ALCOHOL AND WINE IN HEALTH AND DISEASE

ANNALS OF THE NEW YORK ACADEMY OF SCIENCES
Volume 957

ALCOHOL AND WINE IN HEALTH AND DISEASE

Edited by
Dipak K. Das and Fulvio Ursini

The New York Academy of Sciences
New York, New York
2002

Library of Congress Cataloging-in-Publication Data

Alcohol and wine in health and disease / edited by Dipak K. Das and Fulvio Ursini.
p. ; cm. — (Annals of the New York Academy of Sciences ; v. 957)
Includes bibliographical references and index.
ISBN 1-57331-376-9 (cloth : alk. paper) — ISBN 1-57331-377-7 (paper : alk. paper)
1. Wine—Health aspects—Congresses. 2. Alcohol—Health aspects—Congresses. [DNLM: 1. Alcoholic Beverages—Congresses. 2. Ethanol—pharmacology—Congresses. 3. Antioxidants—pharmacology—Congresses. 4. Cardiovascular Diseases—diet therapy—Congresses. 5. Cardiovascular Diseases—prevetion & control—Congresses. WB 444 A354 2002]
I. Das, Dipak Kumar, 1946– II. Ursini, F. III. Series.
Q11 .N5 vol. 957
[RM256]
500 s—dc21
[615/ .78

2002005450
CIP

K-M Research/CCP
Printed in the United States of America
ISBN 1-57331-376-9 (cloth)
ISBN 1-57331-377-7 (paper)
ISSN 0077-8923

ANNALS OF THE NEW YORK ACADEMY OF SCIENCES

Volume 957

May 2002

ALCOHOL AND WINE IN HEALTH AND DISEASE

Editors

DIPAK K. DAS AND FULVIO URSINI

This volume is the result of a New York Academy of Sciences conference entitled **Alcohol and Wine in Health and Disease**, held in Palo Alto, California on April 26–29, 2001.

CONTENTS

Part III. Effects of Alcohol/Wine on Cardioprotection and Atherosclerosis

Part IV. Effects of Alcohol/Wine on Degenerative Diseases and Cancer

Poster Papers

Financial assistance was received from:

Major Funding

- MUSHETT FAMILY FOUNDATION
- U.S. DEPARTMENT OF THE ARMY
- WINE INSTITUTE

Supporters

- DRY CREEK NUTRITION INC., MODESTO, CALIFORNIA
- INDENA S.P.A. MILAN, ITALY

Contributors

- THE CALIFORNIA ASSOCIATION OF WINEGRAPE GROWERS
- DEUTSCHE WEINAKADEMIE
- INTERHEALTH NUTRACEUTICALS INCORPORATED, BENICIA, CALIFORNIA
- SUNTORY LTD.

With sponsorship by the

AMERICAN COLLEGE OF NUTRITION

Alcohol and Wine in Health and Disease

Foreword

ARTHUR L. KLATSKY

Senior Consultant in Cardiology, Kaiser Permanente Medical Center, Oakland, California 94611, USA

Drink a glass of wine after your soup, and you steal a ruble from the doctor.
—Russian Proverb

It is no biological accident that we have the ability to metabolize ethyl alcohol. Natural fermentative processes are widespread, and the fermented alcoholic food beverages (most notably wine and beer) were discovered, not invented. Almost universally used in human societies, these beverages have engendered an abundance of folk wisdom about their harms and benefits. Thus, it is no surprise to find that the above practical preventive medical advice comes from a society long plagued by problems related to uncontrolled drinking.

Scientific evidence of benefit by lighter drinking began to accumulate only in the 20th century. It is now clearer than ever that the health effects of alcohol are a double-edged sword, with harm predominating among heavy drinkers and net benefit evident in light drinkers. Since individual effects cannot reliably be predicted, and harm from heavy drinking can be catastrophic, many persons have great difficulty accepting that there can be any benefit at all.

A number of epidemiologic studies in various countries consistently show substantial (at least 30%) protection against coronary heart disease by moderate amounts of alcohol (10–40 g/day). Protection by moderate alcohol drinking also probably extends to ischemic strokes and ischemic damage to the extremities and possibly to diabetes mellitus and other conditions. Total mortality is slightly favorably affected in middle-aged and older moderate drinkers as compared to abstainers.

Mechanisms underlying an antithrombotic effect of alcohol are becoming established. Alcohol consumption significantly and consistently raises the plasma levels of the antiatherogenic HDL cholesterol. Further, alcohol may reduce the tendency to thrombosis (e.g., plasma fibrinogen levels are decreased, while the efficiency of fibrinolytic pathways is improved), improve endothelial function, and reduce insulin resistance.

There is intense interest in additional effects of nonalcoholic components of specific alcoholic beverages. Antioxidant effects of phenolic components of wine (especially red wine) may contribute to the antiatherosclerotic action of this beverage, by modulation of LDL oxidation and reduction of formation, growth, and maturation of the atherosclerotic plaques. Such antioxidant actions might also play a role in reduction of risk of cancer. Furthermore, there are substantial data supporting hypothetical additional benefit by wine via inhibition of thrombosis and improved endothelial function. Some epidemiologic data, including international comparisons and several

Ann. N.Y. Acad. Sci. **957: ix–x (2002).**

prospective studies, support greater benefit from wine drinking than from other beverage types, but there is no epidemiologic consensus on the issue. The health habits and drinking patterns of wine drinkers tend to be favorable, making interpretation difficult. The issue of additional benefit by wine must be considered unresolved.

All of these areas were covered at the New York Academy of Sciences Conference entitled **Alcohol and Wine in Health and Disease**, held in Palo Alto, California from April 26–29, 2001. There was intense and productive exchange of views. A group of experts considered epidemiologic evidence, beneficial mechanisms of ethyl alcohol, data about nonalcohol ingredients in wine, and public health issues. All present felt that the conference was high-level and authoritative. This volume of the *Annals of the New York Academy of Sciences* presents manuscripts from both speakers and poster presenters at this conference and comprises a state-of-the-art summary of this fascinating field.

Preface

Since humans first learned how to produce wine by fermenting grapes, they have been teased by fear of the mysterious drink that produces an unusual and exciting effect on consciousness while being aware that the same drink could also be a beneficial drug for preventing diseases and relief of pain. Popular appreciation of the virtue of moderation emerged, in contrast to the notion of the detrimental effect of an uncontrolled vice. This duality of feeling is still current: some doctors are fully persuaded of the correctness of recommending complete abstinence even when this is not specifically required by clinical status or social, cultural, or religious rules. Describing the positive effects of wine as "paradoxical" is evocative of a culture in which it is not easily accepted that drinking alcohol, persecuted as a vice, can be beneficial as well. Remarkably, the beneficial properties of the wine were better appreciated by the ancients, at a time when medical suggestions relied on observation and intuition and were enforced by religious doctrine and magical suggestions. The ritual role of wine in Judaism and Christianity is apparently drawn from an ancestral notion of its beneficial effects, somehow linking man to God and immortality. The ancient Greek use of "nectar" to indicate wine is itself a good example of the relationship between wine and prevention of disease—the Greek etymology of the word "nectar" means "to escape from death." Since the biblical period, wine has been considered a symbolic and divine nectar for any kind of social occasion. According to Greek mythology, divinity was bestowed upon wine by the Roman wine god Bacchus. Dionysus, the Greek god of wine, showed mortals how to cultivate grapevines and make wine. A large list of gods/goddesses of wine exists in the literature, including the Roman god Liber, the Sumerian goddess Siduri, the Egyptian goddess Sekhmet, the Aztec god Tepoztecatl, the Chinese god Ssu-ma Hsiang-ju, and the Indian goddess Sura. In the New Testament, during the Last Supper, wine was described as the "blood of Christ." In the Koran, the prophet Mohammed described Paradise as the place where there are "rivers of wine, an exquisite delight for drinkers."

The history of Western civilization and medicine from Hippocrates to Galeno through the Middle Ages to the Renaissance is replete with aphorisms, suggestions, and recommendations by doctors, philosophers, and monks about the beneficial effect of moderate wine use. The notion that wine is beneficial to health is not restricted to the Western world; ancient Chinese medicine also reports a pharmaceutical use of fermented drinks, and a medical text written in India (Charak Sanghita, in the 6th century B.C.) describes its beneficial effects. The most remarkable exception is the ban on alcohol and wine by the Islamic world; this was introduced after the death of Mohammed, however, and there are serious arguments in support of the notion that socioeconomic issues rather than medical concerns gave rise to it. The ban introduced in the United States less than a century ago may have had similar underlying causes. Although wine and alcoholic beverages have had plenty of medical endorsements over the centuries, the first fully scientific notion derives from Pasteur, who described wine as the healthiest and most hygienic beverage. Until recently, when food technology made other drinks equally safe from spoilage, the health benefits of drinking wine were mainly the result of its antibacterial properties.

Ann. N.Y. Acad. Sci. **957: xi–xiv (2002).**

Modern medicine is founded on two basic methodological approaches: the statistical/epidemiological method of "evidence-based medicine" and experimental biomedicine. The increased interest in the health effects of wine and alcohol in the last 10 years has been generated by a combination of these approaches, and any correct claim for public health benefits has to rely on the synergistic results from them both. Epidemiological evidence indicates that a moderate dose of alcohol or wine protects against cardiovascular disease, and this has been discovered to be the explanation for the "paradoxical" data indicating that some populations in Southern Europe have much less cardiovascular disease than would be expected from the prevalence of commonly accepted risk factors related to fat intake and plasma cholesterol levels. Although data about the "French paradox" are solid, they cannot fully satisfy the experimental scientist who would like to know the molecular reasons behind this. In other words, the evidence for the mechanism is missing.

Experimental biomedicine has focused on wine as a major source of natural antioxidants. In the last 20 years an enormous amount of experimental work has been done on the alleged protective role of natural antioxidants. We have learned a lot about how antioxidants protect cells from damaging oxidants produced by aerobic life or pollution, and we have in-depth knowledge of how antioxidants interact with each other in the complex network of antioxidant defenses. The information still missing is that from "evidence-based medicine," and, although several promising studies have been published, an unequivocal definition of the health-protective effects of antioxidant supplementation has never been produced. In conclusion, from experimental biology and *in vitro* studies we know how antioxidants could be beneficial, but we still need a confirmation from controlled clinical studies.

When considering wine, and particularly red wine, as major source of nutritional antioxidants, it would be quite tempting to close a syllogism stating that since wine protects against cardiovascular disease and is rich in antioxidants, and since antioxidants mimic *in vitro* a series of protective events, wine must be beneficial because of its antioxidant content. However, unfortunately, an unequivocal convergence between epidemiological and biochemical evidence has yet to come, and there are several lacunae and open questions. Is only wine protective or is just alcohol enough? Do red wines protect better than white wines? What are the effects of different polyphenols? What is the most appropriate intake? All these questions require answers from both experimental and evidence-based medicine.

The controversial issues of wine and alcohol in health and disease were discussed in a systematic manner at a recent conference organized by the New York Academy of Sciences and held in Palo Alto, California from April 26–29, 2001. The aim of the conference was to bring together basic scientists and clinicians from all over the world to discuss in a cohesive manner the possible health benefits that may be achieved by drinking wine and alcohol. Scientists from diverse areas of research discussed how wine and alcohol could not only provide cardioprotection, but also protect other organs as well. A major part of the discussion focused on the role of wine in cancer prevention. A special panel of scientists emphasized the crucial role of the polyphenolic antioxidants present in wine in preventing degenerative diseases.

Several renowned epidemiologists also took part in this discussion and provided epidemiological evidence supporting the idea that drinking wine and alcohol helps prevent disease. They also emphasized the damaging effects of heavy alcohol drinking. A relatively small part of the conference was devoted to the discussion of good versus bad effects of alcohol drinking. While all of the participants agreed on the beneficial role of mild-to-moderate drinking, they agreed that a proper definition of mild-to-moderate alcohol drinking is far from clear. The familiar advertisements like "one or two glasses of wine a day keep the doctors away" may appear to be too simple, but the general public may have to be carefully educated to understand the real meaning of "mild-to-moderate drinking." It will probably only then be possible to blend the social drinking of alcohol with its medicinal value. Presentations in this book represent a comprehensive overview of available experimental data and personal thoughts of the leading scientist in different areas—from epidemiology to experimental biology—related to the protective effects on human health of alcohol and wine. Since only moderate intake was discussed, relevant aspects of abuse are not considered in this volume. We hope that the growing mass of information on the beneficial effects of wine will contribute to unraveling the mystery of this "magic drink," reminding us of Paracelsus's aphorism that a food can be also a drug or a poison and that the difference is just matter of dosage. Finally we also hope that reading this book will help all of us in drinking our glass of wine more consciously.

Dipak K. Das
Fulvio Ursini

Balancing the Risks and Benefits of Moderate Drinking

R. CURTIS ELLISON

Department of Medicine, Boston University School of Medicine, Boston, Massachusetts 02118, USA

ABSTRACT: The risks of excessive alcohol intake are well known, and heavy drinkers should decrease their intake or stop drinking. On the other hand, in comparison with non-drinkers, moderate drinkers are at much lower risk of cardiovascular disease and certain other diseases and have lower total mortality. Thus, middle-aged or older men and post-menopausal women with no contraindications to alcohol use should be informed that they have, on average, net health benefits from the regular consumption of small-to-moderate amounts of alcohol.

KEYWORDS: coronary artery disease; alcohol; wine

INTRODUCTION

Epidemiologists have known for many years of an inverse association between alcohol consumption and coronary artery disease (CAD). Recent studies have confirmed a protective effect of moderate drinking in both Western industrialized countries, where CAD is such a major health problem, and in Asian nations, where CAD occurs somewhat less frequently. Still, CAD and ischemic stroke are two of the most common causes of death throughout the world. Since the risk of both of these is markedly reduced by moderate drinking, it would take a very strong adverse effect on other common causes of death to balance the health benefits for these cardiovascular diseases. Such is not the case, in that most of the harmful health (and societal) effects of alcohol consumption are due to excessive or inappropriate drinking. Thus, the net effect seen in most epidemiologic studies is lower total mortality rates for moderate drinkers in comparison with non-drinkers: on the average, moderate drinkers live longer than abstainers.

WINE, ALCOHOL AND CORONARY ARTERY DISEASE

The epidemiologic data are quite clear: moderate amounts of alcohol markedly reduce the risk of CAD.[1–5] Of the mechanisms of such an effect, an increase in the protective high-density lipoprotein (HDL) cholesterol from alcohol is probably most

Address for correspondence: R. Curtis Ellison, M.D., Department of Medicine, Boston University School of Medicine, 715 Albany Street, Boston, MA 02118, USA. Voice 617-638-8080; fax: 617-638-8076.
Ellison@bu.edu

important.[6] Our studies from the population-based NHLBI Family Heart Study showed that alcohol consumption was the primary lifestyle factor related to HDL-cholesterol level; the effects were much larger for alcohol than for physical activity.[7] In addition to its effects on lipids, the moderate consumption of wine and other alcoholic beverages favorably affects many factors in the clotting and thrombolytic processes.[8,9] These latter effects tend to be transitory, suggesting that *regular* consumption of alcohol may be the preferable pattern of drinking. Jackson *et al.*[10] demonstrated that, even when adjusting for usual consumption among drinkers, consumption of alcohol within the preceding 24 hours reduced the risk of myocardial infarction and cardiac death when compared with no intake during the preceding 24 hours.

There are data indicating that many of the antioxidants and other phenolic compounds in alcoholic beverages (especially red wine) and in foods are better absorbed or become more biologically active in the presence of alcohol. Of particular interest is the report from the Nurses' Health Study[11] showing that while higher levels of folate were associated with less CAD, the protection was much greater among drinkers than among abstainers. Going from the lowest quintile of folate intake to the highest quintile, the risk of CAD was reduced by only about 15% among non-drinkers but by almost 80% among women consuming 15 g/day or more of alcohol. Further, in the Lyon Diet Heart Study,[12] wine consumption predicted blood levels of vitamin E better than the estimated dietary intake of vitamin E. These and other studies suggest that moderate alcohol consumption may enhance the effects of a number of components of a healthy diet.

EFFECTS OF MODERATE ALCOHOL CONSUMPTION ON OTHER HEALTH CONDITIONS

Coronary artery disease is not the only cardiovascular condition affected favorably by moderate alcohol consumption. Recent studies have shown that moderate drinkers are less likely to suffer ischemic stroke[13,14] and peripheral vascular disease.[15] Further, Cooper *et al.*[16] found that subjects with ischemic congestive heart failure who were light-to-moderate drinkers had fewer deaths and other adverse outcomes than those who were abstainers. Walsh *et al.*[17] found in the Framingham Study that the development of congestive heart failure was lower in subjects who consumed moderate amount of alcohol, both those who had a previous myocardial infarction and those who did not. A number of non-cardiovascular diseases are also known to be less frequent in moderate drinkers than in abstainers. These include diabetes,[18,19] osteoporosis,[20] gall bladder disease,[21] and many other conditions. Therefore, when estimating the benefits of moderate drinking in determining the risk/benefit ratio, one should consider a large number of health conditions. Or, better still, one should look at total, or all-cause mortality associated with moderate drinking.

ADVERSE EFFECTS OF ALCOHOL

All are aware of the adverse effects of excessive or irresponsible alcohol consumption. The acute effects of intoxication may lead to accidents and other harmful events for the drinker as well as others. And while alcoholics are at increased risk of cirrhosis and certain upper digestive cancers and liver cancer, these are not diseases that occur in light-to-moderate drinkers. The only cancer that *may* relate to even moderate drinking is breast cancer in women, and some analyses suggest that there may be a 10% increase in risk for each drink per day a woman consumes.[22] However, we recently reported results among the 5,000 women who have been followed in the Framingham Study for 25–45 years.[23] With repeated assessments of the women's drinking habits and presumed complete detection of breast cancers, we found no increased risk of breast cancer over that of lifetime non-drinkers to be associated with the intake of total alcohol or beer, wine, or spirits consumption among these generally light-to-moderate drinking women. There may well be an increase in risk of breast cancer with drinking for some women, but the magnitude of effect is probably small, while the protective effect against the more common cardiovascular diseases is large. Further, while studies such as the Nurses' Health Study[24] have reported an increase in risk of breast cancer with alcohol consumption, recent research has shown that adequate dietary folate intake may protect against this increase in risk.[25]

THE BOTTOM LINE: ALCOHOL CONSUMPTION AND TOTAL MORTALITY

Regardless of increases or decreases in the risk of specific diseases, what is the bottom line? Is someone more likely or less likely to die of any cause during a specified period of time if he or she drinks? We recently had a report by Thun *et al.*[26] from an American Cancer Society study of almost 500,000 people in the United States. For both men and women, the lowest death rates were among individuals reporting 1–2 drinks per day; in these groups, death rates were 21% lower than those of non-drinkers.

In the Copenhagen Heart Study[27] it was calculated that reducing moderate drinking in a population might actually *increase* mortality. In that study, drinkers who stated that they averaged 1–6 drinks/week had about 35% lower death rates than non-drinkers; heavy drinkers had higher death rates. During follow up, there were 2,229 deaths. The investigators looked at the excess deaths in each category of drinking that could be *attributed to alcohol*. They used death rates for individuals with alcohol consumption of 1–6 drinks per week as the referent group, the group with the lowest death rates in this study, and calculated the number of *excess* deaths (of any cause) that occurred in groups in which the alcohol consumption was either *more* or *less* than this amount. For both men and women, the excess deaths that were attributed to excess alcohol were less than those attributed to no alcohol intake (presumably because of a greater risk of cardiovascular diseases in the latter group). The investigators estimated that if all consumers of seven or more drinks per week decreased their consumption to 1–6 per week, there would have been 117 fewer deaths. On the other hand, if the entire population had stopped drinking, they estimated that there would have been 447 *additional* deaths. This suggests that our efforts to reduce alcohol consumption among heavy drinkers, irresponsible drinkers, and alcoholics should focus on abusers, but should not encourage people who are consuming alcohol moderately and responsibly to stop drinking. In this and several other studies, the net health benefits seemed more favorable for wine than for other alcoholic beverages.

RECOMMENDATIONS TO THE PUBLIC

Are we at the point when governments, health agencies, and physicians should start encouraging certain non-drinking adults to begin to consume small amounts of alcohol for the health benefits? Obviously, no one should advise moderate drinking to a former alcoholic, to a pregnant woman, or to others who should not drink or do not wish to drink for any reason. Anyone who has had problems with alcohol abuse, drug abuse, or other such problems but is currently abstaining should be encouraged to continue to not drink; because of their demonstrated inability to control their drinking, starting to drink "for their health" would not be in their best interests. And no one should be encouraged to drink more that the small amounts that have been shown to be protective against certain diseases.

However, I believe that it is important that physicians and health agencies tell the truth about the risks and benefits of alcohol consumption. For individuals who are drinking in excess, binge drinking, or otherwise consuming alcohol in an unhealthy manner, we should advise them to decrease the amount of alcohol they consume and try to help them learn to drink in a more reasonable manner, or not at all. And for appropriate patients who do not drink or do so only occasionally, and who wish to do so, encouraging a glass of wine or other alcoholic beverage with dinner every evening may be the best advice you can give them. By failing to give older adults at risk of CAD scientifically sound, balanced advice about the potential health benefits of moderate use, as well as the adverse effects of excessive or inappropriate use, we may be doing them a disservice.

The most-recent version of the *Report on Sensible Drinking* from the Department of Health in the United Kingdom[28] stated, regarding adults who are non-drinkers, "While some people do not wish to take up drinking, for religious or other reasons, or there may be medical grounds for them not to do so (and such individuals should make other lifestyle changes to improve health and lower the risk of CAD), middle aged or elderly men and post-menopausal women who drink infrequently (less than one unit per day) or not at all may wish to consider the possibility that light drinking might benefit their health."

CONCLUSIONS

The scientific data now indicate strong protection against CAD from light-to-moderate consumption of alcohol. Governmental and health agencies should present scientifically sound and balanced information on alcohol consumption to the public. Physicians should discuss alcohol consumption with all of their patients. Where there is excessive or irresponsible use, consumption in a healthier manner, or not at all, should be recommended. For patients with previous abuse or with religious, ethical, or medical reasons for abstinence, other approaches for the prevention of CAD should be advised. However, for middle-aged or older men and for post-menopausal women who have no contraindications to alcohol use, especially those at increased risk for CAD, physicians should discuss both the adverse and potentially beneficial aspects of moderate drinking with them. In many circumstances, a joint decision to consume small amounts of an alcoholic beverage on a regular basis, especially wine with meals, may be in the patient's best interests.

REFERENCES

1. RIMM, E.B., E.L. GIOVANNUCCI, W.C. WILLETT, *et al.* 1991. Prospective study of alcohol consumption and risk of coronary disease in men. Lancet **338:** 464–468.
2. KANNEL, W.B. & R. C. ELLISON. 1996. Alcohol and coronary heart disease: the evidence for a protective effect. Clin. Chim. Acta **246:** 59–76.
3. KEIL, U., L.E. CHAMBLESS, A. DORING, *et al.* 1997. The relation of alcohol intake to coronary heart disease and all-cause mortality in a beer-drinking population. Epidemiology **8:** 150–156.
4. KITAMURA, A., H. ISO, T. SANKAI, *et al.* 1998. Alcohol intake and premature coronary heart disease in urban Japanese men. Am. J. Epidemiol. **147:** 59–65.
5. VALMADRID, C.T., R. KLEIN, S.E. MOSS, *et al.* 1999. Alcohol intake and the risk of coronary heart disease mortality in persons with older-onset diabetes mellitus. JAMA **282:** 239–246.
6. GAZIANO, J.M., J.E. BURING & J.L. BRESLOW. 1993. Moderate alcohol intake, increased levels of high-density lipoprotein and its subfractions, and decreased risk of myocardial infarction. N. Engl. J. Med. **329:** 1829–1834.
7. ELLISON, R.C., Y. ZHANG, S. KNOX, *et al.* 1999. Lifestyle determinants of HDL-cholesterol: The NHLBI Family Heart Study. Collected abstracts of the 39th Annual Conference on Cardiovascular Disease Epidemiology and Prevention (Orlando, FL), no. P94. American Heart Association. Dallas, TX.
8. RENAUD, S. & M. DE LORGERIL. 1992. Wine, alcohol, platelets, and the French paradox for coronary heart disease. Lancet **339:** 1523–1526.
9. DIMMITT, S.B., V. RAKIC, I.B. PUDDEY, *et al.* 1998. The effects of alcohol on coagulation and fibrinolytic factors: a controlled trial. Blood Coagul. Fibrinolysis **9:** 39–45.
10. JACKSON, R., R. SCRAGG & R. BEAGLEHOLE. 1992. Does recent alcohol consumption reduce the risk of acute myocardial infarction and coronary death in regular drinkers? Am. J. Epidemiol. **136:** 819–824.
11. RIMM, E.B., W.C. WILLETT, F.B. HU, *et al.* 1998. Folate and vitamin B6 from diet and supplements in relation to risk of coronary heart disease among women. JAMA **279:** 359–364.
12. DE LORGERIL, M., S. RENAUD, N. MAMELLE, *et al.* 1994. Mediterranean alpha-linolenic acid-rich diet in secondary prevention of coronary heart disease. Lancet **343:** 1454–1459.
13. TRUELSEN, T., M. GRØNBAEK, P. SCHNOHR, *et al.* 1998. Intake of beer, wine, and spirits and risk of stroke: The Copenhagen City Heart Study. Stroke **29:** 2467–2472.
14. SACCO, R.L., M. ELKIND, B. BODEN-ALBALA, *et al.* 1999. The protective effect of moderate alcohol consumption on ischemic stroke. JAMA **281:** 53–60.
15. DJOUSSE, L., D. LEVY, J.M. MURABITO, *et al.* 2000. Alcohol consumption and risk of intermittent claudication in the Framingham Heart Study. Circulation **102:** 3092–3097.
16. COOPER, H.A., D.V. EXNER & M.J. DOMANSKI. 2000. Light-to-moderate alcohol consumption and prognosis in patients with left ventricular systolic dysfunction. J. Am. Col. Cardiol. **35:** 1753–1759.

17. WALSH, C.R., M. LARSON, J. EVANS, *et al.* 2002. Alcohol consumption and risk of congestive heart failure in the Framingham Heart Study. Ann. Intern. Med. **136:** 181–191.
18. STAMPFER, M.J., G.A. COLDITZ, W.C. WILLETT, *et al.* 1988. A prospective study of moderate alcohol drinking and risk of diabetes in women. Am. J. Epidemiol. **128:** 549–558.
19. RIMM, E.B., J. CHAN, M.J. STAMPFER, *et al.* 1995. Prospective study of cigarette smoking, alcohol use, and the risk of diabetes in men. BMJ **310:** 555–559.
20. FELSON, D.T., Y. ZHANG, M.T. HANNAN, *et al.* 1995. Alcohol intake and bone mineral density in elderly men and women. The Framingham Study. Am. J. Epidemiol. **142:** 485–492.
21. LA VECCHIA, C., A. DECARLI, M. FERRARONI, *et al.* 1994. Alcohol drinking and prevalence of self-reported gallstone disease in the 1983 Italian National Health Survey. Epidemiology **5:** 533–536.
22. LONGNECKER, M.P. 1994. Alcoholic beverage consumption in relation to risk of breast cancer: meta-analysis and review. Cancer Causes Control **5:** 73–82.
23. ZHANG, Y., B.E. KREGER, J.F. DORGAN, *et al.* 1999. Alcohol consumption and risk of breast cancer: the Framingham Study revisited. Am. J. Epidemiol. **149:** 93–101.
24. FUCHS, C.S., M.J. STAMPFER, G.A. COLDITZ, *et al.* 1995. Alcohol consumption and mortality among women. N. Engl. J. Med. **332:** 1245–1250.
25. ROHAN, T.E., M.G. JAIN, G.R. HOWE, *et al.* 2000. Dietary folate consumption and breast cancer risk. J. Natl. Cancer Inst. **92:** 266–269.
26. THUN, M.J., R. PETO, A.D. LOPEZ, *et al.* 1997. Alcohol consumption and mortality among middle-aged and elderly U.S. adults. N. Engl. J. Med. **337:** 1705–1714.
27. GRONBAEK, M., A. DEIS, T.I. SORENSEN, *et al.* 1994. Influence of sex, age, body mass index, and smoking on alcohol intake and mortality. BMJ **308:** 302–306.
28. Sensible Drinking. The Report of an Inter-Departmental Working Group. 1995. Department of Health, United Kingdom. December, 1995, pp. 28–34.

Alcohol and Cardiovascular Diseases

A Historical Overview

ARTHUR L. KLATSKY

Senior Consultant in Cardiology, Kaiser Permanente Medical Center, Oakland, California 94611, USA

ABSTRACT: Studying its history generally provides insights relevant to current understanding of a subject: the health effects of alcohol consumption is no exception to this rule. Perceiving past errors in the hopes of avoiding their repetition is crucial. Because there are clear disparities in the relationships of alcohol drinking to various cardiovascular conditions, attempts to simplify the subject of alcohol and cardiovascular diseases have delayed understanding this area. Thus, the following are considered separately: cardiomyopathy, arsenic and cobalt beer-drinkers' disease, cardiovascular beri-beri, systemic hypertension, cardiac arrhythmias, stroke, atherosclerotic coronary heart disease (CHD), total mortality, and definitions of safe drinking limits. The basic disparity underlying all alcohol-health relations is between effects of lighter and heavier drinking.

KEYWORDS: alcohol; coronary heart disease; alcoholic cardiomyopathy; arsenic beer-drinkers' disease; cobalt beer-drinkers' disease; cardiovascular beri-beri; hypertension; arrhythmias; stroke

"Those who cannot remember the past are condemned to repeat it."
—George Santayana 1905

INTRODUCTION

Study of a subject's history seldom fails to provide insights relevant to current understanding. Most important is perception of past errors, with the potential for avoiding their repetition. Because there are clear disparities in relations of alcohol drinking to various cardiovascular conditions,[1] attempts to simplify the subject of alcohol and cardiovascular diseases have slowed progress in understanding this area. Thus, the following will be considered separately: cardiomyopathy, arsenic and cobalt beer drinkers' disease, cardiovascular beri-beri, systemic hypertension, cardiac arrhythmias, stroke, atherosclerotic coronary heart disease (CHD), total mortality and definitions of safe drinking limits.

The basic disparity underlying all alcohol-health relations is between effects of lighter and heavier drinking. In 1926 Pearl[2] described this relation in a Baltimore, Maryland study of tuberculosis patients and controls, preceding other reports of the

Address for correspondence: Arthur L. Klatsky, M.D., Senior Consultant in Cardiology, Kaiser Permanente Medical Center, 280 West MacArthur Boulevard, Oakland, CA 94611. USA. Voice: 510-752-6538; fax: 510-752-7456.
hartmavn@pacbell.net or alk@dor.kaiser.org

J-shaped alcohol-mortality curve by half a century. "Heavy/steady" drinkers had the highest mortality; "abstainers" were next; and "moderate" drinkers had the lowest mortality. This was the first description of the J-shaped (from abstinence to heavy drinking) relation of alcohol drinking to risk of death. Most recent studies confirm this relation, with the lowered risk of lighter drinkers due preponderantly to lower CHD risk. Pearl did not know about the favorable CHD mortality of moderate drinkers. Furthermore, his report was made during the U.S. Prohibition era. He made the cautious interpretation that moderate drinking was "not harmful." Cultural context influences what research gets done and how it is interpreted, especially if the subject arouses strong feelings. Perhaps Pearl's major contribution was to realize the fallacy in comparing health risks of all drinkers to abstainers. Such comparison masks differences between the risks of heavy and light/moderate drinkers. His words are memorable: "one cannot judge the role of diet by starvation or excess."

MODERATE AND HEAVY DRINKING: DEFINITIONS

Definitions of moderate and heavy drinking are arbitrary. The operational definition used here is based upon the level of drinking in epidemiological studies above which net harm is usually seen. Three or more drinks per day is called "heavy" drinking, and less than three drinks per day is called "lighter" or "moderate" drinking. Individual factors, including sex and age, lower the upper limit of "light/moderate" drinking for some persons and raise it for others. In survey-based data, systematic "underestimation" (lying) tends to lower the *apparent* threshold for harmful alcohol effects, because some heavy drinkers allege lighter drinking.

A standard-sized drink of wine, liquor, or beer contains approximately the same amount of alcohol. This is fortunate, since people think in terms of "drinks," not milliliters or grams of alcohol. It seems best to this author to describe relations of drinking to health in terms of drinks per day or week. When talking with clients, health professionals should always remember the importance of defining the size of drinks.

ALCOHOLIC CARDIOMYOPATHY

Distinguished nineteenth-century physicians noted a relation of chronic heavy drinking and heart disease.[3–7] Bollinger[8] described cardiac dilatation and hypertrophy among Bavarian beer drinkers, the "Münchener bierherz." He stated that in Munich the average yearly consumption of beer was 432 liters, compared to 82 liters per year elsewhere in Germany.

Graham Steell[6] stated in 1893 "not only do I recognize alcoholism as one of the causes of muscle failure of the heart but I find it a comparatively common one." In 1900, an epidemic of heart disease due to arsenic-contaminated beer occurred in Manchester, England; following this Steell[9] wrote "in the production of the combined affection of the peripheral nerves and the heart met with in beer drinkers, arsenic has been shown to play a conspicuous part." In *The Study of the Pulse*, William MacKenzie[10] described heart failure from alcohol and used the term "alcoholic heart disease." In the early twentieth century there was doubt that alcohol had a direct role in producing heart muscle disease early, but some[11] took a strong view in favor of such a relationship.

For several decades after descriptions of cardiovascular beri-beri[12,13] the concept of "beri-beri heart disease" dominated thinking about alcohol and the heart. For the past 50 years increasing interest has been evident in possible direct toxicity of alcohol upon the myocardial cells. The concept of alcoholic cardiomyopathy is now solidly established by the sheer volume of clinical observations, evidence of decreased myocardial function in many heavy chronic drinkers, and a few good controlled studies. The entity is indistinguishable from dilated cardiomyopathy of other causes; absence of diagnostic tests remains a major impediment to epidemiological study. Most cases of dilated cardiomyopathy in 2002 are of unknown cause, with post-viral autoimmune processes and genetic predisposition the leading etiologic hypotheses. The landmark study[14] of Urbano-Marquez *et al.* showed clear relations of lifetime alcohol consumption to structural and functional myocardial and skeletal muscle abnormalities in alcoholics. The equivalent of 120 grams alcohol/day for 20 years was needed to see a relation. Walshe's[3] term "cirrhosis of the heart" was very appropriate.

Only a small proportion of alcoholics develop cardiomyopathy, a fact leading to interest in predisposing traits which might increase risk. In this context, it is appropriate to consider further the arsenic and cobalt beer-drinker episodes and thiamine (cocarboxylase) deficiency—or beri-beri heart disease.

ARSENIC BEER-DRINKERS' DISEASE

In 1900 an epidemic (exceeding 6,000 cases with more than 70 deaths) in and near Manchester, England, was proved due to accidental contamination of beer by arsenic. Manifestation involved the skin, nervous system, gastrointestinal system and, prominently, the cardiovascular system. A superb clinical description[15] was: (a) "cases were associated with so much heart failure and so little pigmentation that they were diagnosed as beri-beri..."; (b) "the principal cause of death has been cardiac failure," and (c) "at post-mortem...the only prominent signs were interstitial nephritis and the dilated flabby heart... ."

It was estimated that the affected beer had 2–4 parts per million of arsenic, an amount not—in itself—likely to cause serious toxicity. Gowers[16] mentioned prescribing for epilepsy ten times the amount of arsenic involved over long periods of time without toxicity ("the amount of arsenic...was not sufficient to explain the poisoning"). Some seemed to have a "peculiar idiosyncrasy," in that "many persons became ill who drank less beer than others who were not affected." The appointed committee's report[17] suggested that "alcohol predisposed people to arsenic poisoning." As best one can determine, no one suggested the converse.

COBALT BEER-DRINKERS' DISEASE

Recognized sixty-five years after the arsenic-beer episode, this condition was similar in some respects. Reports appeared of heart failure epidemics among beer drinkers in Omaha, Minneapolis, Quebec, and Leuven, Belgium. The symptoms developed fairly abruptly in chronic heavy beer drinkers. The North American patients suffered a high mortality rate, but those who recovered did well despite return, by many, to previous beer habits.

The explanation proved to be the addition of small amounts of cobalt chloride by certain breweries to improve the foaming qualities of beer. Widespread use of detergents (new at that time) in taverns had a depressant effect upon foaming. Quebec investigators[18] tracked down the etiology; the condition became justly known as Quebec beer-drinkers' cardiomyopathy. Removal of the cobalt additive ended the epidemic in all locations.

In Belgium, the cobalt concentrations were less and the cardiac manifestations less severe, with more of the usual findings of chronic cobalt use (polycythemia and goiter). However, even in Quebec, where cobalt doses were greatest, 12 liters of contaminated beer provided only about 8 mg of cobalt, less than 20% of the dose sometimes used as a hematinic. The hematinic use had not been implicated as a cause of heart disease, whereas the first cases of this dramatic heart condition occurred 4–8 weeks after cobalt was added to beer,

It seems almost certain that both cobalt and substantial amounts of alcohol were needed to produce this condition. There must have been other factors, since most exposed persons did not develop the condition. Biochemical mechanisms were not established. Viewing the arsenic and cobalt episodes one observer[19] commented: "This is the second known metal induced cardiotoxic syndrome produced by contaminated beer."

Speculative possibilities which have been mentioned as possible cofactors for alcoholic cardiomyopathy include cardiotropic viruses, drugs, selenium, copper, and iron. Deficiencies (zinc, magnesium, protein, and vitamins) have also been suggested as cofactors, but only thiamine deficiency is proven.

CARDIOVASCULAR BERI-BERI

Aalsmeer and Wenckebach[12] defined, in Javanese polished-rice eaters, high-output heart failure resulting from decreased peripheral vascular resistance. Many assumed that heart failure among Western heavy alcohol drinkers was due to associated nutritional deficiency states. A few heart failure cases in North American and European alcoholics suited this hypothesis. Most did not, however, as they had low output heart failure, were well-nourished, and responded poorly to thiamine. Chronicity and ultimate irreversibility of beri-beri, was used by some to explain the situation. Blacket and Palmer[20] sorted the conditions out: "It (beri-beri) responds completely to thiamine, but merges imperceptibly into another disease, called alcoholic cardiomyopathy, which doesn't respond to thiamine." In beri-beri there is generalized peripheral arteriolar dilatation creating a large arteriovenous shunt and high resting cardiac outputs. A few cases of complete recovery with thiamine within one to two weeks have been documented.

Thus, many cases earlier called "cardiovascular beri-beri" would now be called "alcoholic heart disease." Does chronic thiamine deficiency play a role in some cases of alcoholic cardiomyopathy? Currently unpopular, this hypothesis has not been proved or disproved.

HYPERTENSION

In 1916 Lian[21] reported a relation between heavy drinking (mostly as wine) and hypertension (HTN) in WWI French servicemen. Unless these soldiers exaggerated, they were prodigious drinkers, since the HTN threshold appeared to be at least 2 liters/wine per day. It was almost 60 years before further attention was given to this subject. Since the mid-1970s, dozens of cross-sectional and prospective epidemiological studies[22,23] have solidly established an empirical alcohol-HTN link. The apparent threshold for this relation is approximately three drinks/day, The studies involve both sexes, various ages and populations in North American, Europe, Australia, and Japan. Most studies show no increased HTN with lighter drinking; several show an unexplained J-shaped curve in women with lowest pressures in lighter drinkers. The relation is independent from adiposity, salt intake, education, smoking, beverage type (wine, liquor, or beer), and several other potential confounders.

An early clinical experiment in hospitalized hypertensive men showed increases and decreases in blood pressure with 3–4 days of drinking 4 pints of beer or 3–4 days of abstinence respectively.[24] Similar results were later seen in ambulatory normotensives and hypertensives.[21] Other studies have shown that heavier drinking impairs drug treatment of HTN and that moderation or avoidance of alcohol supplements or betters other nonpharmacological interventions such as weight reduction, exercise, or sodium restriction.[22] The experiments do not show evidence that increased alcohol-associated HTN is related to acute alcohol withdrawal. Even without an established mechanism, the intervention studies strongly support a causal hypothesis and make it probable that alcohol restriction might play a major role in HTN management and prevention.

ARRHYTHMIAS

Based on the observation[25] that supraventricular arrhythmias without evident heart disease were more frequent on Mondays or between Christmas and New Year's Day, the term "holiday heart syndrome" was born. An association of heavier drinking with atrial arrhythmias had been suspected for decades, perhaps occurring especially after a large alcohol-accompanied meal. Atrial fibrillation is the commonest manifestation, typically resolving with abstinence, with or without specific treatment. A Kaiser Permanente study found, in 1,322 persons reporting six or more drinks per day, a doubled relative risk of atrial fibrillation, atrial flutter, supraventricular tachycardia, and atrial premature complexes.[26]

STROKE

Imprecise stroke type diagnosis limited earlier studies of alcohol-stroke relations prior to modern imaging techniques. Both old and new studies of alcohol and stroke have to deal with the complex disparate relations of stroke and alcohol to other cardiovascular conditions. Some studies examined only drinking sprees; others failed to differentiate hemorrhagic and ischemic strokes. Several reports suggested that drinking (heavier especially) carried higher risk of stroke.[27]

In most studies heavier drinkers are at higher risk of hemorrhagic stroke. Several studies suggest that regular lighter drinkers may also be at higher risk of hemorrhagic stroke types, but others have not found this. Associations of alcohol drinking to ischemic stroke risk, are clearly different from those for hemorrhagic stroke. It appears likely that lighter drinking carries lower risk of several types of ischemic stroke; binges aside, the relation of heavier drinking is unclear.[27]

Except for an increased risk of hemorrhagic stroke among heavy drinkers, there is no current consensus about relations of alcohol to various types of stroke, with agreement only that this important subject needs more study.

CORONARY HEART DISEASE

The classic description of angina pectoris[28] by William Heberden in 1786 included: "Wine and spirituous liquors and opium...afford considerable relief."

Thus, it was widely presumed that alcohol is a coronary vasodilator.[29,30] Exercise ECG test data[31,32] suggest that alcohol does not improve myocardial oxygen deficiency and may mask symptoms by a sedative/analgesic effect. Thus, symptomatic benefit is likely to be dangerously misleading in patients with angina who drink before exercise. Available data suggest no major immediate effect of alcohol upon coronary blood flow.[33,34] In the early 1900s an inverse relationship between alcohol consumption and atherosclerotic disease (including CHD) was reported.[35–37] A "solvent" action of alcohol was suggested; another explanation offered was that premature deaths in heavier drinkers precluded development of CHD.[38–40] Population and case-control studies in the past 28 years have solidly established an inverse relation between alcohol drinking and fatal or nonfatal CHD. Since data supporting plausible protective mechanisms against CHD by alcohol have also appeared,[33,34] it now seems likely that alcohol drinking protects against CHD.

In 1819 Dr. Samuel Black, an Irish physician interested in angina pectoris and perceptive about epidemiological aspects, wrote[41] what is probably the first commentary about the "French Paradox." Noting much angina in Ireland but observing little discussion in the writings of French physicians, his explanation of the presumed disparity was "the French habits and modes of living, coinciding with the benignity of their climate and the peculiar character of their moral affections." It was to be 160 years until presentation of international comparison data showing less CHD mortality in wine-drinking countries than in countries where beer or liquor drinking predominate.[42] Confirmatory international comparisons plus reports of nonalcoholic antioxidant phenolic compounds and antithrombotic substances in wine, especially red wine,[33,34] have created great interest in this area. However, prospective population studies show no consensus about the wine/liquor/beer issue.[43,44] This question remains unresolved at this time.

THE "SENSIBLE DRINKING LIMIT"

Attempts to define a safe limit are hardly new, since the medical risks of heavier drinking and the relative safety of lighter drinking have long been evident. Probably the most famous such limit has been known for more than 100 years as "Anstie's Rule."[45] The rule suggested an upper limit of approximately three standard drinks daily. In the mid-19th century, the limit was intended to apply primarily to mature men, but Sir Anstie was a distinguished neurologist and public health activist who emphasized individual variability in the ability to handle alcohol. Modern scientific advances have added little; the threshold for net harm in most population studies is exactly where Anstie, using common-sense observation, placed his limit. Now, as then, considerations of age, sex, and individual risks and benefits become the foci of any discussion in which a health practitioner advises his or her client about alcohol drinking.[46,47]

REFERENCES

1. KLATSKY, A.L. 1995. Cardiovascular effects of alcohol. Scientific Am. Sci. Med. **2:** 28–37.
2. PEARL, R. 1926. Alcohol and Longevity. Knopf. New York.
3. FRIEDREICH, N. 1861. Handbuch der speziellen Pathologic und Therapie. 5th Sect. Krankheiten des Herzens, Ferdinand Enke. Erlangen. The Netherlands.
4. WALSCHE, W.H. 1873. Diseases of the heart and great vessels, 4th edit. Smith, Elder. London, England.
5. STRUMPEL, A. 1890. A Textbook of Medicine. Appleton. New York, p. 294.
6. STEELL, G. 1893. Heart failure as a result of chronic alcoholism. Med. Chron. Manchester **18:** 1–22.
7. OSLER, W. 1899. The Principles and Practice of Medicine, 3rd edit. Appleton. New York.
8. BOLLINGER, O. 1884. Ueber die Haussigkeit und Ursachen der idiopathischen Herzhypertrophie in Munchen. Disch. Med. Wochenschr. (Stuttgart) **10:** 180.
9. STEELL, G. 1906. Textbook on diseases of the heart. Blakiston, Philadelphia, PA, p. 79.
10. MACKENZIE, J. 1902. The Study of the Pulse. Y.J. Pentland. Edinburgh and London, U.K., p. 237.
11. VAQUEZ, H. 1921. Maladies du Coeur. Bailliere et Fils. Paris, France, p. 308.
12. AALSMEER, W.C. & K.F. WENCKEBACH. 1929. Herz und Kreislauf bei der Beri-Beri Krankheit. Wien. Arch. Inn. Med. **16:** 193–272.
13. KEEFER, C.S. 1930. The beri-beri heart. Arch. Intern. Med. **45:** 1–22.
14. URBANO-MARQUEZ, A., R. ESTRICH, F. NAVARRO-LOPEZ, *et al.* 1989. The effects of alcoholism on skeletal and cardiac muscle. N. Engl. J. Med. **320:** 409–415.
15. REYNOLDS, E.S. 1901. An account of the epidemic outbreak of arsenical poisoning occurring in beer drinkers in the North of England and the Midland Counties in 1900. Lancet **i:** 166–170.
16. GOWERS, W.R. 1901. In Royal Medical and Chiururgical Society. Epidemic of arsenical poisoning in beer-drinkers in the north of England during the year 1900. Lancet **i:** 98–100.
17. ROYAL COMMISSION APPOINTED TO INQUIRE INTO ARSENICAL POISONING FROM THE CONSUMPTION OF BEER AND OTHER ARTICLES OF FOOD OR DRINK. 1903. "Final Report," Part I. Wyman and Sons. London, England.
18. MORIN, Y. & P. DANIEL. 1967. Quebec beer-drinkers' cardiomyopathy: etiologic considerations. Can. Med. Assoc. J. **97:** 926–928.
19. ALEXANDER, C.S. 1969. Cobalt and the heart. Ann. Intern. Med. **70:** 411–413.
20. BLACKET, R.B. & A. J. PALMER. 1960. Haemodynamic studies in high output beri-beri. Br. Heart J. **22:** 483–501.

21. LIAN, C. 1915. L'alcoholisme cause d'hypertension arterielle. Bull. Acad. Med. (Paris). **74:** 525–528.
22. BEILIN, L.J. & I.B. PUDDEY. 1992. Alcohol and hypertension. Clin. Exp. Hypertens. Theory Pract. **A14** (1 and 2): 119–138.
23. KLATSKY, A.L. 1995. Blood pressure and alcohol intake. *In* Hypertension: Pathophysiology, Diagnosis, and Management, 2nd edit. J.H. Laragh & B.M. Brenner, Eds.: 2649–2667. Raven Press, Ltd. New York.
24. POTTER, J.F. & D.G. BEEVERS. 1984. Pressor effect of alcohol in hypertension. Lancet **1:** 119–122.
25. ETTINGER, P.O., C.F WU, C. DE LA CRUZ, *et al.* 1978. Arrhythmias and the "holiday heart": alcohol-associated cardiac rhythm disorders. Am. Heart J. **95:** 555–562.
26. COHEN, E.J., A.L. KLATSKY & M.A. ARMSTRONG. 1988. Alcohol use and supraventricular arrhythmia. Am. J. Cardiol. **62:** 971–973.
27. VAN GIGN J., M.J., STAMPFER, C. WOLFE, *et al.* 1993. The association between alcohol consumption and stroke. *In* Health Issues Related to Alcohol Consumption. P.M. Verschuren, Ed.: 43–80. ILSI Press. Washington, DC.
28. HEBERDEN, W. 1786. Some account of a disorder of the breast. Med. Trans. R. Coll. Physicians (London) **2:** 59–67.
29. WHITE, P.D. 1931. Heart Disease. Macmillan. New York, p. 436.
30. LEVINE, S.A. 1951. Clinical Heart Disease, 4th edit. Saunders. Philadelphia, PA, p. 98.
31. RUSSEK, H.I., C.F. NAEGELE & F.D. REGAN. 1950. Alcohol in the treatment of angina pectoris. J. Am. Med. Assoc. **143:** 355–357.
32. ORLANDO, J., W.S. ARONOW, J. CASSIDY, *et al.* 1976. Effect of ethanol on angina pectoris. Ann. Intern. Med. **84:** 652–655.
33. RENAUD, S, M.H. CRIQUI, G. FARCHI, *et al.* 1993. Alcohol drinking and coronary heart disease. *In* Health Issues Related to Alcohol Consumption. P.M. Verschuren, Ed.: 81–124. ILSI Press. Washington, DC.
34. KLATSKY, A.L. 1994. Epidemiology of coronary heart disease—influence of alcohol. Alcohol Clin. Exp. Res. **18:** 88–96.
35. CABOT, R.C. 1904. The relation of alcohol to arteriosclerosis. J. Am. Med. Assoc. **43:** 774–775.
36. HULTGEN, J.F. 1910. Alcohol and nephritis: clinical study of 460 cases of chronic alcoholism. J. Am. Med. Assoc. **55:** 279–281.
37. LEARY, T. 1931. Therapeutic value of alcohol, with special consideration of relations of alcohol to cholesterol, and thus to diabetes, to arteriosclerosis, and to gallstones. N. Engl. J. Med. **205:** 231–242.
38. WILENS, S.L. 1947. The relationship of chronic alcoholism to atherosclerosis. J. Am. Med. Assoc. **135:** 1136–1138.
39. RUEBNER, B.H., K. MIYAI & H. ABBEY. 1961. The low incidence of myocardial infarction in hepatic cirrhosis—a statistical artefact? Lancet **ii:** 1435–1436.
40. PARRISH, H.M. & A.L. EBERLY. 1961. Negative association of coronary atherosclerosis with liver cirrhosis and chronic alcoholism—a statistical fallacy. J. Indiana State Med. Assoc. **54:** 341–347.
41. BLACK, S. 1819. Clinical and Pathological Reports. Alex Wilkinson. Newry, England, pp. 1–47.
42. ST. LEGER, A.S., A.L. COCHRANE & F. MOORE. 1979. Factors associated with cardiac mortality in developed countries with particular reference to the consumption of wine. Lancet **1:** 1017–1020.
43. RIMM, E., A.L. KLATSKY, D. GROBBEE, *et al.* 1996. Review of moderate alcohol consumption and reduced risk of coronary heart disease: is the effect due to beer, wine, or spirits? BMJ **312:** 731–736.
44. KLATSKY. A.L., M.A. ARMSTRONG & G.D. FRIEDMAN. 1997. Red wine, white wine, liquor, beer, and risk for coronary artery disease hospitalization. Am. J. Cardiol. **80:** 416–420.
45. ANSTIE, F.E. 1870. On the uses of wines in health and disease. J.S. Redfield. New York, pp. 11–13.
46. FRIEDMAN, G.D. & A.L. KLATSKY. 1993. Is alcohol good for your health? N. Engl. J. Med. **329:** 1882–1883.

47. KLATSKY, A.L. 2001. Should patients with heart disease drink alcohol? (Editorial.) JAMA **285:** 2004–2005.

Alcohol, Type of Alcohol, and All-Cause and Coronary Heart Disease Mortality

MORTEN GRØNBÆK

Centre for Alcohol Research, National Institute of Public Health, Copenhagen, Denmark

ABSTRACT: Many studies from a variety of countries have shown a U- or J-shaped relation between alcohol intake and mortality from all causes. It is now quite well documented from epidemiologic as well as clinical and experimental studies that the descending leg of the curve results from a decreased risk of cardiovascular disease among those with light-to-moderate alcohol consumption. The findings that wine drinkers are at a decreased risk of mortality from cardiovascular disease compared to non-wine drinkers suggest that substances present in wine are responsible for a beneficial effect on the outcome, in addition to that from a light intake of ethanol. Several potential confounding factors still remain to be excluded, however.

KEYWORDS: coronary heart disease; alcohol intake; flavonoids; wine

A large number of studies from many different countries have shown a U- or J-shaped relation between alcohol intake and mortality from all causes. It is now quite well documented from epidemiologic as well as clinical and experimental studies that the descending leg of the curve is due to a decreased risk of cardiovascular disease among those with light to moderate alcohol intake.[1–9] Some studies have found plausible mechanisms for the beneficial effect of light to moderate drinking. Subjects with a high alcohol intake have a higher level of high density lipoprotein. High density lipoprotein has been found to be a mediator of the effect of alcohol on coronary heart disease. Thus 40–60% of the effect of alcohol on coronary heart disease is likely to be attributable to the effect on high density lipoprotein. Further, drinkers are seen to have a lower level of low density lipoprotein.[9–15] Also, alcohol has a beneficial effect on platelet aggregation, and thrombin level in blood is higher among drinkers than among non-drinkers.[16–18] It remains unsolved whether this apparent protective effect of a light to a moderate alcohol intake is the same for all subsets of the population and for all types of alcoholic beverages.

A few very large prospective studies have shown that only those already at risk of cardiovascular disease seem to benefit from a light to a moderate alcohol intake. A recent study from the American Cancer Society emphasized the different mortality risk functions for different age intervals.[19] Consistent with this, one of the few studies *not* showing a U-shaped relation is one of young (Swedish) men.[20] The potential differences in effect of alcohol in different age groups can be explained by the fact

Address for correspondence: Morten Grønbæk, M.D., Ph.D., Dr. Med. Sci., Centre for Alcohol Research, National Institute of Public Health, Danish Epidemiology Science Centre, Svanemøllevej 25, DK-2100 København Ø, Denmark. Voice: 45-39207777; fax: 45-39273095.

MG@si-folkesundhed.dk

that incidence of different causes of deaths differs from age group to age group. For example, the proportion of death from breast cancer decreases after middle-age, while the relative frequency of death from cardiovascular disease increases dramatically. If alcohol has a carcinogenic effect on the breast and an antiatherogenic effect on the heart, this may explain the different shapes of the curve of the younger and the older populations.

It has been questioned whether pattern of drinking—defined as binge versus steady drinking—also plays a role in the relation between alcohol intake and cardiovascular disease/all cause mortality. In a recent study of 11,511 cases of acute myocardial infarction or fatal coronary heart disease and 6,077 controls in New South Wales, Australia, it was shown that individuals who had a steady small intake of alcohol had lower odds of fatal coronary heart disease, while those who had the same average intake, but consumed their alcohol once or twice per week, did not.[21] Apart from this, the important question of a relation between frequency and amount of drinking has not been examined in observational epidemiologic studies. A very recent study of Americans showed that light to moderate drinkers who had occasions of heavy drinking had a significant higher mortality compared to those who had the same average intake.[22]

Correlational studies suggested that there may be different effects of the different types of alcoholic beverages, by showing that mortality from coronary heart disease was lower in countries where wine was the predominant type of alcohol, than in countries where beer or spirits were the beverages mainly ingested.[18,23,24] In a short report from a cohort study by Klatsky, it was suggested that those who preferred wine were at lower risk of death from cardiovascular disease than beer and spirits drinkers.[25] In the Copenhagen City Heart Study, the differences in the effect of different types of beverages on mortality from cardiovascular disease and all causes were emphasized.[26] It has been suggested that substances present in wine, but not in beer and spirits, are responsible for the lower mortality among wine drinkers. Recent studies from United Kingdom and Sweden and another from Copenhagen, Denmark, supported the above by showing that wine drinkers had lower mortality than beer and spirits drinkers.[27–29]

Substances in wine have been shown to have the platelet aggregation–inhibiting effect also found to pertain to ethanol.[30–33] Further, an inhibiting effect on low density lipoprotein oxidation of unknown factors in wine was shown.[34] Also other mechanisms related to unspecified factors in wine, have been suggested to play a role in the prevention of cardiovascular disease.[35–40] Apart from the effect on platelet and low density lipoprotein oxidation, a vasodilating effect has been proposed,[41] perhaps due to an effect of wine on the endothelium.[42] The suggestions from the experimental studies have only been sparsely reproduced in population studies. Thus, one of the first studies on flavonoid intake in humans, found a lower risk of coronary heart disease among subjects who had a high dietary intake of flavonoids,[43] while recent data from the Nurses Health Study did not support the findings.[38] Several of the components which may have antioxidant properties are present in both fruits and vegetables, as well as in wine.[36,39,44,45]

A high intake of fruit, vegetables and fish and a low intake of saturated fat has been suggested to reduce risk of coronary heart disease morbidity as well as mortality.[45–47] The so-called Mediterranean diet which includes fruits and vegetables, has in six out of ten cohort studies been found to have a weak protective effect on coronary heart disease.[45,47,48] Therefore, diet may play a role in the interpretation of the relation between alcoholic beverage type and coronary heart disease mortality. However, very little attention has been paid to the relation between alcohol beverage choice and diet. In the Danish Diet Cancer and Health Study, a preference for wine was associated with a higher intake of fruit, fish, vegetables, salad, and a higher frequency of use of olive oil for cooking compared with a preference for beer or spirits in both men and women (see Table 5).[49] The association between type of beverage and dietary habits was further illustrated by the odds for a healthy diet by categories of increasing intake of wine.

In conclusion, the findings that wine drinkers are at a decreased risk of mortality from cardiovascular disease compared to non-wine drinkers, suggest that substances present in wine are responsible for a beneficial effect on the outcome, in addition to that from a light intake of ethanol. On the other hand, several potential confounders remains to be excluded.

REFERENCES

1. Yano, K., G.G. Rhoads & A. Kagan. 1977. Coffee, alcohol and risk of coronary heart disease among Japanese men living in Hawaii. N. Engl. J. Med. **297:** 405–409.
2. Hennekens, C.H., B. Rosner & D.S. Cole. 1978. Daily alcohol consumption and fatal coronary heart disease. Am. J. Epidemiol. **107:** 196–200.
3. Kittner, S.J., M.R. Garcia-Palmieri, R. Costas, Jr., *et al.* 1983. Alcohol and coronary heart disease in Puerto Rico. Am. J. Epidemiol. **117:** 538–550.
4. Friedman, L.A. & W. Kimball. 1986. Coronary heart disease mortality and alcohol consumption in Framingham. Am. J. Epidemiol. **124:** 481–489.
5. Shaper, A.G., A.N. Phillips, S.J. Pocock & M. Walker. 1987. Alcohol and ischaemic heart disease in middle aged British men. BMJ **294:** 733–737.
6. Hegsted, D.M. & L.M. Ausman. 1988. Diet, alcohol and coronary heart disease in men. J. Nutr. **118:** 1184–1189.
7. Jackson, R., R. Scragg & R. Beaglehole. 1991. Alcohol consumption and risk of coronary heart disease. BMJ **303:** 211–216.
8. Jackson, R. & R. Beaglehole. 1993. The relationship between alcohol and coronary heart disease: is there a protective effect? Current opinion in lipidology. Lipidology **4:** 21–26.
9. Srivastava, L.M., S. Vasisht, D.P. Agarwal & H.W. Goedde. 1994. Relation between alcohol intake, lipoproteins and coronary heart disease: the interest continues. Alcohol & Alcoholism **29:** 11–24.
10. Dai, W.S., R.E. LaPorte, D.L. Hom, *et al.* 1985. Alcohol consumption and high density lipoprotein cholesterol concentration among alcoholics. Am. J. Epidemiol. **122:** 620–627.
11. Criqui, M.H., L.D. Cowan, H.A. Tyroler, *et al.* 1987. Lipoproteins as mediators for the effects of alcohol consumption and cigarette smoking on cardiovascular mortality: results from the lipid research clinics follow-up study. Am. J. Epidemiol. **126:** 629–637.
12. Puchois, P., N. Ghalim, G. Zylberberg, *et al.* 1990. Effect of alcohol intake on human apolipoprotein A-I-containing lipoprotein subfractions. Arch. Intern. Med. **150:** 1638–1641.
13. Stampfer, M.J., F.M. Sacks, S. Salvini, *et al.* 1991. A prospective study of cholesterol, apolipoproteins, and the risk of myocardial infarction. N. Engl. J. Med. **325:** 373–381.

14. Suh, I., J. Shaten, J.A. Cutler & L.H. Kuller. 1992. Alcohol use and mortality from coronary heart disease: the role of high-density lipoprotein cholesterol. Ann. Int. Med. **116:** 881–887.
15. Gaziano, J.M., J. Buring, J.L. Breslow, *et al.* 1993. Moderate alcohol intake, increased levels of high-density lipoprotein and its subfractions, and decreased risk of myocardial infarction. N. Engl. J. Med. **329:** 1829–1834.
16. Meade, T.W., M.V. Vickers, S.G. Thompson, *et al.* 1985. Epidemiological characteristics of platelet aggregability. BMJ **290:** 428–432.
17. Hillbom, M., M. Kangasaho, M. Kaste, *et al.* 1985. Acute ethanol ingestion increases platelet reactivity: is there a relationship to stroke? Stroke **16:** 19–23.
18. Renaud, S. & M. de Lorgeril. 1992. Wine, alcohol, platelets, and the French paradox for coronary heart disease. Lancet **339:** 1523–1526.
19. Thun, M.J., R. Peto, A.D. Lopez, *et al.* 1997. Alcohol consumption and mortality among middle-aged and elderly U.S. adults. N. Engl. J. Med. **337:** 1705–1714.
20. Andreasson, S., P. Allebeck & A. Romelsjö. 1988. Alcohol and mortality among young men: longitudinal study of Swedish conscripts. BMJ **296:** 1021–1025.
21. McElduff, P. & A.J. Dobson. 1997. How much alcohol and how often? Population based case-control study of alcohol consumption and risk of a major coronary event. BMJ **314:** 1159–1164.
22. Rehm, J., T.K. Greenfield & J.D. Rogers. 2001. Average volume of alcohol consumption, patterns of drinking, and all-cause mortality: results from the US National Alcohol Survey. Am. J. Epidemiol. **153:** 64–71.
23. St. Leger, A.S., A.L. Cochrane & F. Moore. 1979. Factors associated with cardiac mortality in developed countries with particular reference to the consumption of wine. Lancet **1:** 1017–1020.
24. Criqui, M.H. & B.L. Rigel. 1994. Does diet or alcohol explain the French paradox? Lancet **344:** 1719–1723.
25. Klatsky, A.L. & M.A. Armstrong. 1993. Alcoholic beverage choice and risk of coronary artery disease mortality: do red wine drinkers fare best? Am. J. Cardiol. **71:** 467–469.
26. Grønbæk, M., A. Deis, T.I.A. Sørensen, *et al.* 1995. Mortality associated with moderate intake of wine, beer, or spirits. BMJ **310:** 1165–1169.
27. Grønbæk, M., U. Becker, D. Johansen, *et al.* 2000. Type of alcohol consumed and mortality from all causes, coronary heart disease, and cancer. Ann. Intern. Med. **133:** 411–419.
28. Wannamethee, S.G. & A.G. Shaper. 1999. Type of alcoholic drink and risk of major coronary heart disease events and all-cause mortality. Am. J. Public Health **89:** 685–690.
29. Theobald, H., L.O. Bygren, J. Carstensen & P. Engfeldt. 2001. A moderate intake of wine is associated with reduced total mortality and reduces mortality from cardiovascular disease. J. Stud. Alcohol **61:** 652–656.
30. Bierenbaum, M.L., M.L. Struck, A.C. Tomeo, *et al.* 1994. Effect of red and white wine on serum lipids, platelet aggregation, oxidation products and antioxidants. Clin. Res. **42**(2): 157A.
31. Pace-Asciak, C.R., S. Hahn, E.P. Diamandis, *et al.* 1995. The red wine phenolics trans-resveratrol and quercetin block human platelet aggregation and eicosanoid synthesis: implications for protection against coronary heart disease. Clin. Chim. Acta **235:** 207–219.
32. Pace-Asciak, C.R., O. Rounova, S.E. Hahn, *et al.* 1996. Wines and grape juices as modulators of platelet aggregation in healthy human subjects. Clin. Chim. Acta **246**(1-2): 163–182.
33. Renaud, S.C. & J.C. Ruf. 1996. Effects of alcohol on platelet functions. Clin. Chim. Acta **246**(1-2): 77–89.
34. Frankel, E.N., J. Kanner, J.B. German, *et al.* 1993. Inhibition of oxidation of human low-density lipoprotein by phenolic substances in red wine. Lancet **341:** 454–457.
35. Hertog, M.G.L., D. Kromhout, C. Aravanis, *et al.* 1995. Flavonoid intake and long-term risk of coronary heart disease and cancer in the seven countries study. Arch. Intern. Med. **155:** 381–386.

36. Hertog, M.G.L. & P.C.H. Hollman. 1996. Potential health effects of the dietary flavonol quercetin. Eur. J. Clin. Nutr. **50:** 63–71.
37. Knekt, P., R. Järvinen, A. Reunanen & J. Maatela. 1996. Flavonoid intake and coronary mortality in Finland: a cohort study. BMJ **312:** 478–481.
38. Rimm, E.B., M.B. Katan, A. Ascherio, *et al.* 1996. Relation between intake of flavonoids and risk for coronary heart disease in male health professionals. Ann. Intern. Med. **125**(5): 384–389.
39. Nakagami, T., N. Nanaumi-Tamura, K. Toyomura, *et al.* 1995. Dietary flavonoids as potential natural biological response modifiers affecting the autoimmune system. J. Food Sci. **60**(4): 653–656.
40. Ruf, J.-C., J.L. Berger & S. Renaud. 1995. Platelet rebound effect of alcohol withdrawal and wine drinking in rats—relation to tannins and lipid peroxidation. Arterioscler. Thromb. Vasc. Biol. **15:** 140–144.
41. Flesch, M., A. Schwarz & M. Bohm. 1998. Effects of red and white wine on endothelium-dependent vasorelaxation of rat aorta and human coronary arteries. Am. J. Physiol. **275:** H1183–1190.
42. Pendurthi, U.R., J.T. Williams & R.S. Rao. 1999. Resveratrol, a polyphenolic compound forund in wine, inhibits tissue factor expression in vascular cells: a possible mechanism for the cardiovascular benefits associated with moderate consumption of wine. Arterioscler. Thromb. Vasc. Biol. **19:** 419–426.
43. Hertog, M.G.L., E.J.M. Feskens, P.C.H. Hollman, *et al.* 1993. Dietary antioxidant flavonoids and risk of coronary heart disease: the Zutphen Elderly Study. Lancet **342:** 1007–1011.
44. Cao, G., R.M. Russel, N. Lischner & R.L. Prior. 1998. Serum antioxidant capacity is increased by consumption of strawberries, spinach, red wine or vitamin C in elderly women. J. Nutr. **128:** 2383–2390.
45. Potter, J.D. & K. Steinmetz. 1996. Vegetables, fruit and phytoestrogens as preventive agents. IARC Sci. Publ., pp. 61–90.
46. Law, M.R. & J.K. Morris. 1998. By how much does fruit and vegetable consumption reduce the risk of ischaemic heart disease? Eur. J. Clin. Nutr. **52:** 549–556.
47. Ness, A.R. & J.W. Powles. 1997. Fruit and vegetables, and cardiovascular disease: a review. Int. J. Epidemiol. **26**(1): 1–13.
48. Serdula, M.K., T. Byers, A.H. Mokdad, *et al.* 1996. The association between fruit and vegetable intake and chronic disease risk factors. Epidemiology **7**(2): 161–165.
49. Tjønneland, A., M. Grønbæk, C. Stripp & K. Overvad. 1999. Wine intake and diet in a random sample of 48 763 Danish men and women. Am. J. Clin. Nutr. **69:** 49–54.

Wine Phenolics

ANDREW L. WATERHOUSE

Department of Viticulture and Enology, University of California, Davis, California 95616-8749, USA

ABSTRACT: **Wine contains many phenolic substances, most of which originate in the grape berry. The phenolics have a number of important functions in wine, affecting the tastes of bitterness and astringency, especially in red wine. Second, the color of red wine is caused by phenolics. Third, the phenolics are the key wine preservative and the basis of long aging. Lastly, since phenolics oxidize readily, they are the component that suffers owing to oxidation and the substance that turns brown in wine (and other foods) when exposed to air. Wine phenolics include the non-flavonoids: hydroxycinnamates, hydroxybenzoates and the stilbenes; plus the flavonoids: flavan-3-ols, the flavonols, and the anthocyanins. While polymeric condensed tannins and pigmented tannins constitute the majority of wine phenolics, their large size precludes absorption and thus they are not likely to have many health effects (except, perhaps, in the gut). The total amount of phenols found in a glass of red wine is on the order of 200 mg versus about 40 mg in a glass of white wine.**

DEFINITIONS

In order to understand phenolic nomenclature, it is helpful to understand the basis of the general terms used. Phenols or phenolics includes all the substances. *Simple phenols* include those compounds that have a single aromatic ring containing one or more hydroxyl groups, a common example being caffeic acid. *Polyphenolic* compounds are those that have multiple phenol rings within the structure, and examples include catechin and ellagic acid. *Flavonoids* have a very specific three-ring structure. Tannin is a functional term which describes substances that are used to tan hide to leather. Many phenolic compounds have this property, although traditional phenol- containing tannins are natural plant extracts which contain a complex mixture of high molecular weight polyphenolic compounds. *Tannin* is often used loosely to describe high molecular weight phenolic mixtures. The term *condensed tannin* refers to mixtures of polymers of flavonoids and *hydrolyzable tannin* refers to gallic- or ellagic-acid–based mixtures, also called gallotannins or ellagitannins, respectively.

The simple phenols in TABLE 1 are not found in wine, but their nomenclature is a key to understanding terms used for phenolics and more complex names.

This chapter is a modified version of a chapter which previously appeared in *Handbook of Antioxidants*, 2nd Edition, Leslie Packer and Enrique Cadenas, Eds. Marcel Dekker. New York, 2002. Copyright 2002 Andrew L. Waterhouse.

Address for correspondence: Dr. Andrew L. Waterhouse, Department of Viticulture & Enology, University of California at Davis, One Shields Avenue, Davis, CA 95616-8749, USA. Voice: 530-752-4777; fax: 530-752-0382.

alwaterhouse@ucdavis.edu <http://waterhouse.ucdavis.edu>

TABLE 1. Simple phenols

Structure	Names	Notes
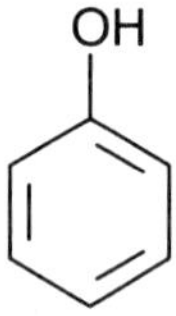 OH (benzene ring)	Phenol (hydroxybenzene)	Rare in natural products but abundant in sore throat sprays
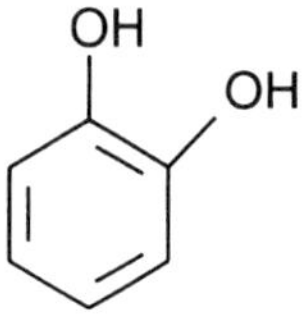 OH, OH (benzene ring)	Pyrocatechol, (1,2 dihydroxybenzene). Note: when a 1,2 dihydroxy functional group is present, it is called a catechol group	Easily oxidized to *o*-quinone. Also called catechol: this same term has been used for catechin, a complex flavonoid described below
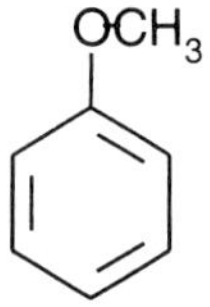 OCH_3 (benzene ring)	Anisole (methoxybenzene)	
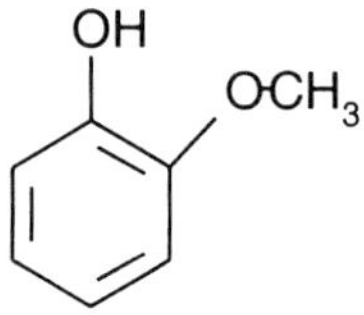 OH, OCH_3 (benzene ring)	Guaiacol (2-methoxyphenol). Note: when a 1-hydroxy-2-methoxy group is present, it is called a guaiacyl group	The methylation eliminates the facile oxidation of the catechol group
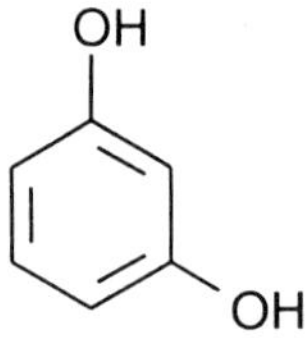 OH, OH (benzene ring)	Resorcinol (1,3 dihydroxybenzene)	Not as easily oxidized as the *ortho* or *para* dihyroxybenzenes
 OH, OH (benzene ring)	1,4 dihydroquinone (1,4 dihydroxybenzene)	Also easily oxidized to quinone, but this form unusual in natural products.
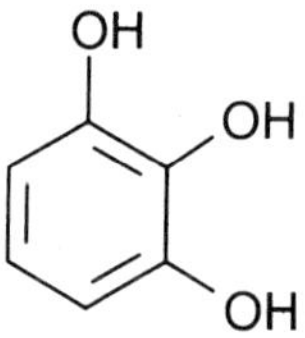 OH, OH, OH (benzene ring)	Pyrogallol 1,2,3-trihydroxy. Note: when a 1,2,3 trihydroxy functional group is present it is call a gallol or galloyl group	Easily oxidized

These names are used in the nomenclature of many different phenolic substances and are helpful to interpret those other names, that is, *res*veratrol contains a resorcinol functional group.

BASIS OF ANTIOXIDANT ACTIVITY

Compounds that are easier to oxidize are often better antioxidants, and that is the case here. The catechol group (1,2 dihydroxy benzene) reacts very readily with oxidants in the form of free radical reactive oxygen species to form a very stable radical anion, the semi-quinone radical. The compounds with catechol or 1,4 dihydroquinone functionality are especially easy to oxidize because the resulting phenoxyl radical can be stabilized by the adjacent oxygen anion. It is stable enough that it does not abstract hydrogens from other substances and will persist long enough to react with another semi-quinone radical, resulting in a disproportionation reaction which yields a quinone and a phenol, in total quenching two radicals. Wine is rich in substances with the catechol group and these compounds impart an antioxidant activity to wine that is its natural preservative. The use of sulfites in wine making enhances this antioxidant property by reducing and recycling the quinone product back to the phenol. If this quinone is not removed, it will eventually react with other nucleophiles, resulting in browning.

NON-FLAVONOIDS

Wine phenolics are grouped into two categories, the flavonoids and non-flavonoids, and the major subgroups will be surveyed. In summary the non-flavonoids are listed with examples and general comments.

Hydroxycinnamic Acids (e.g., caffeic acid)

Ubiquitous in fruits and all plants tissues, and found as esters in fruit–tartrate esters in grapes—, but quinic esters (i.e., cholorgenic acids) in most other fruits

Benzoic Acids (e.g., gallic acid)

Ester forms found in fruit sources, but few free acids are found in fresh sources. Hydrolysis yields the free acids in wine.

Hydrolyzable Tannins (e.g., vescaligin)

Ester-linked oligomers of gallic acid or ellagic acid with glucose or other sugars. These are not present in grapes and only oak-treated wine.

Stilbenes, (e.g., resveratrol)

Hydroxylated stilbenes found as the glycosides in grapes and a few other dietary sources. Produced by the grapevine in all tissues as a phytoalexin in response to fungal attack, such as by *Botrytis cinera*.

HYDROXYCINNAMATES

These are the major phenols in grape juice and the major class of phenolics in white wine. These materials are also the first to be oxidized and subsequently initiate browning, a problem in white wines. There are three common hydroxycinnamates in grapes and wine, those based on coumaric acid (mono 4-hydroxy), caffeic acid (catechol substitution), and ferulic acid (guaiacyl substitution) (see FIGURE 1).

In grape berries the simple hydroxycinnamic acids noted above are not found. Instead these acids exist as esters of tartaric acid. Enologists have adopted trivial names for these compounds—*p*-coutaric acid, caftaric acid, and fertaric acid, respectively. These substances are found in the flesh of the fruit, and thus are found in all grape juices and consequently in all wines. The levels of these compounds vary in grapes, but caftaric acid is by far the predominant cinnamate in grapes, averaging about 170mg/kg in *Vitis vinifera* grapes, while the p-coutaric and fertaric acids occur at about 20 and about 5 mg/kg, respectively, as reported in Ong and Nagel[1] and by Singleton[2] and these relative proportions are maintained in wine.

The naturally occurring esters are susceptible to hydrolysis, and this occurs in the aqueous acidic solution of wine, releasing the simple hydroxycinnamic acids which are readily detected in wine of a few weeks old. In addition, the free acids will partially esterify with the ethanol in wine. The rate of these reactions is variable, depending on the pH of the wine, which typically varies between 3.0 and 3.9. Hydrolysis of the caffeic acids esters of tartaric acid can be catalyzed by an enzyme, hydroxycinnamate ester hydrolyze.[3] Levels of total hydroxycinnamates in finished wine are typically 130mg/L in whites and 60mg/L in reds.

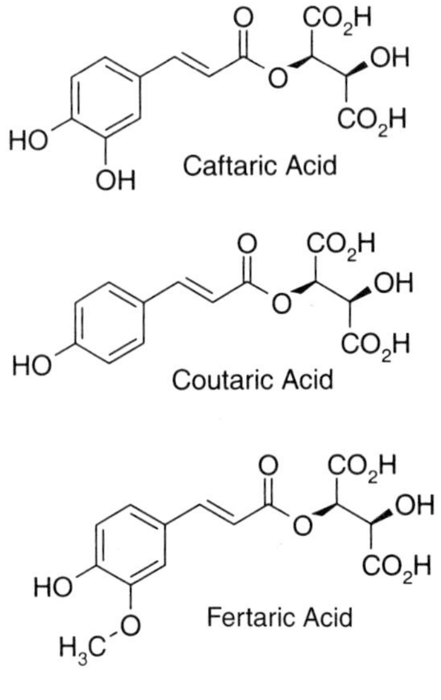

Caftaric Acid

Coutaric Acid

Fertaric Acid

FIGURE 1. Three of the hydroxycinnamates in wine.

In terms of wine sensory qualities, the hydroxycinnamates appear to have no perceptible bitterness or astringency at the levels found in wine. The caftaric and coutaric esters have taste thresholds of 50 and 10–25 ppm in water, with bitterness and astringency being the descriptors,[4] but work by Noble and others[5] showed that in wine the levels of these compounds were below threshold. Hydroxycinnamates have been observed in plasma and urine after the consumption of prunes.[6]

BROWNING

Browning caused by cinnamates is their most important effect in wine making, in particular the color of white wine. When grapes are crushed, polyphenol oxidase enzymes are released, and these enzymes rapidly oxidize the hydroxycinnamates to quinones. Glutathione will quickly react with the quinone, forming a colorless product called grape reaction product (GRP), so juices that have high glutathione have lower browning potential in the short run. Thus the browning potential of musts can be characterized by whether or not they have high relative amounts of GSH present.[7] In raisins, the GSH adduct is not formed and thus they brown quite a bit more.[8] It is possible that the elimination of sulfide aromas by aeration of wine is due largely to the efficient reaction of these quinones with thiols.[9]

Analysis of wines of different ages shows that GRP is slowly hydrolyzed to the GSH-caffeic acid derivative (the tartrate ester is hydrolyzed), but the amides are only partially hydrolyzed.[10] This acid will also be partially converted to the ethyl ester. The specific brown products are not well characterized, but it appears that the hydroxycinnamate quinones react further with flavanols to form the actual pigmented products.[11]

BENZOIC ACIDS

These are a minor component in new wines. Gallic acid appears from the hydrolysis of gallate esters of hydrolyzable tannins and condensed tannin after standing for at least a few months. Gallic acid appears to be stable during aging, as it is one phenolic compound readily visible by chromatographic analysis of older red wines. Its levels in red wine average near around 70 mg/L, while white wines average near 10 mg/L.[12]

HYDROLYZABLE TANNINS

In wine hydrolyzable tannins come from oak and levels are near 100 mg/L for white wines aged for about 6 months in barrel, while red wines will have levels in the range of 250 mg/L after aging two or more years.[13] These phenols are composed of gallic acid and ellagic acid esters with glucose or related sugars (see FIGURE 2). Because of the ester linkage, they are referred to as "hydrolyzable." There are two categories, the gallotannins and the ellagitannins which contain gallic acid or ellagic acid. Hydrolyzable tannins are not found in *V. vinfera*, but can be found in other

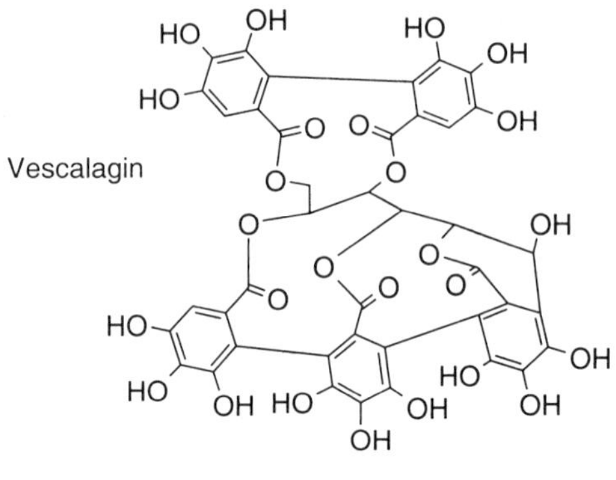

FIGURE 2. An ellagitannin from oak.

fruits, such as raspberries or muscadine grapes. When hydrolyzed in wine from ellagitannins, the ellagic acid will precipitate if it is at high levels, as in muscadine wine. Gallic acid is quite soluble and is found in all aged red wines, and it comes from both hydrolyzable tannins and the condensed tannins found in grape seeds. The sensory impact of oak-derived tannins is controversial.[14]

STILBENES

Stilbenes are another minor class. The principal stilbene in grapes, resveratrol, is produced by vines in response to *Botrytis* infection and other fungal attacks. The actual anti-fungal compounds are the oligomers of resveratrol called the viniferins. Several forms of resveratrol exist including the *cis* and *trans* isomers as well as the glucosides of both isomers. All are found in wine (see FIGURE 3), but in grapes *cis*-resveratrol is absent. Light causes the *cis*/*trans* isomerization.[15] Resveratrol derivatives are found only in the skin of the grape, so much more is found in red wine. The total levels of all forms average about 7 mg/L for reds,[16] 2 mg/L for rosés

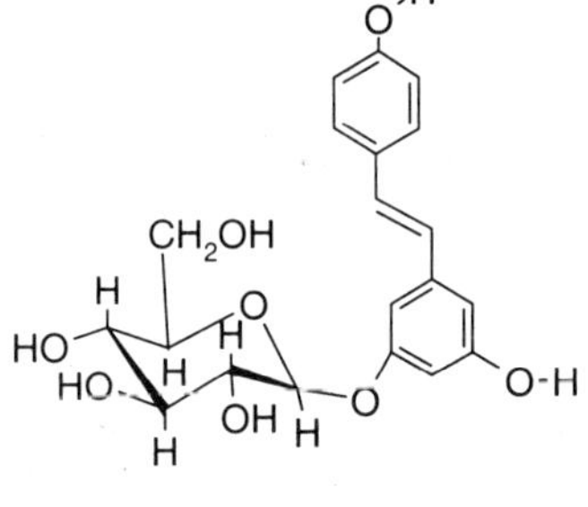

FIGURE 3. *Trans*-piceid, the resveratrol glucoside.

and 0.5 mg/L for white wines.[17] Resveratrol has been implicated as a wine component that may reduce heart disease or cancer, but bioavailability data has not been reported, so it is difficult to assess physiological significance. The interest in the health effects of resveratrol has been greatly stimulated by a report in *Science*[18] implicating anti-cancer activity and in 2001 there were over 200 papers published on resveratrol.

FLAVONOIDS

The wine flavonoids are all polyphenolic compounds, having multiple aromatic rings possessing hydroxyl groups. A specific three-ring system defines flavonoids (see FIGURE 4), there being a central oxygen-containing pyran ring, C ring, of different oxidation states. It is fused to an aromatic ring (A Ring) along one bond and attached to another aromatic ring with a single bond (B ring). The flavonoids found in grapes and wine all have the same hydroxyl substitution groups on ring A, at positions 5 and 7. Differences in the oxidation state and substitution on ring C defines the different *classes* of flavonoids. For instance, a saturated C ring defines the flavans, a keto at position 4 (and unsaturation between 2 and 3) defines the flavones, and the fully aromatic ring, which also has a positive charge, defines the anthocyanidins. The -ol ending further specifies an alcohol substituent on the C ring, as in flavan-3-ol, where the position is distinguished because it could alternatively exist in the 4-position.

The substitution pattern on ring B defines the member of the class. Normal substitution patterns are a hydroxyl at the 4 position with additional oxygen substitution at 3 and/or 5. Those oxygens can be hydroxyls (phenols) or methoxyls at positions 3 and/or 5. Thus the number of class members is relatively short, however, the "free" flavonoid structure can also be substituted further (usually with sugar conjugation on the oxygens); this gives rise to many additional compounds.

The flavonoids comprise the a majority of the phenols in red wine and are derived from extraction of the skins and seeds of grapes during the fermentation process. Since red wines are produced by fermentation of the juice's sugar into alcohol—a good solvent for polyphenol extraction—in the presence of the skins and seeds over a period of 4–10 or more days, there is ample opportunity for extraction of much of the polyphenols into a red wine, and in typical wine making, about half these substance are extracted during the maceration process. The major classes of wine flavonoids are the flavanols, flavonols, and the anthocyanins.

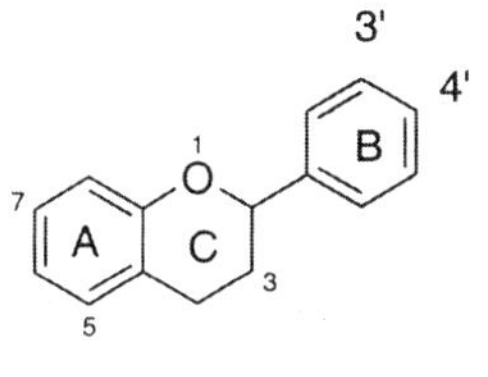

3'
4'
B
1
O
7
A
C
3
5

FIGURE 4. The flavonoid ring system.

Flavan-3-ols (e.g., catechin)

Flavanols make up the most abundant class and include simple monomeric catechins, but most exist in the oligomeric and polymeric proanthocyanidin forms from the skins and seeds.

Flavonols (e.g., quercetin)

Flavonols are found in the berry skin, appear to function as sun screen, and are increased by high sunlight exposure.

Anthocyanins (e.g., malvidin-3-glucoside)

These are the red-colored phenols. While their taste is minimal, better quality red wines in each class often have higher levels of red color. The anthocyanins are converted to other colored forms as a wine ages.

FLAVANOLS

Flavonols are the most abundant class of flavonoids in grapes and wine, and in the grape they are found in both the seed and skin. These are often specifically called the flavan-3-ols to identify the location of the alcohol group on the C ring. The flavan-3-ols are the most reduced form of the flavonoids. Because positions 2 and 3 on the C ring are saturated, stereoisomers exist, and two are found in grapes (see FIGURE 5). The *trans* from is (2R,3S) (+)-catechin and the *cis* form is (2R,3R) (–)-epicatechin. Both catechin and epicatechin have the 3′,4′ catechol substitution on the B ring. The only other B-ring substitution pattern found in wine flavan-3-ols is the 3′4′5′ trihydroxy form, appropriately called the gallo-catechins. Some epigallocatechin is found in grape skin, but gallocatechin is not found in significant amounts. The wine flavan-3-ols are not found as glycosides, typical for the other classes, but instead gallate esters are found, and the gallic acid is esterified at the 3 position of the epi- series only. Epicatechin gallate is a is a small but significant proportion of the flavan-3-ol pool in grapes in the seeds. Thus, for the simple series of monomeric flavanols, there are four different ones found in wine. A few others exist in different foods; of particular note is epigallocatechin gallate in tea. These monomeric flavan-3-ols are sometimes referred to as the "catechins." The levels of total monomeric flavan-3-ols in typical red wine is in the range of 40–120mg/L with the majority usually being

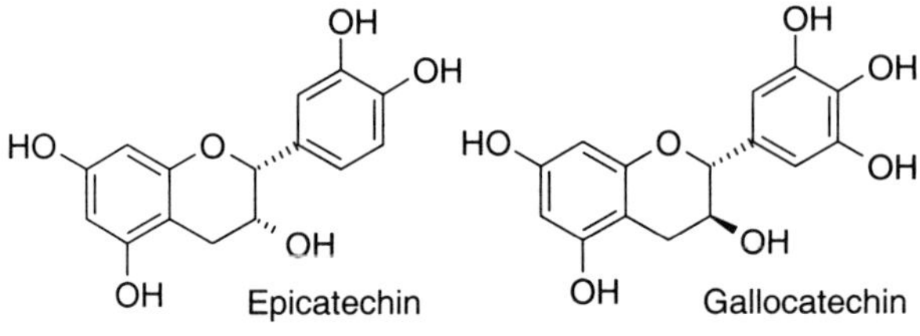

FIGURE 5. *Cis* and *trans* forms of flavan-3-ols.

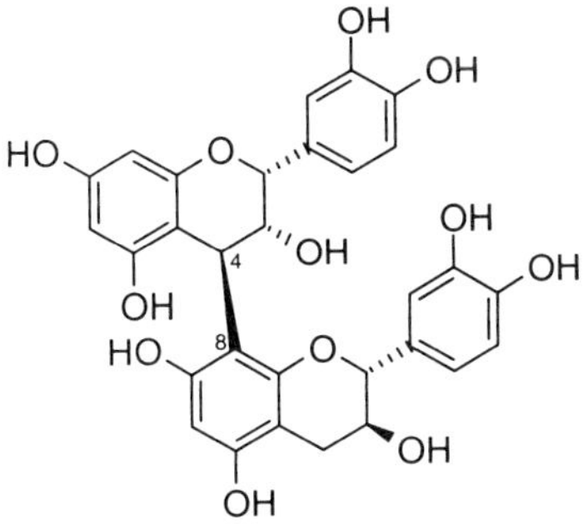

FIGURE 6. Procyanidin B_1.

catechin.[19] The levels are strongly affected by seed extraction techniques and are higher when extended maceration techniques are used.

The majority of phenolic compounds in red wine is from the condensation of flavan-3-ol units to yield the oligomers (proanthocyanidins) and polymers (condensed tannins). The condensation occurs to form covalent bonds between flavan-3-ol units, the most common linkages being 4→8 and 4→6 positions (see FIGURE 6). On average, epicatechin is the predominant unit in condensed tannins from grapes and wine, catechin is the next most abundant (often found at the end or terminal units—those with no bonds at the 4 position). In typical red wines, the amount of polymer plus oligomer is a sizable fraction of the total phenolic level, being in the range of 25 to 50% in new wines and a higher proportion in older wines. Levels are between 0.5 g/L and 1.5 g/L or even higher in some red wines, while in white wine, levels are in the range of 10–50 mg/L and highly dependent on pressing techniques. These substances are unlikely to have many health effects, except in the gut, as they appear to not be absorbed[20] owing to their high molecular weight.[21]

The monomeric catechins are bitter and astringent. In the polymer, the bitterness is minimal, but the astringency remains.[22] Over long aging (many years), a disproportionation reaction can occur, perhaps with some oxidation, so the polymers continue to increase in size until they are no longer soluble in wine and form the precipitate common in older red wines. So, on long aging, the amount of phenols in a wine decreases.

The distribution of the flavanols in grape berries is not the same in all varieties, and in fact has a wide range of differences comparing the seed and skin. The proportion of the flavanols in the seed goes from a low of less than 40% to a high of over 90% as shown in TABLE 2.[23]

FLAVONOLS

Flavonols occur in a wide range of vegetable food sources. This class of compounds is always found in a glycoside form in plants including in grape berries where it is found in grape skin. There are only three forms of the simple flavonoid

TABLE 2. **Distribution of flavan-3-ols within the grape**[23]

Varieties	Monomers (mg/kg)	Percentage in Seeds	Procyanidins (mg/kg)	Percentage in Seeds
Alicante-Bouchet	360	64	287	50
Aramon	213		182	
Syrah	190		176	
Mourvedre	164	58	169	53
Grenache noir	137		125	
Cinsaut	136	37	116	32
Carignan	77	54	73	40
Cabernet Sauvignon	344	83	546	68
Pinot noir	1,165	94	1,609	86

aglycones in grapes, quercetin (see FIGURE 7), myricetin (3′4′5′ trihydroxy) and kaempferol (4′ hydroxy), but since these compounds occur with a diverse combination of glycosidic forms, there are many individual compounds present. The identity of the glycosides has not been established in many different grape varieties, but they have been shown to be mostly the 3-glucosides as well as the 3-glucuronides and small amounts of diglycosides in Cinsault.[24]

A study on Pinot noir has shown that sunlight on the berry skin strongly enhances the levels of the flavonols.[25] Since flavonols absorb UV light strongly at 360 nm, and they appear mostly in the outermost layer of cells in the berry, it appears that the plant produces these compounds as a natural sunscreen. Although this has not been studied in other grape varieties, a study of the levels of phenolic compounds in wine has shown that expensive Cabernet Sauvignon wines expected to come from lower yielding vines with better sun exposure of the fruit has at least three-fold higher levels of flavonols, suggesting that the levels of these compounds in grapes may be a useful indicator of grape sun exposure and perhaps quality. Levels of total flavonols are near 53 mg/L for widely sold Cabernet wine, but over 200 mg/L in more expensive wine.[19]

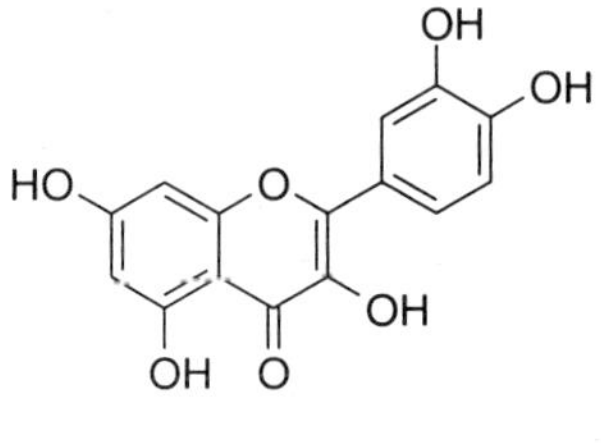

FIGURE 7. Quercetin.

ANTHOCYANINS

Anthocyanins provide the color in red wine and the red and blue colors found in the skins of red or black grapes, but also in many other plants including other foods in the diet. The color is based on the fully conjugated 10 electron A–C ring π-system, with some contribution by the B ring as well. If that is disrupted, the color is lost, as when anthocyanins are bleached by bisulfite. The anthocyanins react with the tannins to produce a "stabilized" anthocyanin or pigmented tannin which persists much longer in wine than the initial form, and it is this stabilized color that persists in most red wines more than a few years old, although in some wines the monomeric forms found in the grape are still found after this time.

The term for the simple flavonoid ring system is anthocyanidin. However, anthocyanidins are never found in grapes or wine, except in trace quantities, because they are unstable. There are five basic anthocyanidins in wine: cyanidin, peonidin, delphindin, petunidin and malvidin, the most abundant in red wines (see FIGURE 8).

"Anthocyanin" implies a glycoside. In *V. vinifera*, this is the single 3-glucoside[1] (see FIGURE 9). In American species and hybrids, the 3,5 diglucoside is found, and its presence is the basis of identifying the use of these grapes in making a wine. The glucose conjugates are further substituted, however. At the six-hydroxyl of the glucose, acyl substitution is also found, with ester linkages connecting either an acetyl group, a coumaryl group, and a small amount of caffeoyl substitution is also observed.

In wine and similar solutions there are several forms of anthocyanins and the proportions strongly affect the color of a solution containing these substances (see FIGURE 10). The charged C ring is an electrophilic center, and it can react with nucleophiles. Common reactions are with water, a pH-dependent reaction, and with bisulfite. In both cases, the red color is lost as the C-ring conjugation is disrupted. The pKa of the flavylium pseudobase form is 2.7. At low pH (less than 1), all forms are converted to the flavylium form, and this treatment is used to assess total

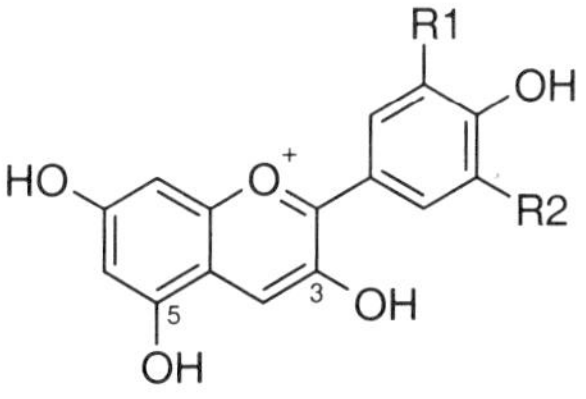

R1, R2
H, OH; cyanidin
H, OCH_3; peonidin
OH, OH; delphinidin
OH, OCH_3; petunidin
OCH_3, OCH_3; malvidin

FIGURE 8. Anthocyanidin structures.

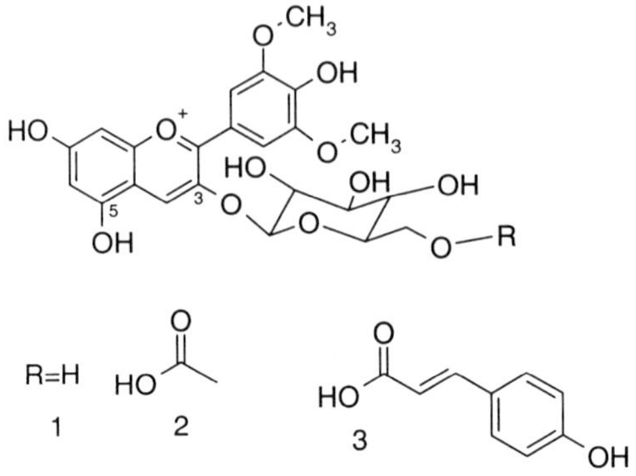

FIGURE 9. Anthocyanidin conjugate forms (anthocyanins).

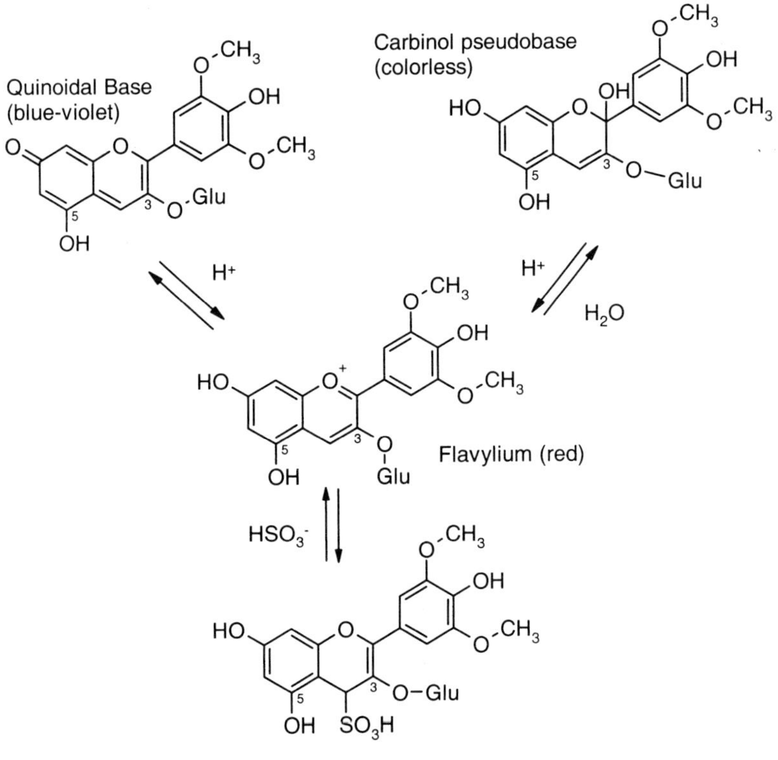

FIGURE 10. Anthocyanin forms present in wine.

anthocyanin content. In addition, there is a quinone form which has a violet hue, and its pKa is 4.7, so it is present in small amounts at high wine pH values.

Anthocyanins interact with other phenolic compounds in solution to create the effect known as co-pigmentation. This is a transient interaction; no chemical bonds are formed. It is a result of the chemical phenomenon called charge transfer complex formation, or π–π interactions.[26] This occurs when there are two aromatic-ring substance in solution that have very different electron densities. The extreme examples are rings with positive charges and those with negative charges. When these differences exist, the rings associate via a weak "bond," and electron density is transferred from the electron-rich ring to the electron-poor ring. Since the flavylium anthocyanin has a positive charge, it is a very good candidate for a charge transfer complex with electron-rich substrates. In addition, the other phenolic compounds in wine are almost all electron rich because the phenol group is a strong electron donor, and thus they are likely to act as partners. As an example, in FIGURE 11 pyrocatechol is used as the electron-rich partner for simplicity, but other phenolic compounds cause the co-pigmentation effect in wine.

The overall effect of co-pigmentation is based on two effects. First, the formation of the π–π- complex causes changes in the spectral properties of the molecules in the flavylium from, in particular increasing absorption (hyperchromic shift) and increasing the wavelength of the absorption (bathochromic shift).[27] But secondly, the stabilization of the flavylium form by the π-complex shifts the equilibrium to better favor the flavylium, thus boosting the proportion of anthocyanin molecules in the red-colored form (FIG. 11). So, the magnitude of the co-pigmentation effect is pH dependent because at very low pH values, all the anthocyanin molecules are already in the flavylium form, and at high pH, the flavylium form is not accessible. There is some speculation that the formation of the π–π-complex enhances the reactions

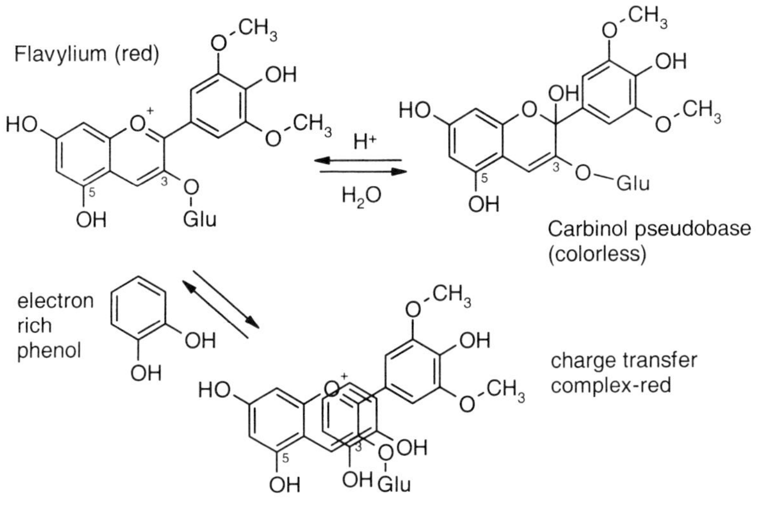

FIGURE 11. Co-pigmentation of wine anthocyanins.

between anthocyanins and the tannins which produce the covalent bonds of pigmented tannins.[28]

Because the effect is bimolecular, it is highly dependent on concentration. It typically occurs in new red wine where the effect can increase apparent color by double. The partners in wine appear to include all other phenols and the anthocyanin itself. Since most of the anthocyanin molecules in a wine solution are in the uncharged, electron-rich pseudo-base form, the anthocyanins act as self-copigmentors.

PHENOLIC LEVELS IN WINES

The levels described here are nominal amounts for wine made from *Vitis vinfera*, the European wine grape, as this grape dominates wine production. There are significant varietal differences within *V. vinifera*, in other words, Cabernet Sauvignon is not the same as Pinot noir, and other grape species will have some different compounds as well as different levels. The levels of the phenolics in wines is also highly variable due to both differences in fruit sources as well as processing. The first and most important factor to understand is that white wines are made by quickly pressing the juice away from the grape solids, while reds are made by fermenting the juice in

TABLE 3. Typical levels of phenolics in red and white table wine[a]

	White Wine		Red Wine	
Phenol Class	Young	Aged	Young	Aged
Non-flavonoids				
Hydroxycinnamates	154	130	165	60
Benzoic Acids	10	15	60	60
Hydrolyzable tannins (from oak)	0	100	0	250
Stilbenes (Resveratrol)	0.5	0.5	7	7
Total mg/L	**164.5**	**245.5**	**232**	**377**
Flavonoids				
Flavanol monomers	25	15	200	100
Proanthocyanidins and condensed tannins	20	25	750	1,000
Flavonols	—	—	100	100
Anthocyanins	—	—	400	90
Others	—	—	50	75
Total mg/L	**45**	**40**	**1,500**	**1,365**
Total all phenols	**209.5**	**285.5**	**1,732**	**1,742**

[a]Young means new wine, less than six months of age, not having been aged or fermented in oak barrels. Aged implies about one year for white, about two years for red and some oak barrel aging (or other oak contact).

the presence of the grape solids (skin and seeds). Since the skins and seeds contain the most of the phenols, red wine is a whole berry extract while white wine is a juice product. In addition, because of the significant sensory effects of these substances on bitterness and astringency, they are controlled by wine makers. The methods of control involve the manipulation of extraction as well as fining with protein to precipitate and remove tannin from finished wine; ultrafiltration is now seeing some use. Typical levels for red and white wines are listed in TABLE 3 based on literature sources noted in the sections above, but in some cases on unpublished data. Keep in mind there is a wide range of variation.

REFERENCES

1. ONG, B.Y. & C.W. NAGEL. 1978. Hydroxycinnamic acid-tartaric acid ester content in mature grapes and during the maturation of white riesling grapes. Am. J. Enol. Viticulture **29:** 277–281.
2. SINGLETON, V.L., J. ZAYA & E. TROUSDALE. 1986. Compostional changes in ripening grapes: caftaric and coutaric acids. Vitis **25:** 107–117.
3. SOMERS, T.C., E. VERETTE & K.F. POCOCK. 1987. Hydroxycinnamate esters of *V. vinifera*: changes during white vinification and effects of exogenous enzyme hydrolysis. J. Sci. Food Agric. **40:** 67–78.
4. OKAMURA, S. & M. WATANABE. 1981. Agric. Biol. Chem. **45:** 2063.
5. VERETTE, E., A.C. NOBLE & C. SOMERS. 1988. Hydroxycinnamates of *Vitis vinifera*: sensory assessment in relation to bitterness in white wine. J. Sci. Food Agric. **45:** 267–272.
6. CREMIN, P., S. KASIM-KARAKAS & A.L. WATERHOUS. 2001. LC/ES-MS detection of hydroxycinnamates in human plasma and urine. J. Agric. Food Chem. **49:** In press.
7. CHEYNIER, V., *et al.* 1990. Must browning in relation to the behavior of phenolic compounds during oxidation. Am. J. Enol. Viticulture **41:** 346–349.
8. SINGLETON, V.L., E. TROUSDALE & J. ZAYA. 1985. One reason sun-dried raisins brown so much. Am. J. Enol. Viticulture **36:** 111–113.
9. CILLIER, J.J.L. & V.L. SINGLETON. 1990. Caffeic acid autoixdation and the effects of thiols. J. Agric. Food Chem. **38:** 1789–1796.
10. CHEYNIER, V.F., *et al.* 1986. Characterization of 2-S-glutathioylcaftaric acid and its hydrolysis in relation to grape wines. J. Agric. Food Chem. **34:** 217–221.
11. RIGAUD, J., *et al.* 1991. Influence of must composition on phenolic oxidation kinetics. J. Sci. Food Agric. **57:** 55–63.
12. WATERHOUSE, A.L. & P.L. TEISSEDRE. 1997. Levels of phenolics in California varietal wine. *In* Wine: Nutritional and Therapeutic Benefits. T. Watkins, Ed.: 12–23. American Chemical Society. Washington, DC.
13. QUINN, K.M. & V.L. SINGLETON. 1985. Isolation and identification of ellagitannins from white oak and an estimation of their roles in wine. Am. J. Enol. Viticulture **36:** 148–155.
14. POCOCK, K.F., M.A. SEFTON & J.P. WILLIAMS. 1994. Taste threshold of phenolic extracts of French and American oakwood: the influence of oak phenols on wine flavor. Am. J. Enol. Viticulture **45:** 429–434.
15. TRELA, B C. & A.L. WATERHOUSE. 1996. Resveratrol: isomeric molar absorptivities and stability. J. Agric. Food Chem. **44:** 1253–1257.
16. LAMUELA-RAVENTÓS, R.M., *et al.* 1995. Direct HPLC analysis of cis- and trans-resveratrol and piceid isomers in Spanish red *Vitis vinifera* wines. J. Agric. Food Chem. **42:** 281–283.
17. ROMERO-PEREZ, A.I. *et al.* 1996. Levels of cis- and trans-resveratrol and their glycosides in white and rose *Vitis vinifera* wines from Spain. J. Agric. Food Chem. **44:** 2124–2128.
18. JANG, M., *et al.* 1997. Cancer chemopreventive activity of resveratrol, a natural product derived from grapes. Science **275:** 218–220.

19. RITCHEY, J.G. & A.L. WATERHOUSE. 1999. A standard red wine: monomeric phenolic analysis of commercial Cabernet Sauvignon wines. Am. J. Enol. Viticulture **50:** 91–100.
20. DONOVAN, J.L., *et al.* 2002. Procyanidins are not bioavailable in rats fed a single meal containing grapeseed extract or the procyanidin dimer B3. Br. J. Nutr. **87:** In press.
21. SCHRAMM, D., *et al.* 1998. Differential effects of small and large molecular weight wine phytochemicals on endothelial cell eicosanoid release. J. Agric. Food Chem. **46:** 1900–1905.
22. ROBICHAUD, J.L. & A.C. NOBLE. 1990. Astringency and bitterness of selected phenolics in wine. J. Sci. Food Agric. **53:** 343–353.
23. BOURZEIX, M., *et al.* 1986. A study of catechins and procyanidins of grape clusters, the wine and other by products of the vine. Bull. l'O.I.V. **59:** 1171–1254.
24. CHEYNIER, V. & J. RIGAUD. 1986. HPLC separation and characterization of flavonols in the skins of *Vitis vinfera* var. Cinsault. Am. J. Enol. Viticulture **37**(4): 248–252.
25. PRICE, S.F., *et al.* 1995. Cluster sun exposure and quercetin in Pinot noir grapes and wine. Am. J. Enol. Viticulture **46:** 187–194.
26. FOSTER, R. 1969. Organic charge-transfer complexes. Academic Press. New York.
27. DANGLES, O. & R. BROUILLARD. 1992. Polyphenols interactions. The copigmentation case: thermodynamic data from temperature and relaxation kinetics. Medium effect. Can. J. Chem. **70:** 2174–2189.
28. MIRABEL, M., *et al.* 1999. Copigmentation in model wine solutions: occurrence and relation to wine aging. Am. J. Enol. Viticulture **50:** 211–218.

Wine, Biodiversity, Technology, and Antioxidants

FULVIO MATTIVI,[a] CHRISTIAN ZULIAN,[a] GIORGIO NICOLINI,[a] AND LEONARDO VALENTI[b]

[a]*Agricultural Institute of San Michele (IASMA), 38010 San Michele all'Adige, Italy*

[b]*University of Milan, DIPROVE, via Celoria 2, 20133 Milan, Italy*

ABSTRACT: Two chemical classes of flavonoids, the flavan-3-ols (catechins and proanthocyanidins) and the anthocyanins, are the natural antioxidants present at the highest concentration in red grape and wine. In the berry, the anthocyanins are localized in the skins, similarly to other highly bioactive phenolics of grape such as the resveratrols and the flavonols, while the flavan-3-ols are contained both in the skins and seeds. During winemaking, only a fraction of the grape flavonoids are selectively extracted into the wine, with a time course and a final yield strongly depending on the grape variety. The knowledge of the diverse and cultivar-specific characteristics of the grape is therefore critical to the appropriate design of the winemaking process. By means of a selective extraction method specifically designed to mimic the winemaking process, it was possible to analyze the "phenolic potential of red grape," thus obtaining quantitative information about amount and localization of the extractable flavonoids in the grape. Twenty-five high-quality red grape cultivars (*V. vinifera*) were studied, including 4 of the worldwide leading cultivars and 21 Italian cultivars with the highest reputation for the production of both young and aged premium red wines. The results clearly indicate that the grape variety plays a central role in determining both the absolute amount of the flavonoids, and the distribution between the berry skin and seeds of the flavan-3-ols. The very high biodiversity of the red grape cultivars in terms of flavonoids indicate a largely under-exploited opportunity to produce a range of diverse premium wines with optimized levels of natural antioxidants.

KEYWORDS: proanthocyanidins; anthocyanins; polyphenols; red wine; grapes; winemaking; antioxidants

INTRODUCTION

In the human diet, about one third of the total intake of polyphenols has been estimated to be accounted by the phenolic acids, while the flavonoids account for the remaining two thirds.[1] The most abundant flavonoids in the diet are the flavan-3-ols (catechins and proanthocyanidins), the anthocyanins and their oxidation products.[1] These two chemical classes of flavonoids are the natural antioxidants present at the highest concentration in red grape and wine.

Address for correspondence: Dr. Fulvio Mattivi, Agricultural Institute of San Michele (IASMA), via E. Mach 1, 38010 San Michele all'Adige, Italy. Voice +39-0461-615259; fax: +39-0461-615288.
fulvio.mattivi@mail.ismaa.it

The anthocyanins and the proanthocyanidins are among the most important compounds for the quality of the red wine, since they are responsible for major characteristics of this beverage, that is, color, bitterness, astringency, and chemical stability towards oxidation. More recently, new attention has been focused on the importance of these components in the human diet, since consistent experimental evidence indicated that they can be at least partially absorbed and metabolized[1–6] and that they can contribute to human health through a number of different mechanisms, only partially elucidated to date.[7]

Red wine represents a concentrated source of anthocyanidins, catechins, and proanthocyanidins, and other phenolics.[1,8–10] In the berry, the anthocyanins are contained only in the skins, with the exception of a few red-juice varieties. Their localization is therefore similar to that of other highly bioactive phenolics of grape such as the resveratrols and the flavonols. The anthocyanins, being highly water-soluble, are dissolved in free forms in the vacuolar sap of skin cells.[11]

The flavan-3-ols are contained in the solid parts of the cluster (berry skins, seeds, and stems). Part of the proanthocyanidins is localized in the berry skin, in different granules dispersed in the vacuolar sap, in forms bound to the proteins of the internal face of the tonoplasts, or bound to the cell wall polysaccharides by osidic bond.[12] A high quantity is contained in the inner part of the grape seeds, under their cuticle and epidermis,[13] and remarkable concentrations are also found in stems. The clusters are normally de-stemmed during the production of red wines; therefore, the berry skins and seeds are the two major contributors of these grape compounds to the wine.

During winemaking, only a fraction of the grape flavonoids is selectively extracted into the wine, with a time course and a final yield strongly depending on the management of the contact of the liquid must with the solid parts of the grape bunches and on the grape variety. Knowledge of the diverse and cultivar-specific characteristics of the grape is therefore critical to the appropriate design of the winemaking process. Unfortunately, the literature provides very little information concerning the contribution to wine of polyphenols from the different parts of grapes.[14]

By means of a selective extraction method specifically designed to mimic the winemaking process, it was possible to analyze the "phenolic potential of the red grape," thus obtaining quantitative information about amount and localization of the flavonoids of the grape which are extractable during the vinification. Twenty-five high-quality red grape cultivars (*V. vinifera*) were studied, including four of the worldwide leading cultivars and twenty-one Italian cultivars with the highest reputation for the production of both young and aged premium red wines.

MATERIALS AND METHODS

Analytical Methods

The analyses of the flavonoids were done by some of the most widely used spectrophotometric methods,[15] carried on under optimized conditions for red wine analysis.[16] A preliminary clean-up of the phenols was performed with a Sep-Pak C-18, 0.5 g (Waters) previously conditioned with 2 mL MeOH followed by 5 mL of 5 mM H_2SO_4. The loading and washing of the polar compounds were made at low pH with diluted sulfuric acid, in order to improve the recovery of acidic phenols, such as gallic acid.

Total Phenols (FC)

The total phenols were assessed by the reduction of phosphotungstic-phosphomolybdic acids (Folin-Ciocalteu's reagent) to blue pigments by phenols in alkaline solution.[17] Concentrations were determined by means of a calibration curve as (+)-catechin, mg/kg of grape.

Proanthocyanidins (PR)

The proanthocyanidins were evaluated by transformation into cyanidin by the method of Di Stefano *et al.*,[15] which uses iron salts as a catalyst to increase the reproducibility of the yield of cyanidin and replaces n-butanol with the optimal percentage of ethanol. Under such conditions, the average yield has been estimated to be 20%, and the PR concentration (mg/kg of grape) can be conventionally expressed as five times the amount of cyanidin formed, by means of a calibration curve with cyanidin chloride ($\varepsilon = 34700$).[15]

Index of Vanillin (VAN)

The catechins and proanthocyanidins reactive to vanillin were analyzed according to the optimized and controlled vanillin-HCl method of Broadhurst and Jones,[18] following the conditions described by Di Stefano *et al.*[15] Concentrations were calculated as (+)-catechin (mg/kg of grape) by means of a calibration curve.

Total Anthocyanins (AT)

An aliquot of 5 mL of red grape extract, diluted (5–20 times, to obtain a final reading in the range 0.3–0.6 AU) with 0.5 M H_2SO_4 was loaded on a conditioned Sep-Pak. The column was washed with 2 mL of 5 mM H_2SO_4 and the red pigments were eluted with 3 mL of MeOH into a 20 mL calibrated flask. A volume of 0.1 mL of concentrated HCl was added and the volume was brought to 20 mL with ethanol:water:HCl (70:30:1). The AT were directly quantified on the basis of their maximal absorbance in the visible range (536–542 nm). The AT content was calculated as malvidin 3-glucoside chloride equivalents (mg/kg of grape), being the experimental value of the molar absorbivity of malvidin 3-glucoside chloride in ethanol:water:HCl (70:30:1), $\varepsilon = 30100$ at 542 nm.

Total and Non-anthocyanin Flavonoids

The total flavonids (FT), the non-anthocyanin flavonoids (FNA), and their difference (ANT), were estimated spectrophotometrically in musts and wines on the basis of their absorbance at 280 nm, according to the literature.[15] Data are given as equivalent of (+)-catechin, mg/L.

Grapes

TABLE 1 reports a detailed list of the grape cultivars studied, the number of samples, their origin and diffusion, and a brief description of some of their main agronomical characteristics. The 122 samples of grapes were harvested in the year 2000. Updated information about the cultivars were derived from the literature.[19–21] The average mechanical data of the berry (average berry weight, percent weight of the seeds and skins), were obtained with the method described below.

Grape Sampling and Extraction

A sample made up of not less than 10 clusters and 3kg of technologically ripe grapes was collected from at least 10 vine plants from the same vineyard. The clusters were sampled at different heights from the soil (low, medium and high) according to the training system, and kept at 4°C until the moment of extraction, which was performed on fresh samples (less than 48h).

The extraction of phenolics from the grape was obtained with a selective extraction method, specifically designed to mimic the winemaking process. A hydroalcoholic solution consisting of EtOH:H_2O (12:88 v/v), containing 100mg/L of SO_2 and 5g/L of tartaric acid, and neutralized by 1/3 with NaOH up to pH = 3.2, was used for the extraction.

After manual separation of the berries from the stems, 200g of berries were randomly sampled, counted and weighed. The skins were manually separated from the seeds of each berry, while the pulp was discarded. Most of the pulp on the inner face of the berry skins was gently removed with the aid of an end-flattened spatula, trying to preserve the skin integrity. The skins were immediately immersed, one by one to avoid oxidations, in a 500mL Erlenmeyer conical glass flask (narrow neck) containing 250mL of hydroalcoholic solution. The flask was weighed before and after the addition of the skins, then a few drops of liquid nitrogen were added to remove the oxygen, and the flask was closed with a plastic film and put in a stove at 30°C for five days. Each flask was stirred by manually rotating it once per day. The seeds were rinsed with cold water, dried with soft laboratory paper, weighed and extracted as described for the skins.

After five days, each Erlenmeyer flasks was cooled, the solution and the solid parts were separated, the skins were gently pressed, and the residual liquid was added to the rest of the solution. The solution was then centrifuged for 9 min at 4,900 rpm to remove the lees. The resulting clear solution was made up to 300mL in a volumetric flask with the hydroalcoholic solution, 250mL were transferred to fill a dark glass bottle, which was sealed, and kept in the dark at 4°C until the analysis.

The precision of the whole method, from sampling in a monoclonal vineyard (Pinot noir) to the analytical result, was evaluated twice by repeatability test ($N = 10$). A major source of variability in the field was found to be the different ripeness of the grapes (evaluated by the degree Brix), with a RSD in the sample of 6.13%. The RSD of the extractable phenolics was for the parameters AT, FC, PR, and VAN in both trials equal to or better than 8.1% for the skins, and better than 13.9% for the seed analysis.

TABLE 1. A detailed list of the grape cultivars investigated, the number of samples, their origin, and a brief description of some of their main agronomical characteristic; all the grapes were harvested in the year 2000 (table continues on following two pages)

Cultivar	Number of samples	Main area of cultivation (Italy)	Area in production (Ha)	Average berry weight (g)[b]	Percent skins by weight[c]	Percent seeds by weight[c]	Ripening[d]	Productivity[e]	Dimension of the cluster[f]	Soluble solids (°Brix) [g]	Acidity (tartaric acid, g/L) [g]	Production of wines
Sagrantino[a]	3	Central (Umbria)	161	1.5	24.6	4.8	L	2	1.5–2	H	M	aged
Tamurro	1	South (Basilicata)	<500	2.3	20.0	2.0	M L	2	1.5	M	L	young
Groppello gentile	1	North (Lombardia)	493	2.4	17.5	3.0	M L	2.5	2.5	M	M	young
Lagrein[a]	3	North (Trentino Alto Adige)	374	2.4	17.7	4.3	L	2.5	2.5	M	M H	young to aged
Rebo	3	North (Trentino)	34	2.1	15.7	5.1	M L	2	2	M H	M	young to aged
Teroldego[a]	23	North (Trentino)	505	2.1	16.6	4.7	M L	2.5	2	M H	M	young to aged
Marzemino[a]	7	North (Trentino, Lombardia)	895	2.1	17.2	5.5	M E	2.5	1.5–3	M H	M	young to aged
Enantio[a]	3	North (Trentino)	2,178	1.5	20.6	4.7	L	2.5	2.5	M L	M H	young
Croatina	3	North West (Lombardia, Piemonte)	4,486	1.8	23.0	2.8	L	3[j]	1.5–3	M	L	aged
Schiava[k]	3	North (Trentino, Lombardia)	1,196	4.0	18.2	3.2	L	3[j]	1.5–3	M	M L	young
Lambrusco salamino	2	North (Emilia, Lombardia)	4,677	1.2	21.6	5.8	M L	2.5	2.5	M L	M H	young

TABLE 1/continued-1.

Cultivar	Number of samples	Main area of cultivation (Italy)	Area in production (Ha)	Average berry weight (g)[b]	Percent skins by weight[c]	Percent seeds by weight[c]	Ripening[d]	Productivity[e]	Dimension of the cluster[f]	Soluble solids (°Brix) [g]	Acidity (tartaric acid, g/L) [g]	Production of wines
Nebbiolo	5	North West (Lombardia, Piemonte)	5,246	2.0	17.6	4.3	L	2.5	2.5	M H	M	aged
Malvasia nera di Lecce	3	South (Puglia)	2,435	2.3	20.9	4.2	M L	2.5	2.5	M H	M H	young
Aglianico[l]	3	South (Campania, Basilicata, Puglia)	13,042	2.1	18.0	4.9	L	2	1.5–2”	M H	M H	aged
Calabrese	2	South Islands (Sicilia)	14,182[h]	2.1	25.5	2.8	M L	2.5	1.5	H	M L	young to aged
Nero d’Avola	2	South Islands (Sicilia)	14,182[h]	2.0	24.7	2.7	M L	2.5	1–2.5	H	M L	young to aged
Primitivo	3	South (Puglia)	17,249	1.9	22.9	3.0	E	2.5	1.5	H	L	young to aged
Montepulciano	3	Central and South (Marche, Abruzzo, Puglia, Sicilia)	31,008	2.6	21.3	2.6	L	3	2.5	M H	M	young to aged
Negroamaro	3	South (Puglia)	31387	2.7	17.8	3.7	L	2.5	2	H	M	young to aged
Barbera	3	North West (Lombardia, Piemonte)	47,120	2.3	17.5	3.3	L	2.5	1–2.5	M H	H	young to aged
Sangiovese	3	Central and North (Toscana, Marche, Umbria, Emilia, Sicilia, Puglia)	86,196	2.4	21.5	4.6	M L	3	1–2.5	M H	M	young to aged

TABLE 1/continued-2.

Cultivar	Number of samples	Main area of cultivation (Italy)	Area in production (Ha)	Average berry weight (g)[b]	Percent skins by weight[c]	Percent seeds by weight[c]	Ripening[d]	Productivity[e]	Dimension of the cluster[f]	Soluble solids (°Brix) [g]	Acidity (tartaric acid, g/L) [g]	Production of wines
Merlot[a]	9	North Central South	31,872 (145,000)[i]	1.7	22.4	4.2	L	2.5	1.5–2.5	H	M L	young to aged
Pinot noir[a]	3	North (Trentino Alto Adige, Veneto, Friuli)	3,538 (37,000)[i]	1.8	20.3	4.5	E	1–1.5	1.5	M H	M	young to aged
Cabernet Sauvignon[a]	25	North Central South	2403 (140,000)[i]	1.6	23.1	4.5	L	2.5	2.5	H	M L	young to aged
Syrah[a]	3	North Central South	102 (35,000)[i]	2.0	17.2	4.0	M L	2.5	2.5	M H	M L	young to aged

[a] Analyzed also in 1998 (data not reported, see discussion).
[b] Experimental average value, vintage 2000.
[c] g per 100 g of berries.
[d] L, late; ML, medium–late; ME, medium–early; E, early.
[e] 1, low; 1.5, medium–low; 2, medium; 2.5, medium–high; 3, high.
[f] 1, small; 1.5, medium–small; 2, medium; 2.5, medium–large; 3, large.
[g] L, low; ML, medium–low; M, medium; MH, medium–high; H, High.
[h] Officially classified as one cultivar.
[i] Estimated area of cultivation, world.
[j] Variable.
[k] Schiava gentile and Schiava grossa.
[l] Aglianico vulture and Aglianico taburno.

Winemaking

Six different wines were produced from three grape samples (two Pinot noir and one Enantio). A single wine was produced out of each grape sample on a semi-industrial scale with a standardized protocol and on an industrial scale with different equipment (TABLE 2). TABLE 2 reports some important conditions affecting the pomace extraction, which were applied in the different processes, after grape crushing and de-stemming, during the skin-contact fermentation and maceration. Samples of must were taken at regular intervals (24 h) and cooled; NaN_3 was added to stop the fermentation and the sample was immediately analyzed for FT, FNA, and ANT.

RESULTS AND DISCUSSION

The data obtained in this study allowed us to obtain a wide comparative survey on twenty-five different red grape cultivars. By means of a selective extraction method specifically designed to mimic the winemaking process, followed by analysis with some simple spectrophotometric assays, it was possible to analyze the "phenolic potential of the red grape," thus obtaining quantitative information about amount and localization of the extractable anthocyanidins and proanthocyanidins in grape. The results are expressed as mg/kg of grape for each class of phenolics (FC, AT, PR, VAN). Since it is possible to obtain about 0.7 L of wine out of 1 kg of grape, by dividing the results in mg/kg of grape by 0.7 it would be possible to estimate the maximal amount of phenolics (mg/L) which could be transferred to wine when the extraction is as good as in the laboratory test and there are no losses due to precipitation, degradation, and absorption by yeasts etc. The comparison of such theoretical values could therefore be considered to correspond to a yield of 100% and could be compared to the actual values obtained in wine to judge the performance of a specific winemaking process.

The data of the extractable phenolics of grape cannot be simply generalized to predict the wine composition, since a high variability in the extraction yield from grape to wine is introduced by the technological factors governing the winemaking process (temperature, duration and intensity of the liquid-solid contact, ethanol concentration, etc.). Moreover, many other chemical (hydrolysis, condensation, oxidation, etc.), microbiological (yeasts, bacteria, fungi), and physical (fining agents, temperature, oxygen, etc.) factors have been demonstrated to further modify the phenolics structure and concentration during the fermentation, fining, and storage of wine.

Nevertheless we observed in validation tests (Mattivi, unpublished data) that the values obtained in grapes have a direct predictive value for wines obtained with a similar winemaking process, and they can provide a general description of the main characteristics of each cultivar, which should then be considered in designing the optimal winemaking process.

The strong similarity of the data for different samples from the same cultivar and the same vintage, even for the few varieties for which a very high number of different samples were considered (TABLE 1), suggested to us that we could simplify the discussion of the very large data set by presenting only the average data for each variety. Great care should be taken in generalizing the average data obtained with the

TABLE 2. **Description of winemaking**

Cultivar	Soluble solids (°Brix)	Titratable acidity of the must (tartaric acid, g/L)	pH	Quantity of grape (kg)	Yeast strain (mg/L)	SO_2 (mg/L)	Systems to increase the extraction	Temperature range (°C)
Pinot noir	20.96	5.8	3.45	65	CAB90 (300)	50	manual cap punching (2 × day)	20–30
				1,500	CAB90 (300)	50	pumping-over 70% of the whole must volume (2 × day)	19–31
Pinot noir	19.96	7.65	3.2	65	CAB90 (300)	50	manual cap punching (2 × day)	22–27
				4,500	GAR26 (500)	40	rotating tank (6 complete rotations × day); pectic enzyme, 30 mg/L	22–28
Enantio	19.82	12.95	2.89	15,000	VL1 (100)	50	automatic cap punching; 12 × 1 day + 8 × 1 day + 3 × 6 days; seed elimination before tank filling	25–31
				100	VL1 (100)	50	manual cap punching (2 × day)	18–25

The table reports the most important conditions affecting the pomace extraction, which were applied in the various processes, after grape crushing and destemming, during the skin-contact fermentations and maceration, both at the semi-industrial (not more than 100 kg) and industrial scale (at least 1,500 kg).

samples in this study to those from grapes grown in different environments, since the levels of phenolics of a given cultivar can be greatly modified by climatic and agronomic conditions.[22]

We observed a large agreement of the values of concentration in 10 varieties which were studied in different vineyards of the same region also in the vintage 1998 (marked with the symbol [a] in TABLE 1, data not shown), and in particular we observed a very close repeatability of the values of localization between skins and seeds.

A remarkable agreement between the present data of localization of the phenolics and the data of wines was observed also for the varieties Enantio and Pinot noir, which were produced with different concentrations of seeds during the winemaking process.[23] The study of these wines allowed also to demonstrate that the PR extracted from the seeds gave a major contribution to the peroxy radical scavenging capacity of red wines.[16] Also the kinetic of extraction of Sagrantino wines can be clearly explained in light of the prevailing localization of the phenolics of this variety in the skins.[24]

Amount of Extractable Phenolics and Anthocyanins in the Grape

The FC assay gives predictable reactions with the compounds bearing one or more phenol groups, thus including all the known classes of phenolics in wine, and provides highly reliable information about their global concentration.[25] Medium-to-high levels generally indicate that a cultivar is good for the production of robust wines able to tolerate long aging; these levels permit more flexible winemaking design. To give an idea of how phenolics contribute to the oxygen capacity of wine, it has been reported that a liter of very rough young wine with 5 g/L of phenolics could consume the oxygen contained in 5 to 10 times its volume of air.[26] Grape cultivars with low to medium-low phenolic levels are usually better for the production of young and fruity wines. They are sometimes used also for the production of medium-aged wines, but

TABLE 3. Mean content of extractable polyphenols (Folin Ciocalteu) in different cultivars of red grape[a]

Range	Cultivar
$1,500 \leq X < 1,800$	Schiava–Barbera–Syrah
$1,800 \leq X < 2,100$	Negroamaro–Nebbiolo–Cabernet Sauvignon
$2,100 \leq X < 2,400$	Calabrese–Groppello–Marzemino–Malvasia nera di Lecce–Rebo–Merlot–Sangiovese–Enantio
$2,400 \leq X < 2,700$	Montepulciano–Nero d'Avola–Tamurro
$2,700 \leq X < 3,000$	Primitivo–Lagrein–Teroldego
$3,000 \leq X < 3,300$	Lambrusco salamino–Pinot noir
$3,300 \leq X < 3,600$	Croatina
$3,600 \leq X < 3,900$	Aglianico
$3,900 \leq X < 4,200$	Sagrantino

[a]Data as (+)-catechin, mg/kg of grape.

in this case they require a more careful winemaking design process and/or blending with stronger cultivars to obtain adequate levels of phenolics.

TABLE 3 ranks the cultivars in 9 groups sorted by increasing amounts of FC: there is a variability factor of 2.7 in the richness of extractable phenolics between the values of the varieties with the extreme average concentrations. Considering the individual samples, the lowest value was found in the Schiava variety (1358 mg/kg) whereas the highest was found in a Sagrantino sample (4628 mg/kg).

The AT parameter is a direct estimate of the red pigments, including both the monomers and the high-molecular mass phenolics containing one or more flavylium cation moiety in the structure.

At the molecular level, the composition of the anthocyanins that accumulate in the grape skin during ripening is quite variable, since it is the synthesis of these secondary metabolites under genetic control at the cultivar level. As a consequence, each cultivar has a peculiar pattern of these compounds, which is a distinguishing chemometric character useful for the classification of grape varieties[27] and of the relevant wines.[28] Also the richness in AT is a characteristic associated with the cultivar. Given the good extractability of the anthocyanins, and in light of the known quantitative contribution of the anthocyanins to the color of wine,[29] we can expect a value over 500 mg/kg of grape to be in general sufficient to obtain a wine of medium color, while lower values can be sometimes more critical, especially when the grape is at the same time particularly rich in proanthocyanidins, as is the cultivar Pinot noir. High values of AT produce highly appreciated, strongly colored wines, which can be used to improve the color of other cultivars, or to produce dark red monovarietal wines such as for the Lagrein and Teroldego cultivars.

TABLE 4 ranks the cultivars in eight groups sorted by increasing amounts of AT: the variability factor for the richness of extractable AT is 4.8 between the average values of the varieties with extreme concentrations. Considering the individual samples, the lowest value was found in a sample of Primitivo (250 mg/kg) whereas the highest value (9.3×) was found in a Teroldego grape (2,323 mg/kg).

TABLE 4. Mean content of extractable anthocyanins in different cultivars of red grape[a]

Range	Cultivar
$300 \le X < 500$	Primitivo–Schiava–Nebbiolo–Pinot noir
$500 \le X < 700$	Sangiovese–Negroamaro–Nero d'Avola–Tamurro–Calabrese–Malvasia nera di Lecce–Groppello
$700 \le X < 900$	Sagrantino–Aglianico–Lambrusco salamino–Cabernet Sauvignon–Barbera
$900 \le X < 1{,}100$	Syrah–Merlot–Montepulciano
$1{,}100 \le X < 1{,}300$	Croatina–Marzemino–Rebo
$1{,}300 \le X < 1{,}500$	Enantio
$1{,}500 \le X < 1{,}700$	Lagrein
$1{,}700 \le X < 1{,}900$	Teroldego

[a]Data as malvidin 3-monoglucoside chloride, mg/kg of grape.

Amount of Extractable Catechins and Proanthocyanidins in the Grape

The PR is a highly specific assay which provides a sound evaluation of the total amount of proanthocyanidins and appears to be mainly linked to variations of high-molecular mass phenolics corresponding to at least five units.[30] The VAN assay is also connected to the amount of flavan-3-ols, but has a different meaning since it is more sensitive to the low-molecular mass phenolics corresponding to two to four units, and is also sensitive to the monomers.

TABLES 5 and 6 present the average data of these two assays, dividing the cultivars into nine and eight groups, respectively. For extractable PR, there is a variability factor of 2.4 between the average values of the varieties with extreme concentrations (Schiava and Sagrantino). Considering the individual samples, the lowest value was found in a sample of Marzemino (1,317 mg/kg) which was considered not typical since it was much lower than the other six samples belonging to the same variety, while the highest value (3.7×) was found in a Sagrantino sample (4,859 mg/kg).

The variability factor of extractable VAN was 4.7 between the average values of the Barbera and Sagrantino varieties. Considering the extreme single values, the lowest was found in a Syrah sample (865 mg/kg) and the highest (5.5×) in a Sagrantino sample (4091).

The ratio VAN/PR, which is expected to be high when the variety is rich in monomer and/or low-molecular weight phenolics,[28] averaged 0.67 for the whole data set, being particularly low in the Enantio (0.44) and Barbera (0.46) grape and, on the opposite end, very high for the Aglianico (0.92) and Pinot noir (1.11) cultivars.

In conclusion, the PR and VAN data allowed us to obtain a ranking of the absolute richness of the different grape cultivars in extractable flavan-3-ols and to highlight the differences in the reactivity of these compounds which suggest different degrees of polymerization associated with different cultivars. More in-depth studies, using the normal phase HPLC method,[30–32] should permit better quantification of these differences.

TABLE 5. Mean content of extractable proanthocyanidins (PR) in different cultivars of red grape[a]

Range	Cultivar
$1{,}600 \leq X < 1{,}900$	Schiava–Syrah–Barbera
$1{,}900 \leq X < 2{,}200$	Calabrese–Negroamaro
$2{,}200 \leq X < 2{,}500$	Marzemino–Groppello–Rebo
$2{,}500 \leq X < 2{,}800$	Merlot–Sangiovese–Cabernet Sauvignon–Malvasia nera di Lecce
$2{,}800 \leq X < 3{,}100$	Nebbiolo–Nero d'Avola–Enantio–Teroldego–Primitivo–Pinot noir–Montepulciano
$3{,}100 \leq X < 3{,}400$	Lagrein
$3{,}400 \leq X < 3{,}700$	Tamurro–Aglianico
$3{,}700 \leq X < 4{,}000$	Lambrusco salamino–Croatina
$4{,}000 \leq X < 4{,}300$	Sagrantino

[a]Data as cyanidin, mg/kg of grape.

TABLE 6. Mean content of extractable catechins and proanthocyanidins, reactive to vanillin (VAN), in different cultivars of red grape[a]

Range	Cultivar
800 ≤ X < 1200	Barbera–Syrah
1,200 ≤ X < 1,600	Marzemino–Enantio–Schiava–Negroamaro–Malvasia nera di Lecce–Rebo–Cabernet Sauvignon–Groppello
1,600 ≤ X < 2,000	Sangiovese–Calabrese–Montepulciano–Nebbiolo–Merlot–Teroldego–Lagrein–Nero d'Avola
2,000 ≤ X < 2,400	Primitivo
2,400 ≤ X < 2,800	Lambrusco salamino–Croatina–Tamurro
2,800 ≤ X < 3,200	(none)
3,200 ≤ X < 3,600	Aglianico–Pinot noir
3,600 ≤ X < 4,000	Sagrantino

[a]Data as (+)-catechin, mg/kg of grape.

Localization of the Catechins and Proanthocyanidins in the Berry

A main goal accomplished with this experimental approach was the evaluation of the localization of the flavan-3-ols between skins and seeds. This is to our knowledge the first report giving a quantitative evaluation of the contribution of the different parts of the berry to the phenolics extractable during the winemaking. In TABLES 7 and 8 the grape cultivars are ranked in 9 and 10 groups, respectively, on the basis of the increasing contribution (percentage) of the seeds to the total extractable PR and VAN.

TABLE 7. Percentage of extractable proanthocyanidins (PR) which are localized in the seeds, average data for different cultivars of red grape[a]

Range	Cultivar
15 ≤ X < 20	Montepulciano–Croatina–Enantio
20 ≤ X < 25	Nebbiolo–Sagrantino–Syrah
25 ≤ X < 30	Barbera–Groppello–Nero d'Avola–Teroldego
30 ≤ X < 35	Negroamaro–Primitivo–Tamurro–Cabernet Sauvignon–Calabrese
35 ≤ X < 40	Marzemino–Sangiovese–Malavasia nera di Lecce–Lagrein–Merlot
40 ≤ X < 45	Aglianico–Rebo
45 ≤ X < 50	Lambrusco salamino–Schiava
50 ≤ X < 55	(none)
55 ≤ X < 60	Pinot noir

[a]Data as percentage of the total extractable PR.

TABLE 8. Percentage of extractable catechins and proanthocyanidins reactive to vanillin (VAN), which are localized in the seeds[a]

Range	Cultivar
$35 \leq X < 40$	Tamurro–Croatina–Montepulciano
$40 \leq X < 45$	Nebbiolo–Sagrantino–Enantio
$45 \leq X < 50$	(none)
$50 \leq X < 55$	Sangiovese–Syrah
$55 \leq X < 60$	Nero d'Avola–Malvasia nera di Lecce
$60 \leq X < 65$	Cabernet Sauvignon–Primitivo–Groppello–Calabrese–Teroldego–Negromaro
$65 \leq X < 70$	Schiava–Merlot
$70 \leq X < 75$	Barbera–Aglianico–Lambrusco salamino–Lagrein
$75 \leq X < 80$	Rebo–Marzemino
$80 \leq X < 85$	Pinot noir

[a]Average data for different cultivars of red grape, as precentage of the total extractable VAN.

Most PR (at least 80%) is in the skins for the cultivars Montepulciano, Croatina and Enantio, and 10 out of 25 cultivars have less than 30% of their PR in the seeds. For all other varieties, the seeds contribute greatly to the extractable PR of the grape, 30–60% of the total.

The extractable low-molecular weight flavan-3-ols (VAN) are even more concentrated in the seeds. The seeds always contribute more than one-third to the total VAN, representing for 19 out of 25 cultivars more than 50% of the total extractable VAN of the grape.

The skins account on average for 20.1% of the berry weight (range 15.7–25.5%), however the seeds account on average for 4% (range 2.0–5.8%), the remaining 75.9% being represented by the pulp (TABLE 1). This means that, on an absolute basis, the concentration of extractable flavan-3-ols is always higher in the seeds than in the skins.

A General Approach to Winemaking Design

We followed the time course of a few winemaking trials in order to describe the kinetics of extraction of the phenolics.

In the Enantio trials, the ANT reached their maximal value in the must after four days of extraction on both the semi-industrial (see FIGURE 1a) and industrial (FIGURE 1b) scales; after this point they started to drop progressively with time. The extraction of the FNA showed in both scales a first, almost linear and strong increase, up to the fourth day, followed by a long-lasting but quantitatively limited extraction of FNA up to day 9 for the semi-industrial trial, while no increase at all was obtained for the industrial trial, where most of the seeds had been removed from the bottom of the tank during the first hours after crushing. These data clearly indicate that the first strong extraction of FNA, similar to the extraction curve of the anthocyanins, is due to the extraction of the FNA from the grape skins, while the extraction from the seeds prevails in the second part of the extraction, starting after the end of the quick fermentation phase. The higher absolute values obtained in the industrial scale are due to a higher extraction yield of the industrial tank in comparison to the experimental one, carried on under less favorable conditions (TABLE 2). The weak extraction of FNA in the second phase at the semi-industrial wine was predictable in the light of the low contribution of Enantio seeds to the total extractable PR (at most 20%, TABLE 7).

In the case of Pinot noir samples (see FIGURES 2a and 3a) the maximal values of ANT are obtained in three to four days after which they started to drop, similarly to the Enantio wines, while the extraction of FNA appeared to be a continuous process

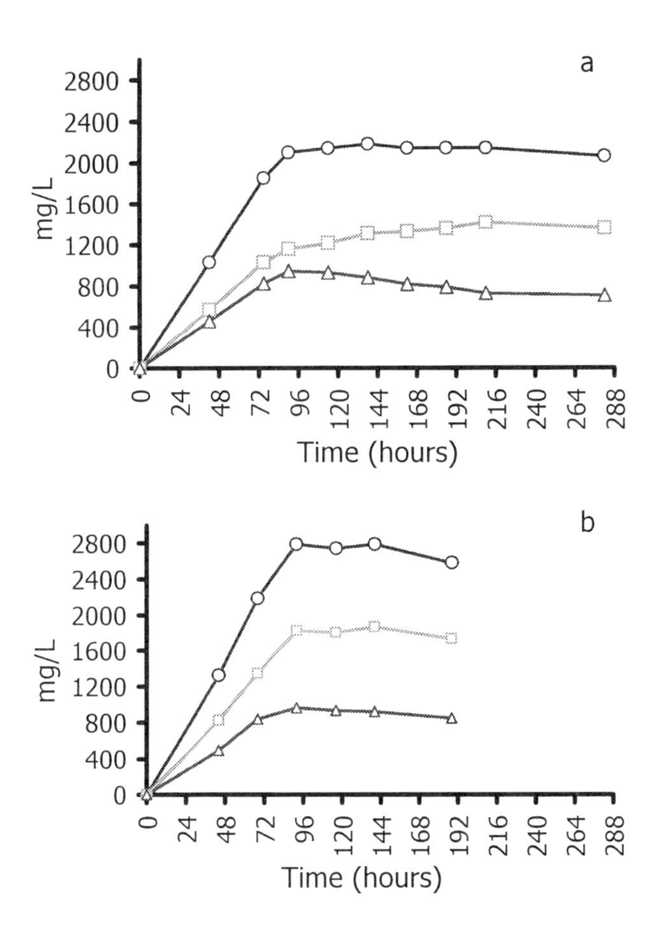

FIGURE 1. Time course of the extraction of the phenolics from the Enantio grape during the winemaking at the semi-industrial (**a**) and industrial (**b**, seeds removed) winery. ○, total flavonoids (FT); FNA, □, non-anthocyanin flavonoids (FNA); △, FT–FNA (ANT); data as (+)-catechin equivalent, mg/L.

which could not be divided into two phases. In the trial shown in FIGURE 2a, the maximal values of FNA were reached on day 9, while for the trial in FIGURE 3a the extraction was still not completed by day 10.

The traditional red-winemaking process applied to the industrial sample (FIG. 2b) permitted a stronger extraction of FNA in a shorter time compared to the semi-industrial trial. A first quick extraction from the skins was followed by a second phase which corresponds to the intense extractions from the seeds, interrupted by the winemaker after six days when the desired levels of FNA were obtained. The industrial trial with the second and different sample of Pinot noir (FIG. 3b) showed a time course similar to the previous one. In this case, the use of a rotating tank permitted the highest yield, possibly because of a more effective soaking of the seeds obtained with such equipment. Also in the case of the industrial Pinot noir where the two extraction phases were distinguishable, the second one gave ca. 50% of the total

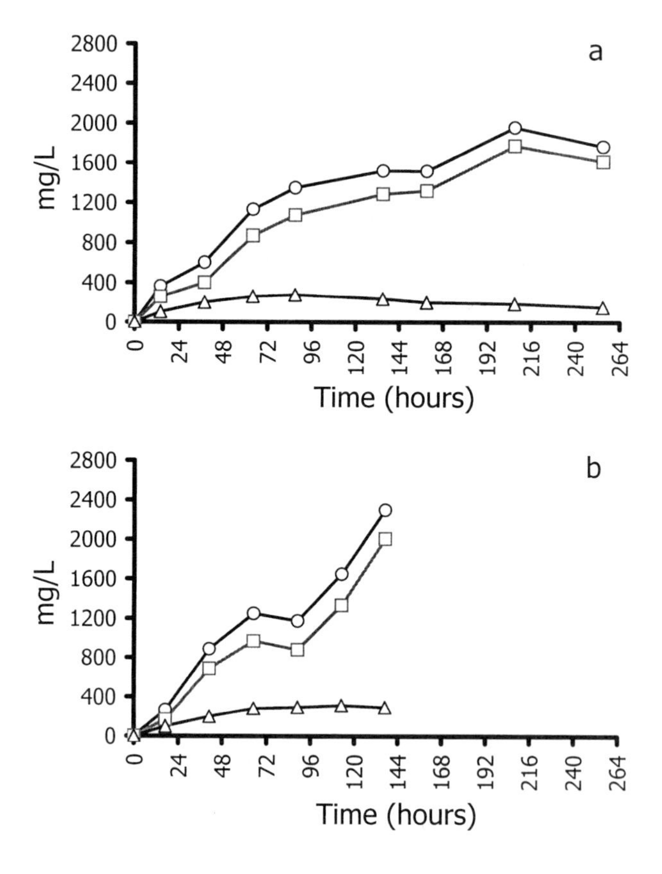

FIGURE 2. Time course of the extraction of the phenolics from the Pinot noir grape during winemaking at the semi-industrial (**a**) and industrial (**b**, traditional tank) winery. ○, total flavonoids (FT); FNA, □, non-anthocyanin flavonoids (FNA); △, FT–FNA (ANT); data as (+)-catechin equivalent, mg/L.

FNA to the must without reaching the end-point of extraction, which was expected in the light of the localization of at least 55% of the PR of this cultivar in the seeds (TABLE 7).

The close agreement between extractable phenolic composition in grapes and the time-course of the winemaking trials demonstrates the usefulness of the information about the localization of the flavan-3-ols in the different solid parts of the grape berry. This information can be easily applied to elucidate and possibly predict the time-course of the winemaking process, thereby adding new and important knowledge very helpful in choosing the optimal technological parameters for red-wine production.

Great care should be taken in designing the specific winemaking process appropriate for each variety. When dealing with compounds that broadly affect the sensory properties of the wine, an approach aimed at maximizing the concentration

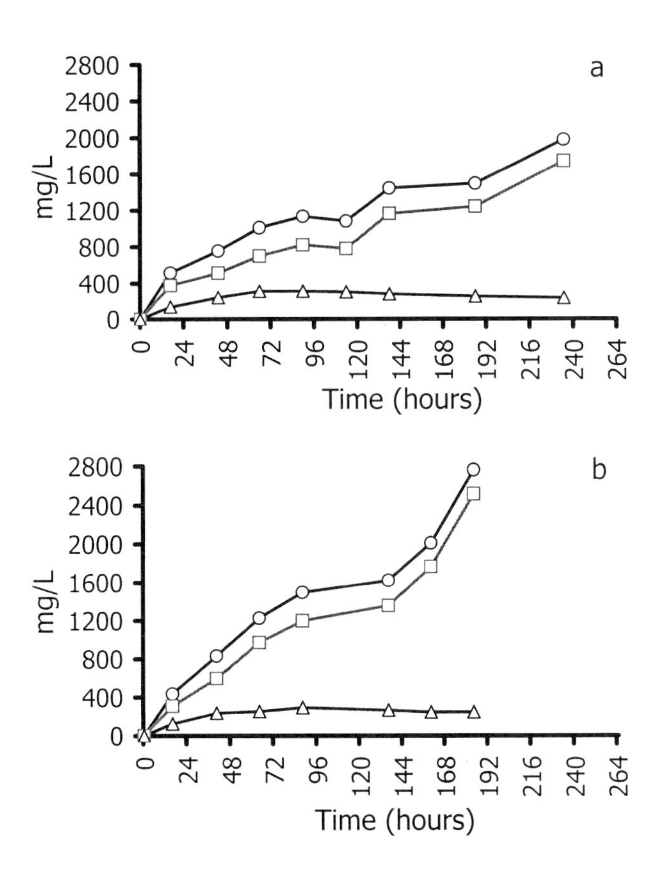

FIGURE 3. Time course of the extraction of the phenolics from the Pinot noir grape during winemaking at the semi-industrial (**a**) and industrial (**b**, rotating tank) winery. ○, total flavonoids (FT); FNA, □, non-anthocyanin flavonoids (FNA); △, FT–FNA (ANT); data as (+)-catechin equivalent, mg/L.

of phenolics is not always appropriate for the production of premium red wines. While on the one hand, high values of AT do not present any particular technological or sensory drawback and are therefore often appreciated, the wines having the highest values of PR and especially of VAN are expected in some cases to be more bitter and astringent and to require a longer maturation time before their consumption as medium-to-long–aged wines. Aged wines have a lower content of monomers, which are expected to be more readily absorbed,[7] and contain more complex structures which lead to the formation of stable semiquinone radicals.[33] All the grape cultivars analyzed in this study have important concentrations of extractable phenolics, when compared to many other fruits and beverages.[10] It should also be taken into consideration that an elegant red wine with a medium and well-balanced phenolic content can be easily consumed with a number of different foods and is therefore more likely to provide a daily contribution of phenolics to the diet.

CONCLUSIONS

The results clearly indicate that the grape cultivar plays a central role in determining both the absolute amount of the extractable flavonoids, and the distribution between the berry skin and seeds of the extractable proanthocyanidins.

In the twenty-five cultivars used for this study, the average distribution of the extractable flavan-3-ols between the seeds and the skins was found to vary in the range 15.9%–57.0% for PR and 37.0%–81.7% for VAN, illustrating differences critical for the appropriate winemaking design. It should be observed that most of the technical equipment for winemaking is designed to manage in the most flexible and efficient way the solid-liquid contact with the skins. In the light of the very important contribution of the grape seeds proanthocyanidins to the phenolics of red wines, as evidenced in this study, the development of winemaking equipment allowing also an improved management of the seeds—either to remove part of them or to better control the extraction, depending on the target composition of the wine—is highly desirable.

The quantitative estimate of the localization of the extractable proanthocyanidins in the different solid parts of the grape can be easily applied on a laboratory scale to elucidate the time-course of the extractions of the flavan-3-ols during the industrial winemaking process. This information contributes knowledge very useful for the choice of the most appropriate winemaking design, keeping in mind that, at least for the proanthocyanidins, the enologist should usually be concerned with the optimization and not necessarily with the maximization of their concentration.

All the grape cultivars analyzed in this study were shown to have high concentrations of extractable phenolics. The average data for the cultivars varied in the range 1,533–4,174 mg/kg (FC), 382–1,830 mg/kg (AT), 1,737–4,241 mg/kg (PR), and 865–3,649 mg/kg (VAN).

Most of the red wines on the world market are made from a few grape varieties. The very high biodiversity of the red grape cultivars in terms of flavonoids described in this study indicates a largely under-exploited opportunity to produce a range of diverse premium wines with optimized levels of natural antioxidants, in order to positively contribute to human health.

ACKNOWLEDGMENTS

This work was in part supported by the Provincia Autonoma di Trento and Consiglio Nazionale delle Ricerche, and by VITIS Rauscedo scrl (Pordenone, Italy). We are grateful to them.

Special thanks are due to Umberto Pichler of Cantine Mezzacorona (Mezzocorona, Italy), to Andreas Prast and Giuliano Cova (IASMA) and to Angelo Divittini, Cesare Bosio, Piero Donna, Marco Tonni and Francesco Cisani (DIPROVE) for their helpful collaboration in this research.

REFERENCES

1. SCALBERT, A. & G. WILLIAMSON. 2000. Dietary intake and bioavailability of polyphenols. J. Nutr. **130**(8S Suppl.): 2073S–2085S.
2. HOLLMAN, P.C.H. & M.B. KATAN. 1998. Absorption, metabolism and bioavailability of flavonoids. *In* Flavonoids in Health and Disease. C.A. Rice-Evans & L. Packer, Eds.: 483–522. Marcel Dekker, Inc. New York.
3. BAGCHI, D., M. BAGCHI, S.J. STOHS, *et al.* 2000. Free radicals and grape seeds proanthocyanidin extract: importance in human health and disease prevention. Toxicology **148**(2–3): 187–197.
4. MIYAZAWA, T., K. NAKAGAWA, M. KUDO, *et al.* 1999. Direct intestinal absorption of red fruits anthocyanins, cyanidin-3-glucoside and cyanidin-3,5-diglucoside, into rats and humans. J. Agric. Food Chem. **3:** 1083–1091.
5. YOUDIM K.A., B. SHUKITT-HALE, S. MACKINNON, *et al.* 2000. Polyphenolics enhance red blood cells resistance to oxidative stress: in vitro and in vivo. Biochim. Biophys. Acta **1523**(1): 117–122.
6. TSUDA, T., F HORIO & T. OSAWA. 2000. The role of anthocyanins as an antioxidant under oxidative stress in rats. Biofactors **13**(1–4): 133–139.
7. GERMAN, B.J. & R.L.WALZEM. 2000. The health benefits of wine. Annu. Rev. Nutr. **20:** 561–593.
8. MATTIVI, F. & G. NICOLINI. 1997. Analysis of polyphenols and resveratrol in Italian wines. BioFactors **6:** 445–448.
9. VRHOVSEK, U. & F. MATTIVI. 1998. Specific substances in wine for the prevention of coronary artery disease. *In* Proceedings of the International Symposium "Cardiovascular diseases," Z. Pajer & D. Stiblar-Martincic, Eds.: 449–463. Lubiana, Slovenia.
10. PAGANGA, G., N. MILLER & C.A. RICE-EVANS. 1999. The polyphenolic content of fruit and vegetables and their antioxidant activities. What does a serving constitute? Free Radical Res. **30:** 153–162.
11. AMRANI-JOUTEI, K. & Y. GLORIES. 1995. Tanins et anthocyanes: localisation dans la baie de raisin et mode d'extraction. Rev. Fr. Oenol. **153:** 28–31.
12. AMRANI-JOUTEI, K., Y. GLORIES & M. MERCIER. 1994. Localization of tannins in grape berry skins. Vitis **33:** 133–138.
13. AMRANI-JOUTEI, K. 1993. Localisation des anthocyanes et des tanins dans les raisin: étude de leur extractibilitè. Ph.D. thesis, University of Bordeaux II, France.
14. DE FREITAS, V.A.P., Y. GLORIES & A. MONIQUE. 2000. Developmental changes of procyanidins in grapes of red *Vitis vinifera* varieties and their composition in respective wines. Am. J. Enol. Vitic. **4:** 397–403.
15. DI STEFANO, R., M.C. CRAVERO & N. GENTILINI. 1989. Methods for the study of wine polyphenols. L'Enotecnico : 83–89.
16. RIGO, A., F. VIANELLO, G. CLEMENTI, *et al.* 2000. Contribution of the proanthocyanidins to the peroxy-radical scavenging capacity of some Italian red wines. J. Agric. Food Chem. **48:** 1996–2002.
17. DI STEFANO, R. & S. GUIDONI. 1989. The analysis of total polyphenols in musts and wines. Vignevini : 47–52.

18. BROADHURST, R. B. & W.T. JONES. 1978. Analysis of condensed tannins using acidified vanillin. J. Sci. Food Agric. **28:** 788–794.

19. AA.VV. 1998. Enotria, il quaderno della vite e del vino 1998. Documenti e statistiche 1997. Suppl. to Il Corriere Vinicolo, 15. Unione Italiana Vini Ed., Milano, Italy.

20. CALÒ A., A. SCIENZA & A. COSTACURTA. 2001. I vitigni d'Italia. Edagricole, Bologna, Italy.

21. AA.VV. 2000. Catalogo dei cloni del nucleo di premoltiplicazione delle Venezie. N.P.V.V. Ed, Pordenone, Italy

22. BERTAMINI, M. & F. MATTIVI. 1999. Meteorological and microclimatic effects on Cabernet Sauvignon from Trentino area. Part II: Flavonoids and resveratrol in wine. *In* Proceedings of GESCO, Sicily, Italy, 6–12 June. **11:** 502–509.

23. NICOLINI, G., F. MATTIVI, R. GIMENEZ MARTINEZ & U. MALOSSINI. 1998. Polyphenols extracted from seeds and characteristics of red wines from Trentino–Italy. Riv. Vitic. Enol. **2:** 31–50.

24. NICOLINI, G. & L. VALENTI. 2001. Survey of the polyphenol content of Sagrantino wines by methods useful in the process control. Riv. Vitic. Enol. **1:** 47–63.

25. SINGLETON, V.L., R. ORTHOFER & R.M. LAMUELA-RAVENTOS. 1999. Analysis of total phenols and other oxidation substrates and antioxidants by means of Folin-Ciocalteu reagent. Methods Enzymol. **299:** 152–178.

26. BOULTON, R.B., V.L. SINGLETON, L.F. BISSON & R.E. KUNKEE. 1996. Principles and Practices of Winemaking. Chapman & Hall. New York..

27. MATTIVI, F., A. SCIENZA, O. FAILLA, *et al.* 1989. *Vitis vinifera*: a chemotaxonomic approach—anthocyanins in the skin. *In* Proceedings of the 5th International Symposium on Grape Breeding. Vitis, Special Issue 1990 : 119–133.

28. MATTIVI, F. & G. NICOLINI. 1997. Analysis of polyphenols and resveratrol in Italian wines. BioFactors **6:** 445–448.

29. MATTIVI, F., H. ROTTENSTEINER & D. TONON. 2001. The tristimulus measurement of the color of hydroalcoholic solutions of anthocyanins. Riv. Vitic. Enol. **2/3:** 51–73.

30. VRHOVSEK, U., F. MATTIVI & A.L. WATERHOUSE. 2001. Analysis of red wine phenolics: comparison of HPLC and spectrophotometric methods. Vitis **40**(2): 87–91.

31. KENNEDY, J.A. & A.L. WATERHOUSE. 2000. Analysis of pigmented high-molecular-mass phenolics using ion-pair, normal-phase high-performance liquid chromatography. J. Chromatogr. A. **866:** 25–34.

32. WATERHOUSE, A.L., S. IGNELZI & J.R. SHIRLEY. 2000. A comparison of methods for quantifying oligomeric proanthocyanidins from grape seed extracts. Am. J. Enol. Vitic. **4:** 383–389.

33. ROSSETTO, M., F. VIANELLO, A. RIGO, *et al.* 2001. Stable free radicals as ubiquitous components of red wines. Free Radical Res. **35:** 933–939.

Chemistry of the Antioxidant Effect of Polyphenols

WOLF BORS AND CHRISTA MICHEL

Institut für Strahlenbiologie, GSF Forschungszentrum für Umwelt und Gesundheit, D-85764 Neuherberg, Germany

ABSTRACT: Most plant-derived polyphenols exhibit strong antioxidant potentials, established by various assay procedures. With pulse radiolysis experiments, absolute scavenging rate constants can be obtained with a variety of oxidizing radicals which allow further structure-activity correlations and, combined with EPR spectroscopy, detailed insight into the mechanisms governing these antioxidant reactions. The most striking difference occurs between regular flavonoids and both condensed and hydrolyzable tannins. The tannins are considered superior antioxidants as their eventual oxidation may lead to oligomerization via phenolic coupling and enlargement of the number of reactive sites, a reaction which has never been observed with the flavonoids themselves.

KEYWORDS: polyphenols; resveratrol; antioxidants; flavonoids

INTRODUCTION

Polyphenols comprise a wide area of natural substances of plant origin. Almost all of them exhibit a marked antioxidant activity. Typical examples in order of increased complexity are hydroxy stilbenes such as resveratrol, an antioxidant in grapes and wine,[1,2] oligomeric catechol structures based on caffeic acid moieties found in several Lamiatae plants (rosmarinic acid, salvianolic acids,[3] yunnaneic acids,[4] etc.), the large group of flavonoids,[5] monomeric and oligomeric flavan-3-ols [derivatives of (+)-catechin or (−)-epicatechin, also known as proanthocyanidins or condensed tannins],[6] or gallo- and ellagitannins (hydrolyzable tannins).[7,8] Structures of typical compounds representing each class are shown in SCHEME 1. In the context of this conference, both flavonoids and flavan-3-ols are ingredients of grape seeds and skin and thus present in red wine especially.[9–13] Both classes of compounds have been suggested to be responsible for the so-called French paradox.[14–16] Gallotannins occur in oak,[17,18] and in addition to being leached into wine during barrique aging,[19–21] it has been suggested that they protect the other flavonoids from oxidation,[22,23] that is, they may act as first line of defense against oxidative degradation of wines in general.

Various antioxidant assays have established the antioxidant potential of most of these substances, prominent among them the so-called TEAC assay (trolox-equivalent antioxidant capacity[24]) and the DPPH assay (bleaching of the absorption

Address for correspondence: Dr. Wolf Bors, Institut für Strahlenbiologie, GSF Research Center, D-85764 Neuherberg, Germany. Voice: +49-89-3187-2508; fax: +49-89-3187-2818. bors@gsf.de of the stable free radical 2,2-diphenyl-1-picryl-hydrazyl[25]). The latter assay, in particular, has been used to screen a wide variety of polyphenolic compounds with ever lower IC_{50}-values with increasing numbers of hydroxy groups in the respective molecules.[26]

In our own studies we employed pulse radiolysis to determine absolute rate constants for scavenging of electrophilic radicals such as hydroxyl (·OH) or azide (·N_3)

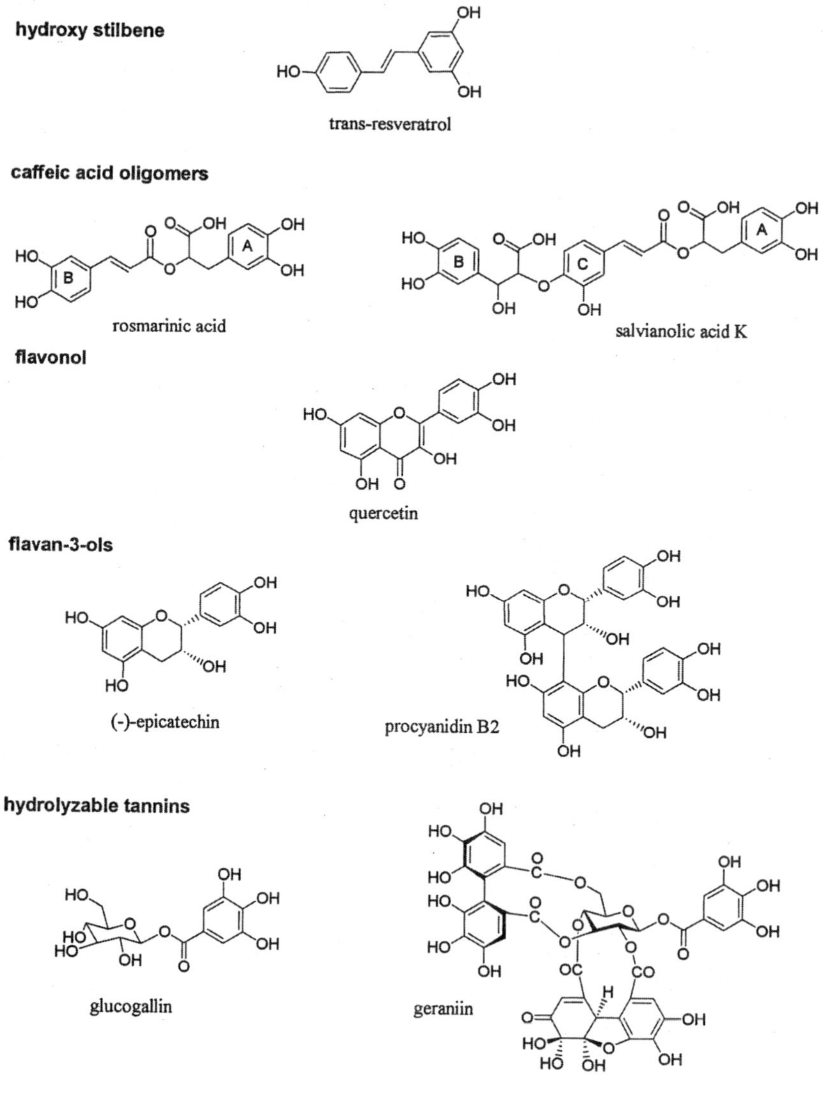

hydroxy stilbene

trans-resveratrol

caffeic acid oligomers

rosmarinic acid

salvianolic acid K

flavonol

quercetin

flavan-3-ols

(-)-epicatechin

procyanidin B2

hydrolyzable tannins

glucogallin

geraniin

SCHEME 1. Typical polyphenols of various structural classes.

radicals, but also superoxide anions ($O_2^{\cdot-}$) and—in some cases—organic peroxyl radicals.[27–31] Aside from the scavenging rate constants, pulse radiolysis also allows the determination of the stability of the antioxidant radicals, as well as the kinetics of interactions with other reactants.[32] These approaches to determining univalent redox potentials[32,33] are crucial to determine the location of the various polyphenolic antioxidants within the lattice of the antioxidant network, whereas the intermittent transient spectra yield insufficient information on the structure of the respective radicals. For the latter purpose EPR spectroscopy is the ideal tool.[34–38] Quantum-mechanical calculations (*ab initio*, density-functional theory) become more and more attractive to obtain other physico-chemical parameters[39,40] and eventually to corroborate structural assignments of the EPR coupling constants (Bors *et al.*, manuscript in preparation).

The present paper attempts to give a short overview on the kinetic, structural and mechanistic properties of the various polyphenolic antioxidants and their respective radicals. As shall be seen, depending on the various structures, quite distinct mechanisms control the antioxidant properties of this large but diverse structural class.

KINETIC PARAMETERS

The relative antioxidant assays TEAC[41,42] and DPPH bleaching[25] (and other related ones[43]) yield limited information on structure-activity relationships (SAR), cf. the above-mentioned correlation of the antioxidant potential with the number of hydroxy groups[26] or the direct correlation observed for the degree of oligomerization for proanthocyanidins with their antioxidant capacity up to a certain limit.[42] In addition, pulse-radiolytic determinations of absolute rate constants of ·OH radicals with flavan-3-ols and gallotannins revealed a linear correlation between rate constants and the number of reactive hydroxy groups (see FIGURE 1).[44,45] The fact that for gallotannins rate constants exceed the diffusion-controlled limits is considered evidence for the attack at several target sites simultaneously, the prime example being penta-galloyl glucose with five accessible galloyl moieties around the central sugar molecule.[44]

For the nominal polyphenolic *trans*-resveratrol (3,5,4′-trihydroxystilbene) a recent pulse radiolysis study verified that only the *para*-hydroxy group acts as a radical target site whereas the *meta*-dihydroxy structure merely participates in the secondary stabilization of the phenoxyl radical.[46] This is another proof for the higher stability of semiquinones derived from catechol (*ortho*-dihydroxy) or hydroquinone (*para*-dihydroxy) as compared to those from *meta*-dihydroxy compounds.[47]

With regard to the interaction of polyphenolic antioxidants and/or their radicals with other substrates, those reactions with reductants such as ascorbate are most pertinent for the determination of univalent redox potentials.[32,33] Owing to the complex kinetic interrelationships of such redox systems, few data are available. Thus, the only conclusions to be drawn at the present are that the flavonoids have too diverse redox potentials to bracket them neatly within the antioxidant network. Rather, depending on their structure [flavones and flavonols vs. flavanones and dihydroflavon(ol)s], they are either capable of oxidizing ascorbate to its radical state or reducing the ascorbate radical.[32]

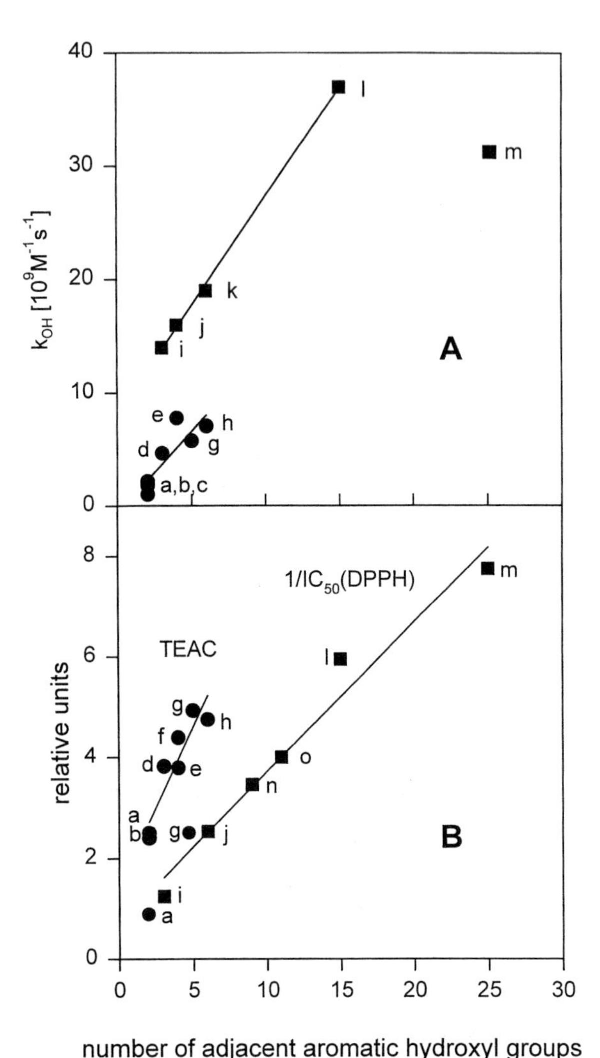

FIGURE 1. Correlation of antioxidant capacity with the number of reactive hydroxy groups (**A**) absolute rate constants from pulse radiolysis experiments (**B**) relative antioxidant capacity from TEAC[41,42] and DPPH[26] assays. (●) flavan-3-ols; a–catechin, b–epicatechin, c–Pycnogenol®, d–epigallocatechin, e–procyanidin B2, f–procyanidin A2, g–epicatechin gallate, h–epigallocatechin gallate. (■) gallo-/ellagitannins; i–propylgallate, j–glucogallin, k–hamamelitannin, l–pentagalloyl glucose, m–tannic acid, n–corilagin, o–geraniin.

Most of the semiquinones or aroxyl radicals derived from the various polyphenolic structures have been observed to decay by second-order processes during the pulse-radiolytic experiments.[28] This disproportionation, in consequence, leads to regeneration of the parent polyphenol and formation of the quinoid oxidation product in equal parts. In fact, the behavior of the antioxidant radicals (phenoxyl, semiquinone or aroxyl radicals) and that of the corresponding quinones is the major determinant for the distinct antioxidant mechanism of these substances, as detailed below.

MECHANISTIC IMPLICATIONS

The various assays for antioxidant potential as well as the pulse-radiolytic determination of diffusion-controlled rate constants merely established their very effective scavenging potential, i.e. their effectiveness as chain-breaking antioxidants.[48,49] In isolated cases such as kaempferol, however, high scavenging rate constants do not translate into a high antioxidant potential since the radical in question is rather unstable and, at least in model lipid matrices, capable of carrying out chain-propagating reactions.[31] In general, the mostly observed second-order decay reactions in aqueous solutions are slow enough that such chain-propagating reactions are unlikely to occur. Furthermore, in a complex biological milieu, second-order processes—i.e., reactions between two radical species at their low steady-state levels—have to compete with interactions with other cellular components, similar to those seen in studies with ascorbate.

Should, however, in a restricted space or by enzymatic catalysis, two aroxyl radicals be capable of disproportionation, then quite distinct differences become apparent between the regular flavonoids and flavan-3-ols. While quinones of neither group have ever been isolated, their presence could be indirectly inferred from trapping experiments and/or formation of secondary products with quinones as obligatory intermediates. The hypothetical quinones and quinone methides of quercetin have been calculated using density-functional theory[40] and could be trapped with glutathione. Such nucleophilic reactions are likewise possible with biological macromolecules (DNA or proteins) with potential cytotoxic consequences. Alternatively, the quinones can undergo futile redox cycling with the formation of reactive oxygen species,[50,51] again an ultimately pro-oxidant effect.

Quite the opposite takes place with the hypothetical quinones derived from flavan-3-ols. The fact, that the oligomeric structures could only be formed via nucleophilic addition of a phenolic substrate to a quinone intermediate (phenolic coupling or S_N2 reaction), has led to numerous attempts to replicate this process by chemical synthesis.[52–59] Surprisingly, except for one example,[60] none of the products isolated corresponded to the major oligomers known from plant materials, i.e. the procyanidin A or B series.[61,62] In our own attempt to follow this phenolic coupling of (−)-epigallocatechin gallate (EGCG) after radical initiation with horseradish peroxidase/hydrogen peroxide (HRP/H_2O_2) or alkaline autoxidation (both processes leading to intermediary formation of EPR-detectable radicals[45]), we also employed NMR spectroscopy. In alkaline solutions of EGCG, mostly epimerization seems to take place.[63] Addition of HRP/ H_2O_2 causes only incomplete coupling reactions. Thus far, concomitant spectral shifts and line broadening are too indistinct to distinguish between the two pathways depicted in SCHEME 2. Proof of the potential formation of theasinensins by 2′/2″ coupling in Oolong tea[64–66] by initial formation of a B-ring semiquinone is therefore still lacking. In any case, phenolic coupling as evident from the presence of the proanthocyanidin oligomers in green tea,[65,67,68] red wine,[9–13] or cocoa,[69–71] has important implications for the antioxidant potential of these substances. As shown by Plumb *et al.*,[42] the linear increase of the TEAC value with the degree of oligomerization up to $n = 6$ is proof for the beneficial consequences of phenolic coupling: in contrast to flavonoid (quercetin) quinones, which are evidently incapable of forming auto-condensates, the coupling process retains the number of the reactive hydroxy groups, in effect doubling them with each dimerization step (see also SCHEME 2). These compounds thus become ever superior antioxidants until a limit is reached probably due to increased insolubility.

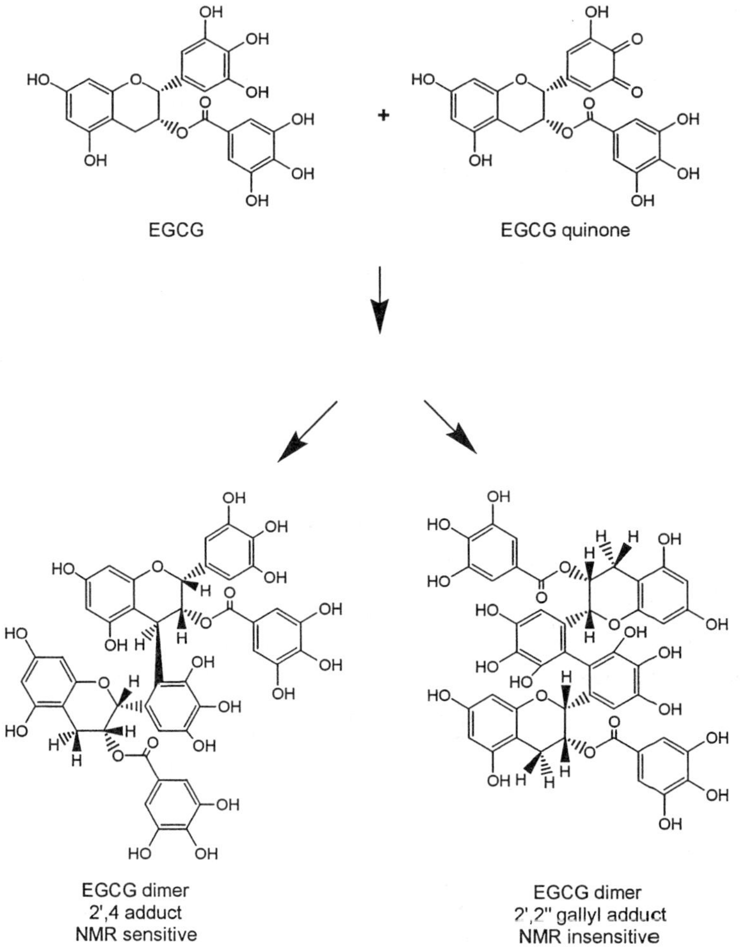

EGCG

EGCG quinone

EGCG dimer
2',4 adduct
NMR sensitive

EGCG dimer
2',2" gallyl adduct
NMR insensitive

SCHEME 2. Phenolic coupling reaction of EGCG with two possible condensation products.

Phenolic coupling is also obviously the reaction responsible for the formation of ellagitannins from gallotannins, even though here again the enzymes in question have never been isolated—except for those catalyzing attachment of galloyl moieties to the sugar ring.[72,73] Starting from pentagalloyl glucose (PGG), we first of all have to account for the distinct pathways governed by the stereo configuration of PGG (see SCHEME 3).[74] In fact, only the axial position of the galloyl moieties eventually leads to the dehydrogenation between 2–4 and 3–6 gallates forming hexahydroxydiphenic acid dimers (HHDP) and further oxidation to dehydrohexahydroxydiphenic acid (DHHDP) moieties.[74] Equatorial positions, in contrast, lead to HHDP couples between the 2–3 and 4–6 gallates and no further oxidation to DHHDP. However, ellagitannins can condense even further, evident from the various structural moieties detected thus far, e.g., gallagyl, valoneoyl, sanguisorboyl, to name just a few.[7,74]

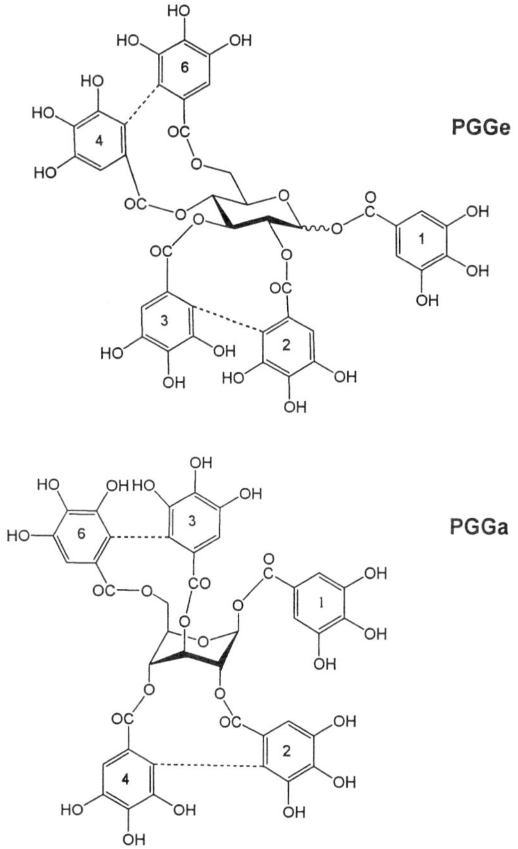

SCHEME 3. Potential condensation routes for PGG in axial and equatorial orientation.

The fact that with PGG this phenolic coupling is an intramolecular process, whereas with EGCG and the other flavan-3-ols it is intermolecular, should enhance the likelihood of this reaction for the gallotannins. In fact, chemical syntheses along these lines were more successful than with the proanthocyanidins.[75–77] NMR studies of autoxidizing PGG in alkaline solutions exhibit line narrowing (in contrast to the line broadening with EGCG) and ester hydrolysis with the liberation of glucose. The catalytic action of HRP/H_2O_2 caused much more extensive coupling reactions than with EGCG, the multiple and diverse products still awaiting identification.[45] To account for the specificity of the biosynthetic coupling reactions of gallotannins to ellagitannins, we might therefore contemplate the participation of some type of dirigent protein. Such a protein has recently been described for lignin biosynthesis[78–80] and may be a necessary adjunct to facilitate the biosynthetic process.

STRUCTURAL INFORMATION AND STRUCTURE–ACTIVITY RELATIONSHIPS (SAR)

The study of *trans*-resveratrol is the simplest example of how pulse-radiolytic studies help to differentiate between the reactivities of the individual hydroxy groups.[46] In the case of flavonoids, we have early on established the critical structural constraints for optimal antioxidant potential as

- *(i)* presence of a catechol group in the B-ring,
- *(ii)* a 2,3-double bond, and
- *(iii)* 3- and 5-hydroxy groups adjacent to the 4-keto structure.[81]

Although for quite a number of studies, these SAR principles could basically be confirmed,[39,82–86] they are not necessarily confined to antioxidant properties.[48] More elaborate quantum-mechanical studies have furthermore pinpointed a crucial role of the 3-hydroxy group of flavonols. On the one hand, its capability of forming hydrogen bonds with the 2′/6′ positions of the B ring was suggested to facilitate the radical formation of the 4′-hydroxy group, resulting in the superior antioxidant potential observed for quercetin.[39] On the other hand, the presence of a free 3-OH group could also be regarded as the nucleus of potential quinoid structures with probable pro-oxidant activities.[40]

The unequivocal linear relationships observed for the antioxidant potential of flavan-3-ols and proanthocyanidins and the number of reactive hydroxy groups—which we define as only those hydroxy groups present in a catechol or pyrogallol structure[44]—clearly point to these structures as the principal (and only) radical target sites. While we find a similar correlation for the simple gallotannins, a surprising dichotomy was observed for the more complex ellagitannins. Despite the very limited number of three compounds investigated, corilagin, geraniin and amariin, both pulse-radiolytic[44] and EPR[38] studies point to a limited accessibility for radical reactions of the DHHDP moiety as compared to the HHDP moiety and the gallate group in C1 position. Yet, in the DPPH assay, where hydrogen transfer to a rather large organic radical is measured, the linearity between reactivity and number of hydroxy groups is still observed.[26] At the present, we have no explanation for this discrepancy.

CONCLUSIONS

The physico-chemical and theoretical approaches (PR, EPR, NMR, DFT) which we have employed in our studies of polyphenolic antioxidants representing different structural classes have allowed us to

(i) determine their effectiveness as radical scavengers or chain-breaking antioxidants,
(ii) identify the structures and stabilities of the intermediary antioxidant radicals,
(iii) draw conclusions regarding the detailed mechanism of the compounds from different structural classes.

Most prominent among these differences is the paradigm shift in the behavior of flavonoid semiquinones and quinones on the one hand and the semiquinones and quinones from tannins, both condensed or hydrolyzable, on the other hand. We thus have now a convincing rationale for the superior antioxidant potential of the latter compounds.

REFERENCES

1. FRÉMONT, L. 2000. Biological effects of resveratrol. Life Sci. **66:** 663–673.
2. HUNG, L.M., J.K. CHEN, S.S. HUANG, *et al.* 2000. Cardioprotective effect of resveratrol, a natural antioxidant derived from grapes. Cardiovasc. Res. **47:** 549–555.
3. LI, L.N. 1998. Biologically active components from traditional Chinese medicines. Pure Appl. Chem. **70:** 547–554.
4. TANAKA, T., A. NISHIMURA, I. KOUNO, *et al.* 1997. Four new caffeic acid metabolites, yunnaneic acids E-H, from Salvia yunnanensis. Chem. Pharm. Bull. **45:** 1596–1600.
5. HARBORNE, J.B., Ed. 1988. The Flavonoids. Advances in Research since 1980. Chapman & Hall, London.
6. FERREIRA, D., E.V. BRANDT, J. COETZEE & E. MALAN. 1999. Condensed tannins. Progr. Chem. Organ. Natural Prod. **77:** 21–67.
7. OKUDA, T. 1999. Novel aspects of tannins—renewed concepts and structure-activity relationships. Curr. Organ. Chem. **3:** 609–622.
8. CLIFFORD, M.N. & A. SCALBERT. 2000. Ellagitannins—nature, occurrence and dietary burden. J. Sci. Food Agric. **80:** 1118–1125.
9. MAFFEI FACINO, R., M. CARINI, G. ALDINI, *et al.* 1994. Free radicals scavenging action and anti-enzyme activities of procyanidines from *Vitis vinifera*—a mechanism for their capillary protective action. Arzneim. Forsch. **44:** 592–601.
10. WILLIAMS, R.L. & M.S. ELLIOTT. 1997. Antioxidants in grapes and wine: chemistry and health effects. *In* Natural Antioxidants. Chemistry, Health Effects, and Applications. F. Shahidi, Ed.: 150–173. AOCS Press. Champaign, IL.
11. BALDI, A., A. ROMANI, N. MULINACCI, *et al.* 1997. The relative antioxidant potencies of some polyphenols in grapes and wines. ACS Sympos. Ser. **661:** 166–179.
12. DE FREITAS, V.A.P. & Y. GLORIES. 1999. Concentration and compositional changes of procyanidins in grape seeds and skin of white *Vitis vinifera* varieties. J. Sci. Food Agric. **79:** 1601–1606.
13. SOUQUET, J.M., B. LABARBE, C. LE GUERNEVÉ, *et al.* 2000. Phenolic composition of grape stems. J. Agric. Food Chem. **48:** 1076–1080.
14. RICHARD, J.L., F. CAMBIEN & P. DUCIMETIÈRE. 1981. Particularités épidémiologiques de la maladie coronarienne en France. Nouv. Presse Medic. **10:** 1111–1114.
15. KLATSKY, A.L. 1997. The epidemiology of alcohol and cardiovascular diseases. ACS Sympos. Ser. **661:** 132–149.
16. ESTRUCH, R. 2000. Wine and cardiovascular disease. Food Res. Int. **33:** 219–226.

17. KÖNIG, M., E. SCHOLZ, R. HARTMANN, *et al.* 1994. Ellagitannins and complex tannins from *Quercus petraea* bark. J. Nat. Prod. **57:** 1411–1415.
18. PUECH, J.L., F. FEUILLAT, J.R. MOSEDALE & C. PUECH. 1996. Extraction of ellagitannins from oak wood of model casks. Vitis **35:** 211–214.
19. SOMERS, C. 1990. An assessment of the 'oak factor' in current winemaking practice. Technical Rev. Aust. Wine Res. Inst. **67:** 3–10.
20. POCOCK, K.F., M.A. SEFTON & P.J. WILLIAMS. 1994. Taste thresholds of phenolic extracts of French and American oakwood—the influence of oak phenols on wine flavor. Am. J. Enol. Viticult. **45:** 429–434.
21. PUECH, J.L., F. FEUILLAT & J.R. MOSEDALE. 1999. The tannins of oak heartwood: structure, properties, and their influence on wine flavor. Am. J. Enol. Viticult. **50:** 469–478.
22. VIVAS, N. & Y. GLORIES. 1996. Role of oak wood ellagitannins in the oxidation process of red wines during aging. Am. J. Enol. Viticult. **47:** 103–107.
23. PEREZ-COELLO, M.S., M.A. SANCHEZ, E. GARCIA, *et al.* 2000. Fermentation of white wines in the presence of wood chips of American and French oak. J. Agric. Food Chem. **48:** 885–889.
24. MILLER, N.J. & C.A. RICE-EVANS. 1996. Spectrophotometric determination of antioxidant activity. Redox Rep. **2:** 161–171.
25. BRAND-WILLIAMS, W., M.E. CUVELIER & C. BERSET. 1995. Use of a free radical method to evaluate antioxidant activity. Food Sci. Technol. **28:** 25–30.
26. YOKOZAWA, T., C.P. CHEN, E. DONG, *et al.* 1998. Study on the inhibitory effect of tannins and flavonoids against the 1,1-diphenyl-2-picrylhydrazyl radical. Biochem. Pharmacol. **56:** 213–222.
27. ERBEN-RUSS, M., W. BORS & M. SARAN. 1987. Reactions of linoleic acid peroxyl radicals with phenolic antioxidants: a pulse radiolysis study. Int. J. Radiat. Biol. **52:** 393–412.
28. BORS, W., W. HELLER, C. MICHEL & M. SARAN. 1992. Structural principles of flavonoid antioxidants. *In* Free Radicals and the Liver. G. Csomos & J. Feher, Eds.: 77–95. Springer, Berlin.
29. BORS, W., C. MICHEL & M. SARAN. 1994. Flavonoid antioxidants: rate constants for reactions with oxygen radicals. Meth. Enzymol. **234:** 420–429.
30. BELYAKOV, V.A., V.A. ROGINSKY & W. BORS. 1995. Rate constants for the reaction of peroxyl free radical with flavonoids and related compounds as determined by the kinetic chemiluminescence method. J. Chem. Soc., Perkin **II:** 2319–2326.
31. ROGINSKY, V.A., T.K. BARSUKOVA, A.A. REMOROVA & W. BORS. 1996. Moderate antioxidative efficiencies of flavonoids during peroxidation of methyl linoleate in homogeneous and micellar solutions. J. Am. Oil Chem. Soc. **73:** 777–786.
32. BORS, W., C. MICHEL & S. SCHIKORA. 1995. Interaction of flavonoids with ascorbate and determination of their univalent redox potentials: a pulse radiolysis study. Free Radical Biol. Med. **19:** 45–52.
33. JOVANOVIC, S.V., S. STEENKEN, M.G. SIMIC & Y. HARA. 1997. Antioxidant properties of flavonoids: reduction potentials and electron transfer reactions of flavonoid radicals. *In* Flavonoids in Health and Disease. C. Rice-Evans & L. Packer, Eds.: 137–161. Marcel Dekker. New York
34. YOSHIOKA, H., K. SUGIURA, R. KAWAHARA, *et al.* 1991. Formation of radicals and chemiluminescence during the autoxidation of tea catechins. Agric. Biol. Chem. **55:** 2717–2723.
35. BORS, W., W. HELLER, C. MICHEL & K. STETTMAIER. 1993. Electron paramagnetic resonance studies of flavonoid compounds. *In* Free Radicals: From Basic Science to Medicine. G. Poli, M. Albano & M.U. Dianzani, Eds.: 374–387. Birkhäuser. Basel.
36. GUO, Q.N., B.L. ZHAO, M.F. LI, *et al.* 1996. Studies on protective mechanisms of four components of green tea polyphenols against lipid peroxidation in synaptosomes. Biochim. Biophys. Acta **1304:** 210–222.
37. BORS, W., C. MICHEL, K. STETTMAIER & W. HELLER. 1997. EPR studies of plant polyphenols. *In* Natural Antioxidants. Chemistry, Health Effects, and Applications. F. Shahidi, Ed.: 346–357. AOCS Press. Champaign, IL.

38. BORS, W., C. MICHEL & K. STETTMAIER. 2000. Electron paramagnetic resonance studies of radical species of proanthocyanidins and gallate esters. Arch. Biochem. Biophys. **374:** 347–355.
39. VAN ACKER, S.A.B.E., M.J. DE GROOT, D.J. VAN DEN BERG, *et al.* 1996. A quantum chemical explanation of the antioxidant activity of flavonoids. Chem. Res. Toxicol. **9:** 1305–1312.
40. BOERSMA, M.G., J. VERVOORT, H. SZYMUSIAK, *et al.* 2000. Regioselectivity and reversibility of the glutathione conjugation of quercetin quinone methide. Chem. Res. Toxicol. **13:** 185–191.
41. SALAH, N., N.J. MILLER, G. PAGANGA, *et al.* 1995. Polyphenolic flavanols as scavengers of aqueous phase radicals and as chain-breaking antioxidants. Arch. Biochem. Biophys. **322:** 339–346.
42. PLUMB, G.W., S. DE PASCUAL-TERESA, C. SANTOS-BUELGA, *et al.* 1998. Antioxidant properties of catechins and proanthocyanidins: effect of polymerisation, galloylation and glycosylation. Free Radical Res. **29:** 351–358.
43. BORS, W., C. MICHEL, M. SARAN & K. STETTMAIER. 1999. Determination of antioxidant activities: relative vs. absolute methods. *In* Different Pathways through Life. Biochemical Pathways of Plant Biology and Medicine. A. Denke, K. Dornisch, F. Fleischmann, *et al.*, Eds.: 9–34. LINCOM Europe, Munich, Germany.
44. BORS, W. & C. MICHEL. 1999. Antioxidant capacity of flavanols and gallate esters: pulse radiolysis studies. Free Radical Biol. Med. **27:** 1413–1426.
45. BORS, W., L.Y. FOO, N. HERTKORN, *et al.* 2001. Chemical studies of proanthocyanidins and hydrolyzable tannins. Antiox. Redox Signal **3:** 995–1008.
46. STOJANOVIC, S., H. SPRINZ & O. BREDE. 2001. Efficiency and mechanism of the antioxidant action of trans-resveratrol and its analogues in the radical liposome oxidation. Arch. Biochem. Biophys. **391:** 79–89.
47. MUSSO, H. & H. DOEPP. 1967. Autoxidationsgeschwindigkeit und Redoxpotential bei Hydrochinon-, Brenzcatechin- und Resorcinderivaten. Chem. Ber. **100:** 3627–3643.
48. BORS, W., W. HELLER, C. MICHEL & K. STETTMAIER. 1996. Flavonoids and polyphenols: chemistry and biology. *In* Handbook of Antioxidants. E. Cadenas & L. Packer, Eds.: 409–466. Marcel Dekker. New York.
49. BORS, W., W. HELLER & C. MICHEL. 1997. The chemistry of flavonoids. *In* Flavonoids in Health and Disease. C. Rice-Evans & L. Packer, Eds.: 111–136. Marcel Dekker. New York.
50. MIURA, Y.H., I. TOMITA, T. WATANABE, *et al.* 1998. Active oxygens generation by flavonoids. Biol. Pharm. Bull. **21:** 93–96.
51. METODIEWA, D., A.K. JAISWAL, N. CENAS, *et al.* 1999. Quercetin may act as a cytotoxic prooxidant after its metabolic activation to semiquinone and quinoidal product. Free Radical Biol. Med. **26:** 107–116.
52. DELCOUR, J.A., D. FERREIRA & D.G. ROUX. 1983. Synthesis of condensed tannins. Part 9. The condensation sequence of leucocyanidin with (+)-catechin and with the resultant procyanidins. J. Chem. Soc., Perkin **I:** 1711–1717.
53. HEMINGWAY, R.W. & P.E. LAKS. 1985. Condensed tannins: a proposed route to 2R,3R-(2,3-cis)-proanthocyanidins. Chem. Comm. 746–747.
54. SINGLETON, V.L. & J.J.L. CILLIERS. 1995. Phenolic browning: a perspective from grape and wine research. ACS Sympos. Ser. **600:** 23–48.
55. CHEYNIER, V., H. FULCRAND, S. GUYOT, *et al.* 1995. Reactions of enzymically generated quinones in relation to browning in grape musts and wines. ACS Sympos. Ser. **600:** 130–143.
56. ESCRIBANO-BAILON, T., O. DANGLES & R. BROUILLARD. 1996. Coupling reactions between flavylium ions and catechin. Phytochemistry **41:** 1583–1592.
57. GUYOT, S., J. VERCAUTEREN & V. CHEYNIER. 1996. Structural determination of colourless and yellow dimers resulting from (+)-catechin coupling catalysed by grape polyphenoloxidase. Phytochemistry **42:** 1279–1288.
58. SCHNEIDER, V. 1998. Must hyperoxidation: a review. Am. J. Enol. Viticult. **49:** 65–73.
59. STEYNBERG, P.J., R.J.J. NEL, H. VAN RENSBURG, *et al.* 1998. Oligomeric flavanoids. Part 27. Interflavanyl bond formation in procyanidins under neutral conditions. Tetrahedron **54:** 8153–8158.

60. Young, D.A., E. Young, D.G. Roux, *et al.* 1987. Synthesis of condensed tannins. Part 19. Phenol oxidative coupling of (+)-catechin and (+)-mesquitol. Conformation of bis-(+)-catechins. J. Chem. Soc., Perkin **I:** 2345–2351.
61. Stafford, H.A. 1990. Pathway to proanthocyanidins (condensed tannins), flavan-3-ols, and unsubstituted flavans. *In* Flavonoid Metabolism. H.A. Stafford, Ed.: 63–99. CRC Press. Boca Raton, FL.
62. Ferreira, D. & X.C. Li. 2000. Oligomeric proanthocyanidins: naturally occurring O-heterocycles. Nat. Prod. Rep. **17:** 193–212.
63. Foo, L.Y. & L.J. Porter. 1983. Synthesis and conformation of procyanidin diastereoisomers. J. Chem. Soc., Perkin **I:** 1535–1543.
64. Hashimoto, F., G.I. Nonaka & I. Nishioka. 1988 Tannins and related compounds. LXIX. Isolation and structure elucidation of B,B′-linked bisflavanoids, theasinensins D-G and oolongtheanin from Oolong tea. Chem. Pharm. Bull. **36:** 1676–1684.
65. Harbowy, M.E. & D.A. Balentine. 1997. Tea chemistry. CRC Crit. Rev. Plant Sci. **16:** 415–480.
66. Ho, C.T. & N. Zhu. 2000. The chemistry of tea. ACS Sympos. Ser. **754:** 316–326.
67. Hashimoto, F., G.I. Nonaka & I. Nishioka. 1992. Tannins and related compounds. CXIV. Structures of novel fermentation products, theagallinin, theaflavonin and desgalloyl theaflavonin from black tea, and changes of tea leaf polyphenols during fermentation. Chem. Pharm. Bull. **40:** 1383–1389.
68. Kiehne, A., C. Lakenbrink & U.H. Engelhardt. 1997. Analysis of proanthocyanidins in tea samples. 1. LC-MS results. Z. Lebensm. Unters. Forsch. **205:** 153–157.
69. Osakabe, N., M. Yamagishi, C. Sanbongi, *et al.* 1998. The antioxidative substances in cacao liquor. J. Nutr. Sci. Vitaminol. **44:** 313–321.
70. Adamson, G.E., S.A. Lazarus, A.E. Mitchell, *et al.* 1999. HPLC method for the quantification of procyanidins in cocoa and chocolate samples and correlation to total antioxidant capacity. J. Agric. Food Chem. **47:** 4184–4188.
71. Wollgast, J. & E. Anklam. 2000. Review on polyphenols in *Theobroma cacao*: changes in composition during the manufacture of chocolate and methodology for identification and quantification. Food Res. Int. **33:** 423–447.
72. Gross, G.G. 1992. Enzymatic synthesis of gallotannins and related compounds. Rec. Adv. Phytochem. **26:** 297–324.
73. Grundhöfer, P. & G.G. Gross. 2000. Purification of tetragalloylglucose 4-O-galloyltransferase and preparation of antibodies against this key enzyme in the biosynthesis of hydrolyzable tannins. Z. Naturforsch. **55c:** 582–587.
74. Haslam, E. & Y. Cai. 1994. Plant polyphenols (vegetable tannins): gallic acid metabolism. Nat. Prod. Rep. : 41–66.
75. Feldman, K.S. & S.M. Ensel. 1994. Ellagitannin chemistry, preparative and mechanistic studies of the biomimetic oxidative coupling of galloyl esters. J. Am. Chem. Soc. **116:** 3357–3366.
76. Quideau, S. & K.S. Feldman. 1996. Ellagitannin chemistry. Chem. Rev. **96:** 475–503.
77. Feldman, K.S., M.D. Lawlor & K. Sahasrabudhe. 2000. Ellagitannin chemistry. Evolution of a three-component coupling strategy for the synthesis of the dimeric ellagitannin coriariin A and a dimeric gallotannin analogue. J. Org. Chem. **65:** 8011–8019.
78. Davin, L.B., H.B. Wang, A.L. Crowell, *et al.* 1997. Stereoselective bimolecular phenoxy radical coupling by an auxiliary (dirigent) protein without an active center. Science **275:** 362–366.
79. Davin, L.B. & N.G. Lewis. 2000. Dirigent proteins and dirigent sites explain the mystery of specificity of radical precursor coupling in lignan and lignin biosynthesis. Plant Physiol. **123:** 453–461.
80. Lewis, N.G. & L.B. Davin. 2000. Phenolic coupling in planta: dirigent proteins, dirigent sites and notions beyond randomness. Part 1. Polyphenol Actual. **20:** 18–26.
81. Bors, W., W. Heller, C. Michel & M. Saran. 1990. Flavonoids as antioxidants: determination of radical scavenging efficiencies. Meth. Enzymol. **186:** 343–354.
82. Krol, W., Z. Czuba, S. Scheller, *et al.* 1994. Structure-activity relationship in the ability of flavonols to inhibit chemiluminescence. J. Ethnopharmacol. **41:** 121–126.

83. CAO, G.H., E. SOFIC & R.L. PRIOR. 1997. Antioxidant and prooxidant behavior of flavonoids: structure-activity relationships. Free Radical Biol. Med. **22:** 749–760.
84. RICE-EVANS, C.A. & N.J. MILLER. 1997. Structure-antioxidant activity relationships of flavonoids and isoflavonoids. *In* Flavonoids in Health and Disease. C. Rice-Evans & L. Packer, Eds.: 199–219. Marcel Dekker. New York.
85. DUARTE SILVA, I., J. GASPAR, G. GOMES DA COSTA, *et al.* 2000. Chemical features of flavonols affecting their genotoxicity. Potential implications in their use as therapeutical agents. Chem. Biol. Interact. **124:** 29–51.
86. DUGAS, A.J., J. CASTAÑEDA-ACOSTA, G.C. BONIN, *et al.* 2000. Evaluation of the total peroxyl radical-scavenging capacity of flavonoids: structure-activity relationships. J. Nat. Prod. **63:** 327–331.

Bioflavonoid-Rich Botanical Extracts Show Antioxidant and Gene Regulatory Activity

KISHORCHANDRA GOHIL[a] AND LESTER PACKER[b]

[a]*Department of Internal Medicine, University of California, Davis, Davis, California 95616, USA*

[b]*Department of Molecular Pharmacology and Toxicology, School of Pharmacy, University of Southern California, Los Angeles, California 90033, USA*

ABSTRACT: Reactive oxygen and nitrogen metabolites are obligatory and essential products of metabolism. Unregulated increase in their production is associated with a number of chronic illnesses. Diets rich in fruits, vegetables, and wines are implicated in the prevention of chronic diseases. Molecular mechanisms by which fruits and vegetables confer their disease-preventive actions are poorly defined. However, recent developments in the fields of genomics and bioinformatics provide powerful tools to investigate the mechanisms by which botanicals affect cellular functions. This monograph illustrates the potential of large-scale messenger RNA analysis to unravel the role of transcription in mediating the effects of botanical extracts with antioxidant properties. The application of microarrays and oligonucleotide arrays shows multiple effects of antioxidant extracts on the expression of a broad spectrum of genes.

KEYWORDS: global gene expression; signal transduction

LIVING GENERATES TOXIC, DISEASE CAUSING, REACTIVE OXYGEN METABOLITES

Antioxidants are necessary to prevent the toxic effects of reactive oxygen metabolites that are generated by essential, life sustaining metabolism. Biochemical processes that produce and utilize reducing equivalents such as reduced nicotinamide adenine dinucleotide (NADH) or nicotinamide adenine dinucleotide phosphate (NADPH) produce reactive oxygen species such as superoxide[1,2] and nitric oxide radicals[3] and highly reactive secondary products such as peroxynitrite formed by interaction between superoxide and nitric oxide radicals.[4,5] These reactive products are generated by electron transport chains that are assembled and organized in biological membranes. For example, endoplasmic reticulum is the primary subcellular compartment that holds cytochrome P-450 systems that generate superoxide radicals during the hydroxylation reactions of xenobiotics and endogeneously generated steroids. The NADPH oxidase system of phagocytic leukocytes[6] plays a vital immune function because it is specialized to generate copious amounts of superoxides to kill engulfed bacteria. Isoforms of NADPH oxidase are present on plasma membranes where they are activated by cytokines and growth factors. Mitochondria produce

Address for correspondence: Dr. Kishorchandra Gohil, Department of Internal Medicine, University of California, Davis, Davis, CA 95616, USA.
kgohil@ucdavis.edu

superoxide and hydrogen peroxide[1] during the oxidation of tricarboxylic acids to generate ATP for maintaining cellular functions. It has been postulated that an imbalance in the production and removal of these reactive metabolites can cause chronic degenerating diseases such as cancers, atherosclerosis, and Alzheimer's disease.[7,8]

DIETARY ANTIOXIDANTS PREVENT CHRONIC DISEASES

The consumption of botanical extracts for prevention and treatment of human diseases is common in all human cultures. Recent epidemiological research has given a new impetus to explore the therapeutic potential of botanical extracts. Particularly noteworthy are the results reporting lowered risk of a variety of cancers in human populations consuming diets with more fruits and vegetables.[9] Also remarkable are the epidemiological studies that suggest lowered susceptibility to coronary heart disease in populations that consume moderate amounts of red wine.[10] These observations have fueled the search for the disease preventive components within a complex mixture of botanical extracts. Polyphenolic and terpenoid components of natural extracts such as *Ginkgo biloba* are of particular interest because of their potential activities in regulating cell proliferation and growth. A synthetic flavonoid, flavopiridol, has been shown to inhibit a number of protein kinases including cyclin-dependent kinases and thereby regulate cell cycle and proliferation.[11] Flavopiridol is currently in clinical trials for the treatment of cancers. Another bioflavonoid extract, silymarin, has been shown to inhibit activation of NF-κB by lipopolysaccharide in a human hepatoma cell line[12] and in keratinocytes[13] and impair receptor tyrosine kinase–signaling pathways in a human epidermoid cell line.[14]

GINKGO BILOBA EXTRACT, EGB 761, AN HERBAL ANTIOXIDANT

Since one of the established activities of *Ginkgo biloba* extract is its antioxidant properties,[15] it is very likely that the flavonoids and terpenoids present in the extract produce their cellular effects by regulating the status of endogenous reactive oxygen metabolites. It is therefore possible that flavonoids present in fruits and vegetables may play an important function in the preventive actions of diets rich in fruits and vegetables against chronic human diseases.

Changes in rates of cell proliferation and death play an important role in initiation and progression of chronic disease and utilize constitutive and inducible metabolic and signaling pathways. Reactive oxygen metabolites have been shown to increase cell proliferation.[16] These effects are likely to be due to changes in the expression of multiple gene families. In the past decade, the techniques for simultaneous monitoring of changes in the expression of a large number of genes, possibly the entire expressed genome of a cell, at the mRNA and protein levels has undergone a revolution. It is now possible to monitor changes in the entire expressed genome of a eukaryotic cell using gene chip technology.[17] We have recently described the effect of *Ginkgo biloba* extract on the gene expression profile of a human bladder cancer cell line and showed a large transcriptional response.[18] We have also shown that the EGb 761 has neuromodulatory activities *in vivo*.[19] In the following pages, we

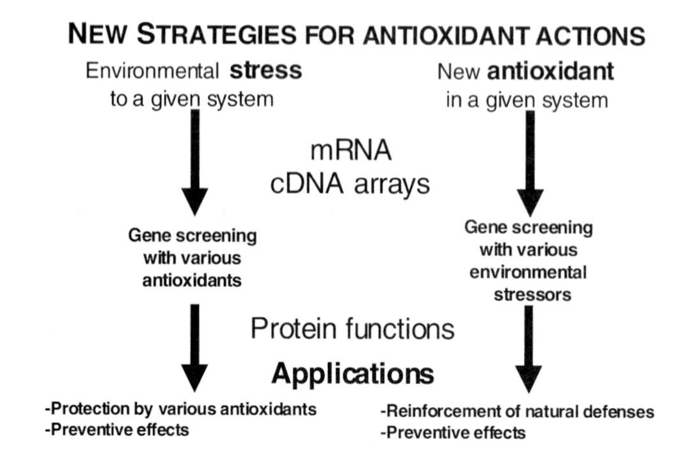

FIGURE 1. Novel strategies for the discovery of antioxidant actions on gene expression are illustrated.

describe the results of applications of different techniques to define changes in the expression of *Ginkgo biloba* extract–sensitive genes in rat neutrophils and in human cancer cells. An analytical strategy such as shown here could be useful for unraveling the molecular basis of the effects of wine (see FIGURE 1).

SELECTIVE INDUCTION OF BCL2 AND TNF-α IN RAT NEUTROPHILS

Neutrophils are endowed with a highly regulated NADPH-oxidase system which generates superoxide radicals and hydrogen peroxide by reduction of molecular oxygen. Superoxide burst is activated in neutrophils as a defense response to the presence of a foreign particle (such as a bacterium) or by nanomolar amounts of phorbol 12-myristate 13-acetate (PMA), an activator of protein kinase C. During the production of superoxide the phagocytic cell is vulnerable to oxidative stress and is known to undergo apoptosis following its activation. Activation of rat neutrophils in the presence of EGb 761 inhibits superoxide production. It has been shown that the apoptotic pathway is regulated by a large number of proteins including bcl-2, bcl-xl, and bax. We therefore examined the effect of EGb 761 on expression of mRNAs for pro-apoptotic protein bax and anti-apoptotic proteins bcl-2 and bcl-xl in rat neutrophils activated by PMA. FIGURE 2 shows that the expression of bax, an apoptotic gene, and bcl-2 and bcl-xl, which code for anti-apoptotic proteins, were unaffected by protein kinase C activator, PMA.

However, EGb 761 doubled the expression of anti-apoptotic bcl-2 without any effect on bax or bcl-xl after 18h following PMA stimulation. EGb 761 also augmented the transcript for TNF-α, an activator of NF-κB, a transcription factor which regulates a large number of cytokine genes and the apoptotic pathways. These observations show that the extract of *Ginkgo biloba* leaves can selectively affect the

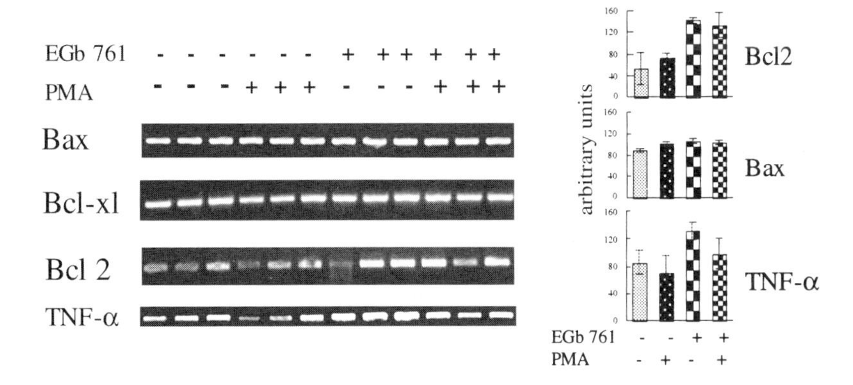

FIGURE 2. EGb 761 upregulates Bcl 2 and TNF-α mRNA in neutrophils. Rat "neutrophils" (10^7) were stimulated with PMA (10 ng/ml) for 15 min and then incubated for 18 h. Total RNA was extracted and subjected to RT-PCR.

transcription of mRNAs important in the regulation of the apoptotic process of neutrophils, which play an important function in the defense of the organism and in remodeling of the tissues.

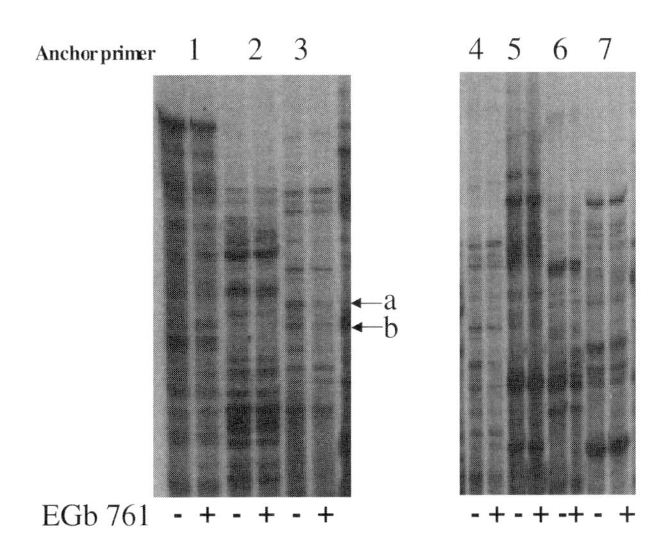

FIGURE 3. PCR-based amplification of a large number of genes can show differential effects of a natural antioxidant extract in the expression of genes in a human bladder cancer cell line.

DIFFERENTIAL DISPLAY OF mRNAs IDENTIFIES A BROAD EFFECT OF EGB 761 ON TRANSCRIPTION

Liang and Pardee[20] described a PCR-based method that enabled simultaneous evaluation of upregulated and down-regulated expression of mammalian genes in a pair of differently treated cells. The method took advantage of random oligonucleotide primers, designed to hybridize to all the cDNAs at sites distal to the oligo-dT tails prepared from mRNAs with four different oligo-dNT primers. mRNA isolated from human bladder cancer cells, T-24 cells, exposed to either the vehicle or EGb 761 showed changes in about 40 different transcripts after screening about 30% of the expressed genome. FIGURE 3 shows a selected region of the differential display autoradiogram.

Several amplified bands (a and b) showed a decrease in expression in EGb 761-treated cells. These changes in the transcript profile need further characterization because the differential display technique suffers from relatively high incidence of false positives. In spite of this limitations the changes described in FIGURE 3 show that large numbers of mRNAs from the T-24 cells were altered by the presence of EGb 761. The identities of these transcripts remain to be defined and are likely to be the mRNAs identified by cDNA arrays on nylon filters and oligonucleotide microarrays described below.

cDNA ARRAYS OF KNOWN GENES IDENTIFY EFFECTS OF EGB 761 ON CYTOKINE AND APOPTOSIS GENES

PCR-amplified cDNA dotted on nylon membranes offers the advantage of rapidly identifying the altered expressions of mRNAs in the presence of the extract. In addition, a good selection of function-specific cDNA arrays are commercially available. One such cDNA array of human cancer genes was used to screen the changes in the gene expression of human bladder cancer cells in the presence of EGb 761. The array screened for 15 different functional classes of genes, including cell cycle regulators, growth regulators, apoptosis, oncogenes, and DNA damage response and repair genes. Radiolabeled cDNA prepared from mRNA isolated from T-24 cells hybridized to 131 of the 588 (22% of the genes on the cDNA array) complimentary gene fragments on the cDNA array.

An example of differential gene expression in human bladder cancer cells in response to the extract of *Ginkgo biloba*, EGb 761, is shown in FIGURE 4. The figure shows that in the group of apoptosis-related genes, mRNA for X compared to that of Y was down-regulated by EGb 761, whereas the expression of cytokine gene Z was upregulated by the presence of EGb 761 for 72h in the cell culture medium. These data show that EGb 761 alters the expression of broad functional groups of genes and support the data obtained by differential display of mRNA discussed above.

DNA Microarray Chips

DNA microarray chips offer a more comprehensive assessment of the transcriptional response of a cell. The technique enables simultaneous screening of changes in the entire expressed genome (up to 30,000 mRNAs) of a cell in response

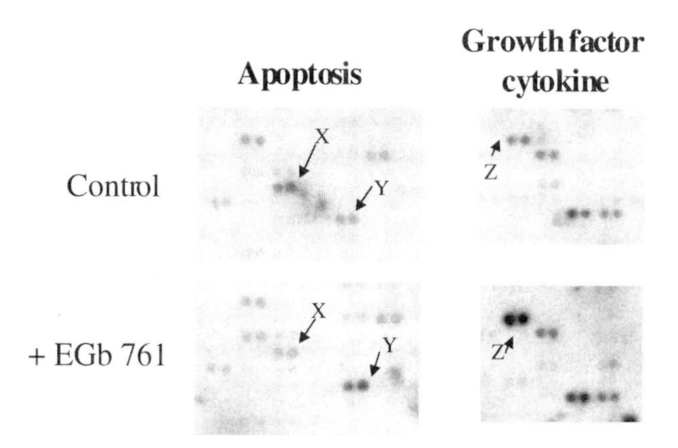

FIGURE 4. This figure shows the differential effects of EGb 761 on the expression of genes encoding proteins for apoptosis, cytokines and growth factors in human bladder cancer cells.

to a variable such as a DNA-damaging agent,[21] or to serum repletion[22] or to a botanical extract.[18] Currently there are two distinct techniques available for monitoring changes in gene expression, and each technique generates a large volume of data of qualitative and quantitative changes in the gene expression profile of the cell under investigation.

One gene profiling technique pioneered by Patrick Brown's and David Botstein's teams at Stanford University,[17,23] utilizes PCR-amplified fragments of cloned DNA (cDNA) dotted, robotically, onto a glass microscope slide, ($1'' \times 3''$). In a typical DNA microarray chip up to 6,000 different cDNAs can be dotted per slide; each dot is about 200 μm in diameter containing about 10 ng of DNA. The source of the cDNA determines the comprehensiveness of the data for the transcriptional response of a selected cell to a defined stimulus. If DNAs are randomly selected from the whole of human genome data bank of known human genes or expressed sequence tags then a typical experiment can screen about 10% of the expressed genome in the selected cell type. However, the response of almost complete (at least 90%) expressed genome to a selected stimulus can be screened if the cDNAs dotted onto the glass slides are obtained from the cDNA library prepared from the specific cell under investigation. The latter approach, therefore, requires an additional step of construction of the cell-specific cDNA library that represents every mRNA expressed in that cell. A DNA genechip generated by the dotting technique is used as microarrayed probes to define changes in two populations of target cDNAs each of which is specifically tagged with a distinct fluorescently labeled nucleotide. The fluorescently labeled target cDNAs from control and test cells, immobilized by specific hybridization to their complementary probes on the gene array, are visualized by a confocal laser scanner. The fluorescent images obtained from the scanner are analyzed by mathematical and statistical algorithms to obtain quantitative information on the number and amount of different mRNAs present in the sample to generate the cell-specific gene expression profiles.

The major difference in the other microarray technique[24] is the manner in which the DNA gene chip is prepared and in the nature of the probes. The probes are oligonucleotides, 23 mers, with gene-specific sequence and are synthesized on a glass surface, 1.28 cm × 1.28 cm, by photolithographic techniques. Up to 300,000 oligonucleotides can be present on a single genechip to analyze changes in about 7,000 different mRNAs (an entire expressed genome of yeast!). The construction of the microarray is such that each target mRNA is probed for specific binding by 20 different oligonucleotides that span from the 5′ end of the message to the 3′ end. Non-specific binding of the target mRNA is measured by hybridization to oligonucleotide probes with mismatched nucleotide in the thirteenth position. Target RNA labeled with a biotinylated nucleotide is prepared by three steps: total or mRNA is isolated from the cells under investigation; double stranded DNA is prepared with a specific T7 primer; and target RNA (cRNA) is amplified in the presence of biotinylated ribose nucleotides by *in vitro* transcription with T7 RNA polymerase. The resulting cRNA is fragmented to obtain about 200 base fragments, which are allowed to hybridize to the probes on the chip by 16-h incubation at the selected temperature.

The primary aim of the application of GeneChips was to evaluate the potential molecular targets of a botanical extract, EGb 761, with therapeutic benefits. Analytical strategies that utilize well-defined *in vitro* assays of cellular responses can generate useful information for designing the more challenging *in vivo* studies. This is best illustrated by the Developmental Therapeutics Program of the National Cancer Institute, which has completed *in vitro* screening of more than 70,000 compounds, including botanical extracts. The application of high density oligonucleotide chips to define the changes in the mRNA profile of a cancer cell under oxidative stress has revealed a novel action of *Ginkgo biloba* extract. The transcripts for heme-oxygenase-1 (HO-1) and mitochondrial superoxide dismutase (MnSOD) were induced as were their respective proteins.[18] Selective expression of genes related to antioxidant functions showed a "signature" of an adaptive response to oxidative stress. Analysis of the gene expression profile of the human cancer cells in response to the extract uncovered a coordinated transcriptional strategy that is utilized to enhance the cell's ability to tolerate oxidative stress. Many of the genes identified in the antioxidant/heat shock/mitochondrial category have antioxidant response elements (ARE) in their promoter regions; AREs are also known as electrophile responsive elements and nuclear respiratory factor element-2 (Nrf2).

The transcriptional response described here partially defines the potential of the flavonoid- and terpenoid-containing extract of *Ginkgo biloba* leaves for regulating the expression of genes encoding antioxidants, mitochondria, the Golgi system, and for the repair and synthesis of DNA in a human cancer cell. The application of reliable techniques that measure global, cellular and molecular responses of selected cells *in vitro* can begin to delineate the specific biological effects of natural extracts and offer a more rational approach to design *in vivo* studies.

REFERENCES

1. Chance, B., H. Sies & A. Boveris. 1979. Hydroperoxide metabolism in mammalian organs. Physiol. Rev. **59:** 527–605.
2. Chakraborti, T., S. Das, M. Mondal, *et al.* 1999. Oxidant, mitochondria and calcium: an overview. Cell Signal **11:** 77–85.

3. HATTORI, R., K. SASE, H. EIZAWA, *et al.* 1994. Structure and function of nitric oxide synthases. Int. J. Cardiol. **47:** S71–S75.
4. GROVES, J.T. 1999. Peroxynitrite: reactive, invasive and enigmatic. Curr. Opin. Chem. Biol. **3:** 226–235.
5. PATEL, R.P., J. MCANDREW, H. SELLAK, *et al.* 1999. Biological aspects of reactive nitrogen species. Biochim. Biophys. Acta **1411:** 385–400.
6. DAHLGREN, C. & A. KARLSSON. 1999. Respiratory burst in human neutrophils. J. Immunol. Methods **232:** 3–14.
7. PACKER, L. 1992. Interactions among antioxidants in health and disease: vitamin E and its redox cycle. Proc. Soc. Exp. Biol. Med. **200:** 271–276.
8. AMES, B.N., M.K. SHIGENAGA & T.M. HAGEN. 1993. Oxidants, antioxidants, and the degenerative diseases of aging. Proc. Natl. Acad. Sci. USA **90:** 7915–7922.
9. STEINMETZ, K.A. & J.D. POTTER. 1996. Vegetables, fruit, and cancer prevention: a review. J. Am. Diet Assoc. **96:** 1027–1039.
10. JOSHIPURA, K.J., F.B. HU, J.E. MANSON, *et al.* 2001. The effect of fruit and vegetable intake on risk for coronary heart disease. Ann. Intern. Med. **134:** 1106–1114.
11. BRUSSELBACH, S., D.M. NETTELBECK, H.H. SEDLACEK & R. MULLER. 1998. Cell cycle-independent induction of apoptosis by the anti-tumor drug Flavopiridol in endothelial cells. Int. J. Cancer **77:** 146–152.
12. SALIOU, C., B. RIHN, J. CILLARD, *et al.* 1998. Selective inhibition of NF-kappaB activation by the flavonoid hepatoprotector silymarin in HepG2. Evidence for different activating pathways. FEBS Lett. **440:** 8–12.
13. SALIOU, C., M. KITAZAWA, L. MCLAUGHLIN, *et al.* 1999. Antioxidants modulate acute solar ultraviolet radiation-induced NF-kappa-B activation in a human keratinocyte cell line. Free Radical Biol. Med. **26:** 174–183.
14. ZI, X., A.W. GRASSO, H.J. KUNG & R. AGARWAL. 1998. A flavonoid antioxidant, silymarin, inhibits activation of erbB1 signaling and induces cyclin-dependent kinase inhibitors, G1 arrest, and anticarcinogenic effects in human prostate carcinoma DU145 cells. Cancer Res. **58:** 1920–1929.
15. MARCOCCI, L., L. PACKER, M.T. DROY-LEFAIX, *et al.* 1994. Antioxidant action of Ginkgo biloba extract EGb 761. Methods Enzymol. **234:** 462–475
16. IRANI, K., Y. XIA, J.L. ZWEIER, *et al.* 1997. Mitogenic signaling mediated by oxidants in Ras-transformed fibroblasts. Science **275:** 1649–1652.
17. DERISI, J.L., V.R. IYER & P.O. BROWN. 1997. Exploring the metabolic and genetic control of gene expression on a genomic scale. Science **278:** 680–686.
18. GOHIL, K., R.K. MOY, S. FARZIN, *et al.* 2000. mRNA expression profile of a human cancer cell line in response to Ginkgo biloba extract: induction of antioxidant response and the Golgi system. Free Radic. Res. **33:** 831–849.
19. WATANABE, C.M., S. WOLFFRAM, P. ADER, *et al.* 2001. The in vivo neuromodulatory effects of the herbal medicine ginkgo biloba. Proc. Natl. Acad. Sci. USA **98:** 6577–6580.
20. LIANG, P., W. ZHU, X. ZHANG, *et al.* 1994. Differential display using one-base anchored oligo-dT primers. Nucleic Acids Res. **22:** 5763–5764.
21. TUSHER, V.G., R. TIBSHIRANI & G. CHU. 2001. Significance analysis of microarrays applied to the ionizing radiation response. Proc. Natl. Acad. Sci. USA **98:** 5116–5121.
22. IYER, V.R., M.B. EISEN, D.T. ROSS, *et al.* 1999. The transcriptional program in the response of human fibroblasts to serum. Science **283:** 83–87.
23. DERISI, J.L. & V.R. IYER. 1999. Genomics and array technology. Curr. Opin. Oncol. **11:** 76–79.
24. LIPSHUTZ, R.J., S.P. FODOR, T.R. GINGERAS & D.J. LOCKHART. 1999. High density synthetic oligonucleotide arrays. Nat. Genet. **21:** 20–24.

Vasodilating Procyanidins Derived from Grape Seeds

DAVID F. FITZPATRICK,[a] BETTYE BING,[a] DAVID A. MAGGI,[a]
RICHARD C. FLEMING,[b] AND REBECCA M. O'MALLEY[b]

[a]*Department of Pharmacology, College of Medicine, and* [b]*Department of Chemistry, University of South Florida, Tampa, Florida 33612-4799, USA*

ABSTRACT: We have shown in previous work that extracts of grape seeds (GSE) and skins, grape juice, and many red wines exhibit endothelium-dependent relaxing (EDR) activity *in vitro*. This EDR activity involves endothelial nitric oxide (NO) release and subsequent increase in cyclic GMP levels in the vascular smooth muscle cells. The NO/cyclic GMP pathway is known to be involved in many cardiovascular-protective roles. The current study focuses on the isolation and identification of EDR-active compounds (procyanidins) from GSE. Crushed Concord grape seeds were extracted with methanol and the extract was separated into seven fractions (A–G) on a Toyopearl TSK-HW-40 column. EDR-active fractions (D–G) were further separated into 25 individual compound peaks by HPLC, 16 of which were EDR active (threshold for relaxation ranged between less than 0.5 μg/mL and greater than 4 μg/mL). Procyanidin identification was accomplished by electrospray-ion trap mass spectrometry (ES-ITMS), MS/MS, and by tannase treatment and acid thiolysis, followed by HPLC and ES-ITMS of the products. Activity of isolated procyanidins tended to increase with degree of polymerization, epicatechin content, and with galloylation. These EDR-active compounds (many of which also possess antioxidant activity), individually or in the form of wines, juices, or nutritional supplements, may be useful in preventing or treating cardiovascular diseases.

KEYWORDS: procyanidins; nitric oxide; endothelium-dependent relaxing activity

INTRODUCTION

The antioxidant properties of various plant flavonoids, including procyanidins, are well known.[1–3] In addition to these properties, procyanidins have recently been shown to possess endothelium-dependent relaxing (EDR) activity in blood vessels *in vitro*.[4–6] The original finding that red wines, grape juice, and other grape products exhibited EDR activity was accompanied by strong evidence that this activity was due to stimulation of nitric oxide (NO) production by the endothelial cells which form the lining of all blood vessels.[4] Vasorelaxation induced by grape extracts, wines etc., was reversed by NO synthase inhibitors, and vasorelaxation could be restored by exposure of the vessel to L-arginine, the normal substrate for NO synthase. Furthermore, EDR was shown to be accompanied by increased vessel levels of cyclic

Address for correspondence: Dr. David F. Fitzpatrick, Department of Pharmacology, College of Medicine, MDC Box 9, and University of South Florida, Tampa, FL 33612-4799, USA. Voice: 813-974-9927; fax: 813-974-2565.
dfitzpat@hsc.usf.edu

GMP, the vascular smooth muscle cell (VSMC) messenger through which NO acts. Cyclic GMP then induces VSCM relaxation by mechanisms that are still not clearly defined, but include reduction of free Ca^{2+} levels within the cell. Subsequently it was shown that aqueous extracts of a wide range of commonly consumed food plants (fruits, vegetables, nuts, herbs) also exhibited EDR activity.[5] More recently the vasodilating compounds have been shown to be phenolics of the proanthocyanidin type.[6]

Our knowledge of the importance of NO has mushroomed since the Nobel Prize-winning work of Furchgott, Murad, and Ignarro who, working independently, showed that the endothelium-derived relaxing factor (EDRF, discovered by Furchgott), was in fact NO, and demonstrated the role and significance of the NO/NOS system in numerous physiological functions of the body. In blood vessels, NO not only resists vasoconstrictor influences, but also decreases platelet aggregation[7] and adherence of platelets and leukocytes to endothelium,[8] inhibits oxidation of low-density lipoproteins (LDL),[9] and diminishes VSMC proliferation.[10] The importance of the NO/NO synthase system is underscored by the finding that a dysfunctional NO system can contribute to several diseases, including atherosclerosis.[11,12] Thus it would appear that consumption (and absorption) of NO-stimulating compounds in the diet, or in the form of dietary supplements, could contribute to prevention or halting the progress of atherosclerosis, and possibly other chronic age-related diseases, or conditions known to involve failure of the NO/NO synthase system, for example, erectile dysfunction.

The exact nature of the plant-derived compounds responsible for EDR activity is not clear. We found in an earlier study that epicatechin-gallate had weak EDR activity. Andriambeloson *et al.*[13] reported that leucocyanidol, a 4-hydroxylated catechin, could produce EDR. It now appears that oligomeric and polymeric procyanidins are the most potent EDR-active compounds.[6] In the current study, we attempted to isolate and identify one or more lower molecular weight EDR-active procyanidins.

METHODS

Grape Seed Extraction and Preliminary Fractionation

Concord grape seeds (provided by Welch Foods, Inc.) were crushed and extracted into methanol, concentrated, filtered, then applied to a column (35×2.5 cm, i.d.) filled with Toyopearl TSK HW-40 resin. Elution was carried out with methanol. Seven fractions were collected, evaporated, and redissolved in water (for EDR testing, HPLC analysis, and tannase treatment) or methanol (for mass spectrometry and acid thiolysis).

Analytical HPLC

A Waters HPLC system was employed and consisted of a U6K injector, two 510 pumps, and a 481 UV/Vis detector, in conjunction with a Radial Pak reverse-phase NovaPak C18 column, protected by a guard column of the same material. The gradient for most analytical runs consisted of: Mobile phase A–water; Mobile phase B –10% acetic acid in water, and the gradient ran from 25% B up to 75% B over the first 47 min; from 85 to 100% B over 47 to 50 min; and 100% B isocratic over 50 to 55 min. Thereafter the gradient was returned to mobile phase A to prepare for the next run. Flow rate was 0.7 mL/min and detection was made at 280 nm. A different gradient was used for HPLC of acid thiolysis–treated samples: mobile phase A–2.5% acetic acid; mobile phase B–40% acetonitrile in A. The gradient ran from 10% B to 50% B in 20 min; up to 100% B at 25 min; isocratic at 100% B to 45 min; then back to 100% A for the next run.

Electrospray-Ion Trap Mass Spectrometry (ES-ITMS)

Fractions eluted from the Toyopearl column and individual peaks were examined by mass spectrometry using a Bruker-Esquire ITMS with an electrospray ionization source and run in the negative ion mode.[14] The electrospray matrix was 80% MeOH: 20% H_2O. A syringe pump was used to deliver the samples to the needle with a flow rate of 25 µL/h.

Chemical Analysis of Procyanidins

For determining whether or not EDR-active procyanidins were galloylated, purified samples in water were incubated with tannase (Sigma) at 35° for varying lengths of time, followed by HPLC analysis of the products. The acid thiolysis method was based on the method of Rigaud *et al.*[15] The purified procyanidin sample in methanol was incubated with an equal amount of benzyl mercaptan solution (12% benzyl mercaptan in 0.4 HCl, made up in methanol) at 50° for varying lengths of time, followed by HPLC analysis of the products.

Aortic Ring Preparation and Bioassay of Fractions and Peaks

The procedure for preparation of rat aortic rings and general aspects of determining mechanical activity has been previously described.[4] Briefly, male Sprague-Dawley rats (200–250 g) were euthanized with an overdose of sodium pentobarbital (100 mg/kg, i.p.), bled, and the thoracic aorta excised, cleaned, and rings (3–4 mm in length) were cut, taking care not to disturb the endothelium. In some instances the endothelium was deliberately removed by gently rubbing the lumen with a curved forceps. The rings were suspended in tissue baths containing a physiological salt solution with the following composition (in millimolar): 118 NaCl; 4.7 KCl; 25 $NaHCO_3$; 1.2 $MgSO_4$; 1.2 NaH_2PO_4; 0.026 EDTA; 1.5 $CaCl_2$; 11 glucose. The solution was bubbled continuously with O_2/CO_2 (95%/5%), and maintained at 37°C. Activity was recorded on a Grass polygraph. After equilibration for at least one hour under 1.5 g of tension, tissues were contracted submaximally (approximately 80% of E_{max}) with 1 µM phenylephrine, and then 3 µM acetylcholine, a known EDR-active compound, was added to the bath to test for intactness of the endothelium. This concentration of acetylcholine is sufficient to produce maximum endothelium-dependent relaxation in intact rings. Rings were washed with physiological salt solution three times over the next 45 min prior to the next sequence.

Screening of extracts, Toyopearl fractions, and HPLC peaks was conducted as follows. Aortic rings were contracted by addition of phenylephrine, and cumulative additions of each sample made, beginning with a concentration determined in preliminary experiments to be below the threshold for relaxation, and increased until a relaxation of approximately 15% (relative to the relaxation induced by 3μM acetylcholine) was achieved. The concentration of sample required to produce this degree of relaxation (15%) was arbitrarily set as the "threshold" for demonstrating relaxation potency for the purpose of rapidly screening the many samples. Subsequently, full concentration-response curves were generated for peak compounds exhibiting the greatest relaxing activity. To test for endothelium dependence, denuded aortic rings were used. Successful endothelium removal was established by a lack of relaxation response to 3μM acetylcholine. Upon testing of the fractions and peaks, none exhibited any relaxing activity using de-endothelialized rings.

RESULTS AND DISCUSSION

Toyopearl Fractionation

Toyopearl TSK HW-40S fractionation of grape seed polyphenolics using methanol as eluent yielded 7 fractions, labeled A through G (see FIGURE 1). The compounds present in the early fractions were easily identified by HPLC by their co-elution with available standards, e.g., fraction A contained primarily gallic acid, fraction B contained (+)-catechin and (−)-epicatechin, and fraction C contained (−)-epicatechin gallate plus other compounds subsequently identified as flavan-3-ol dimers. These three fractions exhibited very little EDR activity when bioassayed,

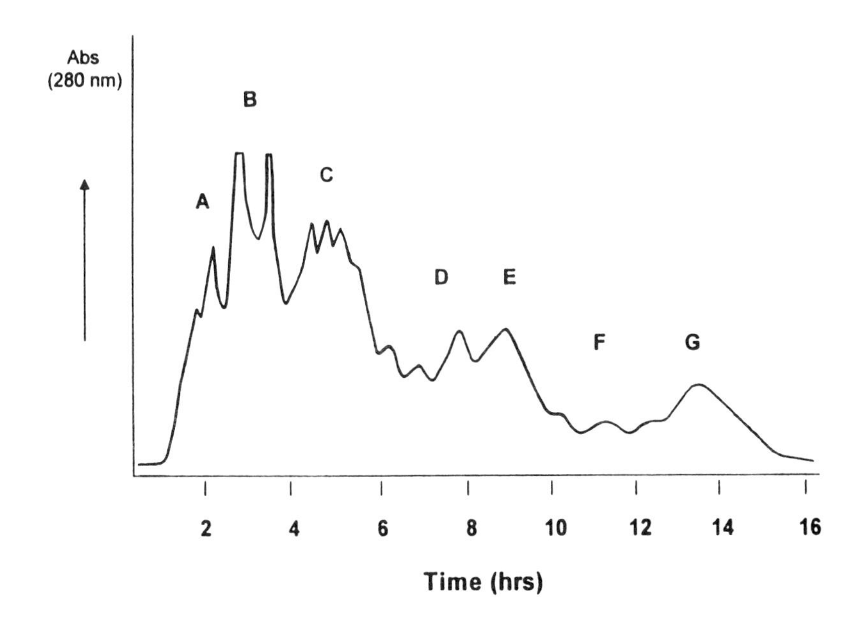

FIGURE 1. Toyopearl TSK HW(40) elution profile of grape seed extract eluted with methanol, absorbance measured at 280 nm. (Reprinted from Fitzpatrick *et al.*[6] with permission from the American Chemical Society.)

whereas the remaining Toyopearl fractions containing proanthocyanidins displayed varying degrees of vasorelaxing potency (see below).

EDR Activity of Toyopearl Fractions and HPLC Peaks

TABLE 1 summarizes the relaxation and ES-ITMS results for most of the Toyopearl fractions and for their constituent peaks. Some peaks were numbered initially, but were present in concentrations too small to collect and test for EDR activity, therefore are not listed in the table. Little or no relaxing activity was observed in fractions A, B, or C, which contained only phenolic acids, monomeric,

TABLE 1. Relaxation data and mass spectrometry information on Toyopearl fractions and HPLC peaks

Toyopearl Fraction	HPLC Peak	EDR Threshold[a] (μg/ml)	ES-ITMS Peak Compounds
Fraction A	—	—	gallic acid; other phenolic acids
Fraction B	—	—	flavanol monomers (catechin, epicatechin)
Fraction C	—	>4	monomer-Gal. (ECG); dimers
Fraction D	D 1	—	trimer
	D 3	>4	trimer
	D 4	—	trimer
	D 6	2–3	dimer-Gal.
Fraction E	E 1	1–2	tetramer
	E 2	—	trimer
	E 3	1	trimer-Gal.
	E 4	1	tetramer
Fraction F	F 3	1–2	tetramer
	F 4	1	dimer-Gal.
	F 5	1–2	tetramer
	F 7	0.5–1	trimer
Fraction G	G 2	1–2	pentamer
	G 3	2–4	tetramer
	G 4	<0.5	tetramer-Gal.
	G 5	<0.5	tetramer-Gal.
	G 6	<0.5	trimer-Gal.
	G 7	<0.5	pentamer

Modified from Fitzpatrick *et al.*[6] with permission from the American Chemical Society.
[a]Amount of fraction or peak material required to produce 15% relaxation.

and dimeric flavanol compounds, respectively, although fraction C (which also contains epicatechin-gallate) showed some relaxing activity at concentrations greater than 4 μg/mL. Relatively more activity (indicated by the "threshold") was seen in subsequent fractions, with activity generally increasing from fraction D through fraction G. EC_{50} values (with confidence intervals) of active Toyopearl fractions (fractions D–G) were as follows: Fraction D, 4.37 (1.30–14.60); Fraction E, 2.97 (1.65–5.38); Fraction F, 2.55 (1.75– 3.72); and Fraction G, 1.20 (0.84–1.71). Fractions eluting later than fraction G (using 70% acetone in water) were quite active but were not pursued since they were higher molecular weight compounds which would probably not be bioavailable and thus would be of less interest to us for future *in vivo* studies.

HPLC peaks derived from fractions E, F, and G that exhibited the greatest EDR activity are indicated in TABLE 2, along with their EC_{50} values. These values ranged from approximately 0.6μg/mL to 2.6μg/mL. The most active compounds include proanthocyanidins larger than dimers (trimers, tetramers and pentamers), and their gallates. Galloylation appeared to increase activity at any given molecular size, for example, the activity of trimer gallates was greater than that of most trimers, and dimer gallate activity was greater than that of the mostly inactive dimers. There also seem to be differences in EDR activity among the members of the isomeric families, for example, all compounds identified as tetramers were not equally active, suggesting that the specific monomeric makeup of the compounds, and possibly the order of the monomeric components within the oligomer, are important for activity.

TABLE 2. EC_{50} values of the most active peaks

Peak Number	Compound	EC_{50} (C.I.)[a]
E1	Tetramer	2.59 (2.49–2.69)
E3	Trimer-G	1.55 (1.28–1.89)
E4	Tetramer	2.25 (2.14–2.37)
F3	Tetramer	1.54 (1.27–1.87)
F4	Dimer-G	1.25 (0.90–1.73)
F6	Trimer	1.17 (0.96–1.43)
G4	Tetramer-G	0.93 (0.83–1.04)
G5	Tetramer-G	0.57 (0.49–0.67)
G6	Trimer-G	1.00 (0.92–1.09)
G7	Pentamer	1.05 (0.85–1.29)

Adapted from Fitzpatrick *et al.*[6] with permission from the American Chemical Society.

[a]Mean concentration of peak material (μg catechin equivalents/ml) required to produce 50% relaxation (C.I., 95% confidence interval).

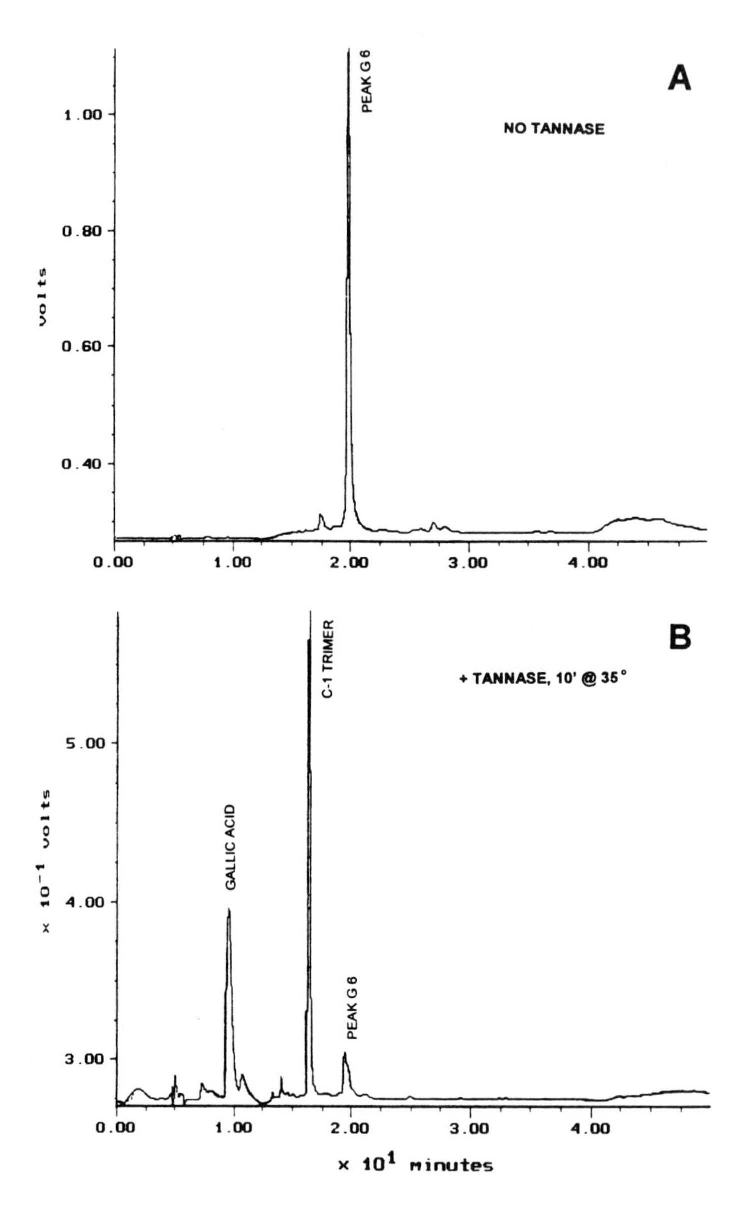

FIGURE 2. Effect of tannase treatment on peak G6. The original peak G6 (**A**) was converted to gallic acid and apparently trimer C1 (**B**), based on retention time.

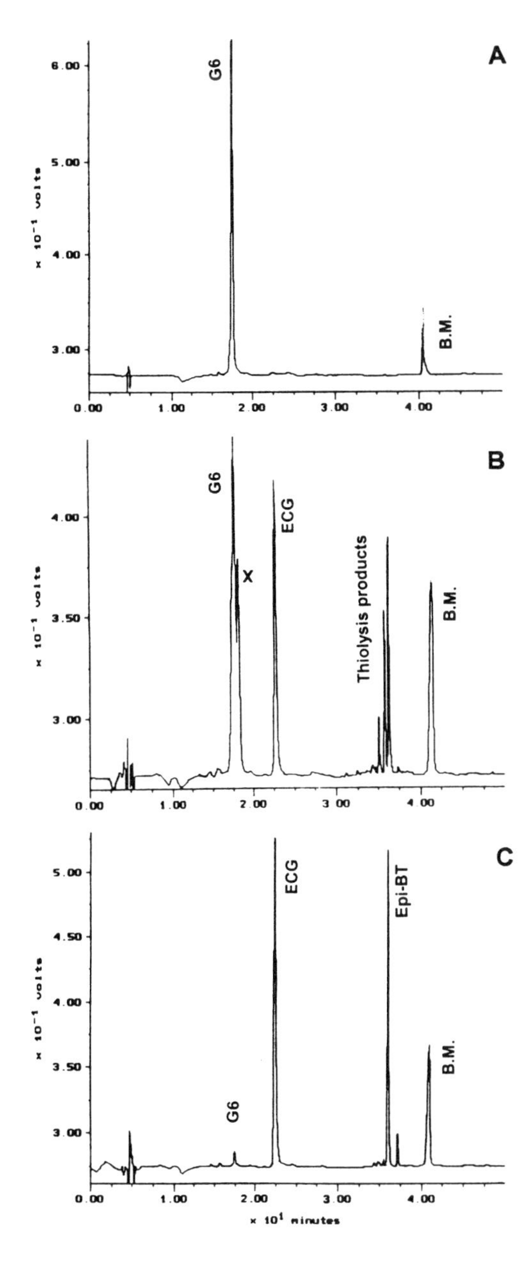

FIGURE 3. Acid thiolysis of peak G6 in the presence of benzyl mercaptan (B.M). The original peak G6 at 0 time (**A**) was converted to the intermediate products (**B**) by acid thiolysis for 1 min at 50° C. The only final products after 15 min at 50° C (**C**) were epicatechin-gallate (ECG) and epicatechin-benzylthioether (Epi-BT). Thus, it is concluded that peak G6 is epicatechin-epicatechin-epicatechin-gallate (procyanidin trimer C1-gallate).

Chemical Analysis of Peak G6

Peak G6 was selected for more detailed analysis because it is one of the smallest of the EDR-active compounds (and more likely to be bioavailable), and is more abundant in these seeds than are other smaller EDR-active procyanidins. This compound was determined to be a trimer-gallate, according to ES-ITMS (TABLES 1 and 2). This was confirmed by treatment with tannase, which resulted in two HPLC peaks (see FIGURE 2), gallic acid and another peak, tentatively identified as trimer C1 (epi-epi-epi).

Partial acid thiolysis of peak G6 with benzyl mercaptan (1 min incubation at 50° C) yielded several intermediate products including epicatechin-gallate, an unidentified peak with a retention time of about 18 min (probably epicatechin-epicatechin-gallate, and two prominent peaks plus a small one at 35–36 min (see FIGURE 3 A and B). The latter were determined to be benzylthioethers (BT's). Complete acid thiolysis (15 min at 50°C) yielded only two peaks, identified as epicatechin-gallate and epicatechin-BT (FIG. 3 C). These results indicate that Peak G6 is comprised only of epicatechins, with a gallate attached to one of the epicatechins. FIGURE 4 shows the most probable scheme for the breakdown products obtained, and that the compound is, in fact epicatechin-epicatechin-epicatechin-gallate, with the gallate attached to the terminal epicatechin (gallate attachment on either of the other two epicatechins would yield entirely different breakdown patterns than that described). The only question is the type of linkage between monomers, i.e., 4-8 or 4-6. The tannase results yielded

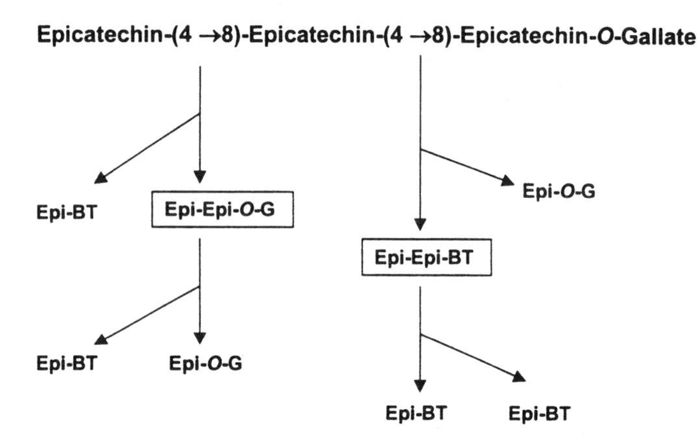

FIGURE 4. Scheme showing intermediate and final products of acid thiolysis of a procyanidin trimer gallate (C1-gallate) in the presence of benzyl mercaptan. "Lower" units (to the right in this illustration) are released by hydrolysis, whereas "upper" units (to the left) yield benzythioethers (-BTs). The intermediate dimer products released after 1 min at 50° C (epi-epi-O-G and epi-epi-BT) are subsequently broken down by thiolysis to yield the final products: epi-O-G and epi-BT.

a peak that appears to be C1 trimer (epi-epi-epi), based on retention time, therefore, the linkage is most likely 4-8. The G6 compound structure, epicatechin-(4-8)-epicatechin-(4-8)-epicatechin-gallate, is shown in FIGURE 5.

Although this is only one example, these results support our previous conclusion[6] that EDR activity is enhanced as the content of (–)-epicatechin in the oligomer increases, and that the presence of gallate(s) augments activity as well. The EC_{50} values of some of the more active of the compounds (TABLE 2) were in the low μg/mL and sub-μg/mL range. It is not known whether these levels of the EDR-active procyanidins (or their metabolites) are achieved by consumption of a standard diet or even a diet supplemented with extra sources of procyanidins (e.g., red wine, GSE supplements), but there is indirect evidence that they are, indeed, increased to pharmacologically relevant levels: Yamakoshi *et al.*[16] have shown that a proanthocyanidin-rich GSE preparation attenuates the development of aortic atherosclerosis in cholesterol-fed rabbits. Wollny *et al.*[17] have shown that oral administration of red wine prevents experimental thrombosis in rats, an effect that is blocked by L-NAME, an NO synthase inhibitor. In a study by Stein *et al.*[18] endothelium-dependent, flow-mediated vasodilation was improved significantly by purple grape juice in coronary artery disease (CAD) patients with impaired endothelial function. While the latter two studies

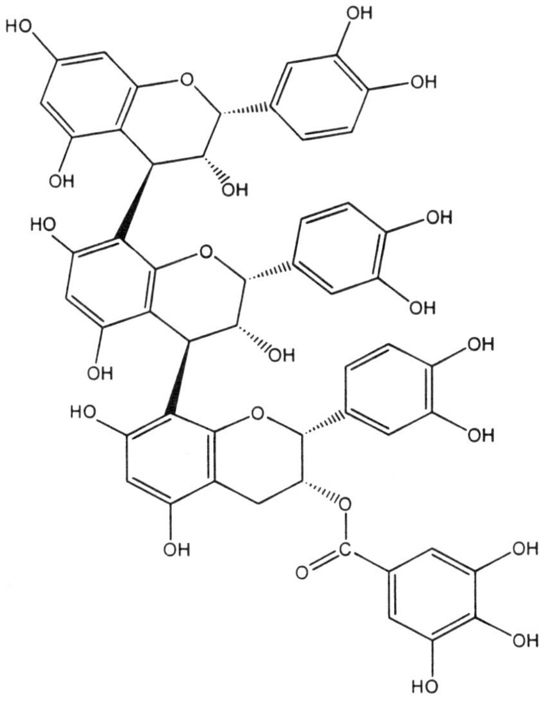

FIGURE 5. Structure of the peak G6 compound identified as epicatechin-(4-8)-epicatechin-(4-8)-epicatechin-gallate (C1-gallate).

clearly indicate endothelium-dependent responses to the grape products (red wine/purple grape juice), the compounds responsible for the responses were not ascertained. It is hoped that the present results will lead to similar types of studies using individual EDR-active compounds

CONCLUSIONS

These results demonstrate that individual compounds with biological activity can be purified and identified by the methods utilized here: solvent extraction, preliminary fractionation by Toyopearl chromatography, analysis by HPLC, ES-ITMS, tannase treatment, and acid thiolysis, with bioassay at each step, for example, using the rat aortic ring preparation for determining EDR activity. The most active EDR-active compounds appear to be galloylated procyanidins containing a preponderance of (–)-epicatechins. It may be possible to isolate such compounds from grape seeds on a large scale, or possibly synthesize them in bulk quantities for *in vivo* studies.

REFERENCES

1. Frankel, E.N., A.L. Waterhouse & P.L. Teissedre. 1995. Principal phenolic phytochemicals in selected California wines and their antioxidant activity in inhibiting oxidation of human low density lipoproteins. J. Agric. Food Chem. **43:** 890–894.
2. Rice-Evans, C.A., N.J. Miller & G. Paganga. 1997. Antioxidant properties of phenolic compounds. Trends Plant Sci. **2:** 152–159.
3. Koga, T., *et al.* 1999. Increase of antioxidative potential of rat plasma by oral administration of proanthocyanidin-rich extract from grape seeds. J. Agric. Food Chem. **47:** 1892–1897.
4. Fitzpatrick, D.F., S.L. Hirschfield & R.G. Coffey. 1993. Endothelium-dependent vasorelaxing activity of wine and other grape products. Am. J. Physiol. **265:** H774–H778.
5. Fitzpatrick, D.F., *et al.* 1995. Endothelium-dependent vasorelaxation caused by various plant extracts. J. Cardiovasc. Pharmacol. **26:** 90–95.
6. Fitzpatrick, D.F., *et al.* 2000. Isolation and characterization of endothelium-dependent vasorelaxing compounds from grape seeds. J. Agric. Food Chem. **48:** 6384–6390.
7. Yao, S., *et al.* 1992. Endogenous nitric oxide protects against platelet aggregation and cyclic flow variations in stenosed and endothelium-injured arteries. Circulation **86:** 1302–1309.
8. Provost, P., *et al.* 1994. Endothelium-derived nitric oxide attenuates neutrophil adhesion to endothelium under arterial flow conditions. Arterioscler. Thromb. **14:** 331–335.
9. Hogg, N., *et al.* 1993. Inhibition of low-density lipoprotein oxidation by nitric oxide. FEBS **334:** 170–174.
10. Scott-Burden, T. & P. Vanhoutte. 1993. The endothelium as a regulator of vascular smooth muscle proliferation. Circulation **87:** V51–V55.
11. Quyyumi, A., *et al.* 1995. Contribution of nitric oxide to metabolic coronary vasodilation in the human heart. Circulation **92:** 320–326.
12. Cannon, R.O. 1998. Role of nitric oxide in cardiovascular disease: focus on the endothelium. Clin. Chem. **44:** 1809-1819.
13. Andriambeloson, E., *et al.* 1997. Nitric oxide production and endothelium-dependent vasorelaxation induced by wine polyphenols in rat aorta. Brit. J. Pharmacol. **120:** 1053–1058.

14. FLEMING, R.C., *et al.* 1999. Identification of procyanidin compounds using nanospray and electrospray ion trap mass spectrometry. Proceedings of the 47th ASMS Conference on Mass Spectrometry and Allied Topics, Dallas, Texas.
15. RIGAUD, J., *et al.* 1991. Micro method for the identification of proanthocyanidin using thiolysis monitored by high-performance liquid chromatography. J. Chromatogr. **540:** 401–405.
16. YAMAKOSHI, J., *et al.* 1999. Proanthocyanidin-rich extract from grape seeds attenuates the development of aortic atherosclerosis in cholesterol-fed rabbits. Atherosclerosis **142:** 139–149.
17. WOLLNY, T., *et al.* 1999. Modulation of haemostatic function and prevention of experimental thrombosis by red wine in rats: a role for increased nitric oxide production. Br. J. Pharmacol. **127:** 747–755.
18. STEIN, J.H., *et al.* 1999. Purple grape juice improves endothelial function and reduces the susceptibility of LDL cholesterol to oxidation in patients with coronary artery disease. Circulation **100:** 1050–1055.

Inhibition by Wine Polyphenols of Peroxynitrite-Initiated Chemiluminescence and NADH Oxidation

ALBERTO BOVERIS, LAURA VALDEZ, AND SILVIA ALVAREZ

Laboratory of Free Radical Biology, School of Pharmacy and Biochemistry, University of Buenos Aires, Buenos Aires, Argentina

ABSTRACT: Peroxynitrite ($ONOO^-$) is a powerful oxidant produced by neutrophils, macrophages, and lymphocytes as a signaling and cytotoxic molecule from their primary production of nitric oxide (NO) and superoxide anion (O_2^-). In the vascular space, $ONOO^-$ will likely oxidize lipoproteins and promote atherogenesis. Pure wine flavonoids (catechin, epicatechin, myricetin), hydroxycinnamates (caffeic acid, ferulic acid, chlorogenic acid), and plain Argentine red wines were assayed as $ONOO^-$ scavengers in two assays: (a) $ONOO^-$-initiated chemiluminescence and (b) $ONOO^-$-dependent oxidation. The assayed polyphenols as well as the red wines were effective inhibitors of the $ONOO^-$-driven oxidation reactions. Fifty percent of the pure substances were observed in the range of 30–300 μM and in the case of red wines with the equivalent of 80–120 μM of flavonoids. The amphipatic nature of wine polyphenols will lead to their accumulation at the lipoprotein surface, according to the Gibbs adsorption equation, where they are likely to prevent $ONOO^-$-induced tyrosine nitration and LDL modification.

KEYWORDS: polyphenols; red wine; NADH oxidation; $ONOO^-$ chemiluminescence

INTRODUCTION

The Biological Role of Peroxynitrite

Peroxynitrite ($ONOO^-$) is both an intracellular and an extracellular metabolite in the human body. On the one hand and concerning the intracellular generation of this species, it has been recently reported that $ONOO^-$ is formed in isolated mitochondria[1] and in mitochondria *in vivo*.[2] Mitochondria have been recognized as an effective source of nitric oxide (NO)[3] and superoxide anion (O_2^-)[4] under physiological conditions. The rate of peroxynitrite formation linked to mitochondrial metabolism can be estimated as accounting for about 2% of the oxygen taken up by the organs.[1,5] On the other hand, concerning the extracellular or vascular production of $ONOO^-$, a series of cell types—most of them immunologically competent, such as macrophages,[6] neutrophils,[7] Kupffer cells,[8] mononuclear leukocytes,[9] and

Address for correspondence: Dr. Alberto Boveris, Facultad de Farmacia y Bioquímica, Universidad de Buenos Aires, Junín 956, 1113 Buenos Aires, Argentina. Voice: 54-11-4964-8245/44; fax: 54-11-4508-3646
aboveris@ffyb.uba.ar

cultured endothelial cells[10]—produce and release $ONOO^-$ vectorially into their extracellular medium, the systemic vascular space. Peroxynitrite is formed in both cases, in the intracellular and the vascular spaces through the diffusion-controlled reaction of NO and O_2^- ($k = 1.9 \times 10^{10}\ M^{-1}\ s^{-1}$).[11] This free radical termination reaction, where both reactants have one unpaired electron in their external orbitals, was proposed by Beckman ten years ago as the molecular mechanism of ischemic injury.[12] Now it is also recognized as the main pathway of intramitochondrial NO utilization,[1] and as one of the molecular strategies which are responsible for the cytotoxicity of specialized immunologic cells.[6–9]

Peroxynitrite is a powerful oxidant that can accept one or two electrons, according to the following chemical equations[13,14]:

Univalent electron transfer:

$$ONOO^- + 1e^- + 2H^+ \rightarrow H_2O + \cdot NO_2 \qquad (E'^{\circ} = +1.6\text{–}1.7\ V).$$

Bivalent electron transfer:

$$ONOO^- + 2e^- + 2H^+ \rightarrow H_2O + NO_2^- \qquad (E'^{\circ} = +1.3\text{–}1.37\ V).$$

At physiologic pH, $ONOO^-$ protonates to yield peroxynitrous acid (ONOOH; pKa = 6.8), which rearranges itself to yield nitrate (NO_3^-) or to decompose producing radical species such as hydroxyl radical ($\cdot OH$) and nitric dioxide ($\cdot NO_2$) with a global half life of about one second.[14] The mechanisms of $ONOO^-$-dependent oxidations are complex and not completely understood at present.[14] However, it has been experimentally well documented that $ONOO^-$ readily oxidizes the sulfhydryl group of cysteine and glutathione (GSH),[15] the sulfur atom of methionine,[16] ascorbate,[17] the purine and pirimidine bases of DNA,[18] and the reduced forms of nicotinamide nucleotides (NADH and NADPH)[19,20] and ubiquinone.[20] Peroxynitrite is also able to start the lipoperoxidation process in biomembranes and liposomes[21] and in isolated LDL,[22] likely by the generation of $\cdot OH$. In addition, $ONOO^-$ added to biological material nitrates protein tyrosine residues forming 3-nitrotyrosine,[23] a process understood to involve hydrogen abstraction by $\cdot OH$ followed by an $\cdot NO_2$ addition eaction.

Wine and Plant Phenolic Compounds: Flavonoids and Hydroxycinnamates

Plant polyphenols constitute a large class of compounds, ubiquitous in plants, containing a number of phenolic hydroxyl groups that are the chemical basis of their antioxidant activity.[24] Red wines contain 800–2000 mg/L of polyphenols,[25,26] expressed as gallic acid equivalents that corresponds to a 5–13 mM gallic acid and to 15–39 mM phenolic groups. Wine polyphenols constitute a large series of compounds that can be grouped in families; the total number of chemical species with antioxidant properties is estimated to be about 300 (according to the currently used HPLC analysis systems). These polyphenols are reducing agents that function as antioxidants by virtue of the hydrogen-donating properties of their phenolic hydroxyl groups[25,26] as well as by their transition metal-chelating abilities.[27,28] The aromatic hydroxyl groups of flavonoids and hydroxycinnamates present an antioxidant pharmacophore comparable to the chroman moiety of tocopherols. Indeed, a high reactivity of flavonoids with radicals such as O_2^-,[29] $\cdot OH$,[25] and NO[30] has been reported. In particular, epicatechin, epigallocatechin and their gallate esters have been shown to

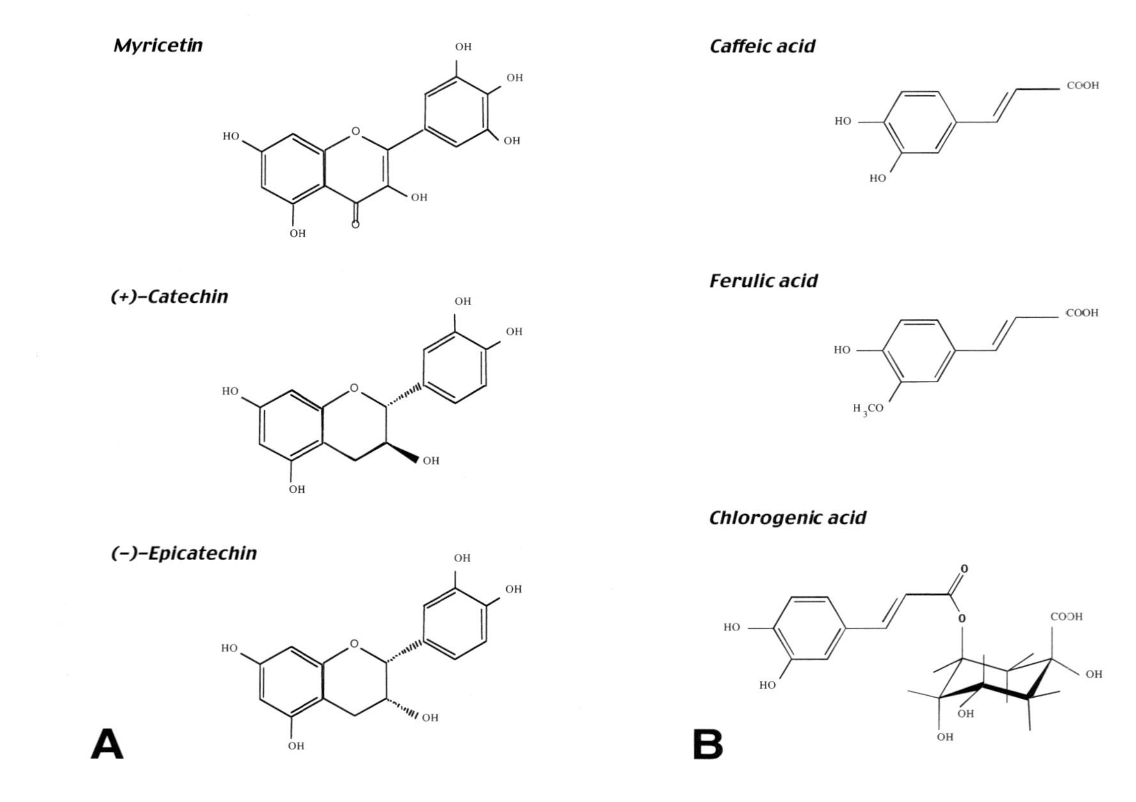

FIGURE 1. Chemical structures of the assayed polyphenols (**A**) and hydroxycinnamates (**B**).

scavenge both aqueous and lipophilic radicals and to protect low-density lipoprotein (LDL) from oxidation by acting as chain-breaking antioxidants.[31] In addition, flavonoids are potent scavengers of $ONOO^-$[30] and red wine exhibits a considerable $ONOO^-$-scavenging activity.[32]

Hydroxycinnamates or phenylpropanoids are widely distributed in plant tissues and are present as dietary phytochemicals at concentrations that are higher than those corresponding to flavonoids.[33] These compounds are present in grape juice and in white and red wines and occur in nature in various conjugated forms. It has been reported that hydroxycinnamates exhibit antimicrobial, antiallergic and anti-inflammatory activities as well as antimutagenic properties.[34,35] They have also been shown to possess peroxyl radical-scavenging properties as measured by their ability to prevent the lipid peroxidation of LDL mediated by heme proteins.[25,36] Recently, it has been reported that catechins and hydroxycinnamates inhibit the $ONOO^-$-induced nitration of tyrosine.[33]

The present chapter is focused in the inhibition by wine polyphenols of two oxidation processes dependent on $ONOO^-$: (a) the $ONOO^-$-initiated chemiluminescence of rat liver homogenate,[37] and (b) $ONOO^-$-dependent NADH oxidation.[38] Pure wine polyphenols, such as the flavonoids (+)-catechin, (–)-epicatechin and myricetin and the hydroxycinnamates caffeic acid, ferulic acid, and chlorogenic acid (see FIGURE 1), grape seed extracts, and 100 samples of Argentine red wines (Fondo Vitivinícola de Mendoza, Mendoza, Argentina) were assayed. Pure polyphenols, grape seed extract, and red wines effectively inhibited both $ONOO^-$-dependent processes.

PEROXYNITRITE-INITIATED HOMOGENATE CHEMILUMINESCENCE

Peroxynitrite was found able to start the lipoperoxidation process in membranes, liposomes and isolated LDL,[21,22] probably by the generation of ·OH. Considering the high sensitivity of the assay of the *tert*-butyl hydroperoxide–initiated homogenate chemiluminescence to assay antioxidants,[39] a similar technique was developed using $ONOO^-$ to provide the radical oxidants for the initiation reactions of the lipoperoxidation process. The details of the experimental procedure are provided in an accompanying paper in this same volume.[37] The assayed flavonoids and the red wine samples inhibited $ONOO^-$-initiated chemiluminescence (see FIGURE 2). The flavonoids showed the following IC_{50}: (+)-catechin, 13 μM; (–)-epicatechin, 7 μM; and myricetin, 20 μM[37] (see FIGURE 3 A).

PEROXYNITRITE-DEPENDENT NADH OXIDATION

Peroxinitrite readily reacts with the reduced forms of pyridine nucleotide producing the corresponding oxidized forms as final products.[19,20] This property and the fluorescence of NADH were recently used by the authors in an assay to evaluate the effectiveness of $ONOO^-$ scavengers.[20,40] A quantitative determination of the effectiveness of wine polyphenols, grape seed extracts, and red wines is given in an accompanying paper in this same volume.[38]

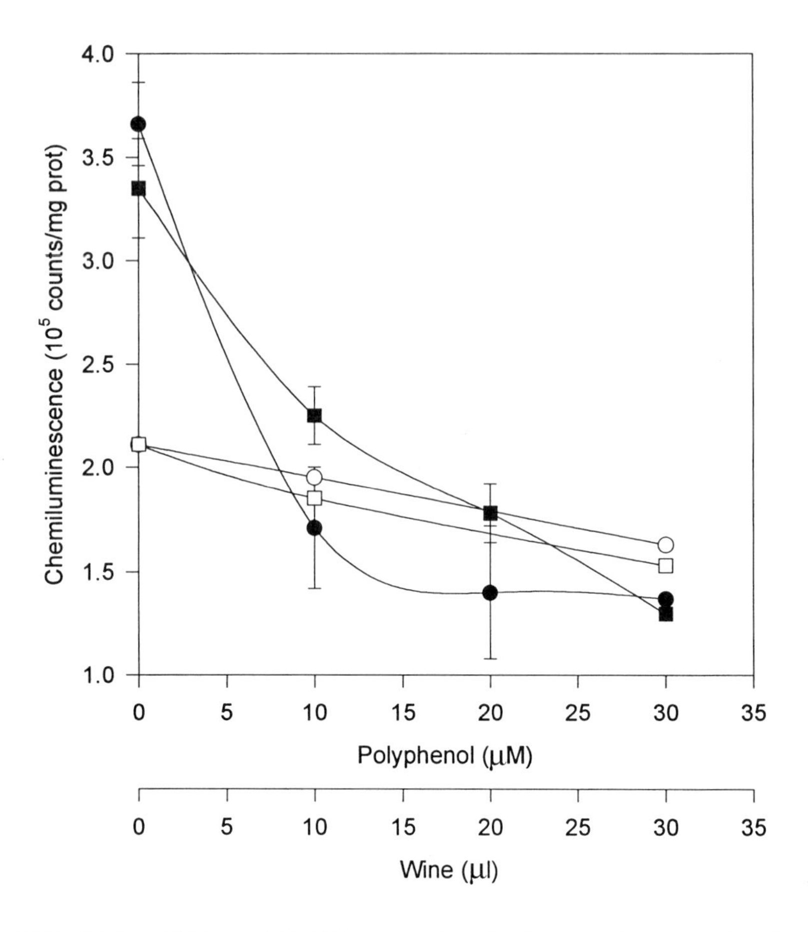

FIGURE 2. Inhibition of ONOOH initiated chemiluminescence by flavonoids and red wines. (●) (+)-catechin; (■) myricetin; (○) malbec and (□) cabernet sauvignon. Wine units are given in mL for 2 mL final assay volume.

The Rate Constants of the Reactions of Flavonoids and Hydroxycinnamates with Peroxynitrite

A simple competition model and the fluorometric determination of NADH, previously described in detail[20,38] were used to determine the apparent rate constants of the reactions of $ONOO^-$ with flavonoids and hydroxycinnamates. This indirect method is sensitive and allows the use of phenolic compounds in the range 20–300 μM, to estimate the apparent second order rate constants for the reaction of $ONOO^-$ and these compounds. The assay is advantageously used under circumstances in which: (a) no direct method or stopped-flow facility is available, (b) pseudo-first-order conditions are not attainable, and (c) high background absorbance is detected.

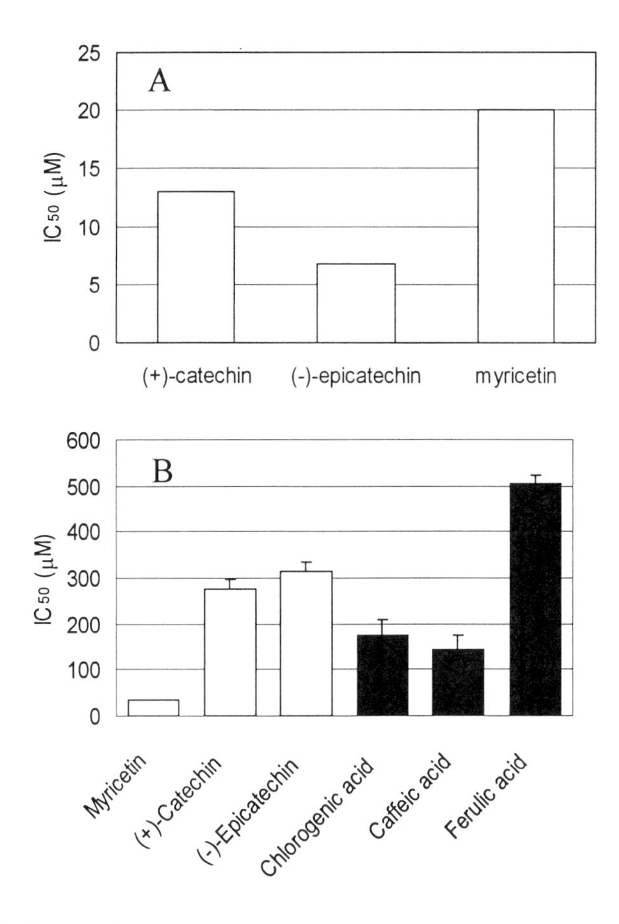

FIGURE 3. IC_{50} of the flavonoids and hydroxycinnamates on $ONOO^-$-dependent oxidations: (**A**) chemiluminescence and (**B**) NADH oxidation.

The chemical structures of the assayed compounds are shown in FIGURE 1. The rate constants were calculated taking into account the second order rate constant for the reaction of NADH and $ONOO^-$ (see Table 1 of Ref. 38). The highest value of the rate constants was the one corresponding to the reaction of $ONOO^-$ with myricetin, with lower rate constants for the other two flavonoids and the hydroxycinnamates. The concentrations inhibiting 50% NAD formation were inversely related and in agreement with the given second order reaction constants (FIG. 3B). The IC_{50} of the flavonoids on the NADH oxidation by $ONOO^-$ were 1.7–35 times higher than the IC_{50} of the same flavonoids on $ONOO^-$ initiated chemiluminescence (FIG. 3). This huge discrepancy indicates that the inhibition of both $ONOO^-$-dependent processes by the same molecule occurs through different mechanisms. The process of lipoperoxidation, which is central to chemiluminescence, occurs in an hydrophobic environment, whereas that NADH oxidation occurs in the aqueous phase.

Inhibition of Peroxynitrite-Dependent NADH Oxidation by Grape Seed Extract and Red Wines

Grape seed and *Ginkgo biloba* extracts, this latter assayed here for comparative purposes, contain a large number of flavonoids that are considered responsible for the free radical scavenging properties of the extracts. When the effect of these two extracts on NADH oxidation by $ONOO^-$ was studied, the estimated half-inhibition concentration, determined from the graphs shown in FIGURE 4, was lower for *Ginkgo biloba* extract than for grape seed extract (see Table 1 of Ref. 38). Considering that the extracts contain a large number of different substances, it follows that both extracts contain very effective $ONOO^-$ scavengers. The units given for IC_{50} inhibition in the referred Table, are µg of extract/ml of reaction medium, which corresponds to similar numbers of flavonoid concentration expressed in µM, considering a content of 30% of flavonoids in the extracts and a molecular weight of 300 Da for the flavonoids. Three selected red wines inhibited NADH oxidation by $ONOO^-$ with IC_{50} of 22–44 µL wine/mL of reaction medium (see Table 1 of Ref. 38).

FIGURE 5 shows the correlation of the effects of the 100 assayed red wines on NADH oxidation and on $ONOO^-$-initiated chemiluminescence. It is seen that the correlation is poor, indicating, once again, that both $ONOO^-$-dependent oxidative processes have different molecular mechanisms and that red wine constituents that have inhibitory effects on both processes are randomly distributed.

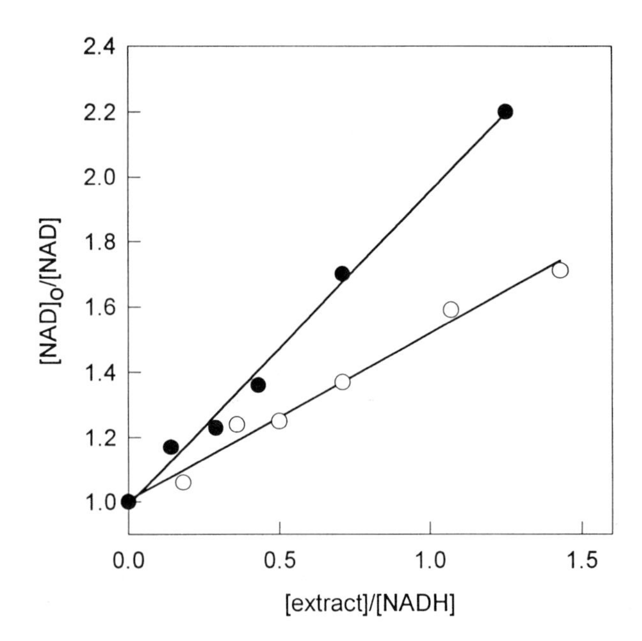

FIGURE 4. Effect of grape seed (○) and *Ginkgo biloba* (●) extracts on the NADH oxidation produced by peroxynitrite. Plot of $[NAD]_0/[NAD]$ as a function of [extract]/[NADH].

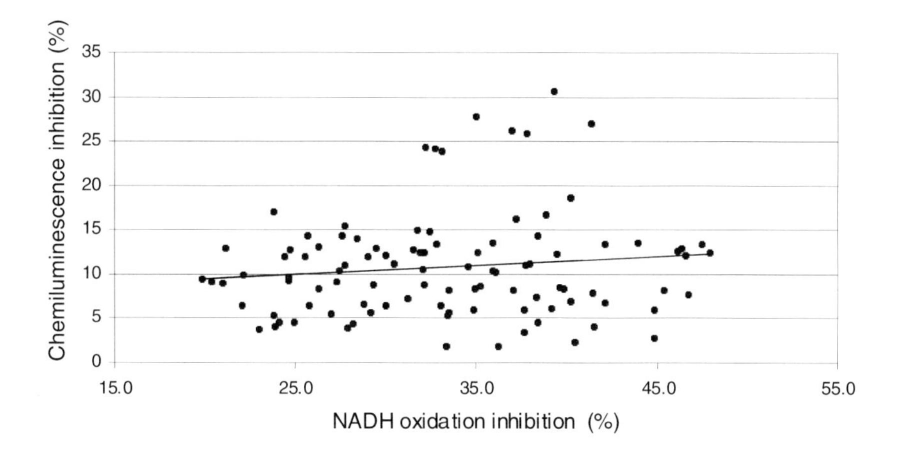

FIGURE 5. Correlation between the chemiluminescence and NADH oxidation assays for the 100 Argentine red wine samples studied.

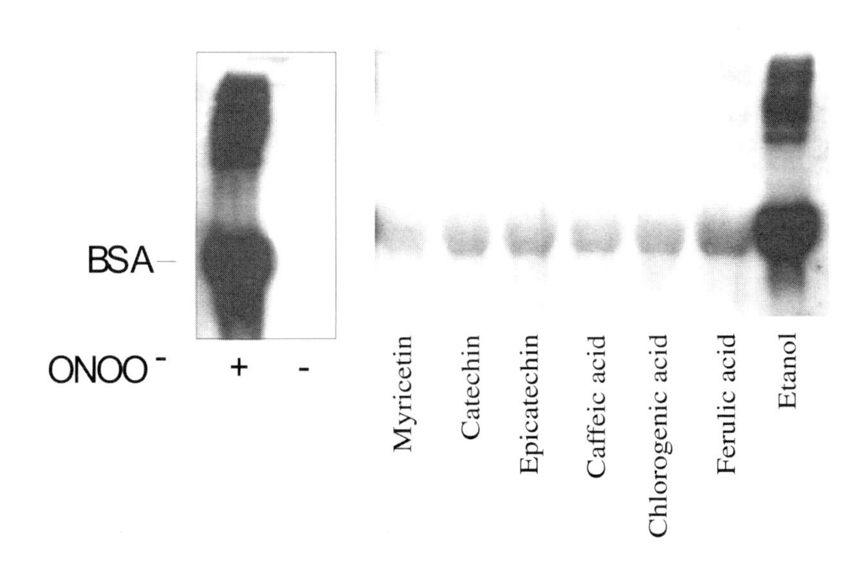

FIGURE 6. Western blot analysis of 3-nitrotyrosine residues. Effect of 200 μM $ONOO^-$ on bovine serum albumin (2 mg/ml) in the absence or presence of 200 μM flavonoids or hydroxycinnamates. Ethanol was assayed as control of myricetin solvent, and it did not protect albumin from nitration.

TYROSINE NITRATION BY PEROXYNITRITE

Tyrosine nitration of bovine serum albumin (BSA) was used as a model of extracellular protein nitration. Bovine serum albumin (2 mg/mL) was exposed to 200 μM $ONOO^-$, in the absence or presence of 200 μM polyphenols. 3-Nitrotyrosine formation was detected by SDS-Page and Western blot analysis (see FIGURE 6). Supplementation with plant phenolic compounds was able to inhibit the tyrosine nitration produced by $ONOO^-$ in the following sequence (according to their effect at the same concentrations of 200 μM): myricetin > (+)-catechin = (−)-epicatechin > caffeic acid = chlorogenic acid > ferulic acid.

The amount of 3-nitrotyrosine was quantified taking as reference for full nitration, the one observed after $ONOO^-$ addition; and for no nitration, the one observed in the absence of $ONOO^-$ (FIG. 6, *left panel*). Flavonoids (myricetin, (+)-catechin and (−)-epicatechin) and disubstituted hydroxycinnamates (caffeic and chlorogenic acid) were found to decrease tyrosine nitration better than monohydroxycinnamates (ferulic acid) (TABLE 1). The observed $ONOO^-$ scavenging effects of flavonoids and hydroxycinnamates are in good correlation with the results reported by Panalla *et al.*[33]

DISCUSSION

A large number of biomolecules constitute potential targets for $ONOO^-$ in its two chemical actions, as an oxidizing and as a nitrating species. However, only few molecules are expected to be preferential targets due to the kinetic conditions, that is, reaction rates and reactant concentrations. The knowledge of the reaction kinetics of $ONOO^-$ with biomolecules not only allows the prediction of the preferential rates of $ONOO^-$ reactions in biological systems, but also provides information about the use of reductants as $ONOO^-$ scavengers in determined situations. There is an understanding that after $ONOO^-$ protonation and formation of ONOOH, this chemical species can: (a) rearrange itself to form nitrate **(1)**; (b) react collisionally with reductants; and (c) dissociate homolitically to form the radicals ·OH and $\cdot NO_2$ **(2)**. An

TABLE 1. Effect of plant polyphenols and hydroxycinnamates on the tyrosine nitration of serum albumin produced by $ONOO^-$

		Tyrosine Nitration (%)
− Peroxynitrite		0
+ Peroxynitrite		100
+ Peroxynitrite + Polyphenols	+ Myricetin	20
	+ (+)Catechin	23
	+ (−)Epicatechin	24
+ Peroxynitrite + Hydroxycinnamates	+ Caffeic acid	25
	+ Chlorogenic acid	27
	+ Ferulic acid	35

interesting possibility to be considered is that the homolytic scission occurs preferentially in the solvent cage including ONOOH and a reactive molecule **(3)**. In such conditions the "caged" ·OH reacts with a neighbor AH molecule—through hydrogen abstraction—leaving a radical (·A), which adds to the neighbor $\cdot NO_2$, to yield the nitrated product ($A\text{-}NO_2$).

$$ONOO^- + H^+ \longrightarrow ONOOH \quad \mathbf{(1)}$$

$$ONOOH \longrightarrow [HO\cdots NO_2] \longrightarrow \cdot OH + \cdot NO_2 \quad \mathbf{(2)}$$

$$ONOOH + AH \longrightarrow \begin{pmatrix} HO\cdots NO_2 \\ H\cdots A \end{pmatrix} \longrightarrow A\text{-}NO_2 + HOH \quad \mathbf{(3)}$$

Wine polyphenols afford the AH reductant, where H indicates the phenolic hydrogen, A the aromatic ·AO and $A\text{-}NO_2$ the polyphenol nitroderivative. The

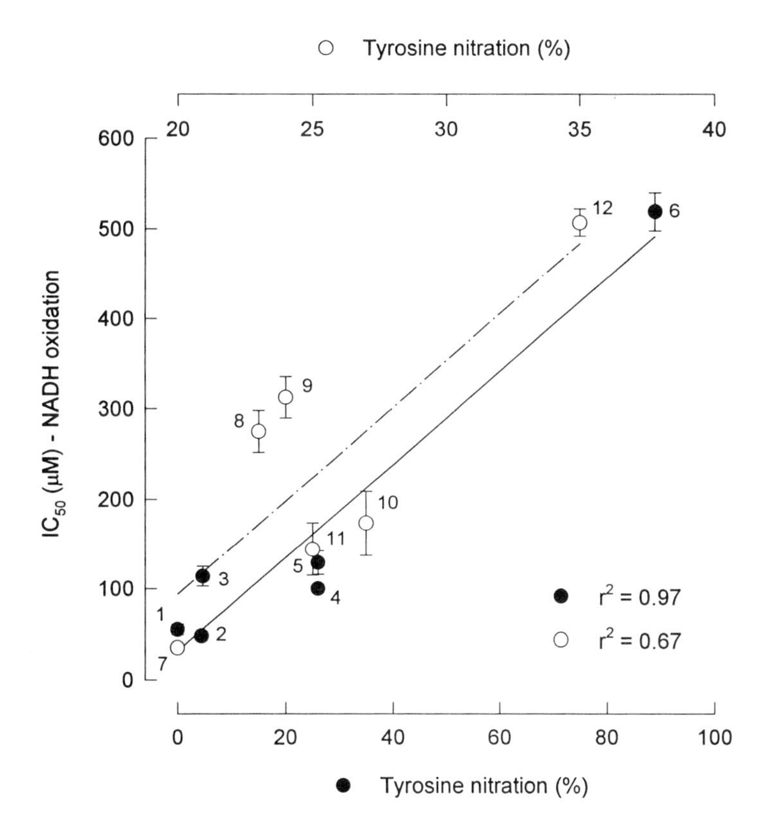

FIGURE 7. Relationships between the inhibition of NADH oxidation, expressed as IC_{50} (μM), and the inhibition of nitration, expressed as percentage of 3-nitrotyrosine. Polyphenols and hydroxycinnamates (○): 7, myricetin; 8, (+)-catechin; 9, (−)-epicatechin; 10, chlorogenic acid; 11, caffeic acid; 12, ferulic acid.

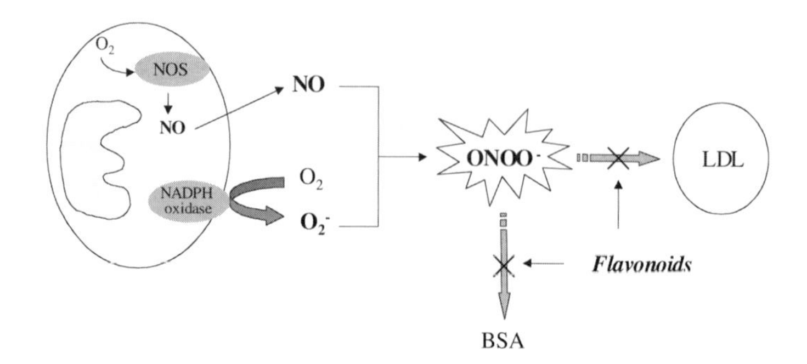

FIGURE 8. Peroxynitrite produced extracellularly by specialized cells (neutrophils, macrophages, lymphocytes, etc.) reaches blood serum albumin and low density lipoproteins. The oxidation and nitration brought about by peroxynitrite can be prevented by bioavailable wine polyphenols.

almost linear relationship found between the inhibition of NADH oxidation and $ONOO^-$ mediated protein nitration by wine polyphenols (see FIGURE 7) appears to indicate that the reductants react or interfere with a common intermediate (ONOOH or $HO \cdots NO_2$), which makes the hypothesis that ONOOH is "activated" in the solvent cage by the co-reactant an appealing one.

Peroxynitrite is released to extracellular and vascular space by the specialized cells during the cytotoxic response[5–8] and by reticulo-endothelial cells.[9] The immunocompetent neutrophils, macrophages and lymphocytes normally circulate in a nonactivated state associated with minimal production of the primary products (NO and O_2^-) and of their secondary product ($ONOO^-$). However, after appropriate stimulation those specialized cells become active sources of $ONOO^-$ with generation rates of 1.0–3.0nmol $ONOO^-$/min × mg protein[8] (see FIGURE 8). Wine polyphenols are interesting and potentially useful $ONOO^-$ scavengers in the vascular space since they are bioavailable and reach plasma concentrations in the 0.3–2.0μM range, as flavonoids, glycosides and conjugates.[41–43] The amphypatic nature of the polyphenols (p) will lead to an accumulation in the interphase between aqueous plasma and LDL surface, according to the Gibbs equation [$\Gamma_p = -a_p/RT\ (d\gamma / da_p)$], which will lead to a more effective protection of the lipoprotein from $ONOO^-$ attack. In summary, wine polyphenols may afford antioxidant and anti-inflammatory actions in the vascular space and in the extracellular fluids by: (a) their general $ONOO^-$ scavenging property; (b) the interfacial excess at the LDL surface, which prevents its oxidation and nitration by $ONOO^-$ and, consequently, they exhibit antiatherogenic effects.

ACKNOWLEDGMENTS

This work was supported by Grants PIP 4110/97 from CONICET, PICT 01608 from ANPCYT, and TB11 from the University of Buenos Aires (Argentina).

REFERENCES

1. PODEROSO, J.J., C.L. LISDERO, F. SCHÖPFER, *et al.* 1999. The regulation of mitochondrial oxygen uptake by redox reactions involving nitric oxide and ubiquinol. J. Biol. Chem. **274:** 37709–37716.
2. BOCZKOWSKI, J., C.L. LISDERO, S. LANONE, *et al.* 1999. Endogenous peroxynitrite mediates mitochondrial dysfunction in rat diaphragm during endotoxemia. FASEB J. **13:** 1637–1647.
3. GIULIVI, C., J.J. PODEROSO & A. BOVERIS. 1998. Purification and characterization of a nitric-oxide synthase from rat liver mitochondria. J. Biol. Chem. **273:** 11044–11048.
4. CHANCE, B., H. SIES & A. BOVERIS. 1979. Hydroperoxide metabolism in mammalian organs. Physiol. Rev. **59:** 527–605.
5. BOVERIS, A., L.E. COSTA & A. BOVERIS. 1999. The mitochondrial production of oxygen radicals and cellular aging. *In* Understanding the Process of Aging. E. Cadenas & L. Packer, Eds.: 1–16. Marcel Dekker, Inc. New York.
6. ISCHIROPOULOS, H., L. ZHU & J.S. BECKMAN. 1992. Peroxynitrite formation from macrophage-derived nitric oxide. Arch. Biochem. Biophys. **298:** 446–451.
7. CARRERAS, M.C., G.A. PARGAMENT, S.D. CATZ, *et al.* 1994. Kinetics of nitric oxide and hydrogen peroxide production and formation of peroxynitrite during the respiratory burst of human neutrophils. FEBS Lett. **341:** 65–68.
8. WANG, J.F., P. KOMAROV, H. SIES & H. DE GROOT. 1991. Contribution of nitric oxide synthase to luminol-dependent chemiluminiscence generated by phorbol-ester activated Kupffer cells. Biochem. J. **279:** 311–314.
9. VALDEZ, L. & A. BOVERIS. 2001. Nitric oxide and superoxide radical production by human mononuclear leukocytes. Antioxidant and Redox Signaling **3:** 505–513.
10. KOOY, N.W. & J.A. ROYALL. 1994. Agonist-induced peroxynitrite production from endothelial cells. Arch. Biochem. Biophys. **310:** 352–359.
11. KISSNER, R., T. NAUSER, P. BUGNON, *et al.* 1997. Formation and properties of peroxynitrite as studied by laser flash photolysis, high-pressure stopped-flow technique, and pulse radiolysis. Chem. Res. Toxicol. **10:** 1285–1292.
12. BECKMAN, J.S. 1990. Ischaemic injury mediator. Nature **345:** 27–28.
13. KOPPENOL, W.H. 1998. The basic chemistry of nitrogen monoxide and peroxynitrite. Free Radical Biol. Med. **25:** 385–391.
14. RADI, R. 1998. Peroxynitrite reactions and diffusion in biology. Chem. Res. Toxicol. **11:** 720–721.
15. RADI, R., J.S. BECKMAN, K.M. BUSH & B.A. FREEMAN. 1991. Peroxynitrite oxidation of sulfhydryls: the cytotoxic potential of superoxide and nitric oxide. J. Biol. Chem. **266:** 4244–4250.
16. PRYOR, W.A., X. JIN & G.L. SQUADRITO. 1994. One- and two-electron oxidations of methionine by peroxynitrite. Proc. Natl. Acad. Sci. U.S.A. **91:** 11173–11177.
17. BARTLETT, K., D.F. CHURCH, P.L. BOUNDS & W.H. KOPPENOL. 1995. The kinetics of the oxidation of L-ascorbic acid by peroxynitrite. Free Radical Biol. Med. **18:** 85–92.
18. KING, P.A., V.E. ANDERSON, J.O. EDWARDS, *et al.* 1992. A stable solid that generates hydroxyl radical upon dissolution in aqueous solution: reactions with proteins and nucleic acid. J. Am. Chem. Soc. **114:** 5430–5432.
19. KIRSCH, M. & H. DE GROOT. 1999. Reaction of peroxynitrite with reduced nicotinamide nucleotides; the formation of hydrogen peroxide. J. Biol. Chem. **274:** 24664–24670.
20. VALDEZ, L., S. ALVAREZ, S. LORES ARNAIZ, *et al.* 2000. Reactions of peroxynitrite in the mitochondrial matrix. Free Radical Biol. Med. **29**(3/4): 349–356
21. RADI, R., J.S. BECKMAN, K.M. BUSH & B.A. FREEMAN. 1994. Peroxynitrite-induced membrane lipid peroxidation: the cytotoxic potential of superoxide and nitric oxide. Arch. Biochem. Biophys. **288:** 481–487.
22. DARLEY-USMAR, V.M., N. HOGG, V.J. O'LEARY, *et al.* 1992. The simultaneous generation of superoxide and nitric oxide can initiate lipid peroxidation in human low density lipoprotein. Free Radical Res. Commun. **17:** 9–20.

23. TIEN, M., B.S. BERLETT, R.L. LEVINE, *et al.* 1999. Peroxynitrite-mediated modification of proteins at physiological carbon dioxide concentration: pH dependence of carbonyl formation, tyrosine nitration and methionine oxidation. Proc. Natl. Acad. Sci. U.S.A. **96:** 7809–7814.
24. RICE-EVANS, C.A., N.J. MILLER, P.G. BOLWELL, *et al.* 1995. The relative antioxidant activities of plant-derived polyphenolic flavonoids. Free Radical Res. Commun. **22:** 375–383.
25. BORS, W., W. HELLER, C. MICHEL & M. SARAN. 1990. Flavonoids as antioxidants: determination of radical-scavenging efficiencies. Methods Enzymol. **186:** 343–355
26. RICE-EVANS, C.A., N.J. MILLER & G. PAGANGA. 1996. Structure-antioxidant activity relationships of flavonoids and phenolic acids. Free Radical Biol. Med. **20:** 933–956.
27. THOMPSON, M., C.R. WILLIAMS & G.E.P. ELLIOT. 1976. Stability of flavonoids complexes of copper (II) and flavonoids antioxidant activity. Anal. Chim. Acta **85:** 375–381.
28. BROWN, J.E., H. KHODR, R.C. HIDER & C.A. RICE-EVANS. 1998. Structural dependence of flavonoid interactions with Cu^{2+} ions: implications for their antioxidant properties. Biochem. J. **330:** 1173–1178.
29. SICHEL, G., C. CORSARO, M. SCALLIA, *et al.* 1991. In vitro scavenger activity of some flavonoids and melanins against O_2^-. Free Radical Biol. Med. **11:** 1–7.
30. HAENEN, G.R.M.M. & A. BAST. 1999. Nitric oxide radical scavenging of flavonoids. Methods Enzymol. **301:** 490–503.
31. SALAH, N., N.J. MILLER, G. PAGANGA, *et al.* 1995. Polyphenolic flavanols as scavengers of aqueous phase radicals and as chain-breaking antioxidants. Arch. Biochem. Biophys. **322:** 339–346.
32. HAENEN, G.R.M.M., J.B.G. PAQUAY, R.E.M. KORTHOUWER & A. BAST. 1997. Peroxynitrite scavenging by flavonoids. Biochem. Biophys. Res. Commun. **236:** 591–593.
33. PANNALA, A.S., S. SINGH & C. RICE-EVANS. 1999. Flavonoids as peroxynitrite scavengers in vitro. Methods Enzymol. **300:** 207–235.
34. SANTOS, A.C., S.A. UYEMURA, J.L.C. LOPES, *et al.* 1998. Effect of naturally occurring flavonoids on lipid peroxidation and membrane permeability transition in mitochondria. Free Radical Biol. Med. **24:** 1455–1461.
35. WOOD, A.W., M.T. HUANG, R.L. CHANG, *et al.* 1982. Inhibition of the mutagenicity of bay-region diol epoxides of polycyclic aromatic hydrocarbons by naturally occurring plant phenols: exceptional activity of ellagic acid. Proc. Natl. Acad. Sci. U.S.A. **79:** 5513–5517.
36. CASTELLUCCIO, C., G.P. BOLWELL, C. GERRISH & C.A. RICE-EVANS. 1996. Differential distribution of ferulic acid to the major plasma constituents in relation to its potential as an antioxidant. Biochem. J. **316:** 691–694.
37. ALVAREZ, S., L. ACTIS-GORETTA & A. BOVERIS. 2002. Polyphenols and red wine as peroxynitrite scavengers: a chemiluminescent assay. Ann. N.Y. Acad. Sci. **957:** this volume.
38. VALDEZ, L., L. ACTIS-GORETTA & A. BOVERIS. 2002. Polyphenols in red wine prevent NADH oxidation induced by peroxynitrite. Ann. N.Y. Acad. Sci. **957:** this volume.
39. GONZALEZ FLECHA, B., S. LLESUY & A. BOVERIS. 1991. Hydroperoxide-initiated chemiluminescence: an assay for oxidative stress in biopsies of heart, liver and muscle. Free Radical Biol. Med. **10:** 93–100.
40. BOVERIS, A., S. ALVAREZ, S. LORES ARNAIZ & L. VALDEZ. 2001. Peroxynitrite scavenging by mitochondrial reductants and plant polyphenols. *In* Handbook of Antioxidants. E. Cadenas & L. Packer, Eds. Marcel Dekker. New York.
41. PAGANGA, G. & C.A. RICE-EVANS. 1997. The identification of flavonoids as glycosides in human plasma. FEBS Lett. **401:** 78–82.
42. MANACH, C., C. MORAND, V. CRESPY, *et al.* 1998. Quercetin is recovered in human plasma as conjugated derivatives which retain antioxidant properties. FEBS Lett. **426:** 331–336.
43. AZIZ, A.A., C.A. EDWARDS, M.E. LEAN & A. CROZIER. 1998. Absorption and excretion of conjugated flavonols, including quercetin-4′-O-β-glucoside and isorhamnetin-4′-O-β-glucoside by human volunteers after the consumption of onions. Free Radic. Res. **29:** 257–269.

The Protective and Anti-Protective Effects of Ethanol in a Myocardial Infarct Model

MAIKE KRENZ,[a] MICHAEL V. COHEN,[b,c] AND JAMES M. DOWNEY[b]

[a]*Department of Molecular and Cardiovascular Biology, The University of Cincinnati, Cincinnati, Ohio 45229, USA*

Departments of [b]*Physiology and* [c]*Medicine, University of South Alabama, College of Medicine, Mobile, Alabama 36688, USA*

ABSTRACT: Recent data indicate that acute alcohol exposure can have a preconditioning-like protective effect on the heart. We investigated the effect of ethanol exposure shortly before regional ischemia in an infarct model. Both in the open-chest rabbit and in the isolated rabbit heart, exposure of the heart to ethanol significantly reduced infarct size, but only if the alcohol were washed out or sufficiently metabolized before the onset of ischemia. If ethanol were still present during ischemia, it could not only prevent its own protective effect, but also abolish protection induced by ischemic preconditioning or the mitochondrial K_{ATP} channel activator diazoxide. In the *in vitro* model, we tested for possible mediators of ethanol-induced protection and made comparisons to the signaling cascade of ischemic preconditioning. Neither adenosine receptor blockade with 8-(p-sulfophenyl) theophylline, scavenging of free radicals with *N*-2-mercaptopropionyl glycine, nor closure of K_{ATP} channels with glibenclamide affected ethanol's protective effect. However, either a PKC inhibitor or a protein tyrosine kinase inhibitor could completely block ethanol-induced infarct size reduction. Both the protective and anti-protective effects of ethanol had a threshold of about 5 mM. Thus, ethanol-induced protection is mediated by protein kinase C and at least one protein tyrosine kinase, but, in contrast to ischemic preconditioning, is not triggered by either adenosine receptors, free radicals, or K_{ATP} channels. Ethanol can only exert its protective effect if it is removed before the onset of ischemia. If still present during ischemia, ethanol has the opposite effect, and inhibits preconditioning by an as yet unidentified mechanism.

KEYWORDS: **myocardial infarction; preconditioning; ischemia; alcohol**

INTRODUCTION

For some time alcohol has been suspected of having a protective effect on the cardiovascular system. While most studies have proposed a long-term adaptation that might slow atherosclerosis or even result in expression of a protective protein in the heart, more recent studies suggest a much more acute effect. Chen *et al.*[1] demonstrated that acute alcohol exposure can have a protective effect on rat hearts. Moreover,

Address for correspondence: James M. Downey, Ph.D., Department of Physiology, MSB 3024, University of South Alabama, College of Medicine, Mobile, AL 36688, USA. Voice: 251-460-6818; fax: 251-460-6464.
jdowney@usouthal.edu

this effect was mediated by protein kinase C-ε. Since protein kinase C-ε has also been shown to play a pivotal role in the signaling cascades of both classical and delayed ischemic preconditioning (IPC),[2,3] it was, therefore, suggested that acute alcohol exposure might be protecting by triggering the same signaling machinery as IPC. As a novel pharmacomimetic of ischemic preconditioning which is available to the general public, alcohol might be a very useful clinical tool. We, therefore, wanted to further characterize its protection in an infarct model of myocardial ischemia/reperfusion. Our aim was to determine the conditions under which alcohol can induce IPC-like protection and to compare the signal transduction pathways of alcohol-induced protection with the signal transduction elements of IPC that have been identified so far.

WASHOUT PRIOR TO ISCHEMIA IS CRITICAL FOR THE ABILITY OF ETHANOL TO PROTECT

In light of the data of Chen *et al.*[1] in rats, we first tested ethanol in an open-chest rabbit model of myocardial infarction.[4] Ethanol (0.35–1.4 g/kg) was infused intravenously over 10 min. Ten minutes after the end of the infusion, a major branch of the left coronary artery was ligated for 30 min followed by three hours of reperfusion. The size of the risk area was determined postmortem in the isolated heart perfused

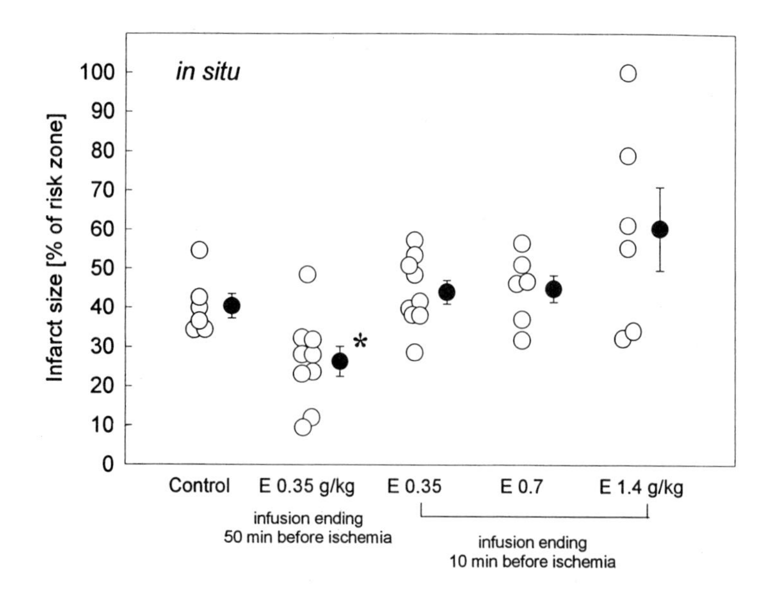

FIGURE 1. Effects of acute ethanol (E) exposure on infarct size in the open-chest rabbit. *Open symbols* represent individual experiments and *closed symbols* the group means ± SEM. Infarct size was not affected if the 10-min ethanol infusion (0.35, 0.7, or 1.4 g/kg) were started 20 min before the onset of ischemia. However, infarct size was significantly reduced in comparison to control (* $p < 0.05$) if the 10-min infusion of 0.35 g/kg ethanol were started one hour prior to ischemia. (Reprinted from Krenz *et al.*,[4] with permission.)

retrogradely through the aortic root by retightening the snare and injecting fluorescent particles into the perfusate. Infarct size was measured by planimetry of necrotic regions of transverse sections of the heart stained with triphenyltetrazolium chloride (TTC). The results of *in situ* experiments are shown in FIGURE 1. Infarct size was not different from control at any dose tested. At the onset of ischemia, ethanol serum levels were 126 ± 11 mg/dL (approximately 26 mM) in the group infused with 1.4 g/kg ethanol, 64 ± 5 g/dL (approximately 14 mM) with 0.7 g/kg, and 56 ± 5 (approximately 12 mM) mg/dL with 0.35 g/kg ethanol. The serum levels were indeed similar to the alcohol concentrations used by Chen *et al.*[1] in their isolated rat heart studies. Two possibilities came to mind. The first was that perhaps ethanol could not trigger preconditioning in the rabbit heart. The second was that it had triggered a preconditioned state but that the continued presence of ethanol during the ischemic insult somehow interfered with its protection. In most previous studies of ethanol and cardioprotection the ethanol had been withdrawn prior to the ischemic test so that the evaluation was performed in alcohol-free conditions. In our *in situ* model, however, the ethanol bathed the heart during ischemia as well as prior to it.

We tested the latter hypothesis in isolated rabbit hearts where ethanol could be given and withdrawn at will. Hearts were perfused with Krebs-Henseleit buffer solution containing 10, 20, or 50 mM ethanol for 5 min. After 10 min of washout, a major branch of the left coronary artery was occluded with a snare for 30 min and then the

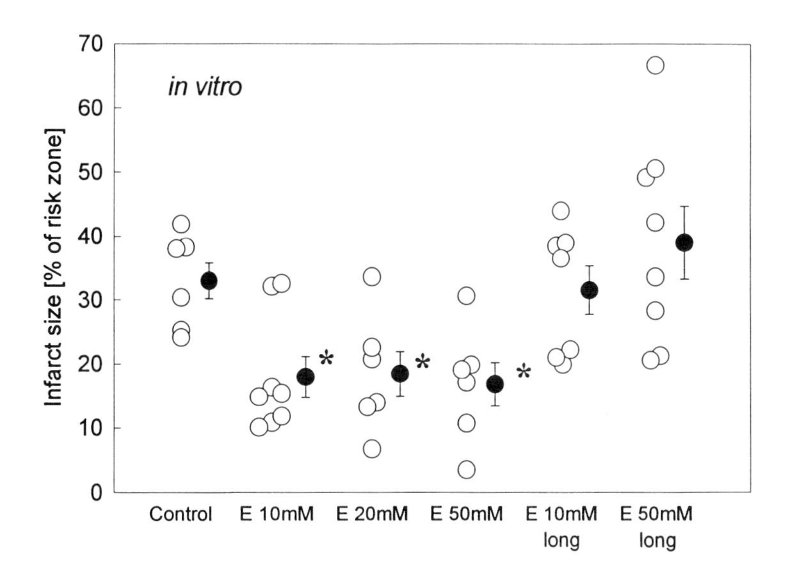

FIGURE 2. Effects of acute ethanol (E) exposure on infarct size in isolated rabbit hearts. *Open symbols* represent individual experiments and *closed symbols* the group means ± SEM. Five min of perfusion with ethanol (50 mM) followed by 10 min of washout prior to ischemia (*Ethanol short*) significantly reduced infarct size in comparison to control (* $p < 0.05$), but not to the same extent as seen in the preconditioned group (PC). If the ethanol exposure (50 mM) were extended for 45 min until the end of ischemia (*ethanol long*), infarct sizes were not different from control. (Reprinted from Krenz *et al.*,[4] with permission.)

hearts were reperfused for 2 h. Risk zone and infarct size were quantified by fluorescente particles and TTC staining as in the *in situ* heart experiments described above. FIGURE 2 (E groups) shows that a five-minute exposure to any of the 3 doses of ethanol followed by 10 min of washout significantly reduced infarct size when compared to untreated hearts. However, it should be noted that the infarct size reduction was not as pronounced as seen with one cycle of IPC in that same model. However, if either the 10 or the 50 mM ethanol exposure were extended to 45 min (lasting until the end of ischemia), hearts were no longer protected (FIG. 2, long groups). Thus it would appear that if ethanol is present during ischemia, it actually blocks its own protection.

We next designed a hybrid model to test whether the hearts in the *in situ* experiments had actually ever been in a protected state. Open-chest rabbits were again given 0.7 g/kg ethanol over 10 min. Ten minutes after the ethanol infusion was stopped, the hearts were rapidly excised, mounted on a Langendorff apparatus, and perfused with Krebs-Henseleit buffer solution to wash out alcohol for 10 min. The hearts then underwent 30 min of regional ischemia and two hours of reperfusion and TTC staining as described above. The hearts in this group showed significantly reduced infarct size (see FIGURE 3), indicating that ethanol infusion in the open-chest rabbit administered shortly before the onset of ischemia can indeed induce a protected state. The

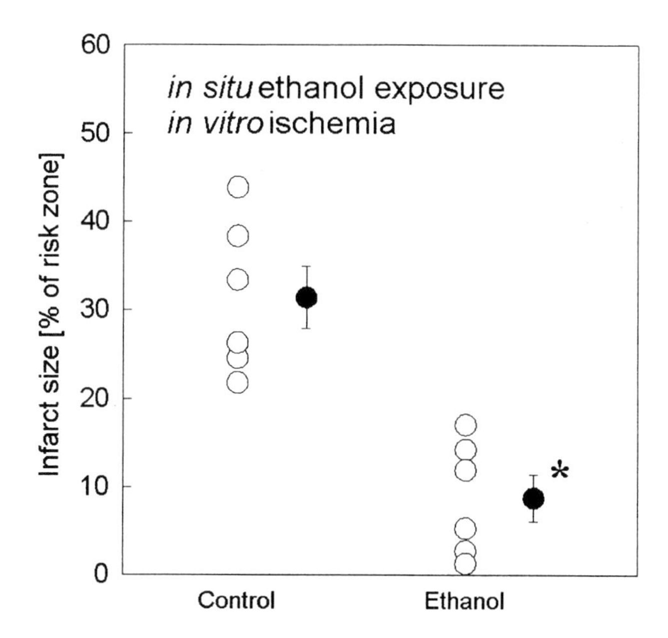

FIGURE 3. Effects of ethanol (E) infusion in the open-chest rabbit followed by regional ischemia in the Langendorff mode (hybrid preparation). *Open symbols* represent individual experiments and *closed symbols* the group means ± SEM. Ethanol exposure (0.7 g/kg) before excision of the hearts resulted in significantly reduced infarct sizes as compared to hearts from untreated rabbits ($*p < 0.05$). (Reprinted from Krenz *et al.*,[4] with permission.)

evidence again indicated that protection by ethanol is dependent upon the absence of ethanol during ischemia. Another possibility was that ethanol somehow could only protect in the buffer-perfused heart and was incapable of protecting in the more clinically relevant *in situ* model.

To rule out the above possibility we went back to the *in situ* model but altered the schedule by giving the ethanol an hour before the ischemic insult, which would allow much of the ethanol to be metabolized prior to ischemia. A 10-min infusion of 0.35 g/kg ethanol was started 1 h before the onset of ischemia. This protocol caused infarct size to be significantly reduced in comparison to that in the untreated control hearts (FIG. 1, "infusion ending 50 min before ischemia" group). The long interval between the end of the ethanol infusion and the onset of ischemia allowed serum levels to drop from 56 ± 5 mg/dL 10 min after the end of the ethanol infusion to 28 ± 2 mg/dL (approximately 6 mM) at the onset of ischemia. This observation further supported the hypothesis that ethanol can be protective, but that it must be washed out or, as demonstrated here, sufficiently metabolized prior to ischemia before protection will be realized.

In summary, in both the isolated rabbit heart and the open-chest rabbit models acute ethanol exposure could significantly reduce infarct size, but only if ethanol were no longer present or, if present, only at very low concentrations during ischemia. Therefore, ethanol has a twofold action on the rabbit heart: treatment prior to ischemia induces protection, whereas continued exposure to ethanol during ischemia inhibits that protection.

Is this twofold action of ethanol unique to rabbits? In most of the previous studies on ethanol and cardioprotection the ethanol was withdrawn prior to testing for resistance to ischemic injury. However, Chen *et al.*[1] reported that creatine kinase released during reperfusion of three isolated rat hearts was still reduced when ethanol was present both before and throughout global ischemia. Thus, ethanol's presence during ischemia may not have an anti-protective effect in rats. On the other hand, in open-chest dogs ethanol infusion 10 min prior to the onset of regional ischemia did not result in a reduction in infarct size.[5] Acute ethanol exposure followed by washout prior to ischemia has not been tested in a dog model, but the data suggest that ethanol may have the same twofold action on dog myocardium as we observed in rabbits. The effect on human myocardium is, of course, unknown.

SIMILARITIES AND DIFFERENCES BETWEEN ALCOHOL-INDUCED PRECONDITIONING AND IPC

To date, the signal transduction pathway of IPC is only partially understood. Adenosine, α-adrenergic, opioid, and bradykinin receptors as well as free radicals have been identified as triggers of the signaling cascade.[6–10] The signal is then further transferred via protein kinase C[11,12] and at least one protein tyrosine kinase.[13–15] The end-effector of IPC has yet to be identified, but possible candidates are phosphorylated proteins that change cytoskeletal stability[16] or energy utilization in the mitochondria.[17] The mitochondrial ATP-sensitive potassium (K_{ATP}) channel has also been proposed as a possible end-effector.[18–20] In order to elucidate the signaling pathway involved in alcohol-induced cardioprotection, a variety of inhibitors of the known steps of IPC were used in our isolated rabbit heart model with short ethanol exposure followed by washout.

First, we tested for the involvement of adenosine receptors and oxygen-derived free radicals. The five-minute perfusion with ethanol (50mM) was bracketed by a 15-min infusion of either the adenosine receptor inhibitor 8-(p-sulfophenyl) theophylline (SPT, 100μM) or the free radical scavenger N-2-mercaptopropionyl glycine (MPG, 300μM). While ethanol was protective (16 ± 3% infarction vs. 33 ± 3% in untreated hearts) neither agent had an effect on ethanol-induced cardioprotection (11 ± 2% in SPT and 15 ± 3% in MPG groups), indicating that, in contrast to IPC, neither adenosine receptors nor free radicals play a role.[21] That finding indicates that ethanol is probably triggering the signal transduction pathway somewhere downstream of the receptors. We did not test for involvement of adenosine receptors or free radicals during the index ischemia.

Next, we investigated the effect of protein kinase C or protein tyrosine kinase inhibition on alcohol-induced cardioprotection. Two inhibitors, chelerythrine and genistein, were used with the same timing and at the same doses as already known to be effective for inhibition of IPC. In combination with the five-minute exposure to 50 mM ethanol, the protein kinase C inhibitor chelerythrine (5 μM) was given over 15 min starting 5 min prior to the onset of ischemia. Chelerythrine inhibited ethanol-induced infarct size reduction: 27 ± 3% infarction which was not significantly different from that in untreated hearts (33 ± 3%) and was larger than the 16 ± 3% seen in the group treated with ethanol alone. When a five-minute ethanol infusion (50mM) was combined with the broad-spectrum protein tyrosine kinase inhibitor genistein (50μM), starting 5 min prior to the onset of ischemia and continuing for 15 min into ischemia, ethanol-induced protection was also abolished (29 ± 4% infarction). Neither chelerythrine nor genistein alone affected infarct size (31 ± 3% and 34 ± 3% infarction, respectively). Thus, just as observed in IPC, both protein kinase C and at least one protein tyrosine kinase are critical signal transduction elements in alcohol-induced reduction of infarct size.[21]

The situation was far less clear regarding the involvement of the mitochondrial K_{ATP} channel. Two different K_{ATP} channel inhibitors, 5-hydroxydecanoate (5HD) and glibenclamide, were used in combination with short ethanol exposure in isolated hearts. 5HD is thought to be a specific inhibitor of the mitochondrial K_{ATP} channel, whereas glibenclamide closes both the mitochondrial and sarcolemmal K_{ATP} channels.[22,23] Both agents have been shown to inhibit IPC in isolated rabbit hearts.[18] The 5-min ethanol perfusion was bracketed by a 15-min perfusion with either 5HD (200μM) or glibenclamide (5μM). To our surprise, the specific inhibitor 5HD blocked ethanol-induced protection (29 ± 4% infarction), whereas the nonspecific inhibitor glibenclamide did not completely block the protection (21 ± 3% infarction).[21] We have no definitive explanation for these findings. We suspect that the blocking effect of 5HD may have been unrelated to its actions on the mitochondrial K_{ATP} channel. 5HD is lipophilic and it may have interfered with ethanol's effect on a membrane structure. Had the block been due to K_{ATP} channel closure, then glibenclamide should have aborted protection as well.

In summary, the signal transduction pathway for alcohol-induced cardioprotection resembles that of IPC, but also displays important differences. Neither adenosine nor free radicals is involved in ethanol-induced cardioprotection. As in IPC, protein kinase C and at least one protein tyrosine kinase are important mediators. The mitochondrial K_{ATP} channel does not seem to be involved in cardioprotection induced by acute ethanol exposure. At first this seems incongruous because it has been suggested the K_{ATP} channel is the end-effector of IPC. However, we have recently presented evidence that K_{ATP} channels may serve as triggers rather than end-effectors of IPC.[24] In that paradigm receptor occupancy appears to result in opening of mitochondrial K_{ATP} channels. Channel opening in turn causes free radical production by mitochondria. We propose that it is the free radicals that activate the kinases. Our data would indicate that ethanol stimulates somewhere downstream of the free radicals but upstream of the kinases.

In rats and guinea pigs, α-adrenoreceptors and adenosine receptors, respectively, have been shown to mediate ethanol-induced protection that resembles delayed preconditioning.[25] In those studies animals were fed a liquid diet containing up to 20% (guinea pigs) or 36% (rats) ethanol over many weeks. The animals were withdrawn from ethanol the night before sacrifice, and then the hearts were isolated and underwent ischemia and reperfusion in the absence of ethanol. In this setting, alcohol-induced cardioprotection strongly resembles delayed preconditioning. We did not test for receptors other than those for adenosine, so we cannot exclude the possibility that some of the other receptors such as opioid, adrenergic or bradykinin, might be involved in the cardioprotection induced by acute ethanol exposure.

The involvement of mitochondrial K_{ATP} channels has been tested in models of chronic feeding. It was reported that 5HD[26] and glibenclamide[27] can abolish protection from chronic alcohol consumption in rats and dogs, respectively. Whether chronic ethanol administration protects hearts by the same mechanism as acute ethanol exposure has not been investigated, but given that classical and delayed preconditioning involve similar but not identical signal transduction pathways,[28,29] we would expect differences. The role of protein kinase C, however, has been evaluated in an acute rat model resembling our isolated heart experiments.[1] In accordance with our data, Chen *et al.*[1] observed that isoform-specific inhibition of protein kinase C-ε abolished ethanol-induced protection. Further studies are needed to fully understand the signaling pathways of preconditioning-like cardioprotection induced by chronic and acute ethanol exposure.

ALCOHOL CAN ACT AS AN INHIBITOR OF IPC IF PRESENT DURING ISCHEMIA

We also tried to address the mechanism of ethanol's anti-protective effect during ischemia. Because protection from ethanol seemed to be the result of signaling that resembled that of IPC, we hypothesized that alcohol would also block IPC's protection. In the open-chest rabbit, IPC was induced by one cycle of five minutes of coronary occlusion followed by 20 min of reperfusion prior to the index ischemia. IPC was combined with a 10 min infusion of ethanol (0.7 g/kg) starting 20 min before the onset of the index ischemia. FIGURE 4 reveals that ethanol completely abolished the protection induced by IPC. We also tested the effect of ethanol exposure during ischemia on pharmacological preconditioning. The mitochondrial K_{ATP} channel opener diazoxide (10 mg/kg) was infused one minute before the start of the 10-min

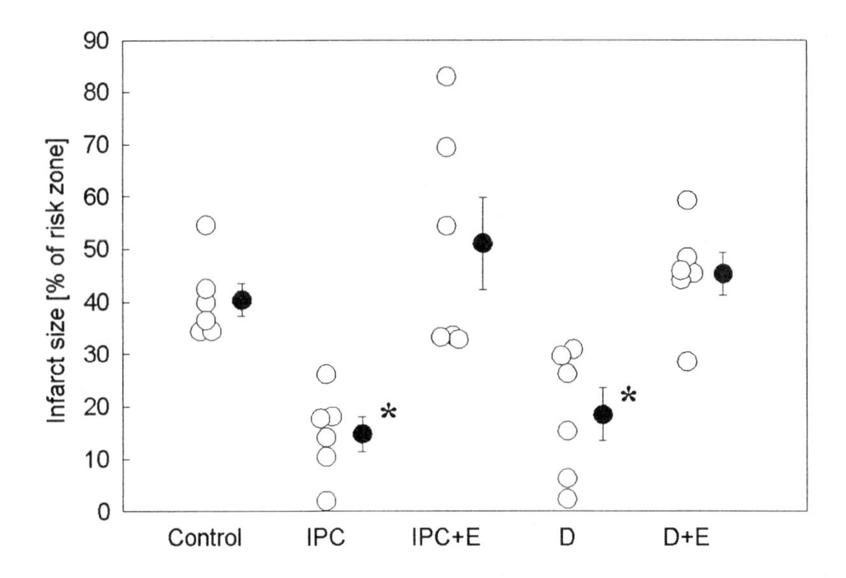

FIGURE 4. Effect of ethanol (E) (0.7 g/kg) on protection induced by ischemic preconditioning (IPC) (5-min occlusion followed by 20-min reperfusion prior to the index ischemia) or diazoxide (D) (10 mg/kg) in the open-chest rabbit. *Open symbols* represent individual experiments and *closed symbols* the group means ± SEM. Ethanol abolished the cardioprotective effect of both ischemic preconditioning and diazoxide. (Reprinted from Krenz *et al.*,[4] with permission.)

ethanol infusion (0.7 g/kg). In this group, in contrast to all other *in situ* ethanol groups, the index ischemia was started immediately after the end of the ethanol infusion. Whereas diazoxide given alone with the same timing significantly reduced infarct size (FIG. 4), diazoxide's protection was again abolished by ethanol.

From the above experiments it was not clear whether ethanol had actually blocked the IPC mechanism or whether the protective effect of ethanol exposure prior to ischemia was counterbalanced by a nonspecific toxic effect of ethanol exposure during ischemia. If this were the case, then we should have been able to unmask the toxic effect in the open-chest model by a combination of ethanol and blockade of the IPC pathway. The most potent inhibition of IPC's protection is effected by combined administration of a protein kinase C inhibitor and a protein tyrosine kinase inhibitor.[30,31] However, even double blockade with staurosporine, 50 μg/kg, and genistein, 5 mg/kg, accompanying ethanol infusion did not increase infarct size above that seen in the control group (see FIGURE 5). This suggests that ethanol's block of protection is a very specific one at some, as yet unidentified, step in the signal transduction pathway of both IPC and ethanol-induced protection.

The question arises whether ethanol can ever be protective in the *in situ* setting? As described above we have shown that a small dose followed by a prolonged washout could produce a window of protection. We went back to the isolated heart model and performed two series of experiments to try to further address this question.[32] In

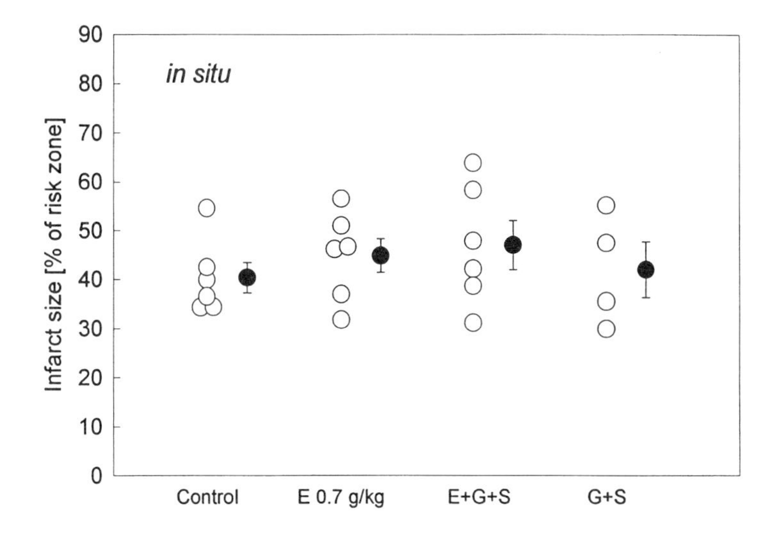

FIGURE 5. Effect of ethanol (E) alone (0.7 g/kg) or in combination with staurosporine (S) (50 µg/kg) plus genistein (G) (5 µg/kg) on infarct size in the open-chest rabbit. *Open symbols* represent individual experiments and *closed symbols* the group means ± SEM. Infarct size was not affected if staurosporine and genistein.

the first we infused ethanol for 5 min followed by 10 min of washout. Concentrations of ethanol from 2.5 to 50 mM ethanol were tested. Protection was lost at 2.5 mM while 5 mM partially protected suggesting a protective threshold near 5 mM. Next, hearts were preconditioned with 5 min of global ischemia followed by 10 min of washout. Ethanol (5–50 mM) was infused for 35 min starting five minutes prior to the onset of the 30-min index ischemia. Protection was present only in the 5-mM group, and absent in hearts exposed to the higher concentrations. Thus ethanol's thresholds for both triggering and blocking the IPC pathway are very similar, and are near 5 mM. This would make it very unlikely that a plasma level could be found that will clearly protect a patient's heart.

In summary, the presence of ethanol during ischemia resulted in a strong anti-protective effect. Ethanol inhibited not only its own preconditioning-like protection, but also abolished protection from IPC and activation of mitochondrial K_{ATP} channels. Since the combination of ethanol with blockade of IPC did not extend the size of infarcts over that seen in the control group, we propose that the anti-protective effect of ethanol is likely to be a very specific action at an as yet unidentified step within IPC's signaling cascade. The anti-protective and protective effects share a similar threshold at about 5 mM.

CLINICAL IMPLICATIONS

A number of recent epidemiological studies have focused on the effects of regular alcohol consumption. Light-to-moderate chronic ethanol consumption was associated with reduced cardiovascular mortality in several studies.[33–36] As potential mechanisms of this beneficial effect, favorable changes in lipid metabolism and hemostasis have been described.[34,37] However, chronic alcohol consumption especially at higher levels has well-known deleterious effects on organs other than the heart, so that changes in drinking habits cannot be generally recommended on the basis of the epidemiological data so far.[38] IPC provides a window of protection that lasts for an hour or two. About 24 hours after the first window of protection subsides, a more long-lasting phase of protection termed the "second window" or "late-phase" preconditioning ensues. While not as robust as the first window of protection, this late-phase protection can last up to four days[39] and is thought to reflect expression of a protective protein.[28] Generally it has been found that almost any stimulus that triggers the first window is also followed by a second-window type protection. Thus it is very likely that a single alcoholic drink each day triggers the first-window pathway and results in a second window response which is reinforced on a daily basis. Thus while protection may be absent for the hour or so after a drink, the individual likely is protected for the remainder of the time.

Whether acute ethanol exposure has a preconditioning-like protective or an antiprotective effect on human myocardium has not been directly tested. However, it has been shown that ethanol ingestion immediately before a treadmill test can aggravate angina and ischemic ECG changes,[40] suggesting that acute ethanol has the same twofold action on human myocardium as we observed in rabbit myocardium. There is good clinical evidence that many patients with acute myocardial infarction enjoy the benefit of IPC through antecedent angina.[41] If alcohol were present in the blood stream during a myocardial infarction, however, it would likely block the beneficial effect of that antecedent IPC. Thus, we would propose that alcohol should not be administered to the patient with acute myocardial infarction under any circumstance until strong evidence to the contrary emerges.

ACKNOWLEDGMENTS

The present study was supported in part by grants from the National Institutes of Health, National Heart, Lung, and Blood Institute (HL 20648 and HL 50688). M. Krenz was sponsored by a grant from the Deutsche Forschungsgemeinschaft (Kr 1432/2-1).

REFERENCES

1. Chen, C.-H., M.O. Gray & D. Mochly-Rosen. 1999. Cardioprotection from ischemia by a brief exposure to physiological levels of ethanol: role of epsilon protein kinase C. Proc. Natl. Acad. Sci. U.S.A. **96:** 12784–12789.

2. Liu, G.S., M.V. Cohen, D. Mochly-Rosen, *et al.* 1999. Protein kinase C-ε is responsible for the protection of preconditioning in rabbit cardiomyocytes. J. Mol. Cell Cardiol. **31:** 1937–1948.

3. Qiu, Y., P. Ping, X.-L. Tang, *et al.* 1998. Direct evidence that protein kinase C plays an essential role in the development of late preconditioning against myocardial stunning in conscious rabbits and that ε is the isoform involved. J. Clin. Invest. **101:** 2182–2198.

4. KRENZ, M., C.P. BAINES, X.-M. YANG, *et al.* 2001. Acute ethanol exposure fails to elicit preconditioning-like protection in *in situ* rabbit hearts because of its continued presence during ischemia. J. Am. Coll. Cardiol. **37:** 601–607.
5. ITOYA, M., J.D. MORRISON & H.F. DOWNEY. 1998. Effect of ethanol on myocardial infarct size in a canine model of coronary artery occlusion-reperfusion. Mol. Cell. Biochem. **186:** 35–41.
6. LIU, G.S., J. THORNTON, D.M. VAN WINKLE, *et al.* 1991. Protection against infarction afforded by preconditioning is mediated by A_1 adenosine receptors in rabbit heart. Circulation **84:** 350–356.
7. BANERJEE, A., C. LOCKE-WINTER, K.B. ROGERS, *et al.* 1993. Preconditioning against myocardial dysfunction after ischemia and reperfusion by an α_1-adrenergic mechanism. Circ. Res. **73:** 656–670.
8. SCHULTZ, J.E.J., E. ROSE, Z. YAO, *et al.* 1995. Evidence for involvement of opioid receptors in ischemic preconditioning in rat hearts. Am. J. Physiol. **268:** H2157–H2161.
9. HARTMAN, J.C., T.M. WALL, T.G. HULLINGER, *et al.* 1993. Reduction of myocardial infarct size in rabbits by ramiprilat: reversal by the bradykinin antagonist HOE 140. J. Cardiovasc. Pharmacol. **21:** 996–1003.
10. BAINES, C.P., M. GOTO & J.M. DOWNEY. 1997. Oxygen radicals released during ischemic preconditioning contribute to cardioprotection in the rabbit myocardium. J. Mol. Cell. Cardiol. **29:** 207–216.
11. LIU, Y., K. YTREHUS & J.M. DOWNEY. 1994. Evidence that translocation of protein kinase C is a key event during ischemic preconditioning of rabbit myocardium. J. Mol. Cell. Cardiol. **26:** 661–668.
12. MITCHELL, M.B., X. MENG, L. AO, *et al.* 1995. Preconditioning of isolated rat heart is mediated by protein kinase C. Circ. Res. **76:** 73–81.
13. DAS, D.K., N. MAULIK, T. YOSHIDA, *et al.* 1996. Preconditioning potentiates molecular signaling for myocardial adaptation to ischemia. Ann. N.Y. Acad. Sci. **793:** 191–209.
14. FRYER, R.M., J.E.J. SCHULTZ, A.K. HSU, *et al.* 1998. Pretreatment with tyrosine kinase inhibitors partially attenuates ischemic preconditioning in rat hearts. Am. J. Physiol. **275:** H2009–H2015.
15. BAINES, C.P., L. WANG, M.V. COHEN, *et al.* 1998. Protein tyrosine kinase is downstream of protein kinase C for ischemic preconditioning's anti-infarct effect in the rabbit heart. J. Mol. Cell. Cardiol. **30:** 383–392.
16. GANOTE, C. & S. ARMSTRONG. 1993. Ischaemia and the myocyte cytoskeleton: review and speculation. Cardiovasc. Res. **27:** 1387–1403.
17. MURRY, C.E., R.B. JENNINGS & K.A. REIMER. 1986. Preconditioning with ischemia: a delay of lethal cell injury in ischemic myocardium. Circulation **74:** 1124–1136.
18. GROVER, G.J. & K.D. GARLID. 2000. ATP-sensitive potassium channels: a review of their cardioprotective pharmacology. J. Mol. Cell. Cardiol. **32:** 677–695.
19. GARLID, K.D., P. PAUCEK, V. YAROV-YAROVOY, *et al.* 1997. Cardioprotective effect of diazoxide and its interaction with mitochondrial ATP-sensitive K^+ channels: possible mechanism of cardioprotection. Circ. Res. **81:** 1072–1082.
20. LIU, Y., T. SATO, B. O'ROURKE, *et al.* 1998. Mitochondrial ATP-dependent potassium channels: novel effectors of cardioprotection? Circulation **97:** 2463–2469.
21. KRENZ, M., C.P. BAINES, G. HEUSCH, *et al.* 2001. Acute alcohol-induced protection against infarction in rabbit hearts: differences from and similarities to ischemic preconditioning. J. Mol. Cell. Cardiol. **33:** 2015–2022.
22. JABŮREK, M., V. YAROV-YAROVOY, P. PAUCEK, *et al.* 1998. State-dependent inhibition of the mitochondrial K_{ATP} channel by glyburide and 5-hydroxydecanoate. J. Biol. Chem. **273:** 13578–13582.
23. SATO, T., N. SASAKI, J. SEHARASEYON, *et al.* 2000. Selective pharmacological agents implicate mitochondrial but not sarcolemmal K_{ATP} channels in ischemic cardioprotection. Circulation **101:** 2418–2423.
24. PAIN, T., X.-M. YANG, S.D. CRITZ, *et al.* 2000. Opening of mitochondrial K_{ATP} channels triggers the preconditioned state by generating free radicals. Circ. Res. **87:** 460–466.

25. MIYAMAE, M., M.M. RODRIGUEZ, S.A. CAMACHO, *et al.* 1998. Activation of ε protein kinase C correlates with a cardioprotective effect of regular ethanol consumption. Proc. Natl. Acad. Sci. U.S.A. **95:** 8262–8267.
26. ZHU, P., H.-Z. ZHOU & M.O. GRAY. 2000. Chronic ethanol-induced myocardial protection requires activation of mitochondrial K_{ATP} channels. J. Mol. Cell. Cardiol. **32:** 2091–2095.
27. PAGEL, P.S., W.G. TOLLER, E.R. GROSS, *et al.* 2000. K_{ATP} channels mediate the beneficial effects of chronic ethanol ingestion. Am. J. Physiol. **279:** H2574–H2579.
28. BOLLI, R. 2000. The late phase of preconditioning. Circ. Res. **87:** 972–983.
29. COHEN, M.V., C.P. BAINES & J.M. DOWNEY. 2000. Ischemic preconditioning: from adenosine receptor to K_{ATP} channel. Annu. Rev. Physiol. **62:** 79–109.
30. VAHLHAUS, C., R. SCHULZ, H. POST, *et al.* 1998. Prevention of ischemic preconditioning only by combined inhibition of protein kinase C and protein tyrosine kinase in pigs. J. Mol. Cell. Cardiol. **30:** 197–209.
31. FRYER, R.M., J.E.J. SCHULTZ, A.K. HSU, *et al.* 1999. Importance of PKC and tyrosine kinase in single or multiple cycles of preconditioning in rat hearts. Am. J. Physiol. **276:** H1229–H1235.
32. KRENZ, M., X.-M. YANG, J.M. DOWNEY, *et al.* 2002. Dose–response relationships of the protective and anti-protective effects of acute ethanol exposure in the isolated rabbit heart. Heart Dis. **4:** in press.
33. CAMARGO, JR., C.A., C.H. HENNEKENS, J.M. GAZIANO, *et al.* 1997. Prospective study of moderate alcohol consumption and mortality in US male physicians. Arch. Intern. Med. **157:** 79–85.
34. SUH, I., B.J. SHATEN, J.A. CUTLER, *et al.* 1992. Alcohol use and mortality from coronary heart disease: the role of high-density lipoprotein cholesterol. The Multiple Risk Factor Intervention Trial Research Group. Ann. Intern. Med. **116:** 881–887.
35. FUCHS, C.S., M.J. STAMPFER, G.A. COLDITZ, *et al.* 1995. Alcohol consumption and mortality among women. N. Engl. J. Med. **332:** 1245–1250.
36. ALBERT, C.M., J.E. MANSON, N.R. COOK, *et al.* 1999. Moderate alcohol consumption and the risk of sudden cardiac death among US male physicians. Circulation **100:** 944–950.
37. DEMROW, H.S., P.R. SLANE & J.D. FOLTS. 1995. Administration of wine and grape juice inhibits in vivo platelet activity and thrombosis in stenosed canine coronary arteries. Circulation **91:** 1182–1188.
38. FAGRELL, B., U. DE FAIRE, S. BONDY, *et al.* 1999. The effects of light to moderate drinking on cardiovascular diseases. J. Intern. Med. **246:** 331–340.
39. BAXTER, G.F., F.M. GOMA & D.M. YELLON. 1997. Characterisation of the infarct-limiting effect of delayed preconditioning: timecourse and dose-dependency studies in rabbit myocardium. Basic Res. Cardiol. **92:** 159–167.
40. AHLAWAT, S.K. & S.B. SIWACH. 1994. Alcohol and coronary artery disease. Int. J. Cardiol. **44:** 157–162.
41. KLONER, R.A., T. SHOOK, E.M. ANTMAN, *et al.* 1998. Prospective temporal analysis of the onset of preinfarction angina versus outcome: an ancillary study in TIMI-9B. Circulation **97:** 1042–1045.

Cardiovascular Protection by Alcohol and Polyphenols

Role of Nitric Oxide

DALE A. PARKS[a,b,d] AND FRANCOIS M. BOOYSE[c]

[a]Departments of Anesthesiology, [b]Physiology & Biophysics, and [c]Medicine and the [d]Center for Free Radical Biology, University of Alabama at Birmingham, Birmingham, Alabama 35233, USA

ABSTRACT: Cardiovascular disease, and in particular coronary heart disease (CHD), remains the leading cause of death in both men and women in the United States. Much epidemiologic evidence indicates that alcoholic beverages, and in particular red wine, results in a reduction in cardiovascular risk factors and decreases mortality; however, the mechanisms of this cardiovascular protection remains elusive. This review discusses evidence to suggest that ·NO plays a critical role in cardiovascular protection and that nitric oxide synthase (NOS) is the responsible cardioprotective protein (see Bolli *et al.* 1998. Basic Res. Cardiol. 93: 325–338).

KEYWORDS: nitric oxide; coronary heart disease; nitric oxide synthase; alcohol; red wine

INTRODUCTION

Cardiovascular disease, and in particular coronary heart disease (CHD), remains the leading cause of death in both men and women in the United States.[1] In patients with cardiovascular risk factors such as hypercholesterolemia, hypertension, or aging,[2,3] endothelial dysfunction predisposes to the development of structural vascular changes.[4,5] Although CHD is a multifactorial disease process, it is clear that the atherosclerotic narrowing of epicardial coronary vessels and thrombosis play a critical role in acute myocardial infarction (MI) and sudden death. The pathogenesis of CHD and the ensuing atherothrombotic complications resulting in MI, are complex and are believed to involve cellular and molecular mechanisms[6] that have been altered through interactions with various environmental and/or systemic factors (i.e., CHD risk factors). Consequently, any systemic factors (such as, alcohol or wine components) that will reduce, minimize or inhibit these induced dysfunctions will be expected to reduce the overall risk for CHD and CHD-related mortality.

Address for correspondence: Dale A. Parks, Ph.D., Department of Anesthesiology, 619 19th Street South, University of Alabama at Birmingham, Birmingham, AL 35233, USA. Voice: 205-934-4665; fax: 205-934-7437.

Dale.parks@ccc.uab.edu

MODERATE ALCOHOL IN CARDIOPROTECTION

There is general consensus that heavy alcohol consumption has long been associated with vascular (hemorrhagic stroke, hypertension) and myocardial (cardiomyopathies, arrhythmias, CHD) complications.[1,7] Paradoxically, a growing body of epidemiological evidence, including both retrospective and prospective studies, indicates that long-term, light-to-moderate alcohol consumption (1–2 drinks/day for men, 1 drink/day for women) is associated with a reduced incidence of coronary artery disease when compared to nondrinkers or heavy drinkers.[8,9] Recent studies indicate that not only do alcoholic beverages decrease the risk of cardiovascular disease by 20–40% compared to that in nondrinkers,[10–13] but they also reduce total mortality by almost one-third.[10] The J- or U-shaped association between alcohol intake and risk for CHD, with the lowest risk observed with moderate drinkers, higher risk for nondrinkers, and highest risk for the heaviest drinkers has been supported by several large prospective cohort studies.[11,12,14,15] This cardioprotection has been attributed to the direct effects of moderate ethanol/polyphenols on increasing high-density lipoproteins, HDL,[16,17] decreasing platelet aggregation,[18] enhancing fibrinolytic activity by upregulation of tissue plasminogen activator, decreasing fibrinogen[19,20] and decreasing ischemia-reperfusion injury.[21] In spite of large numbers of epidemiological studies demonstrating the relationship between the consumption of moderate levels of alcoholic beverages and cardioprotection, the detailed molecular mechanism of this alcohol/polyphenol-induced cardioprotection remains elusive.

POLYPHENOLS IN CARDIOPROTECTION

There is currently no clear-cut evidence that one alcoholic beverage is more protective than another. However, recent evidence from the Zutphen study,[22] the seven countries study,[23] the Copenhagen heart study,[24] and a Finnish study[25] have demonstrated a significant inverse association between flavonoid intake and mortality from coronary heart disease (CHD), MI, and stroke. Wine is an important component of the Mediterranean diet and is postulated to be responsible for the "French paradox" in which some regions of France have about 40% lower mortality rates for CHD than North-European countries with similar CHD risk factors.[26] Red wine has been demonstrated to increase the antioxidant capacity of plasma in humans,[27,28] decrease platelet aggregation,[29,30] increase fibrinolysis,[31] inhibit *ex vivo* LDL oxidation,[32] reduce susceptibility of human plasma to lipid peroxidation,[27,33] increase plasma levels of HDL cholesterol and apolipoprotein A-I,[34] and decrease the atherogenic lipoprotein, Lp(a).[35] Administration of white wine with similar alcohol content, or pure alcohol did not demonstrate a cardioprotective effect. The only consistent difference between red and white wines is that red wine contains about 20-times more polyphenols,[36] in particular flavonoids (e.g., quercetin, catechin, epicatechin) and stilbenes (e.g., resveratrol). The red wine polyphenols have been demonstrated to trigger ·NO-dependent cell signaling, including endothelial-dependent relaxation that is modulated by both $O_2^{-\cdot}$ and SOD, a family of enzymes that catalyze the formation of hydrogen peroxide and molecular oxygen. Quercetin, a flavanol that does not appear in white wines, was shown to scavenge $O_2^{-\cdot}$, singlet oxygen, and lipid peroxy radicals[37] as well as inhibit copper-catalyzed oxidation of LDL.[38] Quercetin pretreatment of vessels revealed that *in vivo* production of ·NO was inversely correlated with $O_2^{-\cdot}$ production.[39] Similar observations have been reported with epicatechin, in which the polyphenol is linked to Ca^{2+} levels and ·NO release.[40] These data indicate that both alcohol and the red wine polyphenols may play a role in the cardiovascular protection associated with alcoholic beverages.[41]

ENDOTHELIAL NOS AND VASCULAR HOMEOSTASIS

Nitric oxide is produced from arginine and oxygen in a variety of mammalian cell types by three distinct NOS isozymes: two constitutively transcribed forms—endothelial (eNOS) and neuronal (nNOS)—and an inducible form (iNOS), which is found in a number of cell types including macrophages and vascular smooth muscle cells.[42–44] Nitric oxide produced from eNOS is a key regulator of vascular homeostasis, including vascular tone[45] and blood pressure,[46,47] and acts as well as an antithrombogenic agent.[48,49] Decreased eNOS protein and inadequate ·NO production is an early and persistent feature of arteriosclerosis and the vascular dysfunction associated with heart failure, diabetes, and hypertension.[50,51] It was also recently demonstrated that eNOS is down regulated in the peripheral vasculature (conduit and resistance arteries) following MI.[52] Decreased ·NO production, as would result from loss of eNOS, could result in impaired vascular function, vasoconstriction, platelet aggregation, smooth muscle cell proliferation, and leukocyte adhesion.[49,53] The pivotal role of eNOS in vascular homeostasis is illustrated by the observation that gene transfer of eNOS (replication-deficient adenovirus that carries cDNA for eNOS) and increased expression of eNOS can improve the vascular dysfunction that is associated with atherosclerosis[54] and MI[52] as well as limiting thrombosis and vascular smooth muscle proliferation in angioplasty-injured vessels.[55] Vascular disorders, such as atherosclerosis, are associated with a progressive loss of endothelial cell function and the accumulation of oxidation products of LDL and other proteins.[56] Reactive oxygen and nitrogen species may mediate the oxidative modification of LDL. This hypothesis is supported by the observation that antibodies directed to a specific oxidized lipid-protein adduct reacts with material in atherosclerotic lesions.[57] Major sources of reactive oxygen species in the vasculature include xanthine oxidase, NO synthase and a membrane-associated NADH/NADPH oxidase expressed by endothelial and vascular smooth muscle cells.[50] In summary, the decreased NO production associated with vascular disease could be a result of either the alteration of endothelial cell signaling and impaired activation of NOS or oxidative inactivation by excessive production of superoxide in the cell wall.

THE ·NO HYPOTHESIS OF CARDIOPROTECTION

The heart reacts to a mild ischemic insult by becoming more resistant to a subsequent ischemic stress, a phenomenon referred to as ischemic preconditioning (PC). Ischemic PC occurs in two phases, an early phase which develops within minutes of the ischemic episode and lasts for 2–3h and a late phase which manifests 12–24h later and lasts for 3–4 days. Late phase PC requires 12–24h to develop requiring upregulation of cardioprotective genes that are responsible for the development of late PC. It has been postulated that nitric oxide synthase (NOS) is the cardioprotective protein responsible for ischemic preconditioning.[58] It is hypothesized that a brief ischemic episode results in production of ·NO and $O_2^{-\cdot}$, which triggers a complex signal transduction cascade that leads to activation of transcription factors, upregulation of cardioprotective genes and ultimately increased activity of NOS.[59] The $O_2^{-\cdot}$ could also react with ·NO to form peroxynitrite ($ONOO^-$), a potent oxidant. The ·NO-dependency of late phase PC is supported by the observation that NOS inhibitors attenuate cardioprotection and that cardioprotection cannot be reproduced in iNOS knock out mice.[58] It has also been postulated that oxidants produced during the initial ischemic period activate endogenous antioxidant defenses (increased MnSOD) 24h later, and that these antioxidants contributes to late PC.[60] These observations of an antioxidant-sensitive mechanism[61] can be reconciled with the concept that increased ·NO is responsible for preconditioning since the increased SOD would decrease $O_2^{-\cdot}$, inhibit the reaction of ·NO with $O_2^{-\cdot}$ thereby increasing ·NO. These studies suggest a mechanistic link between the induction of NOS and late phase ischemic pre-conditioning.

The provocative hypothesis that ischemic pre-conditioning and ethanol-dependent cardioprotection may share common, but as yet undefined, molecular mechanisms has been advanced.[21] Chronic ethanol consumption improved recovery of contractile function and minimized myocyte damage associated with myocardial ischemia-reperfusion, a result similar to that observed with ischemic PC. Chronic ethanol consumption appears to involve changes in gene expression since cardioprotection is not observed for four weeks of ethanol treatment and persists 12–16h after removing ethanol from the diet.[21] The eNOS gene has been identified not only in endothelial cells but also in cardiac myocytes.[62] There may also be changes in the distribution of eNOS in the vasculature with cardiac disease (congestive heart failure), which results in decreased eNOS in the endothelial cell and increased eNOS in the smooth muscle.[63] Therefore, it is reasonable to postulate that ·NO plays a critical role in cardiovascular protection and that nitric oxide synthase (NOS) is the responsible cardioprotective protein.[58]

SUMMARY

In summary, the characteristics of the cardioprotection associated with consumption of alcoholic beverages are indicative of mechanisms resulting in improved functional and metabolic performance in both the myocardium and vasculature. There are a number of actions of ·NO which would be predicted to be cardioprotective, including decreases in cardiac contractility,[49,64] decreasing coronary resistance,[64] diminishing myocardial oxygen consumption,[65] improving metabolic function,[66] inhibiting calcium influx[67] and opening of sarcolemmal, or perhaps mitochondrial, ATP-dependent potassium channels (K^+_{ATP}).[68] We suggest that increased ·NO bioavailability plays an important role in ethanol- and polyphenol-dependent cardioprotective mechanisms through the regulation of antioxidant and NO-producing enzymes.

REFERENCES

1. KLATSKY, A.L. 1996. Alcohol, coronary disease, and hypertension. Annu. Rev. Med. **47:** 149–160.
2. LUSCHER, T.F., F.C. TANNER, M.R. TSCHUDI & G. NOLL. 1993. Endothelial dysfunction in coronary artery disease. Annu. Rev. Med. **44:** 395–418.
3. ZEIHER, A.M., H. DREXLER, B. SAURBIER & H. JUST. 1993. Endothelium-mediated coronary blood flow modulation in humans. Effects of age, atherosclerosis, hypercholesterolemia, and hypertension. J. Clin. Invest. **92:** 652–662.
4. ROSS, R. 1993. The pathogenesis of atherosclerosis: a perspective for the 1990s. Nature **362:** 801–809.
5. FLAVAHAN, N.A. 1992. Atherosclerosis or lipoprotein-induced endothelial dysfunction. Potential mechanisms underlying reduction in EDRF/nitric oxide activity. Circulation **85:** 1927–1938.
6. BOOYSE, F.M. & D.A. Parks. 2001. Moderate wine and alcohol consumption: beneficial effects on cardiovascular disease [Review]. Thromb. Haemost. **86:** 517–528.
7. MOSS, M., B. BUCHER, F.A. MOORE, *et al.* 1996. The role of chronic alcohol abuse in the development of acute respiratory distress syndrome in adults. JAMA **275:** 50–54.
8. KLATSKY, A.L., M.A. ARMSTRONG & G.D. FRIEDMAN. 1997. Red wine, white wine, liquor, beer, and risk for coronary artery disease hospitalization. Am. J. Cardiol. **80:** 416–420.
9. DOLL, R. 1997. One for the heart. Br. Med. J. **315:** 1664–1668.
10. THOMAS, C.E. & D.J. REED. 1990. Radical-induced inactivation of kidney Na^+,K^+-ATPase: sensitivity to membrane lipid peroxidation and the protective effect of vitamin E. Arch. Biochem. Biophys. **281:** 96–105.
11. FAGRELL, B., U. DE FAIRE, S. BONDY, *et al.* 1999. The effects of light to moderate drinking on cardiovascular diseases. [Review] [68 refs]. J. Intern. Med. **246:** 331–340.
12. GAZIANO, J.M., T.A. GAZIANO, R.J. GLYNN, *et al.* 2000. Light-to-moderate alcohol consumption and mortality in the Physicians' Health Study enrollment cohort. J. Am. Coll. Cardiol. **35:** 96–105.
13. BOFFETTA, P. & L. GARFINKEL. 1990. Alcohol drinking and mortality among men enrolled in an American Cancer Society prospective study. [see comments]. Epidemiology **1:** 342–348.
14. THUN, M.J., R. PETO, A.D. LOPEZ, *et al.* 1997. Alcohol consumption and mortality among middle-aged and elderly U.S. adults. N. Engl. J. Med. **337:** 1705–1714.
15. KLATSKY, A.L. 1999. Moderate drinking and reduced risk of heart disease. Alcohol Res. & Health (Journal of the National Institute on Alcohol Abuse & Alcoholism) **23:** 15–23.
16. MCCONNELL, M.V., I. VAVOURANAKIS, L.L. WU, *et al.* 1997. Effects of a single, daily alcoholic beverage on lipid and hemostatic markers of cardiovascular risk. Am. J. Cardiol. **80:** 1226–1228.
17. GAZIANO, J.M., C.H. HENNEKENS, S.L. GODFRIED, *et al.* 1999. Type of alcoholic beverage and risk of myocardial infarction. Am. J. Cardiol. **83:** 52–57.
18. RENAUD, S.C. & J.C. RUF. 1996. Effects of alcohol on platelet functions. Clin. Chem. Acta **246:** 77–89.
19. AIKENS, M.L., H.E. GRENETT, R.L. BENZA, *et al.* 1998. Alcohol-induced upregulation of plasminogen activators and fibrinolytic activity in cultured human endothelial cells. Alcohol. Clin. Exp. Res. **22:** 375–381.
20. RIDKER, P.M., D.E. VAUGHAN, M.J. STAMPFER, *et al.* 1994. Association of moderate alcohol consumption and plasma concentration of endogenous tissue-type plasminogen activator. JAMA **272:** 929–933.
21. MIYAMAE, M., I. DIAMOND, M.W. WEINER, *et al.* 1997. Regular alcohol consumption mimics cardiac preconditioning by protecting against ischemia-reperfusion injury. Proc. Natl. Acad. Sci. USA **94:** 3235–3239.
22. KELI, S.O., M.G. HERTOG, E.J. FESKENS & D. KROMHOUT. 1996. Dietary flavonoids, antioxidant vitamins, and incidence of stroke: the Zutphen study. Arch. Intern. Med. **156:** 637–642.

23. HERTOG, M.G., D. KROMHOUT, C. ARAVANIS, *et al.* 1995. Flavonoid intake and long-term risk of coronary heart disease and cancer in the seven countries study. Arch. Intern. Med. **155:** 381–386.
24. TRUELSEN, T., M. GRONBAEK, P. SCHNOHR & G. BOYSEN. 1998. Intake of beer, wine, and spirits and risk of stroke : the Copenhagen City Heart Study. Stroke **29:** 2467–2472.
25. KNEKT, P., R. JARVINEN, A. REUNANEN & J. MAATELA. 1996. Flavonoid intake and coronary mortality in Finland: a cohort study [see comments]. BMJ. **312:** 478–481.
26. FRANKEL, E.N., J. KANNER, J.B. GERMAN, *et al.* 1993. Inhibition of oxidation of human low-density lipoprotein by phenolic substances in red wine. Lancet **341:** 454–457.
27. SERAFINI, M., G. MAIANI & A. FERRO-LUZZI. 1998. Alcohol-free red wine enhances plasma antioxidant capacity in humans. J. Nutr. **128:** 1003–1007.
28. WHITEHEAD, T.P., D. ROBINSON, S. ALLAWAY, *et al.* 1995. Effect of red wine ingestion on the antioxidant capacity of serum [see comments]. Clin. Chem. **41:** 32–35.
29. PACE-ASCIAK, C.R., S. HAHN, E.P. DIAMANDIS, *et al.* 1995. The red wine phenolics trans-resveratrol and quercetin block human platelet aggregation and eicosanoid synthesis: implications for protection against coronary heart disease. Clin. Chem. Acta **235:** 207–219.
30. BERETZ, A., A. STIERLE, R. ANTON & J.P. CAZENAVE. 1982. Role of cyclic AMP in the inhibition of human platelet aggregation by quercetin, a flavonoid that potentiates the effect of prostacyclin. Biochem. Pharmacol. **31:** 3597–3600.
31. CONSTANT, J. 1997. Alcohol, ischemic heart disease, and the French paradox. Clin. Cardiol. **20:** 420–424.
32. KONDO, K., A. MATSUMOTO, H. KURATA, *et al.* 1994. Inhibition of oxidation of low-density lipoprotein with red wine. Lancet **344:** 1152.
33. FUHRMAN, B., A. LAVY & M. AVIRAM. 1995. Consumption of red wine with meals reduces the susceptibility of human plasma and low-density lipoprotein to lipid peroxidation. Am. J. Clin. Nutr. **61:** 549–554.
34. LAVY, A., B. FUHRMAN, A. MARKEL, *et al.* 1994. Effect of dietary supplementation of red or white wine on human blood chemistry, hematology and coagulation: favorable effect of red wine on plasma high-density lipoprotein. Ann. Nutr. Metab. **38:** 287–294.
35. LECOMTE, E., B. HERBETH, F. PAILLE, *et al.* 1996. Changes in serum apolipoprotein and lipoprotein profile induced by chronic alcohol consumption and withdrawal: determinant effect on heart disease? Clin. Chem. **42:** 1666–1675.
36. SOLEAS, G.J., E.P. DIAMANDIS & D.M. GOLDBERG. 1997. Wine as a biological fluid—history, production, and role in disease prevention. J. Clin. Lab. Anal. **11:** 287–3133.
37. CONQUER, J.A., G. MAIANI, E. AZZINI, *et al.* 1998. Supplementation with quercetin markedly increases plasma quercetin concentration without effect on selected risk factors for heart disease in healthy subjects. J. Nutr. **128:** 593–597.
38. VINSON, J.A. 1998. Flavonoids in foods as in vitro and in vivo antioxidants. Adv. Exp. Med. Biol. **439:** 151–164.
39. HUK, I., V. BROVKOVYCH, V.J. NANOBASH, *et al.* 1998. Bioflavonoid quercetin scavenges superoxide and increases nitric oxide concentration in ischaemia-reperfusion injury: an experimental study. Br. J. Surg. **85:** 1080–1085.
40. HUANG, Y., N.W. CHAN, C.W. LAU, *et al.* 1999. Involvement of endothelium/nitric oxide in vasorelaxation induced by purified green tea (–)epicatechin. Biochem. Biophys. Acta **1427:** 322–328.
41. RIMM, E.B., A. KLATSKY, D. GROBBEE & M.J. STAMPFER. 1996. Review of moderate alcohol consumption and reduced risk of coronary heart disease: is the effect due to beer, wine, or spirits. Br. Med. J. **312:** 731–736.
42. PARKS, D.A., K.A. SKINNER, H.A. SKINNER & S. TAN. 1998. Multiple organ dysfunction syndrome: role of xanthine oxidase and nitric oxide. Pathophysiology **5:** 49–56.
43. BILLIAR, T.R. 1995. Nitric oxide. Novel biology with clinical relevance. Ann. Surg. **221:** 339–349.
44. RUBBO, H., V. DARLEY-USMAR & B.A. FREEMAN. 1996. Nitric oxide regulation of tissue free radical injury. Chem. Res. Toxicol. **9:** 809–820.
45. RADOMSKI, M.W. & S. MONCADA. 1993. Regulation of vascular homeostasis by nitric oxide. Thromb. Haemost. **70:** 36–41.

46. SHESELY, E.G., N. MAEDA, H.S. KIM, *et al.* 1996. Elevated blood pressures in mice lacking endothelial nitric oxide synthase. Proc. Natl. Acad. Sci. USA **93:** 13176–13181.
47. PATEL, R.P., J. MCANDREW, H. SELLAK, *et al.* 1999. Biological aspects of reactive nitrogen species. Biochim. Biophys. Acta **1411:** 385–400.
48. LAMAS, S., D. PEREZ-SALA & S. MONCADA. 1998. Nitric oxide: from discovery to the clinic. Trends Pharmacol. Sci. **19:** 436–438.
49. BECKMAN, J.S. & W.H. KOPPENOL. 1996. Nitric oxide, superoxide, and peroxynitrite: the good, the bad, and ugly. Am. J. Physiol. **271:** C1424–C1437.
50. KOJDA, G. & D. HARRISON. 1999. Interactions between NO and reactive oxygen species: pathophysiological importance in atherosclerosis, hypertension, diabetes and heart failure. Cardiovasc. Res. **43:** 562–571.
51. CHANNON, K.M., M.A. BLAZING, G.A. SHETTY, *et al.* 1996. Adenoviral gene transfer of nitric oxide synthase: high level expression in human vascular cells. Cardiovasc. Res. **32:** 962–972.
52. GABALLA, M.A. & S. GOLDMAN. 1999. Overexpression of endothelium nitric oxide synthase reverses the diminished vasorelaxation in the hindlimb vasculature in ischemic heart failure in vivo. J. Mol. Cell. Cardiol. **31:** 1243–1252.
53. DARLEY-USMAR, V.M., J. MCANDREW, R. PATEL, *et al.* 1997. Nitric oxide, free radicals and cell signalling in cardiovascular disease. Biochem. Soc. Trans. **25:** 925–929.
54. OOBOSHI, H., K. TOYODA, F.M. FARACI, *et al.* 1998. Improvement of relaxation in an atherosclerotic artery by gene transfer of endothelial nitric oxide synthase. Arteriolscler. Thromb. Vasc. Biol. **18:** 1752–1758.
55. WU, K.K. 1998. Injury-coupled induction of endothelial eNOS and COX-2 genes: a paradigm for thromboresistant gene therapy. Proc. Assoc. Am. Physicians **110:** 163–170.
56. ROSS, R. 1999. Atherosclerosis—an inflammatory disease. N. Engl. J. Med. **340:** 115–126.
57. SALONEN, J.T., S. YLA-HERTTUALA, R. YAMAMOTO, *et al.* 1992. Autoantibody against oxidised LDL and progression of carotid atherosclerosis. Lancet **339:** 883–887.
58. BOLLI, R., B. DAWN, X.L. TANG, *et al.* 1998. The nitric oxide hypothesis of late preconditioning. Basic Res. Cardiol. **93:** 325–338.
59. DOWNEY, J.M. & M.V. COHEN. 1997. Arguments in favor of protein kinase C playing an important role in ischemic preconditioning. Basic Res. Cardiol. **92**(Suppl. 2): 37–39.
60. ZHAI, X., X. ZHOU & M. ASHRAF. 1996. Late ischemic preconditioning is mediated in myocytes by enhanced endogenous antioxidant activity stimulated by oxygen-derived free radicals. Ann. N.Y. Acad. Sci. **793:** 156–166.
61. TAKANO, H., X.L. TANG, Y. QIU, *et al.* 1998. Nitric oxide donors induce late preconditioning against myocardial stunning and infarction in conscious rabbits via an antioxidant-sensitive mechanism. Circ. Res. **83:** 73–84.
62. GROSS, S.S. & M.S. WOLIN. 1995. Nitric oxide: pathophysiological mechanisms. Annu. Rev. Physiol. **57:** 737–769.
63. COMINI, L., T. BACHETTI, G. GAIA, *et al.* 1996. Aorta and skeletal muscle NO synthase expression in experimental heart failure. J. Mol. Cell. Cardiol. **28:** 2241–2248.
64. KITAKAZE, M., K. NODE, T. MINAMINO, *et al.* 1996. Role of nitric oxide in regulation of coronary blood flow during myocardial ischemia in dogs. J. Am. Coll. Cardiol. **27:** 1804–1812.
65. OKUBO, T., N. SUTO, S. KUDO, *et al.* 1999. A study on the effects of endogenous nitric oxide on coronary blood flow, myocardial oxygen extraction and cardiac contractility. Fundamental Clin. Pharmacol. **13:** 34–42.
66. KANAI, A.J., S. MESAROS, M.S. FINKEL, *et al.* 1997. Beta-adrenergic regulation of constitutive nitric oxide synthase in cardiac myocytes. Am. J. Physiol. **273:** C1371–C1377.
67. HOVE-MADSEN, L., P.F. MERY, J. JUREVICIUS, *et al.* 1996. Regulation of myocardial calcium channels by cyclic AMP metabolism. Basic Res. Cardiol. **91**(Suppl. 2): 1–8.
68. LIU, Y., T. SATO, B. O'ROURKE & E. MARBAN. 1998. Mitochondrial ATP-dependent potassium channels: novel effectors of cardioprotection? Circulation **97:** 2463–2469.

Cardioprotection with Alcohol

Role of Both Alcohol and Polyphenolic Antioxidants

MOTOAKI SATO, NILANJANA MAULIK, AND DIPAK K. DAS

Cardiovascular Research Center, University of Connecticut School of Medicine, Farmington, Connecticut 06030-1110, USA

ABSTRACT: Both epidemiological and experimental studies indicate that mild-to-moderate alcohol consumption is associated with a reduced incidence of mortality and morbidity from coronary heart disease. The consumption of wine, particularly red wine, imparts a greater benefit in the prevention of coronary heart disease than the consumption of other alcoholic beverages. The cardioprotective effects of red wine have been attributed to several polyphenolic antioxidants including resveratrol and proanthocyanidins. The results of our study documented that the polyphenolic antioxidants present in red wine, for example, resveratrol and proanthocyanidins, provide cardioprotection by their ability to function as *in vivo* antioxidants while its alcoholic component or alcohol by itself imparts cardioprotection by adapting the hearts to oxidative stress. Moderate alcohol consumption induced significant amount of oxidative stress to the hearts which was then translated into the induction of the expression of several cardioprotective oxidative stress-inducible proteins including heat shock protein (HSP) 70. Feeding the rats with red wine extract or its polyphenolic antioxidants as well as alcohol resulted in the improvement of postischemic ventricular function. Additionally, both wine and alcohol triggered a signal transduction cascade by reducing proapoptotic transcription factors and genes such as JNK-1 and c-Jun thereby potentiating an anti-death signal. This resulted in the reduction of myocardial infarct size and cardiomyocyte apoptosis. The results, thus, indicate that although both wine and alcohol alone reduce myocardial ischemic reperfusion injury, the mechanisms of cardioprotection differ from each other.

KEYWORDS: alcohol; wine; polyphenols; flavonoids; antioxidants; free radicals; nitric oxide; heat shock proteins; adaptation; oxidative stress

INTRODUCTION

Recently, a number of studies have been devoted to understanding the cause of the so-called French Paradox, the anomaly which means that in several parts of France and other Mediterranean countries the morbidity and mortality from coronary heart diseases, both absolutely and relative to other causes of death is significantly lower than that is in other developed countries, despite of the high consumption of fat and saturated fatty acids.[1–10] This cardioprotective effect is believed

Address for correspondence: Dipak K. Das, Ph.D., Cardiovascular Research Center, University of Connecticut, School of Medicine, Farmington, CT 06030-1110, USA. Voice: 860-679-3687; fax: 860-679-4606.
ddas@neuron.uchc.edu

Ann. N.Y. Acad. Sci. **957: 122–135 (2002).**

to be in part related to the regular consumption of wine. Wines, especially red wines, have about 1,800–3,000 mg/L of polyphenolic compounds.[11] Many polyphenolic compounds are potent antioxidants capable of scavenging free radicals and inhibiting lipid peroxidation both *in vitro* and *in vivo*. It has been shown that flavonoids as well as non-flavonoids present in wines inhibit the oxidation of low-density lipoproteins, eicosanoid synthesis, and platelet aggregation as well as promote nitric oxide synthesis.[4,12,13]

Over the past several years, various epidemiological studies have strongly suggested that mild-to-moderate alcohol consumption is associated with a reduced incidence of mortality and morbidity from coronary heart disease.[14–19] Proposed mechanisms for such cardioprotective effects have included among others an increase in high-density lipoprotein (HDL) cholesterol,[20] reduction/inhibition of platelet aggregation,[21] reduction in clotting factor concentrations,[4] reduction in thromboxane synthesis,[5] increase in vasodilatory prostacyclin synthesis,[6] inhibition of low-density lipoprotein (LDL) oxidation,[3] and free radical scavenging.[22]

During the past 20 years studies in different ethnic groups starting from an American cohort and including the recently performed analysis in the MONICA-project provided evidence for a decreased morbidity and mortality from coronary heart disease at one to three drinks a day when compared to total abstainers.[23] Although it is comprehensible that antioxidants like flavonoids and polyphenols found in wines may explain the cardioprotective effects of wine consumption, the mechanism of cardioprotection afforded by alcohol alone remains obscure. The present study compared the effects of wine (alcohol-free red wine extract) and alcohol on cardioprotection. The results demonstrated that both wine and alcohol can protect the heart against ischemia/reperfusion injury to similar extent, but by different mechanisms.

MATERIALS AND METHODS

Red Wine Extract

We vacuum evaporated 0.75L of red wine (under 30°C) to obtain 14.5g of jelly-like extract. One gram of this extract was then solubilized in deionized double-distilled water. The resulting fluid was diluted 1,000-fold with double-distilled water. Vacuum evaporation was performed so as to facilitate the exclusive study of the biological effects of red wine constituents other than ethanol without alcohol interference. The composition of red wine (Syrah region, North Italy) extract used in our study: trans-resveratrol, 110ppm; quercitin, 2.9ppm; salycylic acid; non-detectable.

Trans-resveratrol and all other chemicals used were obtained from Sigma Chemical Company (St. Louis, MO), unless otherwise specified.

Animals

Male Sprague-Dawley rats (275–300g, Harlan Sprague-Dawley) were used in this study. They were provided with food and water *ad libitum* up until the start of the experimental procedure. Rats were randomly assigned to one of several groups, A (control, water only), B (red wine extract), C (alcohol), D (proanthocyanidin) and

E (resveratrol). They were given orally (once a day) either water (control) or the above-mentioned experimental compounds for three weeks. All animals received care in compliance with the principles of laboratory animal care formulated by the National Society for Medical Research and *Guide for the Care and Use of Laboratory Animals* prepared by the National Academy of Sciences and published by the National Institute of Health (publication no. NIH 80-23, revised 1978).

Isolated Rat Heart Model

The perfusion buffer used in this study consisted of a modified Krebs-Henseleit bicarbonate buffer (KHB) (in mM: 118 NaCl, 4.7 KCl, 1.2 $MgSO_4$, 1.2 KH_2PO_4, 25 $NaHCO_3$, 10 glucose, and 1.7 $CaCl_2$, gassed with 95% O_2–5% CO_2, pH 7.4) which was maintained at a constant temperature of 37°C and was gassed continuously for the entire duration of the experiment. Rats were anesthetized with sodium pentobarbital (80 mg/kg b.w., i.p. injection, Abbott Laboratories), anticoagulated with heparin sodium (500 IU/kg b.w., i.v. injection, Elkins-Sinn, Inc.). After ensuring sufficient depth of anesthesia, thoracotomy was performed, hearts were excised and immersed in ice-cold perfusion buffer. Both the aorta and the pulmonary vein were cannulated as quickly as possible and the hearts were perfused in the retrograde Langendorff mode at a constant perfusion pressure of 100 cm H_2O for a stabilization period of 10 minutes. Then the circuit was switched to the antegrade working mode as described previously.[24] The hearts pumped against an afterload of 100 cm H_2O perfused with a constant preload of 17 cm H_2O. Aortic pressure was measured using a Gould P23XL pressure transducer connected to a sidearm of the aortic cannula and the signal was amplified using a Gould 6600 series signal conditioner and monitored on a real-time data acquisition and analysis system (CORDAT II.). Heart rate, developed pressure (defined as the difference of the maximum systolic and diastolic aortic pressures) and the first derivative of developed pressure were all derived or calculated from the continuously obtained pressure signal. Aortic flow was measured using a calibrated flowmeter (Gilmont Instruments Inc.) and coronary flow was measured by timed collection of the coronary effluent dripping from the heart. At the end of 10 minutes, after the attainment of steady state cardiac function, baseline functional parameters were recorded and coronary effluent collected for biochemical assays. The circuit was then switched back to the retrograde mode and the hearts were perfused with either KHB (Group A), red wine extract (Group B), or trans-resveratrol (Group C), the latter at a concentration of 10 μM, for a duration of 15 minutes. At the end of this period, hearts were subjected to global ischemia for 30 minutes followed by a two-hour reperfusion. The first 10 minutes of reperfusion was in the retrograde mode to allow for post-ischemic stabilization and thereafter in the antegrade working mode to allow for assessment of functional recovery, which was recorded at 15 minutes, 30 minutes, 60 minutes, and 120 minutes into reperfusion. Coronary effluent samples for biochemical assays were also collected at the above timepoints and stored at −20°C.

Infarct Size Estimation

Hearts to be used for infarct size calculations ($n = 6$ per group) were taken upon termination of the experiment and immersed in 1% triphenyl tetrazolium solution in phosphate buffer (Na_2HPO_4 88 mM, NaH_2PO_4 1.8 mM) preheated to 37°C for 20 minutes and then stored at −70°C for later processing. Frozen hearts (including only ventricular tissue) were sliced in a plane perpendicular to the apico-basal axis into approximately 1 mm thick sections, blotted dry, placed in between microscope slides and scanned on a single pass flat bed scanner (Hewlett-Packard Scanjet 5p).[25] Using the NIH Image 1.6.1 image processing software, each image was subjected to background subtraction, contrast enhancement and grayscale conversion for improved clarity and distinctness. Total as well as infarct zones of each slice were traced and the respective areas were calculated in terms of pixels and cm^2, the respective volumes being obtained in cm^3. The weight of each slice was then recorded to facilitate the expression of total and infarct masses of each slice in grams. The total and infarct masses of each heart were then calculated by taking a weighted average. Infarct size was taken to be the percent infarct mass of total mass for any one heart.

Estimation of Malonaldehyde (MDA) Formation

Malonaldehyde was assayed as described previously, to monitor the development of oxidative stress.[26] In short, coronary effluents were collected in 2 mL of a solution containing 20% trichloroacetic acid, 5.3 mM sodium bisulfite, kept on ice for 10 min, and centrifuged at 3,000×*g* for 10 min; supernatants were then collected, derivatized with 2,4-dinitrophenylhydrazine (DNPH), and extracted with pentane. Aliquots of 25 µl in acetonitrile were injected onto a Beckman Ultrasphere C_{18} (3 µm) column. The products were eluted isocratically with a mobile phase containing acetonitrile:water:acetic acid (40:60:0.1, v/v/v) and measured at three different wavelengths (307 nm, 325 nm and 356 nm) using a Waters M-490 multichannel UV detector. The peak for malonaldehyde was identified by co-chromatography with DNPH derivative of the authentic standard, peak addition, UV pattern of absorption at the three wavelengths, and by GC-MS.

Evaluation of Apoptosis

To examine apoptosis, cardiomyocytes were obtained by well-established methods.[27] Following experiments, a group of hearts were quickly placed into a chilled dissociation buffer containing (mM): NaCl, 137; KCl, 5.4; $CaCl_2$, 1.8; $MgCl_2$, 1.0; KH_2PO_4, 0.44; Na_2HPO_4, 0.34; dextrose, 5.6; HEPES buffer (pH 7.5), 20N; penicillin, 50 U/mL; and streptomycin, 50 µg/mL. The ventricles were cut into 1–2 mm cubes and dissociated by trypsinization (0.05% trypsin-EDTA at 37°C for 10 min). Unfreed cells from the first treatment were discarded, and the sequence was repeated until all tissue was dissociated (approximately five times). Freed cells were collected in cold Dulbecco's modified Eagle's medium (DMEM, GIBCO BRL, Gaithersburg, MD) supplemented with 0.5% fetal calf serum (FCS) and 0.002% DNAse, and washed in the same medium. Immunohistochemical detection of apoptotic cells was carried out using TUNEL[27] assay method in conjunction with an antibody against myosin heavy chain for specifically identifying apoptotic cardiomyocytes.

Western Blot Analysis

To quantify the abundance of p38MAPK, JNK1 and c-Jun protein as well heat shock proteins (HSPs), the heart tissues were homogenized and suspended (5 mg/mL) in sample buffer [10 mM HEPES, pH 7.3, 11.5% sucrose, 1 mM EDTA, 1 mM EGTA, 1 mM diisopropylfluorophosphate (DFP), 0.7 mg/ml pepstatin A, 10 mg/mL leupeptin, 2 mg/mL aprotinin].[28] Proteins were then solubilized with the addition of same amount of 2× Laemmli solution [9% (w/v) SDS; 6% (v/v) β-mercaptoethanol; 10% (v/v) glycerol; and a trace amount of bromophenol blue dye in 0.196 M Tris/HCl (pH 6.7)]. The cellular proteins (50 μL samples) were electrophoresed through 10% SDS-PAGE and then transferred to Immobilon-P membranes (Millipore Corp.) using a semidry transfer system (Bio-Rad). Pre-stained protein standards (Bio-Rad) were run in each gel. The blots were blocked in Tris-buffered saline/Tween-20 (TBS-T containing 20 mM Tris base, pH 7.6; 137 mM NaCl; 0.1% Tween-20) supplemented with 5% BSA for 1 hour, incubated with 1 : 1000 diluted primary rabbit antibodies specifically against either c-Jun or JNK 1 (Santa Cruz Biotech) for two hours, and then with 1 : 10000 diluted secondary antibodies of horseradish peroxidase-conjugated anti-rabbit IgG (Boehringer Mannheim Corp., Inc) for one hour at room temperature. After three washes of five minutes each, blots were treated with Enhanced Chemi-Luminescence (ECL from Amersham) reagents and the JNK1 and c-Jun were detected by autoradiography for variable lengths of time (15 sec to 3 min) with Kodak X-Omat film. Each blot was scanned with a scanning densitometer, and the results are shown as means ± SEM of six different blots per group.

STATISTICAL ANALYSIS

For statistical analysis, a two-way analysis of variance (ANOVA) followed by Scheffe's test was first carried out to test for any differences between groups. If differences were established, the values were compared using Student's *t*-test for paired data. The values were expressed as mean ± SEM. The results were considered significant for $p < 0.05$.

RESULTS

Postischemic Ventricular Function

The effects of alcohol and red wine extract on post-ischemic left ventricular function are summarized in TABLE 1. All data are presented as the mean ± the standard error of the mean. As expected, except for heart rate and coronary flow, postischemic ventricular function was depressed in all three groups. For example, after 60 min of reperfusion AF, LVDP, and LVdp/dt were 12.7 ± 0.5 mL/min, 32 ± 7 mm Hg, and 1454 ± 78 mm Hg/min, respectively, in the control group compared to those at the baseline levels (47.2 ± 2.7 mL/min, 103 ± 6 mm Hg, and 3378 ± 167 mm Hg/sec, respectively). Although similar trends were noticed for other groups, these values were significantly higher for the alcohol group (28.3 ± 0.7 mL/min, 67 ± 4 mm Hg, and 2305 ± 67 mm Hg/min, respectively) and red wine group (30.7 ± 0.9 mL/min, 62 ± 3 mm Hg and 2008 ± 55 mm Hg/sec) compared to control group. This trend remained the same up to 120 min of reperfusion. Thus, both alcohol and red wine displayed significant recovery of post-ischemic myocardial function as compared to control group.

Infarct Size

In this particular study, hearts were arrested by global ischemia for 30 min; therefore, the whole ventricle was considered as the area of risk. Control hearts subjected to ischemia and reperfusion (I/R) had an infarct size of 34.9 ± 3.3% (see FIGURE 1) compared to almost zero when a heart was subjected to 2.5 hours perfusion without ischemia/reperfusion (control) Percent infarct size was significantly reduced for alcohol (21.2 ± 1.8%) and red wine (26 ± 2.2%). The red wine polyphenolic components, proanthocyanidin (21.1 ± 1.5%) and resveratrol (13.9 ± 1.0%) also significantly reduced the myocardial infarct size.

TABLE 1. Effects of feeding alcohol and wine on postischemic left ventricular function[a]

			Reperfusion at		
		Baseline	30 min	60 min	120 min
Heart rate (beats/min)	Control	310 ± 11	293 ± 10	293 ± 8	290 ± 9
	Alcohol	312 ± 8	298 ± 8	296 ± 11	293 ± 8
	Wine	309 ± 7	300 ± 10	298 ± 6	290 ± 5
Coronary flow (ml/min)	Control	26.7 ± 1.5	19.5 ± 0.7	16.9 ± 0.8	14.2 ± 1.0
	Alcohol	27.0 ± 1.2	20.6 ± 1.0	18.8 ± 0.9	17.3 ± 1.1
	Wine	26.5 ± 0.8	21.5 ± 1.2	17.4 ± 1.0	15.9 ± 0.8
Aortic flow (ml/min)	Control	47.2 ± 2.7	14.5 ± 0.3	12.7 ± 0.5	9.8 ± 0.2
	Alcohol	50.4 ± 3.1	30.4 ± 1.1*	28.3 ± 0.7*	23.5 ± 1.0*
	Wine	48.7 ± 2.8	32.2 ± 0.6*	30.7 ± 0.9*	25.1 ± 0.5*
LVDP (mm Hg)	Control	103 ± 6	35 ± 4*	32 ± 7*	25 ± 3*
	Alcohol	110 ± 12	70 ± 9*	67 ± 4*	56 ± 7*
	Wine	104 ± 8	65 ± 6*	62 ± 3*	51 ± 5*
$LV_{max}dp/dt$ (mm Hg/sec)	Control	3,378 ± 167	1,634 ± 121	1,454 ± 78	1,234 ± 103
	Alcohol	3,445 ± 123	2,411 ± 89*	2,305 ± 67*	2,223 ± 45*
	Wine	3,389 ± 120	2,255 ± 105*	2,008 ± 55*	1,988 ± 45*

[a]Rats in three groups were fed orally wince, alcohol, or water (control) for three weeks. Isolated rat hearts were subjected to 30 min ischemia followed by two hours of reperfusion at working mode. Ventricular function was measured at baseline and at 30, 60, and 120 min of reperfusion. $^{*}p < 0.05$ versus control.

MDA Formation

The production of malonaldehyde (MDA) is an indicator for lipid peroxidation and development of oxidative stress. As shown in FIGURE 2, the MDA of the heart increased significantly within four hours of alcohol feeding. The MDA content increased further after 12 hours, but came down significantly within 24 hours of alcohol feeding. After 72 hours, the MDA content of the alcohol-fed heart remained at the baseline level. Feeding of red wine also increased the MDA content slightly, which readily came down to the baseline levels. The results, thus, indicate that alcohol causes the development of oxidative stress in the heart; however, such oxidative stress was only transient.

Expression of Cardioprotective HSPs

Alcohol also caused the induction of the expression of HSP 70 as shown in FIGURE 3. Among the three HSPs examined, only HSP 70 protein was increased compared to control. There was no change in the induction of HSP 27 and HSP 90. Wine did not affect any of the HSPs.

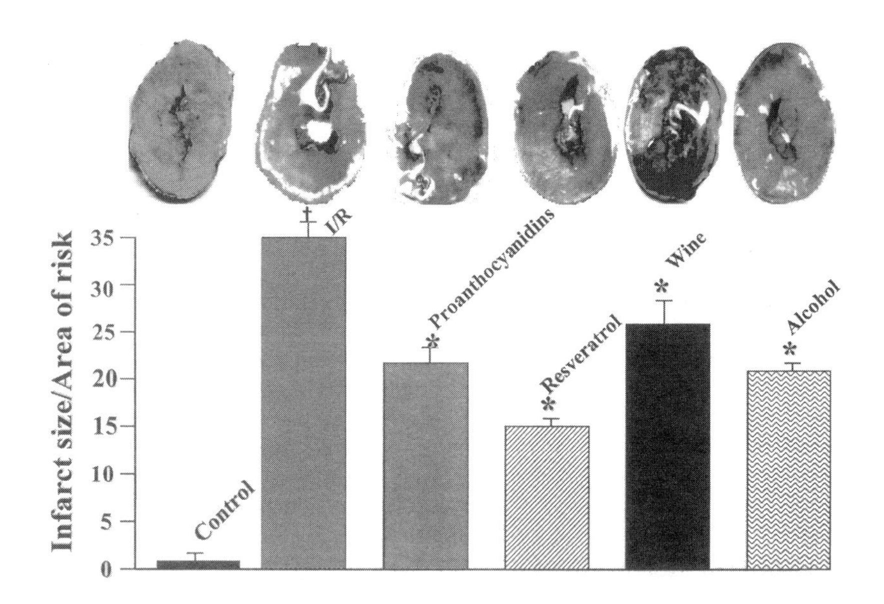

FIGURE 1. Effects of wine, its polyphenolic components, proanthocyanidin and resveratrol, and alcohol on myocardial infarct size. Rats were fed orally with red wine extract, proanthocyanidin, resveratrol or alcohol for three weeks. At the end of three weeks, rats were anesthetized and isolated perfused working hearts were subjected to 30 min of ischemia followed by two hours of reperfusion. The infarct size was measured by TTC staining as described in the METHODS section. Results are expressed as means ± SEM of at least six hearts per group. Representative infarcts are shown (*top*): the light-colored areas (white to yellow stain) represent amount of infarct. The results expressed in a bar graph are shown below. $*p < 0.05$ versus I/R. $\dagger p < 0.05$ versus control.

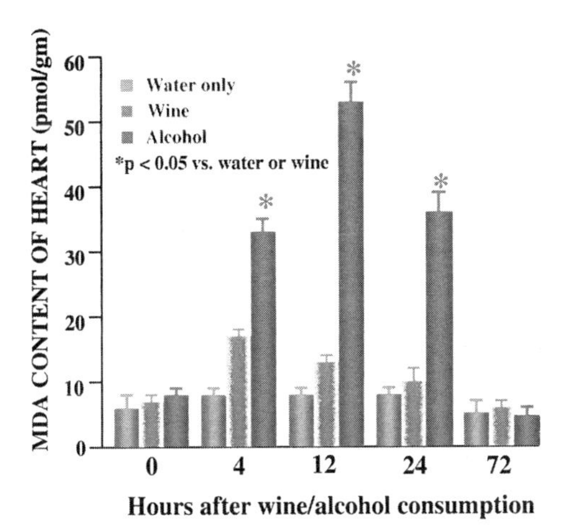

FIGURE 2. Effects of wine, its polyphenolic components, proanthocyanidin and resveratrol, and alcohol on the MDA content of the heart. Rats were fed orally with red wine extract, proanthocyanidin, resveratrol, or alcohol for three weeks. At the end of three weeks, rats were anesthetized and isolated perfused working hearts were subjected to 30 min ischemia followed by two hours of reperfusion. The MDA content was measured by HPLC as described in the METHODS section. Results, means ± SEM of at least six hearts per group, are expressed in a bar graph. $^*p < 0.05$ versus baseline (0 hour) or wine.

Cardiomyocyte Apoptosis

The number of apoptotic cells were significantly higher (19.5%) in the ischemic/reperfused myocardium compared to the control hearts (see FIGURE 4). Wine and its polyphenolic components proanthocyanidin and resveratrol as well as alcohol all were able to reduce the number of apoptotic cardiomyocytes compared to ischemic/reperfused myocardium.

JNK1 and c-Jun Expression

Thirty minutes of ischemia followed by two hours of reperfusion significantly enhanced the amount of the protein levels of p38 MAPK, JNK1, and c-Jun (see FIGURE 5). The ischemia/reperfusion–mediated enhancement of c-Jun and JNK-1 protein levels, but not p38 MAPK, were significantly reduced by wine or alcohol treatment.

DISCUSSION

In this report, we demonstrated that an alcohol-free red wine extract as well as alcohol can protect the hearts from detrimental effects of ischemia reperfusion injury, as evidenced by improved post-ischemic ventricular function and reduced myocardial infarction. Red wine extract and its polyphenolic components, proanthocyanidin and resveratrol reduced the amount of oxidative stress in the heart, as indicated by decreased MDA formation. Alcohol, on the other hand, initially increased the MDA in the heart. However, the MDA content decreased significantly after 24 hours and reduced to baseline level after 72 hours. Both wine and alcohol triggered a signal transduction cascade resulting in the inhibition of proapoptotic factors Jnk and c-Jun, leading to the reduction of cardiomyocyte apoptosis. Myocardial infarct size was also reduced in all groups of hearts. Alcohol, but not the red wine or the polyphenols, induced the expression of cardioprotective HSP 70 protein.

The results of this study suggest that the mechanism of cardioprotection afforded by wine and alcohol is quite different. While, polyphenolic antioxidants including resveratrol and proanthocyanidin are responsible for cardioprotection found in wine, alcohol protects the heart by adapting this organ to oxidative stress. Alcohol consumption appears to induce oxidative stress which is subsequently translated into oxidative stress-inducible proteins. Indeed, alcohol feeding resulted in the

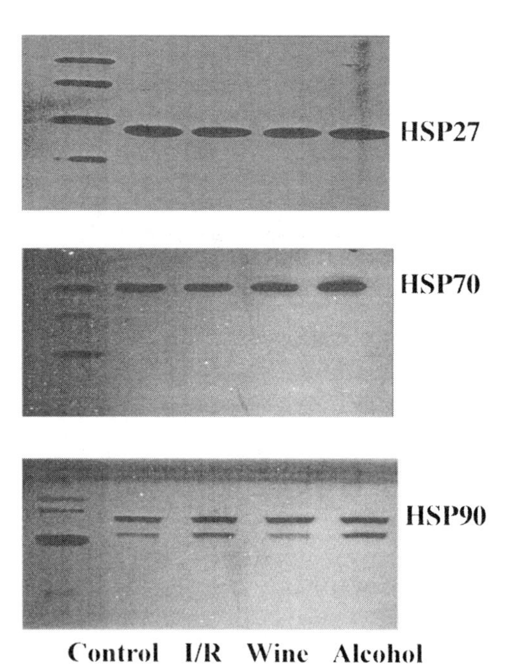

FIGURE 3. Effects of wine, its polyphenolic components, proanthocyanidin and resveratrol, and alcohol on the induction of the expression of HSP 27, HSP 70, and HSP 90 protein content of the heart. Rats were fed orally with red wine extract, proanthocyanidin, resveratrol or alcohol for three weeks. At the end of three weeks, rats were anesthetized and isolated perfused working hearts were subjected to 30 min ischemia followed by two hours of reperfusion. The HSPs was measured by Western blot analysis as described in the METHODS section. Results are representative of at least six hearts per group.

induction of the expression of cardioprotective HSP 70 protein. A large number of evidence exists in the literature indicating cardioprotective abilities of HSPs including HSP 70.[29–32] Currie and colleagues were probably the first group to demonstrate enhanced postischemic ventricular recovery after subjecting the heart to heat stress.[29] The results of many recent studies including our own now support the earlier observation by Currie and further indicate that the same heat shock protein, HSP 70, can also be induced by other stresses such as oxidative stress as shown in the present study.[33,34] The expression of HSP 70 was associated with improved postischemic ventricular function, and decreased infarct size and cardiomyocyte apoptosis.

The mechanism of cardioprotection by red wine is likely to be mediated by its polyphenolic components, resveratrol and proanthocyanidin. As shown in this study wine as well as resveratrol and proanthocyanidin decreased myocardial infarct size and reduced apoptotic cell death to the same extent. Antioxidants have long been known to protect against the damaging effects of free radical–mediated tissue injury, especially ischemia reperfusion injury of the heart and other organs.[35,36] Substantial evidence exists to support the notion that ischemia and reperfusion generate oxygen free radicals which contribute to the pathogenesis of ischemic reperfusion injury.[37,38] The presence of reactive oxygen species were confirmed directly by estimating free radical formation and indirectly by assessing lipid peroxidation and DNA breakdown products.[39] Among the oxygen free radicals, superoxide anion

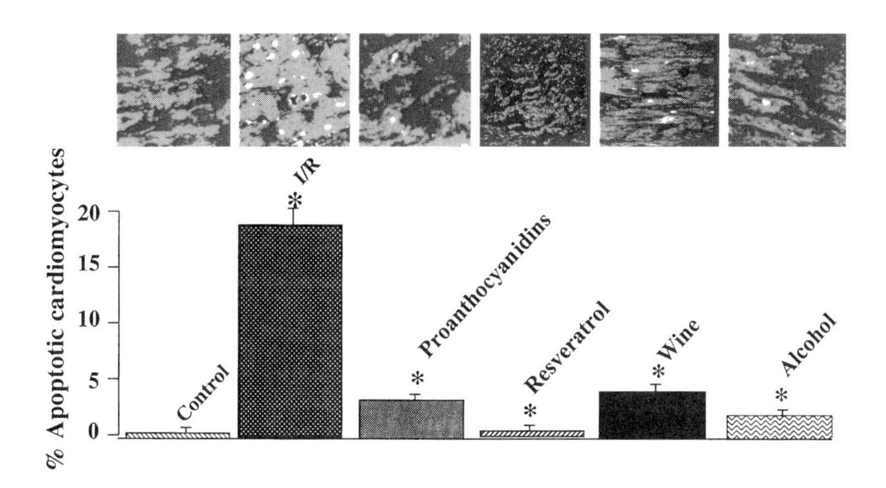

FIGURE 4. Effects of wine, its polyphenolic components, proanthocyanidin and resveratrol, and alcohol on cardiomyocyte apoptosis. Rats were fed orally with red wine extract, proanthocyanidin, resveratrol or alcohol up to a period of three weeks. At the end of three weeks, rats were anesthetized and isolated perfused working hearts were subjected to 30 min ischemia followed by two hours of reperfusion. The apoptotic cardiomyocytes were detected by Tunnel staining in conjunction with a specific antibody against α-myosin heavy chain to specifically stain the cardiomyocytes as described in the METHODS section. Representative apoptotic cells were detected by laser scanning microscopy (*top*). The results (average of at least six/group) expressed in a bar graph are shown below. $*p < 0.05$ versus I/R or control.

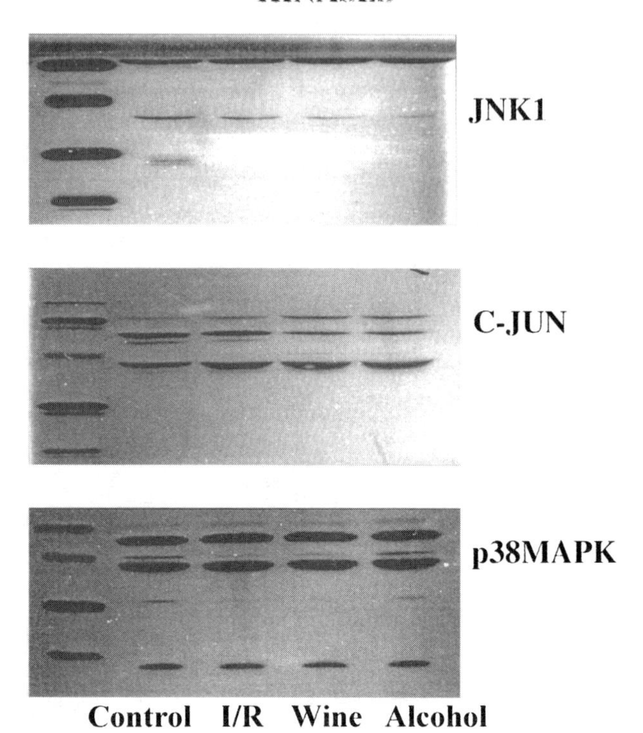

FIGURE 5. Effects of wine, its polyphenolic components, proanthocyanidin and resveratrol, and alcohol on the induction of the expression of p38 MAPK, JNK-1 and c-Fos protein content of the heart. Rats were fed orally with red wine extract, proanthocyanidin, resveratrol or alcohol for three weeks. At the end of three weeks, rats were anesthetized and isolated perfused working hearts were subjected to 30 min of ischemia followed by two hours of reperfusion. P38 MAPK, JNK-1 and c-Fos were measured by Western blot analysis as described in the METHODS section. Results are representative of at least six hearts per group.

(O_2^-) is the most innocent free radical, while the hydroxyl radical (·OH) is the most detrimental to the cells. In spite of their relatively low oxidizing ability compared to ·OH radicals, in biological systems, organic peroxyl radicals could be extremely damaging to the tissues.[40] Tissues like heart are protected from the detrimental actions of peroxyl radicals by the presence of naturally occurring antioxidants such as bilirubin and biliverdin as well as plasma antioxidants.[41] Ascorbic acid and vitamin E comprise the other potent peroxyl radical traps for biological systems.[42] Generally, lipid soluble antioxidants can scavenge chain-carrying lipid peroxyl radicals thereby preventing propagation of lipid peroxidation after the initiation of the lipid peroxidation. Recent study from our laboratory demonstrated that not only are red wine extract and proanthocyanidin and resveratrol potent scavengers of peroxyl radicals, but also they reduced the extent of lipid peroxidation in the ischemic reperfused myocardium.[43–47] These findings seem to be important because these peroxyl radicals are formed *in vivo* in membranes and lipoproteins as intermediate products of lipid peroxidation.

Wine as opposed to other sources of polyphenols and antioxidants is unique in a number of ways. First of all, resveratrol and a few other polyphenols are virtually absent from commonly consumed fruits and vegetables, and thus the consumption of red wine would constitute their only source in the diet. The only other resveratrol source in human consumption is peanuts. Secondly, the various procedures involved in wine production further enrich its polyphenol content. Furthermore, the increased alcohol content as a result of the fermentation process allows for enrichment of total polyphenol content and also for better solubilization of the polyphenols resulting in greater bioavailability than in other foodstuffs.[48] All in all, red wine might possibly be the richest effective source of natural polyphenol antioxidants. Secondary to the antioxidant properties of wines are alcoholic vasodilation, decreases in platelet aggregability, changes in prostacyclin/thromboxane ratios and increased fibrinolytic activities which should be considered as additional benefits caused by mild-to-moderate alcohol consumption.

In summary, the results of the present study indicate that the cardioprotective effects of red wine and alcohol are mediated by two different mechanisms. It appears that wine exerts its cardioprotective abilities through its polyphenolic antioxidants, but alcohol protects the heart from cellular injury by adapting the organ to oxidative stress. Cardioprotective proteins such as HSP 70 produced during the myocardial adaptation to oxidative stress may at least be partially responsible for cardioprotection. This is further supported by the ability of alcohol to trigger a signal transduction cascade potentiating an antideath signal through the reduction of proapoptotic factors JNK anf c-Jun thereby leading to the decrease in cardiomyocyte apoptosis.

ACKNOWLEDGMENTS

This study was supported by NIH HL 34360, HL 22559, HL 33889, HL 56803, and a grant from the California Table Grape Association.

REFERENCES

1. Rimm, E.B., E.L. Giovannucci, W.C. Willett, *et al.* 1991. Prospective study of alcohol consumption and risk of coronary disease in men. Lancet **338:** 464–486.
2. Gaziano, J.M., J.E. Buring, J.L. Breslow, *et al.* 1993. Moderate alcohol intake, increased levels of high-density lipoprotein and its sub-fractions and decreased risk of myocardial infarction. N. Engl. J. Med. **329:** 1829–1834.
3. Frankel, E.N., J. Kanner & J.B. German. 1993. Inhibition of human low-density lipoprotein by phenolic substances in red wine. Lancet **341:** 454–457.
4. Ridker, P.M., D.E. Vaughan, M.J. Stampfer, *et al.* 1994. Association of moderate alcohol consumption and plasma concentration of tissue-type plasminogen activator. JAMA **272:** 929–933.

5. MIKHAILIDIS, D.P., J.Y. JEREMY, M.A. BARRADAS, *et al.* 1983. Effect of ethanol on vascular prostacyclin synthesis, platelet aggregation and platelet thromboxane release. Br. Med. J. **287:** 1495–1498.
6. LANDOLFI, R. & M. STEINER. 1984. Ethanol raises prostacyclin in vivo and in vitro. Blood **64:** 679-682.
7. SAIJA, A., M. SCALESE, M. LAIZA, *et al.* 1995. Flavonoids as antioxidant agents: importance of their interaction with biomembranes. Free Radical Biol. Med. **19:** 481–486.
8. FANCONNEAU, B., P. WAFFO-TEGNO, F. HUGNET, *et al.* 1997. Comparative study of radical scavenger and antioxidant properties of phenolic compounds from vitis vinifera cell cultures using in vitro tests. Life Sci. **61:** 2103–2110.
9. RENAUD, S. & M. DE LORGERIL. 1992. Wine, alcohol, platelets and the French Paradox for coronary heart disease. Lancet **339:** 1523–1526.
10. KLATSKY, A.L., M.A. ARMSTRONG & G.D. FRIEDMAN. 1986. Relations of alcoholic beverage use to subsequent coronary artery disease hospitalization. Am. J. Cardiol. **58:** 710–714.
11. GOLDBERG, D., E. TSANG, A. KARUMANCHIRI, *et al.* 1996. Method to assay the concentrations of phenolic constituents of biological interests in wines. Anal. Chem. **68:** 1688–1694.
12. GRYGLEWSKI, R.J., R. KORBUT, J. ROBAK & J. SWIES. 1987. On the mechanism of antithrombotic action of flavonoids. Biochem. Pharmacol. **36:** 317–322.
13. RENAUD, S.C., A.D. BESWICK, A.M. FEHILY, *et al.* 1992. Alcohol and platelet aggregation: the Caerphilly prospective heart disease study. Am. J. Clin. Nutr. **55:** 1012–1017.
14. ST. LEGER, A.S., A.L. COCHRANE & F. MOORE. 1979. Factors associated with cardiac mortality in developed countries with particular reference to the consumption of wine. Lancet **1:** 1017–1020.
15. HERTOG, M.G.L., E.J.M. FESKENS & D. KROMHOUT. 1997. Antioxidant flavonols and coronary heart disease risk. Lancet **349:** 699.
16. KLATSKY, A.L., M.A. ARMSTRONG & G.D. FRIEDMAN. 1990. Risk of Cardiovascular mortality in alcohol drinkers, ex-drinkers and non-drinkers. Am. J. Cardiol. **66:** 1237–1242.
17. GOLDBERG, D.M., S.E. HAHN & J.G. PARKS. 1995. Beyond alcohol: beverage consumption and cardiovascular mortality. Clin. Chim. Acta **237:** 155–187.
18. KANNEL, W.B. & R.C. ELLISON. 1996. Alcohol and coronary heart disease: the evidence for a protective effect. Clin. Chim. Acta **246:** 59–76.
19. CRIQUI, M.H. 1996. Alcohol and coronary heart disease: consistent relationship and public health implications. Clin. Chim. Acta **246:** 51–57.
20. FUHRMAN, B., A. LAVY & M. AVIRAM. 1995. Consumption of red wine with meals reduces the susceptibility of human plasma and low-density lipoprotein to lipid peroxidation. Am. J. Clin. Nutr. **61:** 549–554.
21. GALLI, C., S. COLLI & G. GIANFRANCESHI. 1984. Acute effects of ethanol, caffeine, or both on platelet aggregation, thromboxane formation and plasma free fatty acids in normal subjects. Drug-Nutr. Interact. **3:** 6107.
22. SAIJA, A., D. MARZULLO, M. SCALESE, *et al.* 1995. Flavonoids as antioxidant agents: importance of their interaction with biomembranes. Free Radical Biol. Med. **19:** 481–486.
23. CRIQUI, M.H. & B.L. RINGEL. 1994. Does diet or alcohol explain the French Paradox? Lancet **344:** 1719–1723.
24. ENGELMAN, D., M. WATANABE, R. ENGELMAN, *et al.* 1995. Hypoxic preconditioning preserves antioxidant reserve in the working rat heart. Cardiovasc. Res. **29:** 133–140.
25. YOSHIDA, T., N. MAULIK, R.M. ENGELMAN, *et al.* 1997. Glutathione peroxidase knockout mice are susceptible to myocardial ischemia reperfusion injury. Circulation **96**(11): 216–220.
26. CORDIS, G.A., N. MAULIK & D.K. DAS. 1995. Detection of oxidative stress in heart by estimating the dinitrophenylhydrazine derivative of malonaldehyde. J. Mol. Cell. Cardiol. **27:** 1645–1653.
27. MAULIK, N., H. SASAKI, S. ADDYA & D.K. DAS. 2000. Regulation of cardiomyocyte apoptosis by redox-sensitive transcription factors. FEBS Lett. **485:** 7–12.

28. MAULIK, N., R.M. ENGELMAN, J.A. ROUSOU, *et al.* 1999. Ischemic preconditioning reduces apoptosis by upregulating anti-death gene Bcl-2. Circulation **100**(II): 369–375.
29. CURRIE, R.W. 1987. Effects of ischemia and reperfusion on the synthesis of stress-induced (heat shock) proteins in isolated and perfused rat hearts. J. Mol. Cell. Cardiol. **19:** 795–808.
30. DAS, D.K., R.M. ENGELMAN & Y. KIMURA. 1993. Molecular adaptation of cellular defences following preconditioning of the heart by repeated ischemia. Cardiovasc. Res. **27:** 578–584.
31. HUTTER, M.M., R.E. SIEVERS, V. BARBOSA & C.L. WOLFE. 1994. Heat shock protein induction in rat hearts. A direct correlation between the amount of heat-shock protein induced and the degree of myocardial protection. Circulation **89:** 355–360.
32. KNOWLTON, A.A., P. BRECHER & C.S. APSTEIN. 1991. Rapid expression of heat shock protein in the rabbit after brief cardiac ischemia. J. Clin. Invest. **87:** 139–147.
33. LIU, X., R.M. ENGELMAN, I.I. MORARU, *et al.* 1992. Heat shock: a new approach for myocardial preservation in cardiac surgery. Circulation **86** (Suppl. 2): 358–363.
34. MAULIK, N., X. LIU, G.A. CORDIS, *et al.* 1994. The reduction of myocardial ischemia reperfusion injury by amphetamine is linked with its ability to induce heat shock. Mol. Cell. Biochem. **137:** 17–24.
35. DAS, D.K. & N. MAULIK. 1994. Evaluation of antioxidant effectiveness in ischemia reperfusion tissue injury methods. Methods Enzymol. **233:** 601–610.
36. DAS, D.K. & N. MAULIK. 1995. Protection against free radical injury in the heart and cardiac performance. *In* Exercise and Oxygen Toxicity. C.K. Sen, L. Packer, O. Hanninen, Eds.: 359–388. Elsevier Science. Amsterdam.
37. ARROYO, C.M., J.H. KRAMER, B.F. DICKENS & W.B. WEGLICKI. 1987. Identification of free radicals in myocardial ischemia/reperfusion by spin trapping with nitrone DMPO. FEBS Lett. **221:** 101–104.
38. TOSAKI A., D. BAGCHI, D. HELLEGOUARCH, *et al.* 1993. Comparisons of ESR and HPLC methods for the detection of hydroxyl radicals in ischemic/reperfused hearts. A relationship between the genesis of oxygen-free radicals and reperfusion-induced arrhythmias. Biochem. Pharmacol. **45:** 961–969.
39. CORDIS, G.A., G. MAULIK, D. BAGCHI, *et al.* 1998. Detection of oxidative DNA damage to ischemic reperfused rat hearts by 8-hydroxydeoxyguanosine formation. Mol. Cell. Cardiol. **30:** 1939–1944.
40. CHANCE, B., H. SIES & A. BOVERIS. 1979. Hydroperoxide metabolism in mammalian organs. Physiol. Res. **59:** 527–540.
41. DAS, D.K., N. MAULIK & I.I. MORARU. 1995. Gene expression in acute myocardial stress. Induction by hypoxia, ischemia, reperfusion, hyperthermia and oxidative stress. J. Mol. Cell. Cardiol. **27:** 181–193.
42. STOCKER, R. & E. PETERHANS. 1989. Synergistic interaction between vitamin E and the bile pigments bilirubin and biliverdin. Biochim. Biophys. Acta **1002:** 238–243.
43. SATO, M., G. MAULIK, P.S. RAY, *et al.* 1999. Cardioprotective effects of grape seed proanthocyanidin against ischemic reperfusion injury. J. Mol. Cell. Cardiol. **31:** 1289–1297.
44. SATO, M., P.S. RAY, G. MAULIK, *et al.* 2000. Myocardial protection with red wine extract. J. Cardiovasc. Pharmacol. **35:** 263–268.
45. SATO, M., D. BAGCHI, A. TOSAKI & D.K. DAS. 2001. Grape seed proanthocyanidin reduces cardiomyocyte apoptosis by inhibiting ischemia/reperfusion-induced activation of JNK-1 and c-Jun. Free Radical Biol. Med. **16:** 729–739.
46. RAY, P.S., G. MAULIK, G.A. CORDIS, *et al.* 1999. The red wine antioxidant resveratrol protects isolated rat hearts from ischemia reperfusion injury. Free Radical Biol. Med. **27:** 160–169.
47. PATAKI, T., I. BAK, P. KOVACS, *et al.* Grape seed proanthocyanidins reduced ischemia/reperfusion-induced injury in isolated rat hearts. Am. J. Clin. Nutr. In press.
48. VRHOVSEK, U., S. WENDELIN & R. EDER. 1997. Effects of various vinification techniques on the concentration of cis- and trans-resveratrol and resveratrol glucose isomers in wine. Am. J. Viticul. **48:** 214–220.

Wine, Diet, Antioxidant Defenses, and Oxidative Damage

DRUSO D. PÉREZ,[a] PABLO STROBEL,[a] ROCÍO FONCEA,[a] M. SOLEDAD DÍEZ,[a] LUIS VÁSQUEZ,[a] INÉS URQUIAGA,[a] OSCAR CASTILLO,[b] ADA CUEVAS,[b] ALEJANDRA SAN MARTÍN,[a] AND FEDERICO LEIGHTON[a]

[a]*Department of Cellular and Molecular Biology and* [b]*Department of Nutrition, Catholic University of Chile, Santiago, Chile*

ABSTRACT: Oxidative stress is a central mechanism for the pathogenesis of ischemic heart disease and atherogenesis, for cancer and other chronic diseases in general, and it also plays a major role in the aging process. Dietary antioxidants constitute a large group of compounds that differ in mechanism of action, bioavailability and side effects. A systematic analysis of the role of the various antioxidants in chronic diseases is hampered by the difficulty of employing death or clinical events as end points in intervention studies. Therefore, valid markers for oxidative stress, which show dose response and are sensitive to changes in dietary supply of antioxidants, are potentially of great value when trying to establish healthy dietary patterns, or when one component, like red wine, is evaluated specifically. To evaluate potential oxidative stress markers we have studied the effect of different diets plus wine supplementation on antioxidant defenses and oxidative damage. In three experimental series, four groups of young male university students, one of older men and other of older women, 20–24 volunteers each, received Mediterranean or occidental (high-fat) diets alone or supplemented with red wine, white wine, or fruits and vegetables. Measurements included, leukocyte DNA 8-OH-deoxyguanosine (8OHdG), plasma 7β-hydroxycholesterol, TBARS and well-characterized antioxidants, and plasma and urine polyphenol antioxidants. In all experimental groups that received red wine, consumption resulted in marked decrease in 8OHdG. The changes observed in 8OHdG correlate positively with the other markers of oxidative damage, and shows a clear inverse correlation with the plasma level of well established antioxidants and with measurements of total antioxidant capacity. Urinary total polyphenol content as well as the sum of some specific plasma species also correlate inversely with 8OHdG. In conclusion, the results identify 8OHdG as a very promising general marker of oxidative stress in nutrition intervention studies in humans, and red wine shows a remarkable protective effect.

KEYWORDS: 8-OH-deoxyguanosine; oxidative stress; antioxidants

Address for correspondence: Federico Leighton, M.D., Faculty of Biological Sciences, Catholic University of Chile, Casilla 114-D, Santiago, Chile. Voice: 56-2-222 2577; fax: 56-2-222 2577.
fleighto@genes.bio.puc.cl

Ann. N.Y. Acad. Sci. **957: 136–145 (2002).**

INTRODUCTION

Human organisms are necessarily exposed to a certain degree of oxidative damage, a consequence of aerobic life and energy metabolism. The intensity of this oxidative stress will depend on the rate of generation of free radicals and other active oxygen and nitrogen oxidative species, and on antioxidant defenses. Diet composition influences both, oxidative damage and antioxidant mechanisms and this explains, at least in part, the relationship among diet and some chronic diseases like atherosclerosis and cancer.[1–3] Diets rich in fruits and vegetables are associated with decreased risk of cancer and cardiovascular disease. Biomarkers of oxidative DNA damage and lipid peroxidation can be used to establish the role of antioxidant in this protection and the optimal intake of those antioxidants. Oxidative DNA damage is a significant contributor to the age-related development of some cancers. Also lipid peroxidation play a key role in the development of cardiovascular disease. This is the base of the oxidative hypothesis of atherogenesis that proposed oxidized LDL as the principal agent of damage, being the endothelial cell the main target for the oxidized LDL-mediated damage.[4,5] Many studies have evaluated the effectiveness of antioxidant vitamins supplementation.[1,2] Also the lower incidence of coronary artery disease associated to diets rich in fruits, vegetables and wine, could be largely explained by the antioxidant content of these diets.[3,6] To evaluate the prooxidant and antioxidant effect of currently used diets and the eventual contribution of moderate wine consumption, volunteers were studied, choosing oxidative leukocyte DNA damage (DNA 8-OH-deoxyguanosine content) as a marker of systemic oxidative stress,[7,8] together with TBARS and 7β-hydroxycholesterol as markers of lipid oxidative damage,[9,10] plasma antioxidants levels[11] and TAR and TRAP[12] as markers of antioxidant status, and endothelial function[13,14] as a marker of oxidative stress and atherogenic risk. The results show a close correlation among plasma antioxidant and oxidative damage, especially with plasma levels of vitamin C and polyphenols. This confirm the hypothesis that the antioxidants present in wine, fruits and vegetables; vitamin C, polyphenols and others, contribute to the healthy effect of some diets.

METHODS

Six groups of 20–24 volunteers each, participated in three diet and wine intervention studies, in the years 1998, 1999, and 2000. Subjects were healthy male university students aged 20–28 years, and older men and women aged 53–73 years from a gerontology center, who signed an informed consent. The Faculty of Medicine ethics committee approved the study protocols. For the 1998 and 2000 experimental protocols, the volunteers received for a period of 90 or 60 days, respectively, either a high-fat (occidental diet) or a diet rich in fruits and vegetables (Mediterranean diet).[15] In the 1999 experimental protocol, we supplied for 90 days two high fat diets (40% of calories from fat), which differed in the vegetable oil employed: sunflower oil (PUFA diet, rich in polyunsaturated fatty acids) and olive oil (MUFA diet, rich in monounsaturated fatty acids). They received 14.5 or 4.2 percent calories from PUFA and 12.6 or 22.2 percent calories from MUFA. During the wine or fruit and vegetables supplementation periods, that lasted three or four weeks each, men received 240 mL/day of white wine or red wine and women 120mL/day, or alternatively a supplement of fruits and vegetables, keeping constant the total caloric intake. No other alcoholic beverage was allowed during the entire study.

Blood samples were collected after overnight fasting. Plasma, serum and leukocytes samples were stored following common laboratory procedures until they were analyzed.

The endothelial function was measured noninvasively as flow-mediated vasodilation of the brachial artery, as previously described.[15]

The 8-OHdG and 2-deoxyguanosine (dG) in DNA from peripheral blood leukocytes were measured by HPLC with electrochemical (EC) and diode array (DAD) detection.[16] Briefly, DNA (100–150μg) was solubilized in 20mM acetate buffer (pH 5.3), digested with nuclease P1 (1mg/mL) and hydrolyzed with alkaline phosphatase (1.5 units). The hydrolysate was filtered and DNA bases were separated by HPLC using a Supelco LC-18 column with 50mM KH_2PO_4 (pH 5.5) containing 10% methanol (1 mL/min) as eluent. The 8-OHdG levels were quantitated with an electrochemical detector and compared to the amount of dG quantitated by DAD. The amount of 8-OHdG in the sample was expressed as 8-OHdG per 10^5 dG.

Vitamin E (as α-tocopherol), β-carotene, lycopene, and ubiquinol were measured in plasma using HPLC with electrochemical detection.[11] Serum vitamin C (ascorbate) was detected by a spectrophotometric procedure based on the reduction of ferric chloride.[17] Total plasma antioxidant capacity was determined as TAR (Total Antioxidant Reactivity) and TRAP (Total Reactive Antioxidant Potential) from luminol-enhanced chemiluminescence measurements.[12] Results were expressed as μM Trolox equivalent.

Plasma polyphenols were analyzed by HPLC using electrochemical and diode array detection and results were expressed as μM rutin equivalent.[18] The Folin-Ciocalteau method was used to determine total polyphenols in urine and results expressed as gallic acid equivalents.[19]

Lipid peroxidation was assessed by measuring the levels of thiobarbituric acid reactive substances (TBARS),[20] and plasma 7β-hydroxycholesterol by gas chromatography method.[21]

Results are presented as mean ± SEM. The measurements within the same group, at different times, were compared by paired *t*-test and analysis of variance (ANOVA) for repeated measurements with Bonferroni adjustment. Bivariant correlation analysis was performed according to Pearson. Values $p < 0.05$ were considered statistically significant.

RESULTS AND DISCUSSION

Oxidative Damage Markers

Although free radicals and reactive oxygen species (ROS) are normally generated during cell metabolism they are implicated in the pathogenesis of several human diseases.[7,8] ROS can interact with biological molecules, such as DNA, lipids and proteins, with potentially deleterious consequences. One of the products of DNA oxidation, 8-hydroxy-2-deoxyguanosine (8-OHdG), has been extensively investigated because it can be measured with high sensitivity, and its levels in target tissues are correlated with oxidative stress and with the incidence of cancer.[22,23] Thus 8-OHdG is a useful and biological relevant marker for the study of oxidative stress. Often studies are performed on DNA isolated from blood leukocytes, and it is assumed that changes reflect oxidative stress in organisms.[1]

Lipid peroxidation *in vivo* is likely to contribute the development and progression of atherosclerosis and other chronic diseases.[4,6,9] Lipid oxidation in human can be assessed by measurement of lipid oxidation products, such as 7β-hydroxycholesterol, a major oxidation product of cholesterol, and lipid hydroperoxides as thiobarbituric acid-reactive substances (TBARS). Both apparently are markers for cardiovascular risk.[9,10]

Oxidative stress markers, or biomarkers, are needed to investigate which diet or particular dietary constituents are capable of reducing the oxidative damage to DNA, lipids and proteins *in vivo*.[1]

The present study summarizes the effects of different diets and of wine supplementation on oxidative damage and plasma antioxidants in healthy volunteers. The oxidative damage associated to different diets and the eventual protective effect of red wine supplementation were evaluated with biomarkers, such as 8-OHdG content in human DNA leukocytes, plasma TBARS and 7β-hydroxycholesterol. The values for 8-OHdG in leukocytes correlate significantly with TBARS and with 7β-hydroxycholesterol, 0.215 ($p = 0.002$) and 0.152 ($p = 0.012$), respectively (see FIGURE 1 and TABLE 1). These results support the validity of these markers to assess oxidative damage in humans.

TABLE 1. Correlation of 8-OHdG with oxidative damage, plasma antioxidants and antioxidant capacity; all studies combined

	R	*p*	*n*
TBARS	0.215	0.002	211[a]
7 β-OH cholesterol	0.152	0.012	269[b]
Vitamin C	−0.236	<0.001	363
Vitamin E	−0.158	0.003	363
β-carotene	−0.425	<0.001	363
Lycopene	−0.362	<0.001	363
Ubiquinol	−0.336	<0.001	363
Cat + prot + gallic.	−0.448	<0.001	266[b]
Polyphenol ur.	−0.183	0.001	348
TAR	−0.113	0.032	363
TRAP	−0.240	<0.001	363

[a]Data from 1998 and 2000 studies.
[b]Data from 1998 and 1999 studies.

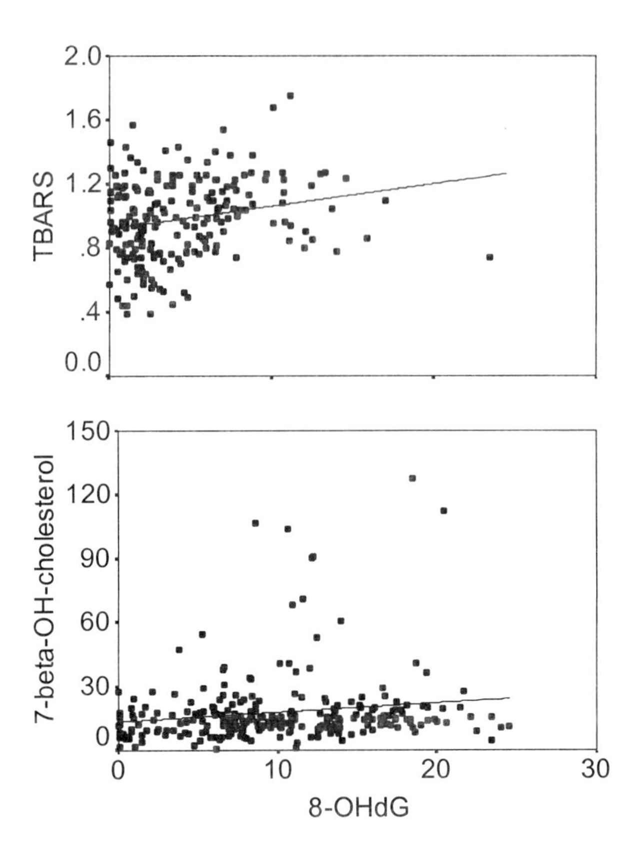

FIGURE 1. Oxidative damage markers. Correlation between 8-OHdG content in DNA leukocytes (8-OHdG per 10^5dG) and plasma concentration of TBARS and 7β-hydroxycholesterol.

Plasma Antioxidants

Several dietary antioxidants have been characterized in terms of their effectiveness, bioavailability and recommended dietary allowances. The main ones are vitamin C, vitamin E, β-carotene, lycopene, and ubiquinol. The changes in the levels of these plasma antioxidants will be reported elsewhere, together with other biochemical parameters. As shown in TABLE 1 and FIGURES 2 and 3, the content of 8-OHdG in leukocytes, in the three experimental groups combined, correlates negatively with each of the plasma antioxidants in a statistically significant fashion. The Pearson correlation coefficients between the level of 8-OHdG in leukocytes and the sum of some plasma polyphenols was −0.448 ($p < 0.001$) the highest value followed by β-carotene, −0.425 ($p < 0.001$); lycopene, −0.362 ($p < 0.001$); ubiquinol, −0.336 ($p < 0.001$); vitamin C, −0.236 ($p < 0.001$) and vitamin E −0.158 ($p < 0.001$). So the correlation among the concentration of catechin, protocatechuic acid, and gallic acid appears at least as effective as the other antioxidants.[24] These antioxidants were supplied by vegetable foods, fruits and also red wine.[25]

The results indicate that 8-OHdG levels are closely correlated with plasma antioxidants. Vitamin C and vitamin E, the main exogenous plasma antioxidants, are capable of reducing the risk of atherogenesis at various levels. Ascorbate is essential and acts directly as an antioxidant while vitamin E, which is also essential, acts via ascorbate, which are both capable of reducing oxidized vitamin E and also of controlling the potential prooxidant capacity of vitamin E.[26] It has been reported that long-term effects of vitamin E, and the combination of vitamins E and C reduced serum 7β-hydroxycholesterol by 50.4% ($p = 0.013$) and 44.0% ($p = 0.041$), respectively; and enhanced the oxidation resistance of isolated lipoproteins and total serum lipids.[27]

There are some epidemiological studies that have evaluated the association between β-carotene and the risk of cardiovascular disease and cancer. Apparently, intervention studies do not show benefits in the prevention of cardiovascular

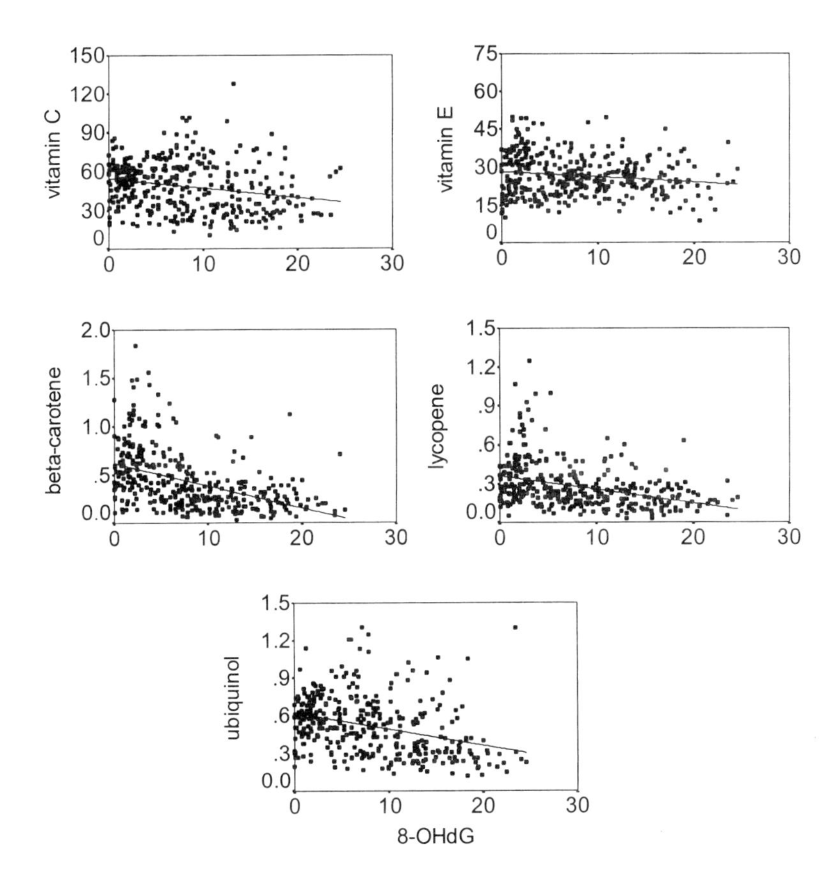

FIGURE 2. Oxidative DNA damage and antioxidants. Correlation between 8-OHdG content in DNA leukocytes (8-OHdG per 10^5dG) and plasma concentration of vitamin C, vitamin E, β-carotene, lycopene and ubiquinol.

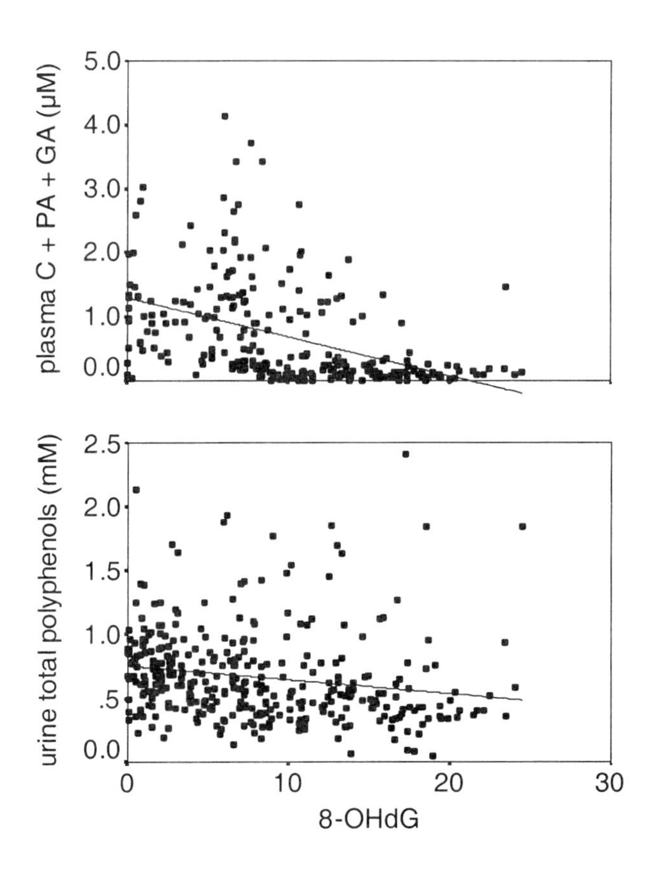

FIGURE 3. Oxidative DNA damage and polyphenols in plasma and urine. Correlation between 8-OHdG content in DNA leukocytes (8-OHdG per 10^5 dG) and plasma concentration of some free polyphenols, represented by the sum of catechin, protocatechuic acid, and gallic acid (µM rutin equivalents), and urine total polyphenols (mM gallic acid equivalents).

disease.[28] Some studies show that high blood levels of carotenoids are associated with a decreased incidence of certain forms of cancer, yet recent results have been contradictory.[29] The role of lycopene, an acyclic form of β-carotene, has been less evaluated with regard to the risk of cardiovascular disease. In a recent study, low levels of plasma lycopene were associated with a 17.8% increment in intima-media thickness of the common carotid artery wall ($p = 0.003$), in men. The authors conclude that low plasma lycopene concentrations are associated with early atherosclerosis, evaluated as increased intima-media thickness of the common carotid artery wall.[30] Also epidemiological studies suggest that high lycopene levels are associated with a decreased risk of prostate cancer.[31] Absorption of dietary carotenoids is incomplete and depends on food mixture and food processing. Lycopene is poorly absorbed from raw tomatoes, but more is taken up from cooked tomatoes or tomato paste.[32]

Extensive use of 8-OHdG as marker of oxidative damage has validated its use with this purpose. Characteristically, different laboratories and different experimental series show differences in basal values or in the magnitude of the changes detected; however there is growing consensus in the explanation for these changes which, with adequate controls, do not invalidate 8-OHdG as a biomarker for oxidative stress.[33]

The DNA protection observed with diets rich in fruits and vegetables as well as with red wine, and to some extent even white wine, is presumably due to the supply of bioavailable antioxidants. Obviously ascorbate is present in fruits and vegetables, but in addition, polyphenol antioxidants should play a major role; they are characteristically abundant in red wine, 1 to 4 g/L, and white wine has 1/10 to 1/5 of this amount.[34] Phytochemicals are considered as valid tools in the prevention of coronary heart disease.[35]

There is considerable epidemiological evidence that diets rich in fruits, grains and vegetables are protective against several human diseases, especially cardiovascular disease and some types of cancer. The vitamins C and E contribute to this protective effect, but there are other constituents that exert additional antioxidants effects, or protect like carotenoids and plant polyphenols.[1]

Plasma Antioxidant Capacity

The total plasma antioxidant capacity values were measured as total antioxidant reactivity (TAR) and total radical antioxidant potential (TRAP). TAR measures the concentration of free radicals that can be initially trapped in the sample. It depends on the quantity and quality of the antioxidants present in the sample. TAR detects hydrosoluble antioxidants, particularly urate and ascorbate. Other components contribute to TAR but have not yet been identified. TRAP measures the total quantity of free radicals that can be trapped in the sample. It depends on the quantity of the antioxidants present in the sample and also detects hydrosoluble antioxidants.[36] Our results (TABLE 1) show that both TAR and TRAP have a significant negative correlation with leukocyte 8-OHdG. This suggests that the plasma antioxidant capacity effectively evaluates protection for oxidative damage.

Polyphenols in Plasma and Urine

Plasma polyphenol levels and urine total polyphenols as presented in FIGURE 3 show a significant inverse correlation with leukocyte 8-OHdG content. The plasma polyphenols illustrated represents the sum of catechin, protocatechuic acid and gallic acid expressed as μM rutin equivalents. These three compounds were selected after a preliminary evaluation of the changes in concentration observed. Other compounds are currently being studied.

The total polyphenols content in urine was measured with a simple method and it could be consider as a procedure for measuring polyphenol metabolism and an index of antioxidant protection status.

Among the large number of polyphenols present in food and wine, some have been detected in plasma and urine;[37] and there is more work to be done in relation to polyphenol metabolism and bioavailability.

Increased levels of plasma polyphenols would protect from oxidative damage, as suggested by our results. Other evidence also supports the hypothesis that wine, as well as fruits and vegetables, decreases oxidative cellular damage to DNA and lipids.[38] In conclusion, the high content of polyphenol antioxidants in red wine contributes to reduce cancer and cardiovascular risk[39] thanks to their antioxidant properties.

ACKNOWLEDGMENT

The work described was supported by Project PBMEC-PUC 1997–2000.

REFERENCES

1. HALLIWELL, B. 1999. Establishing the significance and optimal intake of dietary antioxidants: the biomarker concept. Nutr. Rev. **57:** 104–113.
2. HALLIWELL, B. & J.M.C. GUTTERIDGE. 1999. Free Radicals in Biology and Medicine. Oxford University Press. U.K.
3. RENAUD, S. & M. DE LORGERIL. 1992. Wine, alcohol, platelets, and the French paradox for coronary heart disease. Lancet **339:** 1523–1526.
4. STEINBERG, D., S. PARTHASARATHY, T.E. CAREW, *et al.* 1989. Beyond cholesterol. Modifications of low-density lipoprotein that increase its atherogenicity. N. Engl. J. Med. **320:** 915–924.
5. STEINBERG, D. 2000. Is there a potential therapeutic role for vitamin E or other antioxidants in atherosclerosis? Curr. Opin. Lipidol. **11:** 603–607.
6. AVIRAM, M. 2000. Review of human studies on oxidative damage and antioxidant protection related to cardiovascular diseases. Free Radical Res. **33:** S85–S97.
7. JANSSEN, Y.M.W., B. VAN HOUTEN, P.J.A. BORM & B.T. MOSSMAN. 1993. Cell and tissue responses to oxidative damage. Lab. Invest. **69:** 261–274.
8. AMES, B.N. 1989. DNA damage, ageing and cancer. Free Radical Res. Commun. **7:** 121–128.
9. SALONEN, J.T. 2000. Markers of oxidative damage and antioxidant protection: assessment of LDL oxidation. Free Radical Res. **33:** S41–S46.
10. SALONEN, J.T., K. NYYSSONEN, R. SALONEN, *et al.* 1997. Lipoprotein oxidation and progression of carotid atherosclerosis. Circulation **95:** 840–845.
11. MOTCHNICK, P.A., B. FREI & B. AMES. 1994. Measurement of antioxidants in human blood plasma. Methods Enzymol. **234:** 269–279.
12. LISSI, E.A., M. SALIM-HANNA, C. PASCUAL & M.D. DEL CASTILLO. 1995. Evaluation of total antioxidant potential (TRAP) and total antioxidant reactivity from luminoenhanced chemiluminescence measurements. Free. Radical Biol. Med. **18:** 153–158.
13. VITA, J.A., C.B. TREASURE, E.G. NABEL, *et al.* 1990. Coronary vasomotor response to acetylcholine relates to risk factors for coronary artery disease. Circulation **81:** 491–497.
14. ADAMS, M.R., S. KINLAY, G.J. BLAKE, *et al.* 2000. Atherogenic lipids and endothelial dysfunction: mechanisms in the genesis of ischemic syndromes. Annu. Rev. Med. **51:** 149–167.
15. CUEVAS, A.M., V. GUASCH, O. CASTILLO, *et al.* 2000. A high-fat diet induces and red wine counteracts endothelial dysfunction in human volunteers. Lipids **35:** 143–148.
16. FRAGA, C.G., M.K. SHIGENAGA, J.W. PARK, *et al.* 1990. Oxidative damage to DNA during aging: 8-hydroxy-2′-deoxyguanosine in rat organ DNA and urine. Proc. Natl. Acad. Sci. USA **87:** 4533–4537.
17. DAY, B.R., D.R. WILLIAMS & C.A. MARSH. 1979. A rapid manual method for routine assay of ascorbic acid in serum and plasma. Clin. Biochem. **12:** 22–26.

18. LEIGHTON, F., A. CUEVAS, V. GUASCH, *et al.* 1999. Plasma polyphenols and antioxidants, oxidative DNA damage and endothelial function in a diet and wine intervention study in humans. Drugs Exp. Clin. Res. **25:** 133–141.
19. DUTHIE, G.G., M.W. PEDERSEN, P.T. GARDNER, *et al.* 1998. The effect of whisky and wine consumption on total phenol content and antioxidant capacity of plasma from healthy volunteers. Eur. J. Clin. Nutr. **52:** 733–736.
20. SCHMEDES, A. & G. HOLMER. 1989. A new TBA method for determining free MDA and hydroperoxides selectively as a measure of lipid peroxidation. JAOCS **66:** 813–817.
21. PARK, S.W. & P.B. ADDIS. 1985. Capillary column gas-liquid chromatographic resolution of oxidized cholesterol derivates. Anal. Biochem. **149:** 275–283.
22. HALLIWELL, B. & O.I. ARUOMA. 1991. DNA damage by oxygen-derived species. Its mechanism and measurement in mammalian systems. FEBS Lett. **281:** 9–19.
23. LODOVICI, M., C. CASALINI, R. CARIAGGI, *et al.* 2000. Levels of 8-OHdG as a marker of DNA damage in human leukocytes. Free Radical Biol. Med. **28:** 13–17.
24. URQUIAGA, I. & F. LEIGHTON. 2000. Plant polyphenol antioxidants and oxidative stress. Biol. Res. **33:** 55–64.
25. CROZIER, A., J. BURNS, A.A. AZIZ, *et al.* 2000. Antioxidant flavonols from fruits, vegetables and beverages: measurements and bioavailability. Biol. Res. **33:** 79–88.
26. CARR, A.C., B.Z. ZHU & B. FREI. 2000. Potential antiatherogenic mechanisms of ascorbate (vitamin C) and alpha-tocopherol (vitamin E). Circ. Res. **87:** 349–354.
27. PORKKALA-SARATAHO, E., J.T. SALONEN, K. NYYSSONEN, *et al.* 2000. Long-term effects of vitamin E, vitamin C, and combined supplementation on urinary 7-hydro-8-oxo-2′-deoxyguanosine, serum cholesterol oxidation products, and oxidation resistance lipids in nondepleted men. Arterioscler. Thromb. Vasc. Biol. **20:** 2087–2093.
28. HENNEKENS, C., J. BURING, J. MANSON, *et al.* 1996. Lack of effect of long-term supplementation with beta carotene on the incidence of malignant neoplasms and cadiovascular disease. N. Engl. J. Med. **334:** 1145–1149.
29. COOK, N.R., I.M. LE, J.E. MANSON, *et al.* 2000. Effects of beta-carotene supplementation on cancer incidence by baseline characteristics in the Physicians' Health Study (United States). Cancer Causes Control **11:** 617–626.
30. RISSANEN, T., S. VOUTILAINEN, K. NYYSSONEN, *et al.* 2000. Low plasma lycopene concentration is associated with increased intima-media thickness of the carotid artery wall. Arterioscler. Thromb. Vasc. Biol. **20:** 2677–2681.
31. GIOVANNUCCI, E., A. ASCHERIO, E.B. RIMM, *et al.* 1995. Intake of carotenoids and retinol in relation to risk of prostate cancer. J. Natl. Cancer Inst. **87:** 1767–1776.
32. STAHL, W. & H. SIES. 1996. Lycopene: a biologically-important carotenoid for humans? Arch. Biochem. Biophys. **336:** 1–9.
33. LUNEC, J., K.E. HERBERT, G.D.D. JONES, *et al.* 2000. Development of a quality control material for the measurement of 8-oxo-7,8-dihydro-2′-deoxyguanosine, an in vivo marker of oxidative stress, and comparison of results from different laboratories. Free Radical Res. S27–S31.
34. FRANKEL, E.N., A.L. WATERHOUSE & P.L. TEISSEDRE. 1995. Principal phenolic phytochemicals in selected California wines and their antioxidant activity in inhibiting oxidation of human low-density lipoproteins. J. Agric. Food Chem. **43:** 890–894.
35. VISIOLI, F., L. BORSANI & C. GALLI. 2000. Diet and prevention of coronary heart disease: the potential role of phytochemicals. Cardiovasc. Res. **47:** 419–425.
36. PEREZ, D.D., F. LEIGHTON, A. ASPEE, *et al.* 2000. A comparison of methods employed to evaluate antioxidant capabilities. Biol. Res. **33:** 71–77.
37. DE VRIES, J.H., P.C. HOLLMAN, S. MEYBOOM, *et al.* 1998. Plasma concentrations and urinary excretion of the antioxidant flavonols quercetin and kaempferol as biomarkers for dietary intake. Am. J. Clin. Nutr. **68:** 60–65.
38. THOMPSON, H.J., J. HEIMENDINGER, A. HAEGELE, *et al.* 1999. Effect of increased vegetable and fruit consumption on markers of oxidative cellular damage. Carcinogenesis **20:** 2261–2266.
39. GRONBAEK, M., U. BECKER, D. JOHANSEN, *et al.* 2000. Type of alcohol consumed and mortality from all causes, coronary heart disease, and cancer. Ann. Intern. Med. **133:** 411–419.

Wine Flavonoids Protect against LDL Oxidation and Atherosclerosis

MICHAEL AVIRAM AND BIANCA FUHRMAN

The Lipid Research Laboratory, Technion Faculty of Medicine, The Rappaport Family Institute for Research in the Medical Sciences and Rambam Medical Center, Haifa, Israel

ABSTRACT: We have previously shown that consumption of red wine, but not of white wine, by healthy volunteers, resulted in the enrichment of their plasma LDL with flavonoid antioxidants such as quercetin, the potent free radicals scavenger flavanol, which binds to the LDL via a glycosidic ether bond. This phenomenon was associated with a significant three-fold reduction in copper ion-induced LDL oxidation. The ineffectiveness of flavonoid-poor white wine could be overcome by grape's skin contact for 18 hours in the presence of alcohol, which extracts grape's skin flavonoids. Recently, we observed that the high antioxidant potency of Israeli red wine could be related to an increased content of flavonols, which are very potent antioxidants and their biosynthesis is stimulated by sunlight exposure. To find out the effect (and mechanisms) of red wine consumption on atherosclerosis, we used the apo E deficient (E^0) mice. In these mice, red wine consumption for two months resulted in a 40% decrement in basal LDL oxidation, a similar decrement in LDL oxidizability and aggregation, a 35% reduction in lesion size, and a marked attenuation in the number and morphology of lesion's macrophage foam cells. Red wine consumption resulted in accumulation of flavonoids in the mouse macrophages and these cells oxidized LDL and took up LDL about 40% less than macrophages from placebo-treated mice. Finally, the activity of serum paraoxonase (which can hydrolyze specific lipid peroxides in oxidized LDL and in atherosclerotic lesions) was significantly increased following consumption of red wine by E^0 mice. Red wine consumption thus acts against the accumulation of oxidized LDL in lesions as a first line of defense (by a direct inhibition of LDL oxidation), and as a second line of defense (by paraoxonase elevation and removal of atherogenic lesion's and lipoprotein's oxidized lipids).

KEYWORDS: red wine; flavonoids; polyphenols; antioxidants; lipid peroxidation; LDL; oxidized-LDL; aggregated-LDL; paraoxonase; atherosclerosis

OXIDATIVE STRESS AND ATHEROSCLEROSIS

Atherosclerosis is the leading cause of morbidity and mortality in the western world. The early atherosclerotic lesion is characterized by arterial foam cells, which are derived from cholesterol-loaded macrophages.[1,2] Most of the accumulated cholesterol in foam cells originates from plasma low-density lipoprotein (LDL), which is internalized into the cells via the LDL receptor. However, native LDL does not induce cellular cholesterol accumulation, since the LDL receptor activity is down

Address for correspondence: Prof. Michael Aviram, D.Sc., The Lipid Research Laboratory, Rambam Medical Center, Haifa, Israel, 31096. Voice: 972-4-8542970; fax: 972-4-8542130. aviram@tx.technion.ac.il.

regulated by the cellular cholesterol content.[3,4] LDL has to undergo some modifications, such as oxidation, in order to be taken up by macrophages at enhanced rate via the macrophage scavenger receptors, which are not subjected to down regulation by increased cellular cholesterol content.[5–7]

The oxidative modification hypothesis of atherosclerosis proposes that LDL oxidation plays a pivotal role in early atherogenesis.[8–16] This hypothesis is supported by evidence that LDL oxidation occurs *in vivo*[13,17] and contributes to the clinical manifestation of atherosclerosis. Oxidized LDL (Ox-LDL) is more atherogenic than native LDL, since it contributes to cellular accumulation of cholesterol and oxidized lipids, and to foam cell formation.[5,7,17,18] The process of LDL oxidation appears to occur within the arterial wall, and all major cells of the arterial wall, including endothelial cells, smooth muscle cells and monocyte-derived macrophages can oxidize LDL.[19–23] Macrophage-mediated oxidation of LDL is a key event during early atherosclerosis and requires the binding of LDL to the macrophage LDL receptor under oxidative stress.[19] The interaction of LDL with macrophages under oxidative stress activates cellular oxygenases, which can then produce reactive oxygen species (ROS) capable of oxidizing LDL.[20,24] Under oxidative stress not only LDL and the other plasma lipoproteins are oxidized, but cells' lipids peroxidation also takes place, including lipids in arterial macrophages.[25] Such cells, which are enriched with lipid peroxides, (oxidized macrophages) can easily oxidize LDL.[26]

FLAVONOIDS AND CORONARY ARTERY DISEASE

Consumption of flavonoids in the diet was previously shown to be inversely associated with morbidity and mortality from coronary heart disease.[27] The polyphenolic flavonoids consist of two phenylbenzene (chromanol) rings linked through a pyran ring. Different classes of flavonoids are present in different fruits, vegetables, and beverages, such as tea and wine. Flavonoids may prevent coronary artery disease by inhibiting LDL oxidation, macrophage foam cell formation, and atherosclerosis.[28–33] Flavonoids can reduce LDL lipid peroxidation by acting as free radicals scavengers, metal ion chelators, or by sparing LDL-associated antioxidants. Flavonoids can also reduce macrophage oxidative stress by inhibition of cellular oxygenases such as the NADPH oxidase and also by activating cellular antioxidants such as the glutathione system.[34,35] The effect of flavonoids on cell-mediated oxidation of LDL is determined by their accumulation in the lipoprotein on one hand and in arterial cells, such as macrophages, on the other hand.[24,34,36,37] The antioxidant activity of the flavonoids is related to their chemical structure.[38,39]

RED WINE POLYPHENOLIC FLAVONOIDS

The "French Paradox," (i.e., low incidence of cardiovascular events in spite of diet high in saturated fat), was attributed to the regular drinking of red wine in Southern France.[40] Wine has been part of the human culture for over 6,000 years, serving dietary and socio-religious functions.

The beneficial effect of red wine consumption against the development of atherosclerosis was attributed to the antioxidant activity of its polyphenols. In addition to ethanol, red wine contains a wide range of polyphenols derived from the skin of the grape, with important biological activities.[41,42] Red wine contains the flavonols quercetin and myricetin, (10–20 mg/L), the flavanols catechin and epi(gallo)catechin (up to 270 mg/L), gallic acid (95 mg/L), condensed tannins [catechin and epicatechin polymers (2,500 mg/L)], and also polymeric anthocyanidins.

In previous studies, red wine-derived phenolic acids,[43,44] resveratrol,[45] flavonols (quercetin, myricetin),[46–48] catechins,[49,50] and grape extract[51,52] have all been shown to possess very potent antioxidant properties.

WINE FLAVONOIDS CONTENT CONFERS ITS ANTIOXIDANT CAPACITY AGAINST LDL OXIDATION

In Vitro *Studies*

The antioxidative properties of red wine against *in vitro* LDL oxidation[44,45,53–55] could be attributed to its phenolic substances.[54,55] Confirmation that the active antioxidant components in red wine are phenolic compounds was further supported by the finding that ethanol and wine stripped of its phenols no longer inhibited LDL oxidation.[56] Further evidence that the polyphenols content in wine confer its antioxidant capacity comes from the observations that white wine, which is very poor in polyphenols in comparison to red wine, exhibits very limited antioxidant protection against LDL oxidation, when studied *in vitro*, as well as *in vivo*.[57–63] Red wine contains a much higher concentration of polyphenols than white wine,[62] as a result of the longer period of time (weeks) of contact between the juice and the grape skins during the fermentation process. Unlike red wine, white wine is made from the free running juice, with no contact with the grape skins, resulting in a low polyphenol content. Recently, we have demonstrated the production of white wine with red wine–like antioxidant characteristics by increasing the polyphenol content of the white wine.[64] This was achieved by imposing contact with the grape skins for a short period of time in the presence of added alcohol, in order to augment the extraction of the skin polyphenols into the wine. We have analyzed the antioxidant capacity of white wine samples obtained from whole squeezed grapes that were stored for increasing periods of time before skin removal or from whole squeezed grapes to which increasing concentrations of alcohol were added. White wine obtained from whole squeezed grapes that were incubated for 18 hours with 18% alcohol contained a much higher concentration of polyphenols than untreated white wine and exhibited a significant antioxidant capacity, almost like that of red wine (see FIGURE 1). A maximal free radical–scavenging capacity (as measured by a 79% reduction in the optical density at 517 nm of a 1,1 diphenyl-2-picryl-hydrazyl [DPPH] solution) was induced by the white wine sample derived from whole squeezed grapes, which were preincubated with 18% alcohol. This effect was similar to the free radical–scavenging capacity exhibited by a similar concentration of red wine (FIG. 1A). Furthermore, a maximal inhibition by 87% of copper ion-induced LDL oxidation was induced by wine sample derived from whole squeezed grapes that were

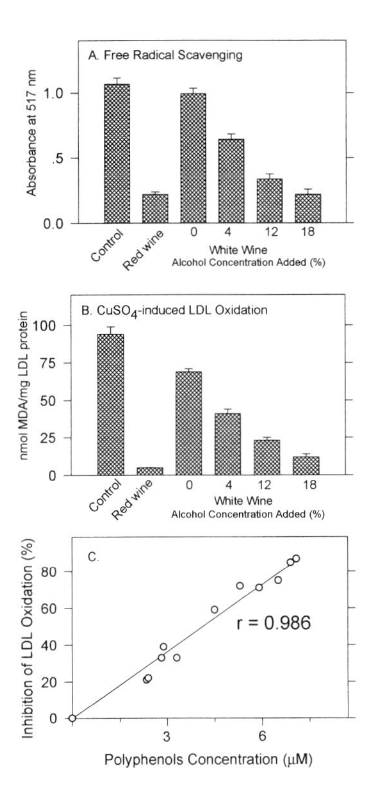

FIGURE 1. The effect of alcohol concentration added to whole squeezed grapes on the polyphenol content of wine and on its antioxidant capacity. Whole squeezed Muscat grapes were incubated for 18 hours with increasing concentrations of alcohol up to 18%, after which the juice was separated from the grape skins and let to ferment into wine. **A.** Aliquots of 25 µl/ml from each wine sample were added to DPPH solution (0.1 mmol/L) and the optical density at 517 nm was recorded after five minutes. Results are expressed as mean ± SD ($n = 3$). **B.** Wine samples at a final concentration of 2 µL/mL were added to LDL (100 mg of protein/L) and incubated with 5 µmol/L $CuSO_4$ for two hours at 37°C. LDL oxidation was measured by the thiobarbituric acid reactive substances (TBARS) assay. Results are expressed as mean ± SD ($n = 3$). **C.** Linear regression analysis of the total polyphenols concentration in wine and the wine-induced inhibition of LDL oxidation.

preincubated with 18% alcohol, a very similar inhibition (94%) to that exhibited by red wine (FIG. 1B). The antioxidant capacity of the above white wine was directly proportional to its polyphenols content (FIG. 1C), and these results are in agreement with previous reports.[58,63]

In Vivo *Studies*

We have previously demonstrated that LDL derived from human male subjects that consumed red wine (400 mL/day for 2 weeks) was more resistant to lipid peroxidation than their LDL obtained before wine consumption.[65] On the contrary, the resistance to oxidation of LDL derived from subjects that consumed the same volume of white wine showed no significant change, in comparison to baseline LDL oxidation rates.[57,65] The administration of 400 mL of red wine to healthy human volunteers for a period of 2 weeks resulted in a substantial prolongation of the lag phase required for the initiation of LDL oxidation (by as much as 130 minutes) (see FIGURE 2A), whereas consumption of a similar volume of white wine had no significant effect on LDL oxidation (FIG. 2B). In parallel, volunteers' LDL, obtained after red wine consumption, had a reduced propensity for copper ion-induced lipid peroxidation in comparison to LDL obtained at baseline, as measured by a 72% decrement in the content of the lipoprotein-associated lipid peroxides (FIG. 2C), whereas after white wine consumption no significant effect was observed (Fig. 2D). The effects of red wine consumption on LDL oxidation could be related to an elevation in the total polyphenols content in the plasma and in the LDL particle. Thus, polyphenolic substances, present in red wine, but not in white wine are absorbed, bind to plasma LDL, and protect the lipoprotein from oxidation.[65]

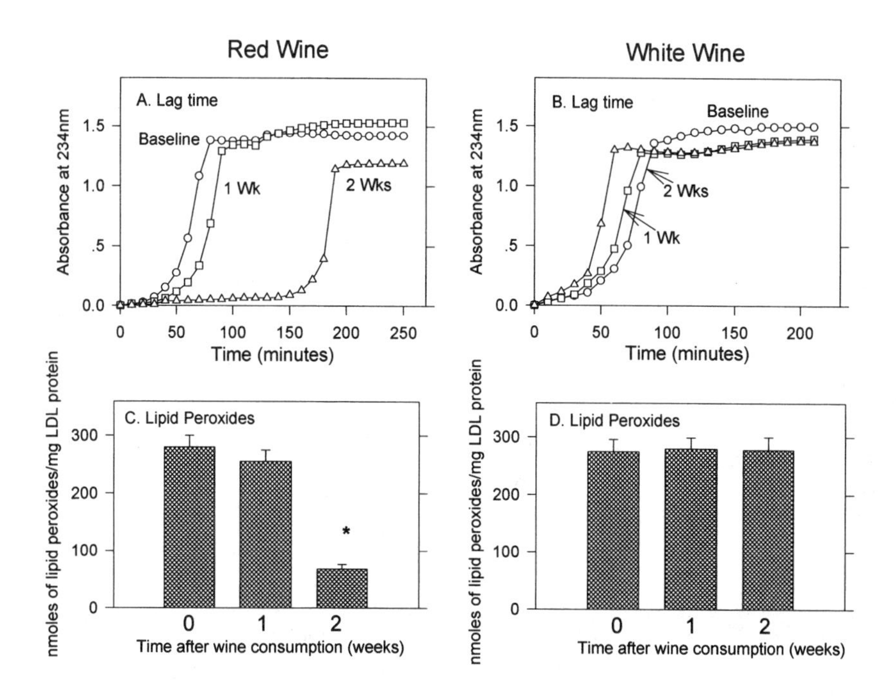

FIGURE 2. The antioxidative effects of red wine versus white wine consumption on LDL oxidation *ex vivo*. LDL (200 μg of protein/ml) isolated before ("0 time", baseline) or after one or two weeks of red wine or white wine consumption, was incubated with $CuSO_4$ (10 μmol/L). **A, B:** LDL oxidation was kinetically monitored by continuously monitoring the absorbance at 234 nm. **C, D:** LDL oxidation was determined by measuring the formation of lipid peroxides. Results are expressed as mean ± SD ($n = 3$). * $p < 0.01$ (vs. time zero).

WINE FLAVONOIDS COMPOSITION AFFECTS ITS ANTIOXIDANT CAPACITY AGAINST LDL OXIDATION

The polyphenol content in white wine obtained from whole squeezed grapes that were incubated for 18 hours with 18% alcohol, even though three-fold greater than untreated normal white wine, was still 4.4-fold less than that found in red wine. However, this polyphenol-rich white wine exhibited antioxidant activity to an extent similar to that of red wine (FIG. 1). These results suggest that not only the quantity of the wine polyphenols is an important determinant of the wine antioxidant capacity, but the diversity of polyphenol types also plays an important role in this effect. Consumption of red wine was shown to inhibit LDL oxidation *ex vivo* in some,[65–67] but not in all human intervention studies,[68,69] and the discrepancies in the results may be related to the variations in polyphenol composition of the wines used. In a recent study by Nigdikar *et al.*,[66] consumption of 400 mL of red wine for two weeks was shown to increase plasma and LDL-associated polyphenols and to protect LDL against copper ion–induced oxidation, as we have previously shown.[65] In the study of Nigdikar *et al.* an increased lag time of only 31% was noted, in comparison to a 290% increase in the lag time obtained in our study.[65] We thus hypothesize that the wine polyphenol composition determines the ability of the wine to inhibit LDL oxidation. Comparison of the composition of the red wine used in our studies[65] with the composition of the wine used in the study performed in UK[66] (both red wines from Cabernet Sauvignon cultivar, one grown in Israel[65] and the other in France[66]) revealed that although total polyphenol content was similar (1,650 and 1,800 mg/L, respectively), the wine from Israel had a six-fold higher content of flavonols (glycosides and aglycones) and of some monomeric anthocyanins.[70]

There is a wide variation in the flavanol content of different red wines throughout the world[71] and a major determinant of this variation is the amount of sunlight to which the grapes are exposed during cultivation.[72] The flavanol synthesis in the skin of the grape is increased in response to sunlight so as to act as a yellow filter against the harmful effect of UV light. Thus, the climatic conditions under which grapes are grown could explain the increased content of flavonols in the Israeli red wines in comparison to the wine studied by Nigdikar *et al.*[66]

ANTIATHEROSCLEROTIC EFFECT OF RED WINE

The direct effect of red wine consumption on the development of atherosclerotic lesions was further studied in the atherosclerotic apolipoprotein E deficient (E^0) mice.[73] These mice were supplemented in their drinking water for a period of six weeks with placebo (1.1% alcohol), or with 0.5 mL of red wine/day per mouse, or with the purified polyphenols catechin or quercetin (50 μg/day/mouse), which are major polyphenols in red wine.[73,74] The atherosclerotic lesions areas in E^0 mice that consumed red wine, catechin or quercetin, were significantly reduced by 40%, 39%, or 46%, respectively, in comparison to the lesion areas measured in the placebo-treated E^0 mice (see FIGURE 3). The mechanisms responsible for reduced atherosclerotic lesion development in E^0 mice that consumed red wine, catechin, or quercetin, can be possibly related to red wine–induced inhibition cellular uptake of lipoproteins, to reduction in oxidative stress, and/or to inhibition of LDL oxidation.

Effect on LDL Uptake by Macrophages

We have investigated macrophage uptake of LDL derived from E^0 mice that consumed red wine, or red wine–derived polyphenols.[73] Incubation of J-774 A.1 macrophages for five hours at 37°C with 10μg of protein/ml of plasma LDL from E^0 mice derived after consumption of catechin, quercetin or red wine, resulted in 31%,

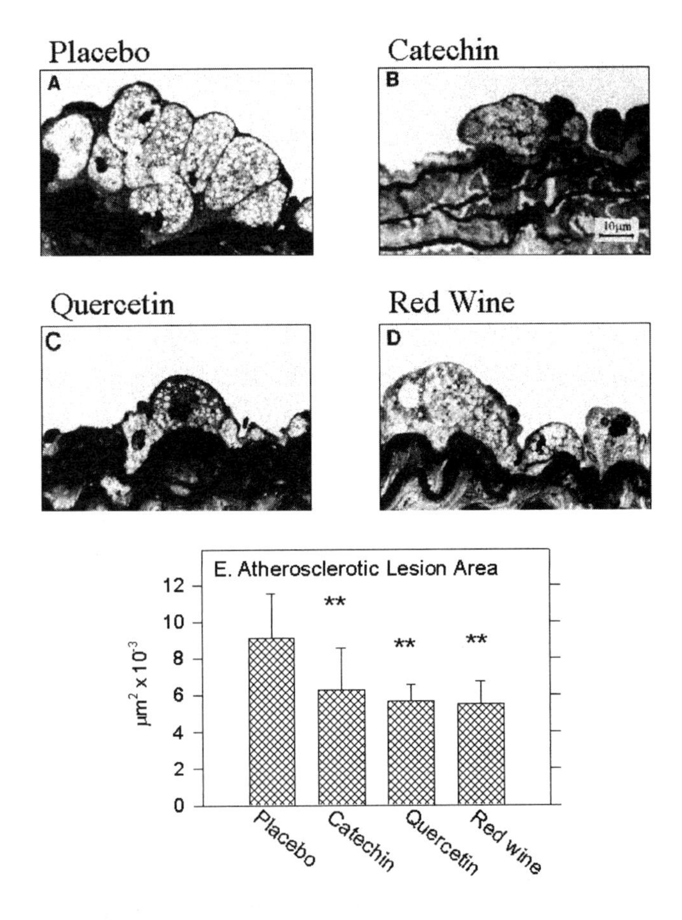

FIGURE 3. Effects of catechin, quercetin or red wine consumption by E^0 mice, on the size of their aortic arch atherosclerotic lesion area. The aortic arch derived from E^0 mice that consumed placebo, catechin, quercetin, or red wine for 6 weeks was analyzed. Photomicrographs of a typical atherosclerotic lesions of the aortic arch following treatment with placebo (**A**), catechin (**B**), quercetin (**C**), or red wine (**D**) are shown. The sessions were stained with alkaline toludine blue. All micrographs are at the same magnification. **D:** The lesion area is expressed in square micrometers ± SD. * $p < 0.05$ versus placebo.

40% or 52% reduced LDL-induced cellular cholesterol esterification respectively, in comparison to LDL derived from the placebo-treated group (see FIGURE 4A). These results suggest that the reduced atherosclerotic lesion formation in E^0 mice that consumed polyphenols may be attributed to reduced uptake of their LDL by arterial macrophages, and hence to attenuation of foam cell formation and atherosclerosis.

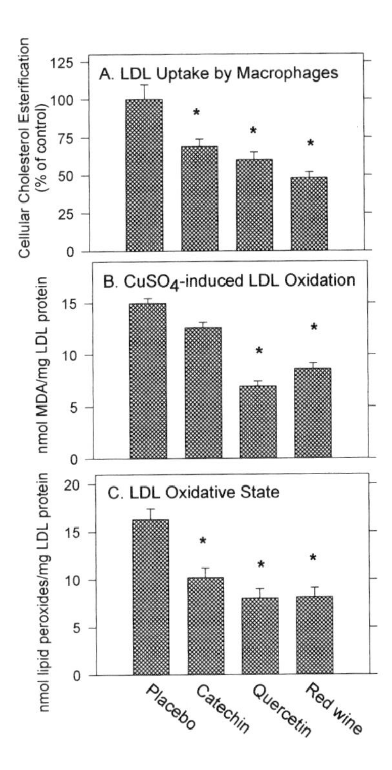

FIGURE 4. Mechanisms responsible for anti-atherosclerotic effects of red wine consumption by E^0 mice. **A:** Macrophage uptake of LDL derived from E^0 mice that consumed catechin, quercetin or red wine. LDL (25 µg of cholesterol/ml) derived from mice that consumed placebo, catechin, quercetin or red wine for six weeks, was incubated for 18 hours at 37°C with J-774 A.1 macrophages. During the last two hours of incubation, [^{3}H]-oleate complexed with albumin was added to medium. The rate of [^{3}H]-cholesteryl oleate formation was determined in the lipid extract of the cells after separation by thin layer chromatography. Results are expressed as mean ± S.D. ($n = 3$). $p < 0.01$ versus placebo. **B:** The effect of catechin, quercetin, or red wine consumption by E^0 mice on the susceptibility of their LDL to oxidation. LDL (100 µg of protein/mL) derived from E^0 mice that consumed placebo, catechin, quercetin or red wine for six weeks, was incubated for two hours at 37°C with 10 µM $CuSO_4$. LDL oxidation was determined by measurement of the LDL-associated thiobarbituric acid reactive substances (TBARS). Results represent mean ± S.D. values ($n = 3$). $p < 0.01$ versus placebo. **C:** The basal oxidative state of LDL derived from E^0 mice that consumed catechin, quercetin or red wine. Lipid peroxides levels were measured in LDL samples (100 µg of protein/mL) derived from E^0 mice that consumed placebo, catechin, quercetin, or red wine for six weeks. Results are expressed as mean ± S.D. ($n = 3$). **$p < 0.01$ versus placebo.

Effect on the Susceptibility of LDL to Oxidation

Since enhanced cellular uptake of LDL is also associated with lipoprotein oxidation, we have further investigated the effect of polyphenols consumption on LDL susceptibility to oxidation. LDL (100μg of protein/ml) derived from E^0 mice, that had their diet supplemented for six weeks with placebo, catechin, quercetin or red wine, was incubated with copper ions (10μM), with the free radical initiator AAPH (5mM), for two hours at 37°C, or with J-774A.1 cultured macrophages under oxidative stress (2μM $CuSO_4$). Copper ion–induced oxidation of LDL derived from E^0 mice after consumption of red wine or quercetin, was delayed by 120 minutes, whereas the onset of lipid peroxidation in LDL derived from E^0 mice that consumed catechin was retarded by only 40 minutes, in comparison to LDL from the placebo group. Furthermore, quercetin or red wine consumption resulted in a 54% and 43% reduction in copper ion–induced oxidation, measured as TBARS (FIG. 4B), or as formation of lipid peroxides, respectively. Similarly, a 83% and 81% reduction in AAPH-induced oxidation, and a 33% and 30% inhibition in macrophage-mediated oxidation, respectively, were noted. These results suggest that consumption of red wine or its flavanol quercetin substantially inhibit LDL oxidative modification by various modes of oxidation inducers.

Effect on LDL Oxidative State

Using the E^0 mice, which are under oxidative stress (contain lipid peroxides in their blood even with no induction of oxidation with some oxidants), we have also analyzed the effect of dietary supplementation of polyphenols on the oxidative state of the mice LDL under basal conditions (not induced by copper ions or by AAPH). LDL derived from E^0 mice that consumed catechin, quercetin or red wine for six weeks, was found to be less oxidized in comparison to LDL isolated from mice that consumed placebo. This was evidenced by a 39%, 48%, and 49% reduced levels of LDL-associated lipid peroxides, respectively (FIG. 4C).

MECHANISMS FOR THE ANTIOXIDATIVE EFFECT OF POLYPHENOLS AGAINST LDL OXIDATION

Scavenging of Free Radicals

To determine the antioxidant capability of red wine, catechin, or quercetin, we performed the DPPH assay. Addition of increasing concentrations of red wine, up to 20μL/mL, to the DPPH solution, decreased the optical absorbance at 517nm from 1.018 to 0.160 O.D. within three minutes. Addition of 100μM of quercetin to the DPPH solution decreased the optical absorbance at 517nm from 1.047 to 0.133 O.D. within eight minutes, whereas during a similar period of time, catechin reduced the optical density of the DPPH solution only to 0.676 O.D units. These results suggest that red wine possesses free radical–scavenging capacity, and that quercetin is more potent in this respect than catechin. These characteristics may contribute to the higher potency of red wine and quercetin, in comparison to catechin, in inhibiting LDL oxidation.

Flavonoids Binding to LDL

To find out whether the antioxidative protection offered to LDL by the red wine-derived polyphenols catechin or quercetin is due to their binding to the lipoprotein, we have preincubated LDL (1 mg of protein/mL) for 18 h at 37°C with 50 μM of the pure polyphenols, or with 10% red wine, followed by removal of the unbound flavonoids (by dialysis). Using a reverse phase-HPLC with UV detection, no significant levels of either catechin or quercetin could be detected in the LDL samples.

As polyphenols interact with surface components of LDL such as fatty acids (ester bond) or sugar residues (ether bond), we performed an alkaline or acidic hydrolysis of the LDL samples prior to the HPLC analysis of the polyphenols. The alkaline hydrolysis of the LDL samples, did not result in the identification of the above flavonoids. However, when acidic hydrolysis was performed on the LDL samples, prior to the HPLC analysis, both catechin (0.35 nmole/mg LDL protein) and quercetin (1.00 nmole/mg LDL protein) were identified, suggesting a glycosidic bond between the flavonoids and the LDL surface protein/lipids components. Using this technique with acidic hydrolysis of the LDL samples, we found that LDL derived after red wine consumption by E^0 mice contained 3.65 nmole of catechin and 3.00 nmole of quercetin/mg LDL protein. No flavonoids could be measured in LDL samples derived from the placebo-treated mice.

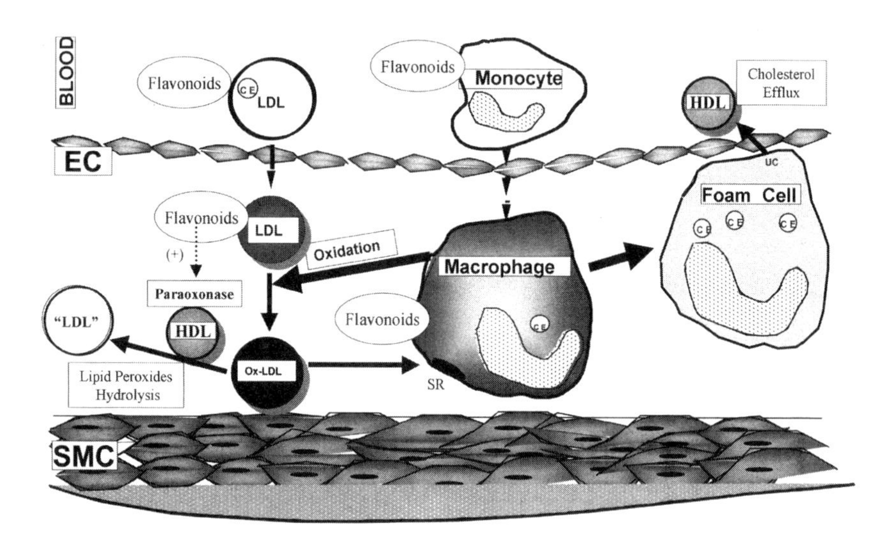

FIGURE 5. Effect of wine flavonoids on LDL oxidation and foam cell formation. Flavonoids can associate directly with LDL, resulting in the inhibition of LDL oxidation. Flavonoids can also associate with arterial cells such as monocytes/macrophages, resulting in the inhibition of macrophage-mediated oxidation of LDL. Furthermore, flavonoids preserve the activity of the enzyme paraoxonase, which further protects lipoproteins from oxidation. Altogether, these effects lead to a reduced formation of macrophage-foam cells, and thus attenuate the development atherosclerotic lesion.

RED WINE AND SERUM PARAOXONASE ACTIVITY

Paraoxonase and Lipid Peroxidation

Human serum paraoxonase (PON 1) is an esterase, which is physically associated with HDL, and also distributed in tissues, such as liver, kidney, and intestine.[75,76] Activities of PON1, which are routinely measured, include hydrolysis of organophosphates, such as paraoxon (the active metabolite of the insecticide parathion), hydrolysis of arylesters, such as phenyl acetate, and also lactonase activities. Paraoxonase has been recently implicated in the protection of LDL from oxidation.[77,78] Human serum paraoxonase activity has been shown to be inversely related to the risk of cardiovascular disease,[79,80] as shown in hypercholesterolemic and diabetic patients.[81–83] HDL-associated PON1 has recently been shown to protect LDL, as well as the HDL itself, against oxidation induced by copper ions or by generators of free radicals,[77,78] and this effect could be related to the hydrolysis of the lipoproteins' oxidized lipids. Inhibition of HDL oxidation by PON1 was shown to preserve the antiatherogenic effect of HDL in reverse cholesterol transport, as shown by its beneficial effect on HDL-mediated macrophage cholesterol efflux.[77]

Effect of Red Wine Consumption on Paraoxonase Activity

Paraoxonase inactivation was shown under oxidative stress[79] and hence, potent antioxidants may preserve the enzyme activity. Serum paraoxonase activity in E^0 mice (whose plasma is oxidized and highly susceptible to oxidation) was found to be lower by 27%, in comparison to the activity observed in control mice (23 ± 3 units/mL in E^0 mice versus 33 ± 2 units/mL in control mice). In serum derived from E° mice after two weeks of catechin, quercetin or red wine consumption, serum paraoxonase activity (measured as arylesterase activity) was higher by 14%, 113%, and 75%, respectively, in comparison to serum obtained from the placebo-treated E^0 mice.[73] The increased levels of serum paraoxonase in E^0 mice that consumed red wine flavonoids can contribute to the reduction in LDL oxidation, secondary to a hydrolysis of LDL-associated lipid peroxides, by paraoxonase.

Red Wine and LDL Aggregation

In addition to LDL oxidation, aggregation of LDL is also an atherogenic modification of the lipoprotein, since aggregated LDL is taken up by macrophages by phagocytosis at increased rate, leading to foam cell formation.[84–87] There are several lines of evidence indicating that LDL aggregation occurs in the arterial wall,[88,89] but little is known about the mechanisms responsible for this modification *in vivo*. It has been shown that extensive oxidation of LDL can lead to its aggregation.[88] LDL aggregation can result from interactions between the lipoprotein hydrophobic domains.[90] Flavonoids, which are multidentate ligands, are able to bind simultaneously to more than one molecule on the lipoprotein surface,[91] and thus can reduce the susceptibility of LDL to aggregation. Dietary supplementation of licorice, or its isoflavan glabridin, or of gingerol, (25 μg flavonoids/day/mouse, for three months) to E^0 mice, resulted in a marked decrease in the susceptibility of the mice's LDL to aggregation.[73,74] Consumption of quercetin, red wine or catechin by E^0 mice also resulted in a reduced susceptibility of their LDL to aggregation by 48%, 50%, or 63%, respectively, in comparison to LDL from placebo-treated mice.

SUMMARY

FIGURE 5 demonstrates our current view on the mechanisms involved in protection against foam cell formation and atherosclerosis by wine flavonoids. Following wine ingestion, wine-derived flavonoids bind to LDL particles and protect them against lipid peroxidation. Under conditions of reduced oxidative stress the activity of the enzyme paraoxonase is preserved, resulting in a further protection of lipoproteins (LDL and HDL) from oxidative modification. Wine flavonoids also protect cells in the arterial wall, including macrophages, from lipid peroxidation, thus reducing macrophage atherogenicity (reduced cellular lipoprotein uptake, decreased macrophage-mediated oxidation of LDL). Wine flavonoids thus reduce macrophage foam cell formation and attenuate atherogenesis.

REFERENCES

1. SCHAFFNER, T., K. TAYLOR, E.J. BARTUCCI, *et al.* 1980. Arterial foam cells with distinctive immuno-morphologic and histochemical features of macrophages. Am. J. Pathol. **100:** 57–80.
2. GERRITY, R.G. 1981. The role of monocytes in atherogenesis. Am. J. Pathol. **103:** 181–190.
3. GOLDSTEIN, J.L. & M.S. BROWN. 1990. Regulation of the mevalonate pathway. Nature **343:** 425–430.
4. BROWN, M.S. & J.L. GOLDSTEIN. 1986. A receptor-mediated pathway for cholesterol homeostasis. Science **232:** 34–47.
5. STEINBERG, D., S. PARTHASARATHY, T.E. CAREW, *et al.* 1989. Beyond cholesterol: modifications of low-density lipoprotein that increase its atherogenicity. N. Engl. J. Med. **320:** 915–924.
6. AVIRAM, M. 1993. Modified forms of low density lipoprotein and atherosclerosis. Atherosclerosis **98:** 1–9.
7. AVIRAM, M. 1993. Beyond cholesterol: modifications of lipoproteins and increased atherogenicity. *In* Atherosclerosis Inflammation and Thrombosis. G.G. Neri Serneri, G.F. Gensini, R. Abbate & D. Prisco, Eds.: 15–36. Scientific Press. Florence, Italy.
8. JIALAL, I. & S. DEVARAJ. 1996. The role of oxidized low density lipoprotein in atherogenesis. J. Nutr. **126**(Suppl.): 1053S–1057S.
9. STEINBERG, D. 1997. Low density lipoprotein oxidation and its pathobiological significance. J. Biol. Chem. **272:** 20963–20966.
10. BERLINER, J.A. & J.W. HEINECKE. 1996. The role of oxidized lipoproteins in atherosclerosis. Free Radical Biol. Med. **20:** 707–727.
11. AVIRAM, M. 1995. Oxidative modification of low density lipoprotein and atherosclerosis. Isr. J. Med. Sci. **31:** 241–249.
12. WITZTUM, J.L. & D. STEINBERG. 1991. Role of oxidized low density lipoprotein in atherogenesis. J. Clin. Invest. **88:** 1785–1792.
13. AVIRAM, M. 1996. Interaction of oxidized low density lipoprotein with macrophages in atherosclerosis and the antiatherogenicity of antioxidants. Eur. J. Clin. Chem. Clin. Biochem. **34:** 599–608.
14. KAPLAN, M. & M. AVIRAM. 1999. Oxidized low density lipoprotein: atherogenic and proinflammatory characteristics during macrophage foam cell formation. An inhibitory role for nutritional antioxidants and serum paraoxonase. Clin. Chem. Lab. Med. **37:** 777–787.

15. Parthasarathy, S., N. Santanam & N. Auge. 1998. Oxidized low-density lipoprotein, a two-faced janus in coronary artery disease? Biochem. Pharmacol. **56:** 279–284.
16. Parthasarathy, S. & S.M. Rankin. 1992. The role of oxidized LDL in atherogenesis. Prog. Lipid Res. **31:** 127–143.
17. Herttuala, S.Y. 1998. Is oxidized low density lipoprotein present in vivo? Curr. Opin. Lipidol. **9:** 337–344.
18. Aviram, M. 1991. The contribution of the macrophage receptor for oxidized LDL to its cellular uptake. Biochem. Biophys. Res. Commun. **179:** 359–365.
19. Aviram, M. & M. Rosenblat. 1994. Macrophage mediated oxidation of extracellular low density lipoprotein requires an initial binding of the lipoprotein to its receptor. J. Lipid Res. **35:** 385–398.
20. Aviram, M., M. Rosenblat, A. Etzioni, *et al.* 1996. Activation of NADPH oxidase is required for macrophage-mediated oxidation of low density lipoprotein. Metabolism **45:** 1069–1079.
21. Witztum, J.L. & D. Steinberg. 1984. Modification of low density lipoprotein by endothelial cells involves lipid peroxidation and degradation of low density lipoprotein phospholipids. Proc. Natl. Acad. Sci. USA **81:** 3883–3887.
22. Heinecke, J.W., H. Rosen & A. Chait. 1984. Iron and copper promote modification of low density lipoprotein by human arterial smooth muscle cells in culture. J. Clin. Invest. **74:** 1980–1984.
23. Parthasarathy, S., D.J. Printz, D. Boyd, *et al.* 1986. Macrophage oxidation of low density lipoprotein generates a modified form recognized by the scavenger receptor. Arteriosclerosis **6:** 505–510.
24. Aviram, M. & B. Fuhrman. 1998. LDL oxidation by arterial wall macrophages depends on the antioxidative status in the lipoprotein and in the cells: role of prooxidants vs. antioxidants. Mol. Cell. Biochem. **188:** 149–159.
25. Fuhrman, B., J. Oiknine & M. Aviram. 1994. Iron induces lipid peroxidation in cultured macrophages, increases their ability to oxidatively modify LDL and affect their secretory properties. Atherosclerosis **111:** 65–78.
26. Fuhrman, B., J. Oiknine, S. Keidar, *et al.* 1997. Increased uptake of low density lipoprotein (LDL) by oxidized macrophages is the result of enhanced LDL receptor activity and of progressive LDL oxidation. Free Radical Biol. Med. **23:** 34–46.
27. Hertog, M.G., D. Kromhout, C. Aravanis, *et al.* 1995. Flavonoid intake and long-term risk of coronary heart disease and cancer in the seven countries study. Arch. Intern. Med. **155:** 381–386.
28. Catapano, A.L. 1997. Antioxidant effect of flavonoids. Angiology **48:** 39–44.
29. Rice-Evans, C.A., N.J. Miller, P.G. Bolwell, *et al.* 1995. The relative antioxidant activities of plant-derived polyphenolic flavonoids. Free. Radical Res. **22:** 375–383.
30. Aviram, M. & B. Fuhrman. 1998. Polyphenolic flavonoids inhibit macrophage-mediated oxidation of LDL and attenuate atherogenesis. Atherosclerosis. **137**(Suppl.): S45–S50.
31. Fuhrman, B. & M. Aviram. 2001. Polyphenols and flavonoids protect LDL against atherogenic modifications. *In* Handbook of Antioxidants: Biochemical, Nutritional and Clinical Aspects, 2nd edit., Chap. 16. 303–336.
32. Fuhrman, B. & M. Aviram. 2001. Anti-atherogenicity of nutritional antioxidants. Drugs **4:** 82–92.
33. Fuhrman, B. & M. Aviram. 2001. Flavonoids protect LDL from oxidation and attenuate atherosclerosis. Curr. Opin. Lipidol. **12:** 41–48.
34. Rosenblat, M., P. Belinky, J. Vaya, *et al.* 1999. Macrophage enrichment with the isoflavan glabridin inhibits NADPH oxidase-induced cell mediated oxidation of low density lipoprotein. J. Biol. Chem. **274:** 13790–13799.
35. Elliott, A.J., S.A. Scheiber, C. Thomas & R.S. Pardini. 1992. Inhibition of glutathione reductase by flavonoids. A structure-activity study. Biochem. Pharmacol. **44:** 1603–1608.
36. Hsiech, R., B. German & J. Kinsella. 1988. Relative inhibitory potencies of flavonoids on 12-lipoxygenase of fish gill. Lipids **23:** 322–326.
37. Luiz da Silva, E., T. Tsushida & J. Terao. 1998. Inhibition of mammalian 15-lipoxygenase-dependent lipid peroxidation in low density lipoprotein by quercetin and quercetin monoglucosides. Arch. Biochem. Biophys. **349:** 313–320.

38. RICE-EVANS, C.A., N.J. MILLER & G. PAGANGA. 1996. Structure-antioxidant activity relationships of flavonoids and phenolic acids. Free Radical Biol. Med. **20:** 933–956.
39. VAN ACKER, S.A.B.E., D.J. VAN-DEN BERG, M.N.J.L. TROMP, *et al.* 1996. Structural aspects of antioxidants activity of flavonoids. Free Radical Biol. Med. **20:** 331–342.
40. RENAUD, S. & M. DE LORGERIL. 1992. Wine alcohol, platelets and the French paradox for coronary heart disease. Lancet **339:** 1523–1526.
41. SOLEAS, G.J., E.P. DIAMANDIS & D.M. GOLDBERG. 1997. Wine as a biological fluid: history, production, and role in disease prevention. J. Clin. Lab. Anal. **11:** 287–313.
42. HERTOG, M.G.L., P.C.H. HOLLMAN & B. VAN DE PUTTE. 1993. Content of potentially anticarcinogenic flavonoids of tea infusions, wines, and fruit juices. J. Agric. Food. Chem. **41:** 1242–1246.
43. NARDINI, M., M. D'AQUINO, G. TOMASSI, *et al.* 1995. Inhibition of human low density lipoprotein oxidation by caffeic acid and other hydroxycinnamic acid derivatives. Free. Radical Biol. Med. **19:** 541–552.
44. ABU-AMSHA, R., K.D. CROFT, I.B. PUDDEY, *et al.* 1996. Phenolic content of various beverages determines the extent of inhibition of human serum and low density lipoprotein oxidation in vitro: identification and mechanism of some cinnamic derivatives from red wine. Clin. Sci. **91:** 449–458.
45. FRANKEL, E.N., A.L. WATERHOUSE & J.E. KINSELLA. 1993. Inhibition of human LDL oxidation by resveratrol. Lancet **341:** 1103–1104.
46. DE WHALLEY, C.V., S.M. RANKIN, R.S. HOULT, *et al.* 1990. Flavonoids inhibit the oxidative modification of low density lipoproteins by macrophages. Biochem. Pharmacol. **39:** 1743–1750.
47. MANACH, C., C. MORAND, O. TEXIER, *et al.* 1995. Quercetin metabolites in plasma of rats fed diets containing rutin or quercetin. J. Nutr. **125:** 1911–1922.
48. VINSON, J.A., Y.A. DABBAGH, M.M. SERRY & J. JANJ. 1995. Plant flavonoids, especially tea flavonols, are powerful antioxidants using an in vitro model for heart disease. J. Agric. Food. Chem. **45:** 2800–2802.
49. MANGIAPANE, H., J. THOMSON, A. SALTER, *et al.* 1992. The inhibition of the oxidation of low density lipoprotein by (+)-catechin, a naturally occurring flavonoid. Biochem. Pharmacol. **43:** 445–450.
50. SALAH, N., N.J. MILLER, G. PAGANGA, *et al.* 1995. Polyphenolic flavanols as scavengers of aqueous phase radicals and as chain-breaking antioxidants. Arch. Biochem. Biophys. **322:** 339–346.
51. LANNINGHAMFOSTER, L., C. CHEN, D.S. CHANCE & G. LOO. 1995. Grape extract inhibits lipid peroxidation of human low density lipoprotein. Biol. Pharm. Bull. **18:** 1347–1351.
52. RAO, A.V. & M.T. YATCILLA. 2000. Bioabsorption and in vivo antioxidant properties of grape extract BioVin: a human intervention study. J. Med. Food **3:** 15–22.
53. MIYAGI, Y., K. MIWA & H. INOUE. 1997. Inhibition of human low density lipoprotein oxidation by flavonoids in red wine and grape juice. Am. J. Cardiol. **80:** 1627–1631.
54. FRANKEL, E.N., J. KANNER, J.B. GERMAN, *et al.* 1993. Inhibition of oxidation human low density lipoprotein by phenolic substances in red wine. Lancet **341:** 454–457.
55. FRANKEL, E.N., A.L. WATERHOUSE & P.L. TEISSEDRE. 1995. Principal phenolic phytochemicals in selected Californian wines and their antioxidant activity in inhibiting oxidation of human low-density lipoproteins. J. Agric. Food Chem. **43:** 890–893.
56. KERRY, N.L. & M. ABBEY. 1997. Red wine and fractionated phenolic compounds prepared from red wine inhibit low density lipoprotein oxidation in vitro. Atherosclerosis **135:** 93–102.
57. FUHRMAN B. & M. AVIRAM. 1996. White wine reduces the susceptibility of low density lipoprotein to oxidation. Am. J. Clin. Nutr. **63:** 403–404.
58. LAMUELA-RAVENTOS, R.M. & M.C. DE LA TORRE-BORONAT. 1999. Beneficial effects of white wines. Drugs Exp. Clin. Res. **25:** 121–124.
59. PAGANGA, G., N. MILLER & C.A. RICE-EVANS. 1999. The polyphenolic content of fruit and vegetables and their antioxidant activities. What does a serving constitute? Free Radical Res. **30:** 153–162.
60. RIFICI, V.A., E.M. STEPHAN, S.H. SCHNEIDER & A.K. KHACHADURIAN. 1999. Red wine inhibits the cell-mediated oxidation of LDL and HDL. J. Am. Coll. Nutr. **18:** 137–143.

61. SERAFINI, M., G. MAIANI & A. FERRO-LUZZI. 1998. Alcohol-free red wine enhances plasma antioxidant capacity in humans. J. Nutr. **128:** 1003–1007.
62. VINSON, J.A. & B.A. HONTZ. 1995. Phenol antioxidant index: comparative antioxidant effectiveness of red and white wines. J. Agric. Food Chem. **43:** 401–403.
63. TUBARO, F., P. RAPUZZI & F. URSINI. 1999. Kinetic analysis of antioxidant capacity of wine. BioFactors **9:** 37–47.
64. FUHRMAN, B., N. VOLKOVA & M. AVIRAM. 2001. White wine with red wine-like properties: increased extraction of grape skin polyphenols improves the antioxidant capacity of the derived white wine. J. Agric. Food Chem. **49:** 3164–3168.
65. FUHRMAN, B., A. LAVY & M. AVIRAM. 1995. Consumption of red wine with meals reduces the susceptibility of human plasma and LDL to undergo lipid peroxidation Am. J. Clin. Nutr. **61:** 549–554.
66. NIGDIKAR, S.V., N. WILLIAMS, B.A. GRIFFIN & A.H. HOWARD. 1998. Consumption of red wine polyphenols reduces the susceptibility of low density lipoproteins to oxidation in vivo. Am. J. Clin. Nutr. **68:** 258–265.
67. CHOPRA, M., P.E.E. FITZSIMONS, J.J. SRAIN, *et al.* 2000. Non-alcoholic red wine extract and quercetin inhibit LDL oxidation without affecting plasma antioxidant vitamins and carotenoid concentrations. Clin. Chem. **46:** 1162–1170.
68. DE RIJKE, Y.B., P.N. DEMACKER, N.A. ASSEN, *et al.* 1996. Red wine consumption does not affect oxidisability of low-density lipoproteins in volunteers. Am. J. Clin. Nutr. **63:** 329–334.
69. CACCETTEA, R.A., K.D. CROFT, L.J. BEILIN & I.B. PUDDEY. 2000. Ingestion of red wine significantly increases plasma phenolic acid concentrations but does not acutely affect *ex-vivo* lipoprotein oxidizability. Am. J. Clin. Nutr. **71:** 67–74.
70. HOWARD, A., C. MRIDULA, D.I. THURNHAM, *et al.* 2002. Red wine consumption and inhibition of LDL oxidation. Medical Hypotheses. In press.
71. MCDONALD, M.S., M. HUGHES, J. BURNS, *et al.* 1998. Survey of free and conjugated myricetin and quercetin content of red wines of different geographical origin. J. Agric. Food Chem. **46:** 368–375.
72. PRICE, S.F., P.J. BREEN, M. VALLADAO & B.T. WATSON. 1995. Clusters sun exposure and quercetin in Pinot noir grapes and wines. Am. J. Enol. Viticulture **46:** 187–194.
73. HAYEK, T., B. FUHRMAN, J. VAYA, *et al.* 1997. Reduced progression of atherosclerosis in the apolipoprotein E deficient mice following consumption of red wine, or its polyphenols quercetin or catechin, is associated with reduced susceptibility of LDL to oxidation and aggregation. Arterioscler. Thromb. Vasc. Biol. **17:** 2744–2752.
74. AVIRAM, M., T. HAYEK & B. FUHRMAN. 1997. Red wine consumption inhibits LDL oxidation and aggregation in humans and in atherosclerotic mice. BioFactors **6:** 415–419.
75. MACKNESS, M.I., B. MACKNESS, P.N. DURRINGTON, *et al.* 1996. Paraoxonases biochemistry, genetics and relationship to plasma lipoproteins. Curr. Opin. Lipidol. **7:** 69–76.
76. LA DU, B.N., S. ADKINS, C.L. KUO & D. LIPSIG. 1993. Studies on human serum paraoxonase/arylesterase. Chem. Biol. Interact. **87:** 25–34.
77. AVIRAM, M., M. ROSENBLAT, C.L. BISGAIER, *et al.* 1998. Paraoxonase inhibits high density lipoprotein (HDL) oxidation and preserves its functions: a possible peroxidative role for paraoxonase. J. Clin. Invest. **101:** 1581–1590.
78. AVIRAM, M., S. BILLECKE, R. SORENSON, *et al.* 1998. Paraoxonase active site required for protection against LDL oxidation involves its free sulfhydryl group and is different from that required for its arylesterase/paraoxonase activities: selective action of human paraoxonase allozymes Q and R. Arterioscler. Thromb. Vasc. Biol. **18:** 1617–624.
79. AVIRAM, M. 1999. Does paraoxonase play a role in susceptibility to cardiovascular disease? Mol. Med. **5:** 381–386.
80. LA DU, B.N., M. AVIRAM, S. BILLECKE, *et al.* 1999. On the physiological role(s) of the paraoxonases. Chem. Biol. Interact. **119/120:** 379–388.
81. MACKNESS, M.I., D. HARTY, D. BHATNAGAR, *et al.* 1991. Serum paraoxonase activity in familial hypercholesterolaemia and insulin-dependent diabetes mellitus. Atherosclerosis **86:** 193–197.
82. ABBOTT, C.A., M.I. MACKNESS, S. KUMAR, *et al.* 1995. Serum paraoxonase activity, concentration, and phenotype distribution in diabetes mellitus and its relationship to serum lipids and lipoproteins. Arterioscler. Thromb. Vasc. Biol. **15:** 1812–1818.

83. GARIN, M.C., R.W. JAMES, P. DUSSOIX, *et al.*, 1997. Paraoxonase polymorphism Met-Leu54 is associated with modified serum concentrations of the enzyme. A possible link between the paraoxonase gene and increased risk of cardiovascular disease in diabetes. J. Clin. Invest. **99:** 62–66.
84. HOFF, H.F. & Y. O'NEIL. 1991. Lesion-derived low density lipoprotein and oxidized low density lipoprotein share a lability for aggregation, leading to enhanced macrophage degradation. Arterioscler. Thromb. Vasc. Biol. **11:** 1209–1222.
85. HOFF, H.F., T.E. WHITAKER & Y. O'NEIL. 1992. Oxidation of low density lipoprotein leads to particle aggregation and altered macrophage recognition. J. Biol. Chem. **267:** 602–609.
86. ISMAIL, N.A., M.Z. ALAVI & S. MOORE. 1994. Lipoprotein- proteoglycan complexes from injured rabbit aortas accelerate lipoprotein uptake by arterial smooth muscle cells. Atherosclerosis **105:** 79–87.
87. HURT, E., G. BONDERS & G. CAMEJO. 1990. Interaction of LDL with human arterial proteoglycans stimulates its uptake by human monocyte-derived macrophages. J. Lipid Res. **31:** 443–454.
88. MAOR, I., T. HAYEK, R. COLEMAN & M. AVIRAM. 1997. Plasma LDL oxidation leads to its aggregation in the atherosclerotic apolipoprotein E deficient mice. Arterioscler. Thromb. Vasc. Biol. **17:** 2995–3005.
89. AVIRAM ,M., I. MAOR, S. KEIDAR, *et al.* 1995. Lesioned low density lipoprotein in atherosclerotic apolipoprotein E-deficient transgenic mice and in humans is oxidized and aggregated. Biochem. Biophys. Res. Commun. **216:** 501–513.
90. KHOO, J.C., E. MILLER, P. MCLOUGHLIN & D. STEINBERG. 1990. Prevention of low density lipoprotein aggregation by high density lipoprotein or apolipoprotein A-I. J. Lipid Res. **31:** 645–652.
91. HAGERMAN, A.E. & L.G. BUTLER. 1981. The specificity of proanthocyanidin—protein interactions. J. Biol. Chem. **156:** 4494–4498.

Scientific Aspects That Justify the Benefits of the Mediterranean Diet

Mild-to-Moderate Versus Heavy Drinking

M.I. COVAS, J. MARRUGAT, M. FITÓ, R. ELOSUA, AND C. DE LA TORRE-BORONAT

Lipids and Cardiovascular Research Unit, Municipal Institute for Medical Research (IMIM), 08003 Barcelona, Spain

ABSTRACT: The Mediterranean diet is now recognized as being both limited in toxicity and abundant in nutrient and non-nutrient protective factors. A large body of basic, clinical and epidemiological studies have been developed in recent years to provide evidence of the benefits of the Mediterranean diet or its components on health. Evidence-based medicine ranks randomized controlled clinical trials as providing the highest level of evidence and expert opinions the lowest. On the basis of these criteria, the current state of knowledge about Mediterranean diet in primary and secondary prevention of disease and mortality and morbidity as functions of the amount of alcoholic beverage consumption, is reviewed. Efficacy versus effectiveness, the role of basic and animal research, and bioavailability studies providing evidence is also discussed.

KEYWORDS: Mediterranean diet; evidence-based medicine; alcohol; wine

In 1979, Keys *et al.* provided ecological evidence of a reduced risk for coronary heart disease (CHD) associated with the traditional Mediterranean diet, despite a high intake of monounsaturated fat.[1] Since then, several paradoxes have been described in the Mediterranean countries which merit being taken into account to design primary and secondary disease prevention strategies.

In the 1980s, the well-known French paradox was described as a high ingestion of fat with a low incidence of CHD mortality rate.[2] In 1997, an Albanian paradox was also described[3] as a high adult life expectancy in a very-low income population. Consistent with its economic situation, Albania has one of the highest infant mortality rates in Europe. In contrast, adult mortality, including cardiovascular disease mortality, is similar to that of other Mediterranean countries. Age-standardized CHD mortality in males (41 per 100,000 inhabitants) in Albania in 1990 was similar to that in Italy and less than half of that in the U.K. In 1998, a Spanish paradox was described as a high prevalence of cardiovascular risk factors in a population with low incidence of CHD in Girona (Spain)[4,5] in the context of the REGICOR Study.

The REGICOR Study has been registering all myocardial infarction (MI) patients in Girona[4] (509,628 inhabitants), Spain, since 1988. From 1994 to 1998 we conducted

Address for correspondence: Dr. M.I. Covas, Lipids and Cardiovascular Research Unit, Municipal Institute for Medical Research (IMIM), Carrer Dr. Aiguader, 80, 08003 Barcelona, Spain. Voice: 34-932211009; fax: 34-932213237.
mcovas@imim.es

a cross-sectional study to assess the prevalence of cardiovascular risk factors in this population (1,748 participants, 72% of response rate).[5] A comparison of the cumulated incidence rates for acute MI in MONICA Centres from 1985 to 1994[6] showed that the lowest MI incidence rates, both for men and women, were located in Southern European countries. In the case of Spain, we observed similar results in the MONICA-Spain[6] and in the REGICOR[4] studies. The REGICOR was an associated-member with the MONICA Study and the methods used were similar. Both studies were undertaken in two areas of Catalonia, in North-Eastern Spain. TABLE 1 shows a comparison of the prevalence of cardiovascular risk factors in men between populations of a Minnesota cohort[7] and the REGICOR[5] studies. Despite the higher levels of main cardiovascular risk factors, cholesterol, systolic blood pressure, and smoking habits, the incidence of MI was three times lower in the Girona population. As a possible protective factor, which might explain part of the differences, the higher physical activity levels in the Girona population must be pointed out. The same differences between the two populations could be observed in women, with the exception of smoking habits which were higher in the Minnesota population. The incidence of MI was 5.5 times lower in the Girona female population.

Thus, we could speak of a Southern European Mediterranean paradox. To explain this paradox, besides the concept of risk factors, we need to introduce the concept of protective factors for disease. Although part of this Mediterranean paradox could be explained by genetic differences among populations, factors which appear to be the best candidates to explain it are those linked to lifestyle, such as diet, physical activity, and environmental factors, such as social cohesion.

The term Mediterranean diet refers to traditional dietary patterns found in areas of the Mediterranean countries around fifty years ago.[1] There are several variants of Mediterranean diets but some common components can be identified (see FIGURE 1):

1. high consumption of grains, vegetables, fruits, dry fruits and legumes;
2. olive oil as a main source of fat;
3. moderate consumption of fish, poultry, milk and dairy products (specially in the form of cheese or yoghurt);
4. low consumption of meat and meat products.[8]

TABLE 1. Comparative prevalence of myocardial infarction (MI), physical activity, and cardiovascular risk factors in men aged 25 to 74 years from Minnesota (USA) and Girona (Spain)

	Population	
Variable	Minnesota	Girona
Cholesterol (mmol/L)	5.26	5.85
Systolic blood pressure (mm Hg)	123	133
Smoking (%)	26	32
Physical activity (kcal/day)	298	441
MI incidence (per 100,000)	613	206

NOTE: From McGovern *et al.* 1996. N. Engl. J. Med. **334:** 884–890; and Masiá *et al.* 1998. J. Epidemiol. Community Health **52:** 707–715.

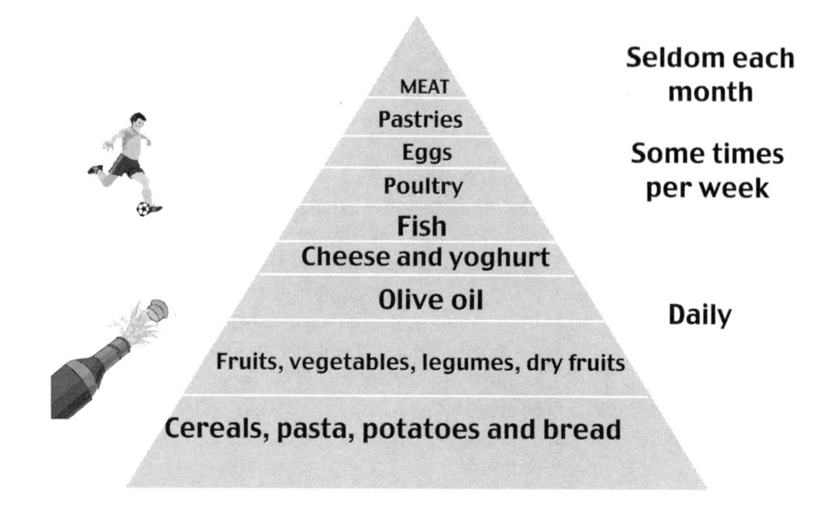

FIGURE 1. Characteristics of the Mediterranean-type diet.

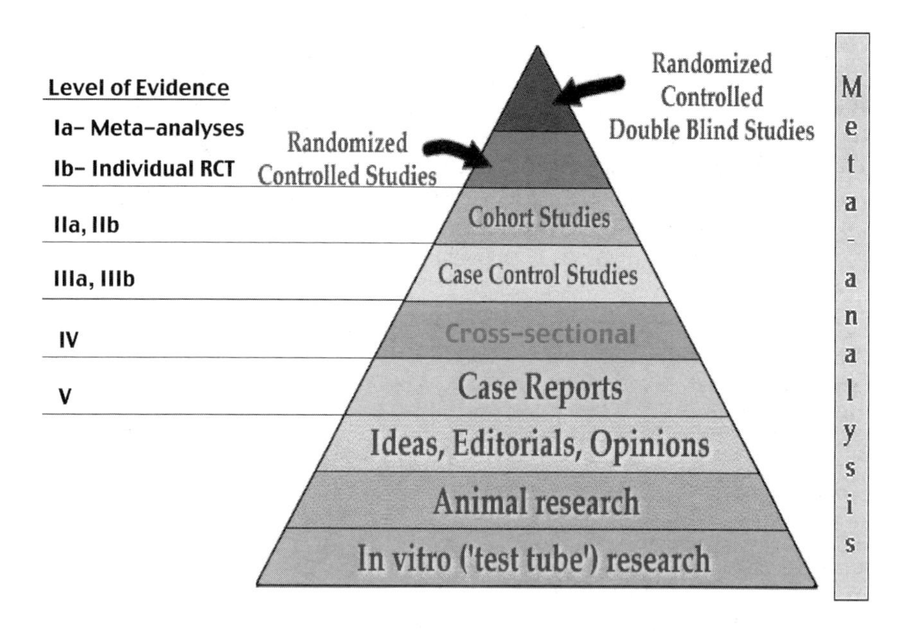

FIGURE 2. Evidence-based medicine. (Adapted from: NHS R&D Centre for Evidence-Based Medicine, U.K.; Canadian Task Force on the Periodic Health Examination; U.S. Preventive Services Task Force; The Swedish Council on Technology Assessment in Health Care (SBU); and the Medical Technologies Evaluation Agency of Catalonia, Spain.)

Associated habits are a high level of physical activity and daily consumption of wine.

Therefore, the question arises as to whether a Mediterranean style diet could be considered to be a Public Health Preventive Intervention. Evaluation of a medical intervention includes first, the assessment of the safety, efficacy, effectiveness and cost-benefit of the procedure, and second, the evaluation of ethical, economic and social consequences from the procedure's adoption.[9]

On the basis of evidence-based medicine, recommendations for the adoption of a medical procedure should stem from the degree of evidence provided by the available scientific information in the literature. Several classifications, based on the scientific rigor of the study design (with close similarities among them), have been proposed. The one presented in FIGURE 2 is based mainly on that of the National Health Service, Research & Development Centre for Evidence-Based Medicine[10] in the U.K. which in turn is based on, and similar to, several others, such as those of the Canadian Task Force on the Periodic Health Examination,[11] the U.S. Preventive Services Tasks Force,[12] the Swedish Council on Technology Assessment in Health Care (SBU),[13] and the Medical Technologies Evaluation Agency from Catalonia, Spain.[9] Grading scientific evidence from I to V is based on the recognized differential capacity of the type of study design to determine scientific evidence, randomized controlled clinical trials (RCTs) being in the upper level of evidence and expert opinions in the lower. Basic research is not included in this classification. Of course, the level of evidence of a particular study depends not only on its type of design, but also on the quality of the study (external and internal validity and statistical power). Typically, the strongest evidence emerges from the agreement of the results of several similarly designed studies. Within each level, meta-analyses are considered to have the highest scientific level, although discrepancies exist about this matter among investigators.

A recommendation for adopting an intervention derives from the scientific evidence, which is classified as adequate, when provided by level I studies, and permitting conclusive recommendations to adopt or not the procedure. Some evidence to adopt a procedure or not is considered to exist when the evidence available only permits the formulation of non-conclusive recommendations. Scientific evidence is classified as insufficient or inadequate when no recommendation can be drawn from the available data.[9,12]

Few studies have tested a Mediterranean-type diet, as a whole in a dietary intervention. We must underline the Lyon Diet Heart Study, which has provided the highest level evidence with a clinical trial in which the Mediterranean type diet not only was shown to reduce CHD events in secondary prevention,[14] but also suggested possible cancer prevention benefits of this diet.[15] Other cohort studies provided evidence about the benefits of a Mediterranean dietary pattern on overall mortality. The Melbourne Study,[16] which replicated the design of a previous Greek study,[17] showed the transferability of the benefits of the Mediterranean diet to other elderly population groups, Anglo-Celts and Greek-Australians. In the Oviedo Study,[18] the adherence to a Mediterranean dietary pattern appeared to be worthwhile in people aged less than 80 years, although it did not provide evidence for over that age. Other cohort studies have provided evidence of the benefits of partial characteristics or some components of the Mediterranean diet. Recently, a cohort study with a large sample size showed that a "prudent dietary pattern," characterized by a high intake of vegetables, fruit, legumes, whole grains, fish and poultry, provided the lowest CHD mortality risk in men.[19]

Given that the first level of evidence is provided by RCTs, one wonders why there are not more RCTs with Mediterranean diet testing? Difficulties in organizing RCTs to test the benefits of the Mediterranean diet, as a whole, on primary end points are:

1. a long follow-up time is needed to assess primary end-points;
2. it is difficult to provide a precise definition of the Mediterranean diet characteristics;
3. a large sample size is also required;
4. compliance of the participants needs to be continuously stimulated, particularly among those whose dietary habits are different from those of the Mediterranean countries.

Alternatives to avoid these limitations have been often used: one of them consists of working with surrogate end points such as lipid level change, lipid or DNA oxidation,[20] and another which consists of testing the effect of individual compounds of the Mediterranean diet (i.e., fruits, olive oil, wine) or nutrients (i.e., fatty acids, vitamins, flavonoids). Nutrients tested have been administered in natural or nutraceutical foods or as isolated compounds. Of course, in dietary intervention studies no double-blind RCTs are possible.

The most closely studied components of the Mediterranean diet are monounsaturated fatty acids (MUFA) and vitamins. Concerning MUFA, as a dietary intervention alone, there are no data on randomized clinical trials with primary end points, such as CHD events. Results of individual cohort studies (level IIb scientific evidence) are discrepant.[21] Several studies, which controlled for a number of potential confounding variables, have reported a protective effect of MUFA against CHD events, i.e., the Nurses Health Study,[22] and the Alpha Tocopherol, Beta-Carotene Cancer Preventive Study.[23] In contrast, some studies, such as the Zuphten Study[24] and the Seven Countries Study,[25] did not find this association, perhaps because they did not control for all possible confounding variables. Results on the protective effect of MUFA intake on cancer development are also controversial or limited.[26, 27] However, a large body of knowledge, including well-controlled RCTs and meta-analyses, have provide evidence of the benefits of MUFA versus saturated fat intake on surrogate cardiovascular risk end points, mainly on serum lipids.[21] There is also growing evidence, although with discrepancies, on the role of MUFA in reducing lipid oxidation.[20,21] Other data studies suggest that MUFA intake may decrease platelet aggregation, increase bleeding time, and increase fibrinolysis, as well as improve insulin sensitivity.[21] Thus, at present, we have adequate evidence that MUFA intake have benefits on cardiovascular risk factors.

Concerning vitamins and β-carotene, the relationship between β-carotene and CHD has been investigated in several observational studies, including ecological, case-control and cohort studies.[28,29] Evidence from case-control studies, level III, supports a role of β-carotene in the prevention of CHD. Controversial results about the protective role of β-carotene in CHD prevention have been obtained in several cohort studies with relative risks ranging from 0.27 to 1.19 for CHD events. Randomized clinical trials with β-carotene supplementation did not show any benefit, on

CHD and cancer prevention.[28,29] The apparent discrepancy between observational studies may be related to consumption of foods rich in β-carotene rather than in β-carotene itself, as foods rich in β-carotene are usually also rich in other antioxidant vitamins. Thus, at present, we have adequate evidence to not recommend isolated β-carotene supplementation and some evidence to recommend the intake of foods rich in β-carotene. A pattern of mixed results applies to vitamin C with insufficient and limited data for recommending its supplementation.[30]

Results from observational cohort studies consistently support an association of vitamin E with reduced CHD risk. Results of large randomized trials are discrepant about the benefits of vitamin E supplementation on CHD and cancer, but suggest that protective effects against CHD could be obtained by supplemental intake levels equal to or greater than 100 IU/d.[31,32] Thus, at present, the evidence is inadequate to recommend vitamin E, supplementation,[31] although some experts differ. In a recent review article Pryor[32] considers that scientific evidence is adequate to recommend a vitamin E supplementation (100–400 IU/d).

Apart from the most closely studied components of the Mediterranean diet, other compounds such as phenolics need further investigation. We are now starting to have evidence about their benefits as antioxidants. However, more information about their benefits on health is needed before making formal daily recommendations. To achieve evidence of the benefits of foods containing phenolic compounds, several possible confounding factors must be taken into account. First of all, non-dietary factors, common to all dietary studies, such as smoking and physical activity. Second, other dietary factors: i.e., the presence of other nutrients or components specific to the tested foods which could mask or enhance the effect (e.g., antioxidant vitamins in vegetables, oleic acid or vitamin E in olive oil, or alcohol in wine). Also, within phenolic compounds themselves several types of phenolics may be responsible for one effect. Some foods contain several groups of phenolic compounds and, in some cases, part of them have not been identified. It is also important to know the bioavailability of phenolic compounds as well as the dose of a specific phenolic compound provided by a food.

On the basis of Bradford Hill criteria[33] to evaluate causality, time sequence of exposure and effect, biological plausibility, and dose-response relationship are some of the criteria which must be tested in observational studies. Here, and particularly with less known dietary compounds and their health benefits, is where the basic science and bioavailability studies become necessary to complete the evidence in order to:

1. establish the plausibility of the hypothesis by analyzing the relationship with end points;
2. check causal relationships in isolated systems, avoiding misinterpretations due to other possible confounding variables;
3. establish the bioavailability of these compounds in humans to support any hypothesis on their possible health benefits; and
4. determine the appropriate dose to be administered.

Food analysis, and its relationship with biological effects, applies particularly to the phenolic compounds because there are few reliable methods to measure them in foods.[30] This fact makes it difficult to accurately estimate the daily intake or the administered doses. As an example, we conducted a study on the bioavailability of tyrosol from ingestion of 50 mL of virgin olive oil in eight healthy volunteers. Urinary tyrosol increased reaching a peak from 0 to 4 hours after olive oil administration.[34] The individual quantities of tyrosol in 24-hour urine of the eight volunteers ranged from 284 to 708 μg. The estimation of tyrosol recovery was quite different depending on the treatment to which olive oil was submitted. Initially, we used a direct phenolic compounds extraction with methanol and water. Levels of tyrosol were 216 μg in 50 mL of virgin olive oil. This result implied a non-realistic urinary tyrosol recovery which ranged from 137% to 327% of the administered dose. However, with the addition of an alcaline and sonication treatments, to mimic intestinal conditions, levels of tyrosol in the olive oil were 1,650 μg. Thus, urinary recovery was estimated to lie between 17% and 43% of the administered dose. These are more realistic recovery levels, similar to those obtained after free tyrosol-enriched olive oil administration in another study.[35]

In consequence, the role of basic science and bioavailability studies in humans is crucial for the interpretation of RCTs and cohort study results.

Some evidence suggests that intake of other nutrients characteristic of the Mediterranean style diet, such as omega-3 fatty acids or folic acid, and foods, such as olive oil, fruits, vegetables, whole grains, and legumes could give protection from the risk of developing CHD, cancer, and other diseases.[20,30,36] Taken together, we could consider that we have some evidence for the benefits of the Mediterranean type diet on health. Some evidence for the beneficial effects of some Mediterranean diet components in CHD and cancer prevention exists, and evidence that intake of these Mediterranean diet components could contribute to reducing the risk factors for these diseases ranges from adequate to some.

Life-style associated with the Mediterranean diet, such as regular physical activity, have also shown very important health benefits.[37] Wine, however, the other daily component of the Mediterranean life-style and diet contains alcohol. We must underline the mortality associated with alcohol-related diseases,[38] as well as the problems associated with its consumption: i.e., accidents or violence.[39] The risks and benefits of alcohol and wine consumption have been extensively discussed and the key issue seems to be how to keep the population within mild-to-moderate alcohol consumption limits.

We have level IIa scientific evidence, i.e., meta-analyses of cohort studies, concerning the relationship between alcohol consumption and all-cause mortality. Results of several meta-analyses summarize the well-known U or J shaped-curves obtained in several cohort studies.[40] One meta-analysis, from data of 16 cohort studies and selected conditions from further 132 epidemiological studies, refers to an increase in the relative risk (RR) of mortality in "non-responsible drinkers," males drinking more than four drinks/day and females drinking more than two drinks/day ($RR > 1.24$).[40] Lower limits for adult males were obtained in another meta-analysis, performed on five cohort studies, in which drinking 43 or more drinks per month (approximately 1.5 drinks/day) was associated with an increased risk of mortality ($RR > 1.45$).[41] Recently, a meta-analysis based on eight cohort studies[39] has reported that doses from one to four drinks/day were associated with a slightly reduced risk of overall mortality and with a reduced risk of CHD.

An increased risk of developing cancer of the upper respiratory and digestive tracts has been reported in two recent meta-analyses, even at a dose of 25 g/day (RR > 1.2).[38,40] Weaker, but significant associations were found for breast cancer in both meta-analyses.[38,40] Data from a recent meta-analysis estimated that the proportion of breast cancer attributable to alcohol intake in USA is 2.1%.[42] Although the opinion of the authors[42] is that this is a small population attributable risk, it should be not be underestimated.

Concerning level I, for ethical reasons it is difficult to perform RCTs including heavy alcohol consumption. RCTs using the opposite, the reduction of alcohol consumption, and its benefits lowering blood pressure both in treated and non-treated hypertensive patients, have been carried out.[43] On surrogated end points a meta-analysis performed on 75 experimental studies showed that a moderate dose of 30 g/day was related to benefits on serum lipids and apolipoproteins and hemostatic factors.[44] The reduction of CHD risk was estimated to be 24.7%.[44]

Concerning the pattern of alcohol consumption, and regarding the current status of alcohol consumption in a Mediterranean life-style country, such as Spain, data from the REGICOR study show that the mean alcohol consumption was 26.6 g/day in men and 5.2 g/day in women in 1995. Concerning men, 53% of them were moderate consumers, drinking less than 40 g/day, and 26% were high consumers, drinking more than or equal to 40 g/day. Concerning women, 36% were moderate consumers, drinking less than a 20 g/day, and only 4% were high consumers, drinking more than or equal to 20 g/day. Alcohol consumption was mainly from wine. These data indicate that in this population with high levels of risk cardiovascular factors[5] and low incidence of CHD,[4] moderate consumption of wine is a usual pattern, particularly in men (see FIGURE 3). These data are in agreement with the results of several cohort studies in which lower risks for CHD and cancer in populations

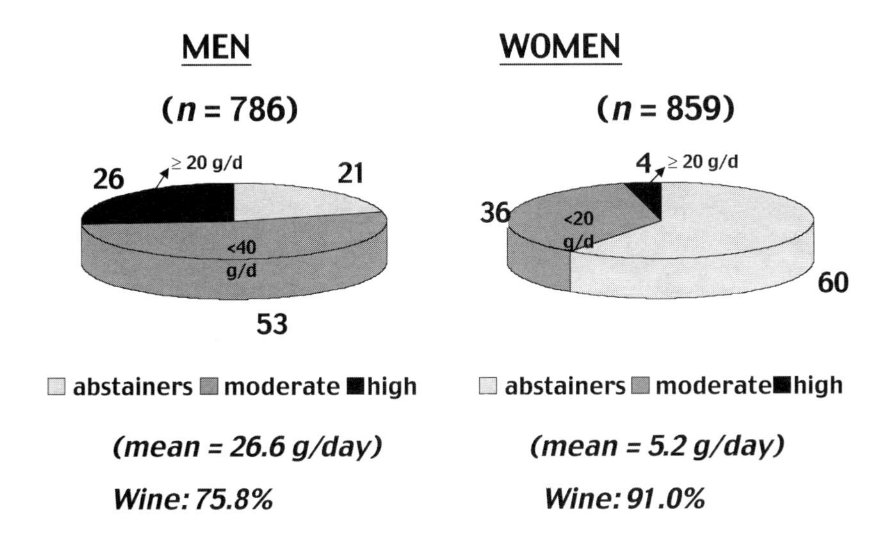

FIGURE 3. Current status of alcohol consumption in a Spanish population—The REGICOR Study.

with a moderate intake of wine have been observed.[45,46] However, other studies included in a meta-analysis did not observe differences associated with the type of alcohol intake on risks or benefits for health.[39]

Given that the pattern of alcohol consumption in this Spanish population was of moderate wine consumption, the interrelationship between this pattern and other risk and protective factors, including dietary factors, was examined. Variables, which were significantly associated in bivariate analyses, were included in logistic regression models. In men, only high MUFA and polyunsaturated fatty acids (PUFA) intakes (i.e., third tercile) were associated with moderate wine consumption. Smoking was not associated after adjustment by age and MUFA and PUFA intakes. In women, only physical activity higher that 150kcal/day was associated with moderate wine consumption. Is physical activity counteracting a non-beneficial effect of moderate wine consumption in women or, is physical activity a confounding factor for the moderate wine consumption protective effect in women? Or, as in the case of men, is moderate wine consumption counteracting the effect of a high fat intake as has been proposed?[47] These questions remain to be elucidated.

Regarding moderate wine consumption one useful approach is to compare its risks and benefits with moderate consumption of other types of alcoholic beverages. Observational studies on this differential effect on primary end points adjusted by possible confounding factors—diet, physical activity, and psychosocial factors—among populations are required. Other approaches may include RCTs with surrogate end points comparing moderate consumption of different types of beverages. Regarding this issue, controversial results on the changes in plasma antioxidant status after red wine, beer and spirits consumption have been described.[48,49] A complementary analysis of the advantages of antioxidant phenolic compounds in wine, may help to differentiate the effects of the antioxidants from the alcoholic content. Several studies have shown an increase in the blood flow through the brachial artery after de-alcoholized red wine ingestion,[50] an inhibition of *in vitro* LDL oxidation after non-alcoholic red wine extract ingestion,[51] and an increased resistance of LDL to oxidation in CHD patients after purple grape juice ingestion, which shows an improvement in the endothelial function.[52]

As a conclusion of this review concerning the scientific assessment of the benefits of the Mediterranean diet, we must underline the recent statements of the American Heart Association Science Advisory, encouraging the realization of RCTs with a Mediterranean-type Step I diet. Literally, it said that "the task at hand is to corroborate the results of the Lyon Diet Heart Study in both primary and secondary prevention models as expediently as possible."[53]

REFERENCES

1. Keys, A., A. Menotti & M.I. Karovene. 1986. The diet and 15-years death rate in the Seven Countries Study. Am. J. Epidemiol. **124:** 903–915.
2. Ducimetiere, P., L. Richard, F. Cambien, *et al.* 1980. Coronary heart disease in middle-aged Frenchmen. Comparisons between Paris Prospective Study, Seven Countries and Pooling Project. Lancet **1:** 1346–1350.
3. Gjonca, A. & M. Bobak. 1998. Albanian paradox, another example of protective effect of Mediterranean lifestyle? Lancet **351:** 835–836.

4. PÉREZ, G., A. PENA, J. SALA, *et al.* 1998. Acute myocardial infarction case fatality, incidence and mortality rates in a population registry in Gerona, Spain, 1990–1992. REGICOR Investigators. Int. J. Epidemiol. **27:** 599-604.
5. MASIÁ, R., A. PENA, J. MARRUGAT, *et al.* 1998. High prevalence of cardiovascular risk factors in Gerona, Spain, a province with low myocardial infraction incidence. REGICOR Investigators. J. Epidemiol. Comm. Health **52:** 707–715.
6. TUNSTALL-PEDOE, H., K. KUULASMAA, M. MAHONEN, *et al.* 1999. Contribution of trends in survival and coronary-event rates to changes in coronary heart disease mortality: 10 year results from 37 WHO MONICA project populations. Monitoring trends and determinants in cardiovascular disease. Lancet **353:** 1547–1557.
7. MCGOVERN, P.G., J.S. PANKOW, E. SHAHAR, *et al.* 1996. Recent trends in acute coronary heart disease-mortality, morbidity, medical care, and risk factors: The Minnesota Heart Survey Investigators. N. Engl. J. Med. **334:** 884–890.
8. TRICHOPOULOU, A. & P. LAGIOU. 1997. Healthy traditional Mediterranean diet: an expression of culture, history, and lifestyle. Nutr. Rev. **55:** 383–389.
9. JOVELL, A.J. & M.D. NAVARRO-RUBIO. 1995. Evaluation of scientific evidence. Med. Clin. (Barc.) **105:** 740–743.
10. SACKETT, D.L. & W.M. ROSENBERG. 1995. On the need for evidence-based medicine. J. Public Health Med. **17:** 330–334.
11. WOLFF, S.M, R.N. BATTISTA, G.M. ANDERSON, *et al.* 1990. Assessing the clinical effectiveness of preventive manoeuvres: analytic principals and systematic methods in reviewing evidence and developing clinical practice recommendations. A report by the Canadian Task Force on the Periodic Health Examination. J. Clin. Epidemiol. **43:** 891–905.
12. U.S. PREVENTIVE TASK FORCE. 1989. Guide to clinical preventive services: an assessment of the effectiveness of 169 interventions. Williams and Wilkins. Baltimore.
13. GOODMAN, C. 1993. Literature Searching and evidence interpretation for assessing health care practices. The Swedish Council of Technology Assessment in Health Care. Stockholm. Sweden.
14. DE LORGERIL, M., P. SALEN, J.L. MARTIN, *et al.* 1999. Mediterranean diet, traditional risk factors and the rate of cardiovascular complications after myocardial infarction. Final report of the Lyon Diet Heart Study. Circulation **99:** 779–785.
15. DE LORGERIL, M., P. SALEN, J.L. MARTIN, *et al.* 1998. Mediterranean dietary pattern in a randomized trial: prolonged survival and possible reduced cancer rate. Arch. Intern. Med. **158:** 1181–1187.
16. KOURIS-BLAZOS, A., C. GNARDELLIS, M.L. WAHLQVIST, *et al.* 1999. Are the advantages of the Mediterranean diet transferable to other populations? A cohort study in Melbourne, Australia. Br. J. Nutr. **82:** 57–61.
17. TRICHOPOULOU, A., A. KOURIS-BLAZOS, M.L.WAHLQVIST, *et al.* 1995. Diet and overall survival in elderly people. Br. Med. J. **311:** 1457–1460.
18. LASHERAS, C., S. FERNANDEZ, A.M. PATTERSON, *et al.* 2000. Mediterranean diet and age with respect to overall survival in institutionalised, non-smoking elderly people. Am. J. Clin. Nutr. **71:** 987–992.
19. HU, F.B., E.B. RIMM, M.J. STAMPFER, *et al.* 2000. Prospective study of major dietary patterns and risk of coronary heart disease in men. Am. J. Clin. Nutr. **72:** 912–921.
20. LAIRON, D. 1998. Mediterranean diet, fats and cardiovascular disease risk: what news? Br. J. Nutr. **82:** 5–6.
21. KRIS-ETHERTON, P.M & THE NUTRITION COMMITTEE. 1999. AHA Science Advisory. Monounsaturated fatty acids and risk of cardiovascular disease. Circulation **100:** 1253–1258.
22. HU, F.B, M.J. STAMPFER, J.E. MANSON, *et al.* 1997. Dietary fat intake and the risk of coronary heart disease in women. N. Engl. J. Med. **337:** 1491–1499.
23. PIETINEN, P., A. ASCHERIO, P. KORHONEN, *et al.* 1997. Intake of fatty acids and risk of coronary heart disease in a cohort of Finnish men: the Alpha Tocopherol, Beta-carotene Cancer Preventive Study. Am. J. Epidemiol. **145:** 876–887.
24. KROMHOUT, D. & C.D.L. COULANDER. 1984. Diet, prevalence and 10-year mortality from coronary heart disease in 871 middle-aged men: the Zuphten Study. Am. J. Epidemiol. **119:** 733–741.

25. KROMHOUT, D., A. MENOTTI, B. BLOEMBERG, *et al.* 1995. Dietary saturated and trans fatty acids and cholesterol and 25-year mortality from coronary heart disease: the Seven Countries Study. Prev. Med. **24:** 308–315.
26. HOLMES, M.D., D.J. HUNTER, G.A. COLDITZ, *et al.* 1999. Association of dietary intake of fat and fatty acids with risk of breast cancer. JAMA **281:** 914–920.
27. BARTSCH, H., J. NAIR & R.W. OWEN. 1999. Dietary polyunsaturated fatty acids and cancers of breast and colorectum: emerging evidence for their role as risk modifiers. Carcinogenesis **20:** 2209–2218.
28. TAVANI, A. & C. LA VECCHIA. 1999. Beta-carotene and risk of coronary heart disease. A review of observational and intervention studies. Biomed. Pharmacother. **3:** 409–416.
29. OMENN, G.S., G.E. GOODMAN, M.D. THORNQUIST, *et al.* 1996. Effects of beta-carotene and vitamin on lung cancer and cardiovascular disease. N. Engl. J. Med. **334:** 1150–1155.
30. VISIOLI, F. 2000. Antioxidants in Mediterranean Diets. *In* Mediterranean Diets. A.P. Simopoulos & F. Visioli, Eds. World Rev. Nutr. Diet. Karger. Basel **87:** 43–55.
31. RIMM, E.B. & M.J. STAMPFER. 2000. Antioxidants for cardiovascular disease. Med. Clin. North Am. **84:** 239–249.
32. PRYOR, W.A. 2000. Vitamin E and heart disease. Basic Science to clinical intervention trials. Free Radical Biol. Med. **28:** 141–164.
33. HILL, A.B. 1953. Observation and experiment. N. Engl. J. Med. **248:** 995–1001.
34. MIRÓ, E., M. FARRÉ, M.I. COVAS, *et al.* 2001. Tyrosol bioavailability in humans after virgin olive oil ingestion. Clin. Chem. **47:** 341–343.
35. VISIOLI, F., C. GALLI, F. BORNET, *et al.* 2000. Olive oil phenolics are dose-dependently absorbed in humans. FEBS Lett. **468:** 159–160.
36. DE LA TORRE-BORONAT, M.C. 1999. Scientific basis for the health benefits of the Mediterranean diet. Drugs Exp. Clin. Res. **XVX:** 1551–1561.
37. MARTÍN , S., R. ELOSUA, M.I. COVAS, *et al.* 1999. Relationship of lipoprotein(a) levels to physical activity and family history of coronary heart disease. Am. J. Public Health **89:** 383–385.
38. CORRAO, G., V. BAGNARDI, A. ZAMBON, *et al.* 1999. Exploring the dose-relationship between alcohol consumption and the risk of several alcohol-related conditions, a meta-analysis. Addiction **94:** 1551–1573.
39. CLEOPHAS, T.J. 1999. Wine, beer and spirits and the risk of myocardial infarction: a systematic review. Biomed. Pharmacother. **53:** 417–423.
40. HOLMAN, C.D., D.R. ENGLISH, E. MILNE, *et al.* 1996. Meta-analysis of alcohol and all-cause mortality: a validation of NHMRC recommendations. Med. J. Aust. **164:** 141–145.
41. LEINO, E.V., A. ROMELSJO, C. SHOEMAKER, *et al.* 1998. Alcohol consumption and mortality. II. Studies of male populations. Addiction **93:** 205–218.
42. TSENG, M., C.R. WEINBERG, D.M. UMBACH, *et al.* 1999. Calculation of population attributable risk for alcohol and breast cancer (United States). Cancer Causes Control **10:** 119–123.
43. BEILIN, L.J., B. PUDDEI & V. BURKE. 1996. Alcohol and hypertension, kill or cure? J. Hum. Hypertens. **10**(Suppl. 2): 51–55.
44. RIMM, E.B., P. WILLIAMS, K. FOSHER, *et al.* 1999. Moderate alcohol intake and lower risk of coronary heart disease: meta-analysis of effects on lipids and haemostatic factors. BMJ **319:** 1523–1528.
45. PRESCOTT, E., M. GRONBAEK, U. BECKER, *et al.* 1999. Alcohol intake and the risk of lung cancer: influence of type of alcoholic beverage. Am. J. Epidemiol. **149:** 463–470.
46. GRONBAEK, M., A. DEIS, T.I.A. SORENSEN, *et al.* 1995. Mortality associated with moderate intakes of wine, beer, or spirits. BMJ **310:** 1165–1169.
47. LEIGHTON, F., A. CUEVAS, V. GUASCH, *et al.* 1999. Plasma polyphenols and antioxidants, oxidative DNA damage and endothelial function in a diet and wine intervention study in humans. Drugs Exp. Clin. Res. **25:** 133–141.
48. DUTHIE, G.G., M.W. PEDERSEN, P.T. GARDNER, *et al.* 1998. The effect of whisky and wine consumption on total phenol content and antioxidant capacity of plasma from healthy volunteers. Eur. J. Clin. Nutr. **52:** 733–736.

49. Van der Gaag, M.S., R. van den Berg, H. van den Berg, *et al.* 2000. Moderate consumption of beer, red wine and spirits has counteracting effects on plasma antioxidants in middle-aged men. Eur. J. Clin. Nutr. **54:** 586–591.
50. Agewall, S., S. Wright, R.N. Doughty, *et al.* 2000. Does a glass of red wine improve endothelial function? Eur. Heart. J. **21:** 74–78.
51. Chopra, M., P.E. Fitzsimons, J.J. Strain, *et al.* 2000. Non-alcoholic red wine extract and quercetin inhibit LDL oxidation without affecting plasma antioxidant vitamin and carotenoid concentrations. Clin. Chem. **46:** 1162–1170.
52. Stein, J.H., J.G. Keevil, D.A. Wiebe, *et al.* 1999. Purple grape juice improves endothelial function and reduces the susceptibility of LDL cholesterol to oxidation in patients with coronary artery disease. Circulation **100:** 1050–1055.
53. Kris-Etherton, P., R.H. Eckel, B.V. Howard, *et al.* 2001. Lyon Diet Heart Study. Benefits of a Mediterranean-Style, National Cholesterol Education Program/American Heart Association Step I dietary pattern on cardiovascular disease. Circulation **103:** 1823–1825.

Antithrombotic Effect of Polyphenols in Experimental Models

A Mechanism of Reduced Vascular Risk by Moderate Wine Consumption

GIOVANNI DE GAETANO,[a] AMALIA DE CURTIS,[b]
AUGUSTO DI CASTELNUOVO,[b] MARIA BENEDETTA DONATI,[a]
LICIA IACOVIELLO,[b] AND SERENELLA ROTONDO[b]

[a]*Centro di Ricerca e Formazione ad Alta Tecnologia nelle Scienze Biomediche, Università Cattolica, 86100 Campobasso, Italy*

[b]*Istituto di Ricerche Farmacologiche Mario Negri, Dipartimento di Medicina e Farmacologia Vascolare, Consorzio Mario Negri Sud, 66030 S. Maria Imbaro (Chieti), Italy*

ABSTRACT: Epidemiological studies have suggested that cardiovascular disease can be decreased by moderate wine consumption, but an overall quantitative estimation of the relationship between wine intake and vascular risk is lacking. A meta-analysis was therefore performed on 19 studies selected on the basis of the availability of specific information on the cardiovascular relative risk (RR) associated with wine consumption. A significant risk reduction (RR: 0.66, 95% CI 0.57–0.75) was associated with moderate (1–2 drinks or 150–300 mL/d) versus no wine consumption. In five studies which excluded ex-drinkers as reference group, the overall RR associated with wine consumption was 0.61 (95% CI 0.57–0.75). A dose-response relation between wine intake and vascular risk resulted in a J-shaped curve, with a significant risk reduction at about 300 mL/d (trend analysis $p = 0.032$). Two studies were also performed to investigate the effects of wine polyphenols on experimental thrombosis in rats. Supplementation for 10 days with alcohol-free red wine—but not white wine or alcohol—induced a significant reduction of stasis-induced venous thrombosis, an effect blunted by NO synthase inhibitor L-NAME. In rats with diet-induced hyperlipidemia, alcohol-free red wine supplementation significantly delayed the thrombotic occlusion of an artificial prosthesis inserted into the abdominal aorta, but did not affect the increased cholesterol and triglyceride levels. TRAP values were significantly higher in animals receiving alcohol-free wine. Altogether these experimental data support an antithrombotic role of polyphenols in the reduced vascular risk associated with moderate wine consumption in man, as shown by our epidemiological studies.

KEYWORDS: thrombosis; cardiovascular disease; wine; alcohol; polyphenols

Address for correspondence: Giovanni de Gaetano, M.D., Ph.D., Via Milano, 56, 66034 Lanciano, Italy. Voice/fax: + 39-0872-714008.
degaetano@cotir.it

Ann. N.Y. Acad. Sci. **957: 174–188 (2002).**

INTRODUCTION

An inverse relationship between light to moderate alcohol consumption and vascular risk has been consistently shown in many epidemiological studies, independently from age, sex, and smoking habits.[1–10] The dose-response relationship between alcohol intake and rate of cardiovascular events and of all-cause mortality has been depicted as a U- or J-shaped curve, in relation to different populations or subgroups, reference groups (abstainers vs. occasional drinkers), range of amounts of alcohol consumed, clinical endpoint.[7,11,12]

The U- or J-shaped relationship between alcohol consumption and clinical events shows that non-drinkers or occasional drinkers have higher incidence and mortality rates than light or moderate drinkers, but rates similar to or lower than heavy drinkers.[13]

Wine intake has been suggested as one possible explanation for the lower than expected coronary heart disease mortality rates in France, despite the prevalence of risk factors (the "French Paradox").[14] Since then, many ecological and observational studies have dealt with the question whether different alcoholic beverages are equivalent in their ability to protect against cardiovascular disease, or a specific beverage, wine in particular, might offer a greater protection, most likely related to its non-alcoholic components.[15–26]

In the first part of this paper we shall summarize the preliminary results of a meta-analysis of epidemiological studies reporting specific data on wine consumption. In the second part, we shall report the experimental results obtained with wine or wine-derived products, to provide evidence for the possible mechanisms underlying specific effects of wine. Other factors that may contribute to the inverse association between moderate wine consumption and cardiovascular risk different from or complementary to the non-alcoholic components will also be briefly discussed.

WINE CONSUMPTION AND VASCULAR RISK REDUCTION

At the present, a rigorous overall estimation of the relative vascular risk associated with any specific alcoholic beverage is lacking.

We have recently performed an overview of the literature[27] aimed at evaluating the relationship between wine consumption and both cardiovascular and cerebrovascular risk and at giving a quantitative estimate of this relationship. More than 50 studies were screened, from which only 19 could be selected that gave quantitative estimation on the relative risk associated with wine consumption (TABLE 1).

The main outcome measure was wine consumption versus the relative risk of morbidity and mortality from cardiovascular disease. The results obtained support the protective role of moderate wine consumption against the risk of vascular events. In fact, pooling data from 13 prospective and case-control studies reporting only relative risk of moderate (one to two drinks per day, 150–300 mL per day) versus no wine consumption, the overall effect was a significant risk reduction of 34% (RR = 0.66; 95% CI 0.57–0.75).

Furthermore, meta-analysis of six additional studies reporting trend analysis for increasing categories of consumption, allowed the definition of a dose-response

TABLE 1A. Characteristics of the studies reporting relative risk of "yes" versus "no" wine consumption

First Author "Yes" versus "No"	Wine intake	Wine intake (mL/d)	RR	95% CI		Type of Cohort	Number of Patients	End Point
Yano, 1977	0 mL/day	0	1			follow up	7,705	CHD total
	≥2	≥2	0.71	0.44	1.16			
Kozararevic, 1980	<1 dr/day	<130	1			follow up	11,121	CHD total
	≥1	≥130	0.68	0.42	1.11			
Friedman, 1986	0 dr/day	0	1			follow up	4,745	CHD mortality
	2	260	0.25	0.06	1.12			
Stampfer, 1988	0 g/day	0	1			follow up	87,526	CHD total
	>5	>54.4	0.4	0.2	0.8			
Rimm, 1991	0 dr/day	0	1			follow up	44,059	CHD total
	2	236	0.98	0.64	1.50			
Klatsky, 1990	0 day/weeks	0	1			follow up	123,840	CHD total
	≥2	≥38.6	0.5	0.4	0.7			
Wannamethee, 1999	<1/7 dr/day	<15.1	1			follow up	7,735	CHD events
	≥1/7	≥15.1	0.92	0.51	1.67			
Rosenberg, 1981	0	0	1			case-control	513/918	MI
	< 4 dr/week	< 74.3	0.5	0.3	0.8			
Kaufman, 1985	0 dr/day	0	1			case-control	2,170/981	MI
	0–2	0–237	1.1	0.6	1.9			

TABLE 1 A/continued.

First Author "Yes" versus "No"	Wine intake	Wine intake (mL/d)	RR	95% CI		Type of Cohort	Number of Patients	End Point
Marques-Vidal, 1996	0	0	1			case-control	561/643	MI
	105.5 mL/day	*105.5 mL/day*	0.86	0.64	1.15			
Sacco, 1999	0 dr/day	0	1			case-control	677/1,139	ischemic stroke
	0.21 dr/day	25.2	0.40	0.23	0.70			
Gaziano, 1999	0 dr/day	0	1			case-control	340/340	MI
	≥0.5	≥59.2	0.58	0.31	1.09			
Thrift, 1999	0 dr/day	0	1			case-control	331/331	intracranial hemorrhage
	>1 dr/month	≥3.5	0.5	0.2	0.9			

NOTES: RR, relative risk; CI, confidence intervals; CHD, coronary heart disease; CAD, coronary artery disease; CVD, cerebrovascular disease; MI, myocardial infarction. When a study did not report the volume of wine (or the amount of ethanol) equivalent to a drink (dr), we considered 1 drink = 130 millilitre (mL) of wine or 12 grams (g) of ethanol.

TABLE 1B. Characteristics of the studies included in the "dose–response" analysis

First Author	Wine intake	Wine intake (mL/d)	RR	95% CI		Type of Cohort	Number of Patients	End Point
Farchi, 1992	22.7 g/day	239.5	1			follow up	1,536	CHD total
	56.4	594.9	0.77	0.34	1.76			
	77.8	820.7	0.67	0.29	1.58			
	108.2	1141.3	1.31	0.64	2.66			
	164.7	1737.3	1.61	0.79	3.31			
Gronbaek, 1995	0 dr/day	0	1			follow up	13,285	CAD + CVD
	1/30	4.2	0.69	0.62	0.77			mortality
	1/7	18.1	0.53	0.45	0.63			
	1–2	126.6–253.2	0.47	0.35	0.62			
	3–5	379.7–632.9	0.44	0.24	0.80			
Truelsen, 1998	0 dr/day	0	1			follow up	13,329	stroke mortality
	1/30	4.2	0.84	0.70	1.02			
	1/7	18.1	0.66	0.50	0.88			
	≥1	≥126.6	0.68	0.45	1.02			

TABLE 1B/continued.

First Author	Wine intake	Wine intake (mL/d)	RR	95% CI		Type of Cohort	Number of Patients	End Point
Renaud, 1999	0 g/day	0	1			follow up	36,250	CAD
	1–21	11.4–240.0	0.88	0.65	1.19			
	22–32	251.4–365.7	0.60	0.45	0.79			
	33–54	377.1–617.1	0.61	0.47	0.79			
	55–98	628.6–1120.0	0.83	0.65	1.07			
	99–131	1131–1497	0.86	0.67	1.19			
	>131	>1497.1	1.13	0.76	1.47			
Tavani, 1996	0 dr/day	0	1			case-control	787/959	MI
	0–2	0–300	1.0	0.7	1.4			
	2–4	300–600	0.9	0.7	1.3			
	4–6	600–900	0.9	0.6	1.4			
	6–7	900–1050	0.7	0.3	1.5			
	>7	≥1050	0.6	0.4	1.0			
Bianchi, 1993	0 dr/day	0	1			case-control	298/685	MI
	0–1	0–150	0.8	0.5	1.2			
	1–2	150–300	1.0	0.6	1.5			
	2–3	300–450	1.7	0.8	3.5			
	>3	>450	2.4	1.0	5.7			

NOTES: RR, relative risk; CI, confidence intervals; CHD, coronary heart disease; CAD, coronary artery disease; CVD, cerebrovascular disease; MI, myocardial infarction. When a study did not report the volume of wine (or the amount of ethanol) equivalent to a drink (dr), we considered 1 drink = 130 millilitre (mL) of wine or 12 grams (g) of ethanol.

relation between wine intake and vascular risk, which resulted in a J-shaped curve, with a statistically significant risk reduction at about 300mL per day (RR = 0.86; 95% CI 0.74–0.99), non-drinkers being the reference group.

The choice of non-drinkers as reference group has often been questioned since this group may include ex-drinkers who have quit because of health problems. Although attention has been drawn in the past years to this possibility, non-drinkers remain, with few exceptions, the baseline comparison group, although various statistical adjustments have been made whenever possible to eliminate the effects of differences in the prevalence of risk factors and pre-existing disease in the various alcohol intake categories. In meta-analyzing the results of the various prospective and case-control studies we had to use the relative risk given by each single study in respect to its own reference group. However, in order to minimize any possible inaccurate evaluation due to the choice of reference group, we performed a sub-analysis, restricted to those studies which excluded ex-drinkers or had other than non-drinkers as reference group. Pooling the results from five such studies,[26,27] the estimated overall risk associated to moderate wine consumption appeared to be significantly reduced by 39% (RR = 0.61; 95% CI 0.57–0.75).

BIOLOGICAL PLAUSIBILITY

Studies showing that alcohol favorably influences hemostatic/thrombotic factors give experimental support to the epidemiological evidence of inverse association between moderate alcohol intake and vascular risk.

The mechanisms through which wine might exert antiatherogenic and antithrombotic effects appear to be distinct from those of alcohol and, at least in part, attributable to biological properties peculiar to its polyphenolic constituents, first of all their recognized antioxidant and radical scavenging properties.[28] Reactive oxygen species produced at site of vascular damage or inflammation can indeed *per se* exacerbate atherogenic and thrombotic processes through the induction of LDL oxidation, stimulation of cell growth and proliferation, and activation of both vascular and blood cells.[29] Different polyphenols have also been shown to modulate the function of cellular components involved in the process of thrombosis in several systems.[18]

Modulation of vascular response, possibly through mechanisms linked to the nitric oxide (NO) pathway, may contribute to the maintenance of blood vessel function.[30] Interference with the arachidonic acid metabolism in both platelets and leukocytes have been reported, which resulted in inhibition of platelet aggregation and reduced synthesis of pro-thrombotic and pro-inflammatory mediators *in vitro* and in experimental models.[18] Furthermore, some polyphenols can modulate specific pathways regulating the expression and activation of genes induced by a variety of agonists; this results in down-regulation of the expression of adhesive molecules and tissue factor activity in both endothelial cells and leukocytes, and ultimately in functional modulation of cell–cell interactions and procoagulant activities.[18,31]

EXPERIMENTAL MODELS: THE ANTITHROMBOTIC ROLE OF WINE POLYPHENOLS

Experimental Venous Thrombosis in Healthy Rats

Bleeding time, *ex vivo* platelet adhesion to collagen, and experimental venous thrombosis were studied in rats after 10 days supplementation in drinking water with either red wine or white wine or alcohol.[32] Red wine supplementation induced significant prolongation of template bleeding time, decrease in platelet adhesion to fibrillar collagen, and reduction in thrombus weight (see FIGURE 1).

Alcohol-free red wine was almost as effective as the original beverage, while white wine was almost ineffective. The NO synthase inhibitor L-NAME prevented the effects of red wine, suggesting a role for alcohol-independent factors interfering with the NO pathway (see TABLE 2). The difference in polyphenolic contents between red wine and white wine and the increased total radical-trapping antioxidant activity (TRAP) found in the plasma of rats given red wine (see FIGURE 2) strongly favor the role of polyphenols rather than of alcohol in the observed effects.

Experimental Arterial Thrombosis in Hyperlipidemic Rats

A second study was performed to investigate the effects of alcohol-free red wine on experimental arterial thrombosis in rats with diet-induced hypercholesterolemia.[33]

A colony of spontaneously normolipidemic rats (FNL) were fed a 2% cholesterol-rich-diet for six months (FNL+D). After five months on this diet, a group of rats was supplemented for one month with alcohol-free red wine (FNL+D+W). Another group of rats was supplemented with alcohol-free red wine starting simultaneously with the cholesterol-rich-diet administration. Cholesterol, triglyceride, coagulation FVII activity, and fibrinogen levels were measured in plasma by standard methods. The thrombotic tendency was estimated measuring the occlusion time (OT) of a prosthesis inserted into the abdominal aorta.[33–35] Antioxidant capacity of rat plasma was measured by TRAP assay.[32]

Five months of a cholesterol-rich diet induced in control rats a dramatic increase in cholesterol and triglyceride levels, with a concomitant shortening of the OT, compared to animals fed a standard diet. Factor VII levels were also increased (see FIGURE 3). Alcohol-free red wine supplementation for one month reversed the protrombotic effect of the diet. Indeed, in FNL+D+W a significant prolongation of OT from 57.6 ± 7.3 to 116 ± 14 h, ($p < 0.01$) was observed as compared to FNL+D. Simultaneous administration of alcohol-free red wine reduced by 36% the shortening

TABLE 2. Effect of NO inhibition by L-NAME (10 g/L/day for 10 days) on the effects of red wine in rats

	BT (sec)	Platelet Adhesion (%)	Thrombus Weight (mg)
Controls	132 ± 13	32 ± 1.3	3.3 ± 0.4
Red wine	213 ± 10**	17 ± 1.5**	1.8 ± 0.3*
Red wine + L-NAME	137 ± 15	34 ± 2.2	3.8 ± 0.8

Mean ± SE; $n = 10$; ** $p < 0.01$; * $p < 0.001$.

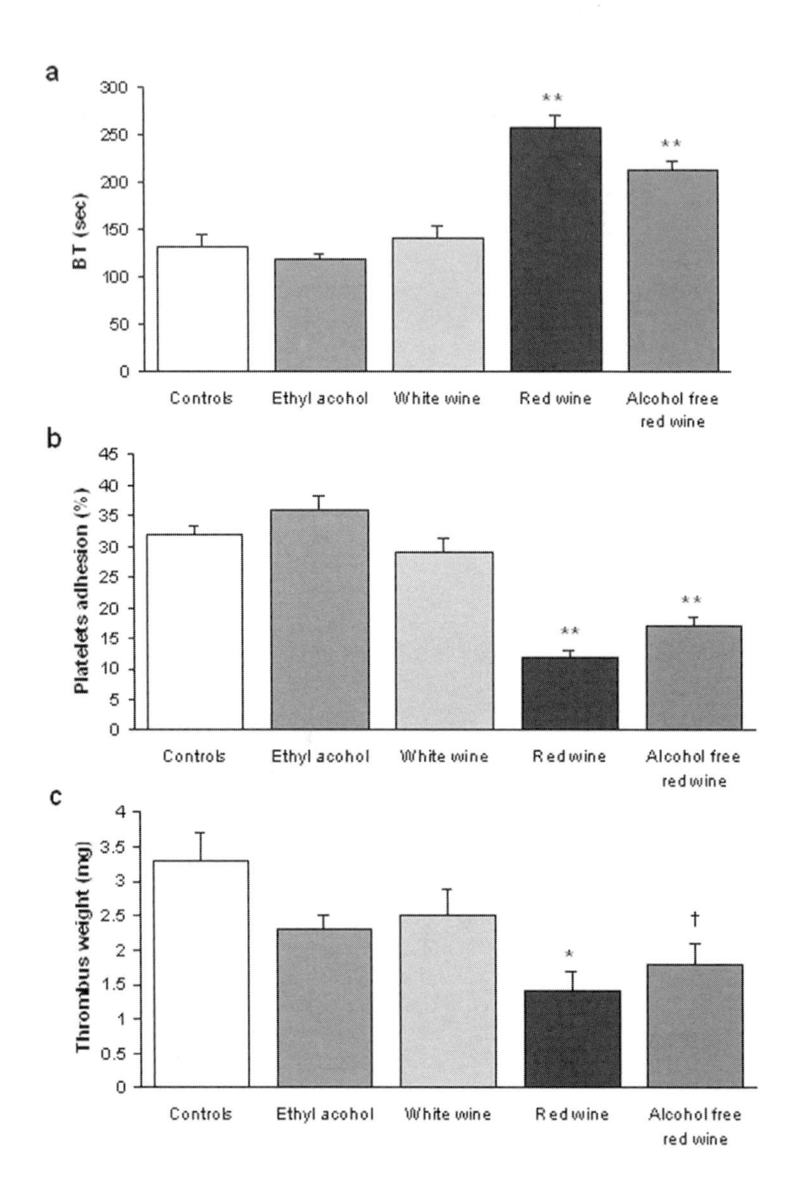

FIGURE 1. Effect of supplementation with different beverages (8.8 ± 0.2 mL/day for 10 days) on bleeding time (BT), platelet adhesion, and thrombus weight in rats. $^{**}p < 0.001$; $^{*}p < 0.01$; $^{\dagger}p < 0.05$.

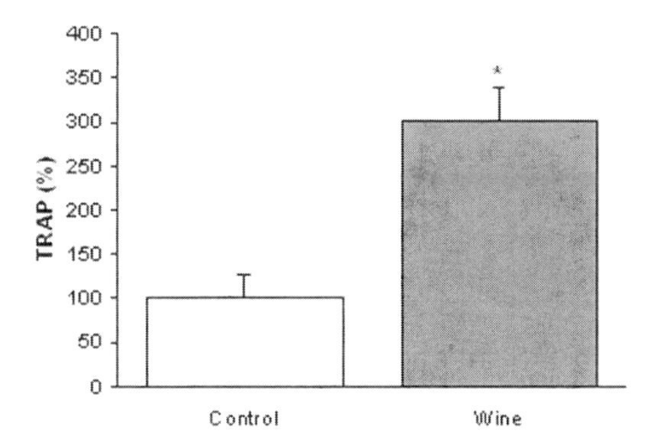

FIGURE 2. Effect of alcohol free red wine (100% intravenous) on plasma antioxidants capacity. $*p < 0.01$.

of OT induced by the diet. Alcohol-free red wine showed a tendency to reduce the increase in cholesterol and triglyceride levels induced by the cholesterol-rich diet. In contrast, neither fibrinogen nor FVII were modified. TRAP values in FNL+D+W were significantly higher as compared to FNL+D (226 ± 73 vs. $100 \pm 10\%$, respectively; $p < 0.05$). Thus, supplementation of alcohol-free red wine reversed the prothrombotic state induced by a cholesterol-rich diet, an effect likely to be correlated to the increased antioxidant capacity of plasma induced by alcohol-free red wine.

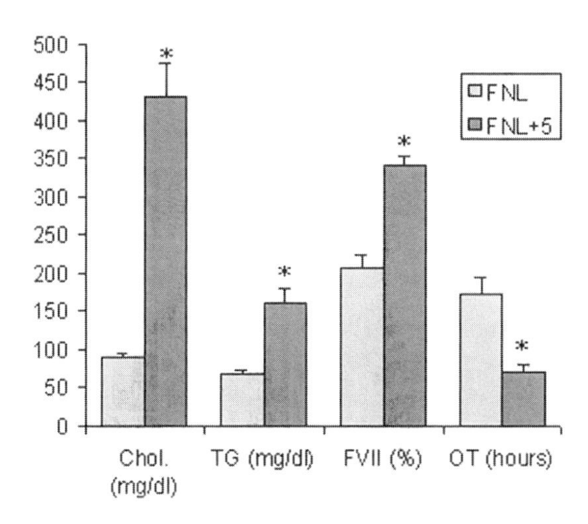

FIGURE 3. Characteristics of rats at standard diet (FNL) and after five months on cholesterol-rich diet (FNL+5). Cholesterol (chol), trygligerides (TG), coagulation factor VII (FVII) and arterial prosthesis occlusion time (OT). $*p < 0.01$.

WINE AND REDUCED CARDIOVASCULAR RISK: A CAUSAL RELATIONSHIP?

From the experimental observations reported here, it seems plausible to hypothesize that absorption of polyphenols from wine or from wine-derived products may exert a number of effects on cellular and plasmatic blood components and on the vascular function that may result in a more favorable hemostatic profile, thus reducing the risk of development and progression of cardiovascular disease.[18,36]

A strong interaction between the alcohol dehydrogenase type 3 (ADH3) genotype and the level of alcohol consumption in relation to the HDL level and the risk of myocardial infarction has recently been observed in a case-control study based on data from the prospective Physicians' Health Study.[37] Moderate alcohol consumption (at least one drink per day) was associated with a decreased risk of myocardial infarction in all three genotype groups; however, when men who were homozygous for the allele inducing a fast rate of alcohol clearance were taken as a reference group, those who were homozygous for the allele determining a slower rate of clearance had the greatest reduction in risk (RR 0.14, 95% CI 0.04–0.45). The latter also had the highest plasma HDL levels.

The interaction among the ADH3 genotype, the level of alcohol consumption and the HDL levels was also found by the same authors in an independent study of postmenopausal women.[37] The finding of an effect of the functional ADH3 polymorphism on the relations between moderate consumption of alcohol and the risk of myocardial infarction lends support to the plausibility of a *causal* interpretation. Indeed, while associations observed in non-randomized epidemiological studies may be influenced by potentially confounding factors, the association between the risk of myocardial infarction and functional variants in a gene that regulates the metabolism of the factors (alcohol) that underlines the apparent benefit observed, adds substantial support to the idea that the exposure to the factors is directly related to causation of the beneficial effect. The authors conclude that the fact that ADH3 metabolizes ethanol and not other compounds suggests that ethanol is responsible for the protective effect. Differences in ethanol metabolism, however, might contribute to differences in the gastrointestinal absorption of other constituents of alcoholic beverages such as the polyphenols.[38]

A key problem in epidemiological studies on dietary habits is that individuals are exposed to complex mixtures of compounds, so that the identification of the specific beneficial (or harmful) compounds may not be possible. Thus, future studies of common variants in metabolic genes and disease will help identifying specific environmental factors as causes of a given disease or of its prevention.[37]

CONCLUDING COMMENTS

Cardiovascular disease is the leading cause of morbidity and death in the industrialized countries, and is increasing in the developing world. The combined effect of a diet that is high in fruit, vegetables, complex carbohydrates, and monounsaturated fat, and low in animal fat and simple sugars, is certainly most important in lowering rates of cardiovascular and possibly other chronic diseases.[39] However, until recently alcohol consumption was frequently overlooked as an important part of the diet.

Many epidemiological studies have suggested that death rates for all causes and cardiovascular disease are lower for people who drink low to moderate amounts of alcohol than for people who do not drink at all and those who drink heavily.

In recent years, several studies have also been performed to determine whether death rates differ for people who drink mostly wine or other alcoholic beverages.

Evidence obtained from the meta-analysis of epidemiological studies as discussed here suggests that there is a consistent, significant beneficial effect from moderate wine consumption. This suggestion is in keeping with experimental data showing that wine contains several substances that may add to the beneficial effect of intake of a light to moderate amount of alcohol.[18]

It has recently been stated[40] that a definite proof could only be obtained by large scale, randomized, clinical end-point intervention trials comparing the effect of drinking wine versus not drinking alcohol, drinking other alcoholic beverages, or consuming wine-based supplements. Such trials however appear to be infeasible for several reasons, including the number of persons needed to observe statistically meaningful clinical endpoints, ethical concerns, and the huge economic investment required. Moreover, it is hard to imagine—at least in Italy or other Mediterranean countries—a controlled trial in which half of a large group of randomized wine-drinking persons would give their informed consent to avoid wine for five years to assess whether the other half of the group continuing to drink would have a reduced chance of developing myocardial infarction. The evidence for the beneficial effect of wine should therefore critically include molecular biology, animal experiments, and observational epidemiological studies.

Although a causal relationship between wine consumption and reduced vascular risk cannot be unequivocally established, it is reasonable to suppose that a moderate and regular wine consumption, together with other healthy lifestyle-related factors, such as a Mediterranean-like diet (that is high in fruit, vegetables, complex carbohydrates, and monounsaturated fat, and low in animal fat and simple sugars) may contribute to the lower rates of cardiovascular disease observed among populations living in Southern Europe.[41] The results obtained in other countries, such as Denmark, Britain and the USA, may allow extension of the beneficial effect of moderate wine consumption to different populations.

Other studies indicate that light-to-moderate alcohol consumption may lower the risk of cardiovascular and all causes events at the same or even greater extent in high risk subgroups, such as hypertensives,[42] diabetics,[43–45] and in patients with a history of cardiovascular disease,[8,46,47] although the latter finding has not been confirmed in a recent study.[48] Whether, and at what extent, the favorable results obtained in high-risk populations also applies to wine-preferrers is not yet established. However, it is worth of mention that light-to-moderate wine consumption by diabetic patients may add to the already ethanol-related positive effect on lipid profile, thrombotic tendency, and insulin levels, by reducing meal-generated oxidative stress.[49]

What should the doctors' advice to their patients be in respect to wine consumption? Abstainers should be informed that, in the absence of contraindications and in the context of healthy eating and lifestyle, low-to-moderate wine consumption may contribute to better cardiovascular health. People who are already low-to-moderate wine consumers should be encourage to continue.

The hazards of excess drinking should always be highlighted and heavy drinkers pushed to cut their consumption to a moderate level.[50]

ACKNOWLEDGMENTS

Work reported here was supported by grants from the European Union (FAIR CT97 3261) and Abruzzo Region (Programma Operativo Multiregionale "Sviluppo locale-Patti Territoriali per l'Occupazione" Obiettivo n. 1 - Sottoprogramma n. 9). Cleo Colombo, Alexandra Cianci, and Rosanna Tucci helped prepare the manuscript.

Thanks are also due to Drs. Annalisa Porto and Michela Vischetti and to Ms. Concetta Amore who performed some of the experimental work mentioned in this review. Drs. Domenico Rotilio, Roberto Lorenzet, Ennio Esposito, Anna Anconitano, and Chiara Cerletti shared with us their thoughts on the role of polyphenols in health and disease and the pleasure of tasting Montepulciano d'Abruzzo wines.

REFERENCES

1. MACLURE, M. 1993. Demonstration of deductive meta-analysis: ethanol intake and risk of myocardial infarction. Epidemiol. Rev. **15:** 328–251.
2. RIMM, E.B., A. KLATSKY, D. GROBBEE & M.J. STAMPFER. 1996. Review of moderate alcohol consumption and reduced risk of coronary heart disease: is the effect due to beer, wine, or spirits. BMJ **312:** 731–736.
3. CLEOPHAS, T.J. 1999. Wine, beer and spirits and the risk of myocardial infarction: a systematic review. Biomed. Pharmacother. **53:** 417–423.
4. CONSTANT, J. 1997. Alcohol, ischemic heart disease, and the French paradox. Coron. Artery Dis. **8:** 645–649.
5. GRØNBAEK, M., A. DEIS, T.I.A. SØRENSEN, *et al.* 1995. Mortality associated with moderate intakes of wine, beer, or spirits. BMJ **310:** 1165–1169.
6. RENAUD, S.C., R. GUEGUEN, J. SCHENKER & A. D'HOUTAUD. 1998. Alcohol and mortality in middle-aged men from eastern France. Epidemiology **9:** 184–188.
7. DOLL, R., R. PETO, E. HALL, *et al.* 1994. Mortality in relation to consumption of alcohol: 13 years' observations on male British doctors. BMJ **309:** 911–918.
8. THUN, M.J., R. PETO, A.D. LOPEZ, *et al.* 1997. Alcohol consumption and mortality among middle-aged and elderly U.S. adults. N. Engl. J. Med. **337:** 1705–1714.
9. GAZIANO, J.M., T.A. GAZIANO, R.J. GLYNN, *et al.* 2000. Light-to-moderate alcohol consumption and mortality in the Physicians' Health Study enrollment cohort. J. Am. Coll. Cardiol. **35:** 96–105.
10. ALBERT, C.M., J.E. MANSON, N.R. COOK, *et al.* 1999. Moderate alcohol consumption and the risk of sudden cardiac death among US male physicians. Circulation **100:** 944–950.
11. CORRAO, G., V. BAGNARDI, A. ZAMBON & S. ARICO. 1999. Exploring the dose-response relationship between alcohol consumption and the risk of several alcohol-related conditions: a meta-analysis. Addiction **94:** 1551–1573.
12. LA VECCHIA, C. 1995. Alcohol in the Mediterranean diet: assessing risks and benefits. Eur. J. Cancer Prev. **4:** 3–5.
13. GRØNBAEK, M., U. BECKER, D. JOHANSEN, *et al.* 2000. Type of alcohol consumed and mortality from all causes, coronary heart disease, and cancer. Ann. Intern. Med. **133:** 411–419.
14. RENAUD, S. & M. DE LORGERIL. 1992. Wine, alcohol, platelets, and the French paradox for coronary heart disease. Lancet **339:** 1523–1526.

15. KLATSKY, A.L., M.A. ARMSTRONG & H. KIPP. 1990. Correlates of alcoholic beverage preference: traits of persons who choose wine, liquor or beer. Br. J. Addict. **85:** 1279–1289.
16. KLATSKY, A.L. & M.A. ARMSTRONG. 1993. Alcoholic beverage choice and risk of coronary artery disease mortality: do red wine drinkers fare best? Am. J. Cardiol. **71:** 467–469.
17. CRIQUI, M.H. & B.L. RINGEL. 1994. Does diet or alcohol explain the French paradox? Lancet **344:** 1719–1723.
18. ROTONDO, S. & G. DE GAETANO. 2000. Protection from cardiovascular disease by wine and its derived products: epidemiological evidence and biological mechanisms. *In* Characteristics of Mediterranean Diet. F. Visioli & A.P. Simopoulos, Eds: 90–113. World Rev. Nutr. Diet, Basel, Karger.
19. TJØNNELAND, A., M. GRØNBAEK, C. STRIPP & K. OVERVAD. 1999. Wine intake and diet in a random sample of 48763 Danish men and women. Am. J. Clin. Nutr. **69:** 49–54.
20. RENAUD, S.C., R. GUEGUEN R,G. SIEST & R. SALAMON. 1999. Wine, beer, and mortality in middle-aged men from eastern France. Arch. Intern. Med. **159:** 1865–1870.
21. WANNAMETHEE, S.G. & A.G. SHAPER. 1999. Type of alcoholic drink and risk of major coronary heart disease events and all-cause mortality. Am. J. Public Health **89:** 85–90.
22. HOMMEL, M. & A. JAILLARD. 1999. Alcohol for stroke prevention? N. Engl. J. Med. **341:** 1605–1606.
23. TRUELSEN, T., M. GRØNBAEK, P. SCHNOHR & G. BOYSEN. 1998. Intake of beer, wine, and spirits and risk of stroke. The Copenhagen city heart study. Stroke **29:** 2467–2472.
24. SACCO, R.L., M. ELKIND, B. BODEN-ALBALA, *et al.* 1999. The protective effect of moderate alcohol consumption on ischemic stroke. JAMA **281:** 53–60.
25. THRIFT, A.G., G.A. DONNAN & J.J. MCNEIL. 1999. Heavy drinking, but not moderate or intermediate drinking, increases the risk of intracerebral hemorrhage. Epidemiology **10:** 307–312.
26. ROTONDO, S., A. DI CASTELNUOVO & G. DE GAETANO. 2001. The relationship between wine consumption and cardiovascular risk: from epidemiological evidence to biological plausibility. Ital. Heart J. **2:** 1–8.
27. DI CASTELNUOVO, A., S. ROTONDO, L. IACOVIELLO, *et al.* 2000. Wine Consumption and Vascular Risk: a Meta-Analysis. Proceed XXVth World Congress of Vine and Wine, Paris, June 19–23, pp 9–13.
28. RICE-EVANS, C., N.J. MILLER & G. PAGANGA. 1997. Antioxidant properties of phenolic compounds. Trends Plant Sci. **2:** 152–159.
29. ABE, J. & B.C. BERK. 1998. Reactive oxygen species as mediators of signal transduction in cardiovascular disease. Trend. Cardiovasc. Med. **8:** 59–64.
30. ANDRIAMBELOSON, E., A.L. KLESCHYOV, B. MULLER, *et al.* 1997. Andriantsitohaina R. Nitric oxide production and endothelium-dependent vasorelaxation induced by wine polyphenols in rat aorta. Br. J. Pharmacol. **120:** 1053–1058.
31. MEZZETTI, A., A. DI SANTO & R. LORENZET. 2000. The polyphenolic compounds resveratrol and quercetin downregulate tissue factor expression by human endothelial and mononuclear cells. Haematologica **85**(Suppl. 5): 54.
32. WOLLNY, T., L. AIELLO, D. DI TOMMASO, *et al.* 1999. Modulation of hemostatic function and prevention of experimental thrombosis by red wine in rats: a role for increased nitric oxide production. Br. J. Pharmacol. **127:** 747–755.
33. DE CURTIS, A., A.L. PORTO, M. VISCHETTI, *et al.* 2001. Effects of alcohol-free red wine on experimental thrombosis in hyperlipidemic rats. (Abstr.) Thromb. Hæmost. Supplement, July 2001.
34. HORNSTRA, G. & A. VENDELMANS-STARRENBURG. 1973. Induction of experimental arterial occlusive thrombi in rats. Atherosclerosis **17:** 369–382.
35. REYERS, I., G. DE GAETANO & M.B. DONATI. 1983. Protective effect of warfarin in arterial thrombosis: evidence from a vascular prosthesis model in rats. Thromb. Hæmost. **50:** 821–823.
36. RIMM, E.B., P. WILLIAMS, K. FOSHER, *et al.* 1999. Moderate alcohol intake and lower risk of coronary heart disease. Meta-analysis of effects on lipids and haemostatic factors. BMJ **319:** 1523–1528.

37. HINES, L.M., M.J. STAMPFER, J. MA, *et al.* 2001. Genetic variation in alcohol dehydrogenase and the beneficial effect of moderate alcohol consumption on myocardial infarction. N. Engl. J. Med. **344:** 549–555.
38. DUTHIE, G.G., M.W. PEDERSEN, P.T. GARDNER, *et al.* 1998. The effect of whisky and wine consumption on total phenol content and antioxidant capacity of plasma from healthy volunteeers. Eur. J. Clin. Nutr. **52:** 733–736.
39. KRAUSS, R.M., R.H. ECKEL, B. HOWARD, *et al.* 2000. AHA Dietary Guidelines. Stroke **31:** 2751–2766.
40. GOLDBERG, I.J., L. MOSCA L, M.R. PIANO & E.A. FISHER. 2001. Wine and your heart: a science advisory for healthcare professionals from the Nutrition Committee, Council on epidemiology and prevention, and Council on cardiovascular nursing of the American Heart Association. Circulation **103:** 472–475.
41. DE LORGERIL, M. 1998. Mediterranean diet in the prevention of coronary heart disease. Nutrition **14:** 55–57.
42. BERGER, K., U.A. AJANI, C.S. KASE, *et al.* 1999. Light-to-moderate alcohol consumption and risk of stroke among U.S. male physicians. N. Engl. J. Med. **341:** 1557–1564.
43. VALMADRID, C.T., R. KLEIN, S.E. MOSS, *et al.* 1999. Alcohol intake and the risk of coronary heart disease mortality in persons with older-onset diabetes mellitus. JAMA **282:** 239–246.
44. AJANI, U.A., C.H. HENNEKENS, A. SPELSBERG & J.E. MANSON. 2000. Alcohol consumption and risk of type 2 diabetes mellitus among US male physicians. Arch. Intern. Med. **160:** 1025–1030.
45. SOLOMON, C.G., F.B. HU, M.J. STAMPFER, *et al.* 2000. Moderate alcohol consumption and risk of coronary heart disease among women with type 2 diabetes mellitus. Circulation **102:** 494–499.
46. MUNTWYLER, J., C.H. HENNEKENS, J.E. BURING & J.M. GAZIANO. 1998. Mortality and light to moderate alcohol consumption after myocardial infarction. Lancet **352:** 1882–1885.
47. COOPER, H.A., D.V. EXNER & M.J. DOMANSKI. 2000. Light-to-moderate alcohol consumption and prognosis in patients with left ventricular systolic dysfunction. J. Am. Coll. Cardiol. **35:** 1753–1759.
48. SHAPER, A.G. & S.G. WANNAMETHEE. 2000. Alcohol intake and mortality in middle aged men with diagnosed coronary heart disease. Heart **83:** 394–399.
49. CERIELLO, A., N. BORTOLOTTI, E. MOTZ, *et al.* 1999. Meal-generated oxidative stress in diabetes. The protective effect of red wine. Diabet. Care **22:** 2084–2085.
50. DI CASTELNUOVO, A., S. ROTONDO, L. IACOVIELLO, *et al.* 2002. A meta-analysis of wine and beer consumption in relation to vascular health. Circulation. In press.

Thioredoxin Superfamily and Thioredoxin-Inducing Agents

KIICHI HIROTA,[a] HAJIME NAKAMURA,[b]
HIROSHI MASUTANI,[a,b] AND JUNJI YODOI[a,b]

[a]*BioMedical Special Research Unit, Human Stress Signal Research Center, National Institute of Advanced Science and Technology (AIST), Ikeda, Osaka 563-8577, Japan*

[b]*Department of Biological Responses, Institute for Virus Research, Kyoto University, Sakyo-ku, Kyoto, Japan 606-8507*

ABSTRACT: Mammalian thioredoxin (TRX) with redox-active dithiol in the active site plays multiple roles in intracellular signaling and resistance against oxidative stress. TRX is induced by a variety of stresses including infectious agents as well as hormones and chemicals. TRX is secreted from activated cells such as HTLV-I–transformed T-cells as a redox-sensitive molecule with cytokine-like and chemokine-like activities. The promoter of the TRX gene contains a series of stress-responsive elements. In turn, TRX promotes activation of transcription factors such as NF-κB, AP-1, and p53. We have reported that natural substances including estrogen, prostaglandins, and cAMP induce mRNA, protein, and secretion of TRX. These agents seemed to exert their physiological functions including cytoprotective actions partly through the induction of TRX without massive oxidative stress, which induces TRX strongly as well as other stress proteins. We report here a new TRX inducer substance, geranylgeranylacetone (GGA), which is originally derived from a natural plant constituent and has been used in the clinical field as an anti-ulcer drug. We have demonstrated that GGA induces the messenger RNA and protein of TRX and affects the activation of transcription factors, AP-1 and NF-κB, and that GGA blunted ethanol-induced cytotoxicity of cultured hepatocytes and gastrointestine mucosal cells. We will discuss a possible novel molecular mechanism of GGA, which is to protect cells via the induction of TRX and activation of transcription factors such as NF-κB and AP-1. Identification of the particular TRX-inducing components may contribute to the elucidation of the molecular basis of the "French Paradox," in which good red wines are beneficial for the cardiovascular system.

KEYWORDS: thioredoxin; induction; redox regulation; cytoprotection

INTRODUCTION

Thioredoxin (TRX) was first described in 1964 by Laurent *et al.* as a small redox protein from *Escherichia coli*.[1] TRX family members have subsequently been shown

Address for correspondence: Dr. Junji Yodoi, Department of Biological Responses, Institute for Virus Research, Kyoto University, 53 Shogoin Kawahara-cho, Sakyo-ku, Kyoto, Japan 606-8507. Fax: +81-75-751-5766.
yodoi@virus.kyoto-u.ac.jp

to be present ubiquitously from arche to man.[2] Human TRX has been identified by us and other investigators under several names: as adult T cell leukemia-derived factor, an interleukin-2 (IL-2) receptor–inducing factor produced by human T lymphotrophic virus type I–infected T cells; as an interleukin-1 (IL-1)-like cytokine produced by Epstein-Barr virus (EBV)-infected B lymphoblastoid cells; and as early pregnancy factor, part of a complex in the serum of pregnant animals that increases the complement-dependent inhibition of lymphocytes binding to heterologous blood cells (rosette formation).[3] TRXs have been implicated in a number of mammalian cell functions. Activity has been found outside the cell (cell growth stimulation and chemotaxis), in the cytoplasm (as an antioxidant and a cofactor), in the nucleus (regulation of transcription factor activity), and in the mitochondria.[3]

Many physicochemical stimuli including UV irradiation, and hydrogen peroxide have been shown to induce TRX expression and secretion. On the other hand, certain substances which do not have so called "oxidative stress" properties also induce TRX expression and secretion to prevent cell injury against oxidative stress.[3,4]

The focus of this review is to describe the biological properties of TRX and induction of TRX by non-toxic substances. The clinical implications of TRX will be also discussed. FIGURE 1 shows the various roles played by TRX.

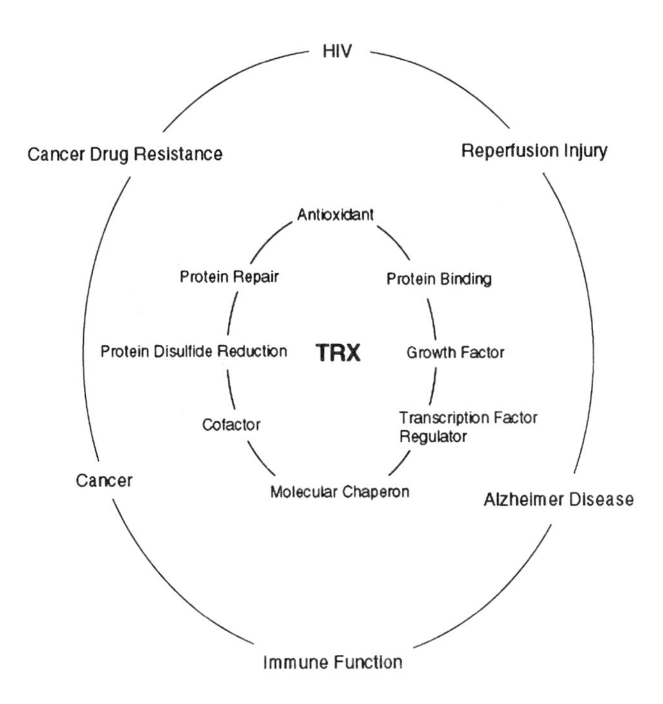

FIGURE 1. TRX play a variety of role in pathophysiogy of major disease states. TRX participates in essential intracellular process (*inner circle*) and is involved in either or phasophysiologic process in severe major disease states (*outer circle*).

TRX FAMILY PROTEINS

Human TRX is a 105-amino-acid protein with a molecular weight of 12 kDa. Chicken, mouse, rat, and bovine TRXs have also been cloned. TRX exists both with and without the *N*-terminal methionine, and numbering is from the *N*-terminal methionine. Human and other mammalian TRXs contain, in addition to the two catalytic site Cys residues -Trp-Cys32-Gly-Pro-Cys35-Lys, three other Cys residues, Cys62, Cys69, and Cys73 (numbers are based on human TRX), not found in bacterial TRXs. The C-terminal Cys73 residue is shown to be involved in dimer formation of TRX and may convey unique biological properties to mammalian TRX (reviewed in Ref. 2).

A second, slightly larger TRX based on electrophoretic mobility was identified in pig heart mitochondria.[5] This second TRX, TRX-2, was cloned from a rat heart library and found to encode an 18 kDa protein with 166 amino acid residues that had a conserved TRX catalytic site, but lacked the other Cys residues found in mammalian TRX. A 60-amino-acid *N*-terminal extension of TRX-2 exhibited characteristics consistent with a mitochondrial translocation signal, and Western blotting has confirmed the mitochondrial localization of TRX-2.[5–7]

A 32-kDa thioredoxin-like cytosolic protein (p32TRXL) has been cloned from a human testis cDNA library.[8] The predicted protein sequence is 289 amino acids with an *N*-terminal TRX domain of 105 amino acids, a conserved TRX active site (-Cys-Gly-Pro-Cys-), and a high degree of homology to human TRX. The sequence of the remaining 184 C-terminal amino acids showed no homology to other proteins in the database. p32TRXL is ubiquitously expressed in human tissues, with the highest levels appearing in stomach, testis, and bone marrow. Neither the full length p32TRXL protein nor the *N*-terminal 105- or 107-amino-acid TRX-like fragments were reduced by TRX reductase and NADPH. However, when reduced by dithioethreitol, the *N*-terminal 107-amino-acid fragment of p32TRXL was able to reduce insulin.[8] A 435-amino-acid redox protein with similarity to TRX but with a -Trp-Cys-Pro-Pro-Cys- catalytic site has been cloned from a mouse YAC library and localized to the nucleus. The protein has been named nucleoredoxin (NRX).[9] Insulin degradation assay using bacterially produced recombinant NRX demonstrated that NRX has a TRX-like activity in terms of disulfide bond cleavage.[9] Matsuo *et al.* reported the molecular cloning of a gene, TMX (transmembrane TRX-like protein) encoding a novel protein of 280 amino acid residues.[10] TMX has one TRX-like domain and a potential transmembrane domain. The TRX-like domain of TMX also has the activity to cleavage the interchain disulfide bond in insulin. Among recent additions to the TRX family are a protein specifically expressed in spermatozoa, named Sptrx (sperm-specific trx)[7] and a protein interacting with protein kinase C, named PICOT (protein kinase C-interacting cousin of TRX).[11]

Glutaredoxin (GRX), also known as thioltransferase, has GSH-disulfide oxidoreductase activity with the redox-active site, -Cys-Pro-Tyr-Cys-.[12–14] GRX reduces low molecular weight disulfides and proteins in concert with NADPH and GSH reductase.

SUBCELLULAR LOCALIZATION OF TRXs

TRX behaves as a soluble protein after disruption of cells. In resting cells, TRX, which does not have a specific localization signal, exists as one of the cytoplasmic proteins. Accumulation in the membrane fraction has been implied by Holmgren *et al.*[13] Moreover, it has been reported that TRX is constitutively present as a surface-associated sulfhydryl protein in plasma membrane of a wide range of cells.[13]

In contrast, under pathophysiological condition, TRX shows specific intracellular localization. We have reported that in cultured human retinal pigment epithelial cells exposed to H_2O_2, TRX is detected in mitochondoria using immunoelectron microscopy. In primary cultured keratinocytes, we first observed that TRX translocates from the cytoplasm into the nucleus quickly after UVB irradiation.[15–18] In addition to UVB irradiation, treatment of cells with PMA, H_2O_2, UV irradiation, hypoxia, the cancer drug cisplatin, and hemin has been reported to cause the translocation of TRX from the cytoplasm to the nucleus.[19–21] The mechanism for this translocation is not known, but it may be the consequence of TRX's being carried along bound to another protein with a nuclear import sequence. This translocation phenomenon is detected in other cell lines and observed on the overexpressed TRX in mouse fibroblast NIH3T3.[16] These might suggest that TRX is one of the stress-sensing molecules of cells. We will describe later the interaction of TRX and transcription factors and transcription factor enhancing molecules as an intranuclear function of TRX. As previously noted, TRX-2 is found in the mitochondria[5] while p32TrxL is a cytosolic protein.[8] The varied subcellular localization of TRXs suggests that TRXs have essential roles in many cellular process. NRX is localized mainly in the nucleus and GRX is found in both the nucleus and the cytoplasm.[9,14]

SECRETION OF TRX

TRX, despite its function as an intracellular disulfide-reducing enzyme and its lack of a signal sequence, has been found to play some extracellular roles. Several laboratories including ours have purified human TRX in the culture supernatant of the cell lines, indicating that TRX could be released by an unknown mechanism from the cells (reviewed in ref. 3). TRX is actively secreted by a variety of normal and transformed cells, including fibroblasts, airway epithelial cells, and activated B and T lymphocytes. Rubartelli *et al.* showed that two drugs, Brefeldin A and Dinitrophenol, which block transport through the exocytotic pathway, do not inhibit secretion of TRX.[22,23] This indicates that secretion of TRX does not follow the classical ER-Golgi route. The secretory mechanism for TRX shares several features with the alternative pathway described for interleukin-1 (IL-1)β. However, unlike IL-1β, TRX is not detected in membrane-bound compartments of secreting cells.[22,23]

Plasma levels of TRX in normal individuals vary between 10 and 80ng/mL.[24] The plasma TRX level is elevated in certain human diseases including HIV infection[24,25] and cancer.[26]

A shorter, 10-kDa, form of TRX has been reported to be secreted and bound to the outer membrane of human U937 cells and MP-6 cells.[27] Although it still has increased eosinophil cytotoxic activity, this truncated TRX does not have insulin disulfide-reducing activity.[28] Some evidence suggests that this is formed by proteolytic cleavage of TRX.

INDUCTION OF TRX

The coding region of the human TRX gene spans 1.3 kb and is organized into five exons.[29] TRX is characterized as a stress-inducible protein. A lot of physicochemical stimuli and agents are found to induce TRX gene expression. The promoter region contains many possible regulatory binding motifs compatible with constitutive expression, including SP1 and with an oxidative stress response element.[30] A variety of stress stimuli increase TRX expression in cells including hypoxia,[31] lipopolysaccharide,[32] O_2,[33,34] hydrogen peroxide,[35,36] phorbol ester, viral infection,[26,37,38] X-ray radiation, and UV irradiation.[36]

TRX is induced in peripheral blood mononuclear cells (PBMC) by phytohemagglutinin, Con A or OKT3 mAb ligation, which is almost completely suppressed by the immunosuppressant FK506. Whereas cyclosporin A also inhibits TRX expression in OKT3 mAb-stimulated PBMC, rapamycin fails to affect it in spite of exhibiting growth inhibition. In addition, exogenous IL-2 could not increase TRX production in FK506-treated PBMC or in PHA blasts.[39]

In the erythroleukemic cell line K562, TRX is induced by hemin (ferriprotoporphyrin X).[18,40] Heat shock factor-2 is indicated as the agent responsible for this activation.[40] We recently showed that hemin activates the TRX gene expression through the antioxidant-responsive element (ARE) and that TRX gene induction is regulated via ARE by the binding of NF-E2 p45/small Maf in an unstimulated condition, of Nrf2/small Maf in hemin stimulation, and of the Jun/Fos families of proteins in PMA stimulation. This is a demonstration of a novel molecular mechanism through the ARE/EpRE by a switch of binding factors including CNC-bZIP/small Maf transcription factors and the Jun/Fos families of proteins, depending on different stimuli.[18]

TRX expression is also induced by stimuli without so-called stress properties. TRX is transcriptionally induced by retinol (vitamin A) in monkey tracheobronchial epithelial cells.[41] Estradiol increases TRX expression in primary cultures of human endometrial stromal cells,[42,43] and estradiol, although not progesterone and testosterone, increases the expression of TRX in bovine artery endothelial cells.[44] Prostaglandin (PG) E_1 has cytoprotective activity against oxidative stress.[45] PGE_1 also increases TRX expression in human retinal pigment epithelial cell line.[45] TRX expression is augmented in a dose-dependent manner when retinal pigment epithelial cells are treated with 10nM–1 M PGE_1 one hour before exposure to hydrogen peroxide. Intracellular cyclic AMP level is elevated by PGE_1 when the cells are simultaneously exposed to hydrogen peroxide. Forskolin, an activator of adenylate cyclase, and dibutylyl cAMP, a cyclic AMP analog, can also induce TRX and extend survival of retinal pigment epithelial cells. On the other hand, TRX induction and cellular protection by PGE_1 was blocked by Rp diastereoisomer of cyclic adenosine 3′, 5′, monophosphorothioate, a competitive inhibitor of cyclic AMP-dependent protein kinase. TRX induction is augmented significantly by pretreatment with PGI_2, a stimulator of the cyclic AMP–dependent signal pathway, while treatment with PGF2α, a stimulator of the inositol phosphate–dependent signal pathway, failed to enhance TRX. These findings indicate that PGE_1 has a cytoprotective activity against oxidative injury, partly through TRX induction via cyclic AMP–dependent pathway.[45] Geranylgeranylacetone (GGA), an acyclic polyisoprenoid, was developed in Japan and is widely used as an ulcer drug. In addition to its protective effect on gastric mucosal cells, GGA also has anti-apoptotic effects against ischemia and reperfusion injury in hepatocytes and intestinal cells. We have demonstrated that GGA induces protein expression of TRX in cultured hepatocytes and gastric mucosal cells and prevents ethanol-induced cytotoxicity in these cells.[46,47] GGA induces messenger RNA of TRX.[47] In addition, ELISA of TRX shows that GGA induces secretion of TRX into the culture media.[46] Notably, PGE_1, dibutylyl cAMP, and GGA are widely used clinically and their cell toxicity has not been reported.

On the other hand, expressional regulation of the other TRX family molecules is currently under investigation.

BIOLOGICAL ACTIVITY OF TRX

As originally reported, TRX is a electron donor for ribonucleotide reductase. TRX is also suggested to be a source of reducing equivalents for vitamin K epoxide reductase, which is necessary for the biosynthesis of plasma clotting factors.

Besides these roles as a cofactor, TRX by itself acts as antioxidant or reactive oxygen species scavenger. TRX by itself can scavenge singlet oxygen, hydroxyl radical and hydrogen peroxide. TRX is also an efficient electron donor to human glutathione peroxide and members of thioredoxin peroxide (peroxiredoxin) superfamily. The potent reducing activity of TRX is the most important implication of its activity as an anti-apoptotic protein.

TRX is also a key player in keeping intracellular protein disulfides and oxidatively modified cysteine residues such as sulfenic acid derivatives reduced. Among many targets of TRX, redox regulation of transcription factors has been most investigated. TRX selectively regulates DNA-binding activity of a number of transcription factors. NF-κB, AP-1, CREB, PEBP2/CBF, Myb, estrogen receptor, glucocorticoid receptor, and p53 are targets of redox regulation of TRX.[14,16,17,19,20,35,48–50] TRX also regulates transcription factor by interaction with redox factor 1 (Ref-1), which

TABLE 1. TRX binding proteins

Protein	Role of TRX	References
ASK1	Reduced TRX binds ASK1 to inhibit downstream signaling	53
P40 Phox	Biological significance unknown	57
Vitamin D3 upregulated protein 1 (VDUP-1)	VDUP-1 inhibits TRX function	54–56
PKC	TRX inhibits autophosphorylation and histone phosphorylation	58
Lipocain	Biological significance unknown	59

has a reducing activity and apurine/apyrimidine endonuclease repair activity.[19,51,52] TRX enhances transcriptional activity of hypoxia-inducible factor 1 (HIF-1) in a unique manner. TRX reduces the cysteine residue of one of its subunits, HIF-1α, via Ref-1 and allows HIF-1 to interact with p300/CBP.[21]

Several TRX-binding proteins are identified by yeast two-hybrid and phage-display screening assay systems[53–59] (see TABLE 1). A protein to which TRX binds is apoptosis signaling-regulating kinase 1 (ASK1), which is an activator of the c-Jun *N*-terminal kinase (JNK) and p38 MAPK kinase pathway and is required for TNF-α–induced apoptosis.[53,60,61] Another TRX-binding protein, TBP-2, which is identical to vitamin D3-upregulated protein1 (VDUP1) has been reported.[54–56] Treatment of HL-60 cells with 1α, 25-dihydroxyvitamin D3 caused an increase of TBP-2/VDUP1 expression and downregulation of the expression and the reducing activity of TRX.[54] It is likely that TRX-TBP-2/VDUP1 interaction plays an important role in the redox regulation of growth and differentiation of the cells sensitive to a variety of inducers, including 1 α, 25-dihydroxyvitamin D3 responses.[54,55]

Extracelluar TRX exerts immunomodulatory properties. In fact, human TRX was found as a secreted protein upregulating the IL-2 receptor α chain and having co-cytokine activity from HTLV-I–transformed T lymphocytes. Extracellular TRX has proinflammatory effects by potentiating cytokine release from fibroblasts as well as monocytes. Recently, TRX has been shown to act as a chemotactic protein.[62] Because TRX does not increase intracellular Ca^{2+} and its activity is not inhibited by pertussis toxin, the chemotactic action of TRX differs from that of known chemokines though G protein-coupled receptors. A series of experiments using recombinant TRX with the mutation of the active site cysteins to serines resulted in loss of chemotactic activity, suggesting that the redox property of TRX is required.[62]

CLINICAL IMPLICATIONS

TRX levels in plasma/serum are elevated in oxidative stress–associated disorders including viral infection and ischemia-reperfusion injury.[63] Especially in HIV infection, plasma TRX levels are inversely correlated with GSH levels in lymphocytes and AIDS patients with elevated TRX levels die more quickly than those with normal plasma levels of TRX.[25] In Hepatitis C virus infection, patients with elevated serum TRX and elevated serum ferritin are more insensitive to interferon therapy.[64] TRX levels in synovial fluid increase in association with the disease progress of rheumatoid arthritis.[65] Thus, secreted levels of TRX are indicative of systemic or local oxidative stress.

Exogenous TRX protects cells from hydrogen peroxide or hypoxia.[66] Intravenous administration of TRX attenuates ischemic-reperfusion injury in animal models. Recently we found that circulating TRX suppresses neutrophil adhesion on endothelial cells and extravasation into inflammatory sites.[67] TRX administration or overexpression attenuates bleomycin-induced or cytokine-induced interstitial pneumonia (Hoshino *et al.*, manuscript in preparation). Further studies are now in progress to clarify the possible clinical application of TRX and TRX-inducers.

IN SUMMARY

In this paper, we reviewed the various biological property of TRX.

Substances in vegetables and other foods inducing TRX should be beneficial for the mucosal epithelial cells as well as other tissue, enhancing the protective activity against noxious stimuli or invasion of microorganisms. Recently Ueda *et al.* reported that baicalin, which is one of major ingredients of Chinese herbal medicine, induces apoptosis in Jurkat cells acting as prooxidants.[68] Thus, the substances such as falvonoids and polyphenols in intoxicating beverages may have a potential utility as redox regulatory drugs. Identification of the particular TRX-inducing components may contribute to the elucidation of the molecular basis of the "French paradox," which seems to indicate that good red wines are beneficial for the cardiovascular system.

ACKNOWLEDGMENT

We thank Ms. Yoko Kanekiyo for secretarial help.

REFERENCES

1. LAURENT, T.C., E.C. MOORE & P. REICHARD. 1964. Enzymatic synthesis of deoxyribonucleotides VI. Isolation of and characterization of thioredoxin, the hydrogen donor from *Escherichia coli* B. J. Biol. Chem. **239:** 3436–3444.
2. HOLMGREN, A. 1985. Thioredoxin. Annu. Rev. Biochem. **54:** 237–271.
3. NAKAMURA, H, K. NAKAMURA & J. YODOI. 1997. Redox regulation of cellular activation. Annu. Rev. Immunol. **15:** 351–369.
4. POWIS, G. & W.R. MONTFORT. 2001. Properties and biological activities of thioredoxin. Annu. Rev. Pharmacol. Toxicol. **41:** 261–295.
5. SPYROU, G., E. ENMARK, A. MIRANDA-VIZUETE & J. GUSTAFSSON. 1997. Cloning and expression of a novel mammalian thioredoxin. J. Biol. Chem. **272:** 2936–2941.
6. MIRANDA-VIZUETE, A., A.E. DAMDIMOPOULOS & G. SPYROU. 2000. The mitochondrial thioredoxin system. Antioxid Redox Signal **2:** 801–810.
7. MIRANDA-VIZUETE, A., J. LJUNG, A.E. DAMDIMOPOULOS, *et al.* 2001. Characterization of Sptrx, a novel member of the thioredoxin family specifically expressed in human spermatozoa. J. Biol. Chem. **276:** 31567–31574.
8. LEE, K.K., M. MURAKAWA, S. TAKAHASHI, *et al.* 1998. Purification, molecular cloning, and characterization of TRP32, a novel thioredoxin-related mammalian protein of 32 kDa. J. Biol. Chem. **273:** 19160–19166.
9. KUROOKA, H., K. KATO, S. MINOGUCHI, *et al.* 1997. Cloning and characterization of the nucleoredoxin gene that encodes a novel nuclear protein related to thioredoxin. Genomics **39:** 331–339.
10. MATSUO, Y., N. AKIYAMA, H. NAKAMURA, *et al.* 2001. Identification of a novel thioredoxin-related transmembrane protein. J. Biol. Chem. **276:** 10032–10038.
11. WITTE, S., M. VILLALBA, K. BI, *et al.* 2000. Inhibition of the c-Jun N-terminal kinase/AP-1 and NF-κB pathways by PICOT, a novel protein kinase C-interacting protein with a thioredoxin homology domain. J. Biol. Chem. **275:** 1902–1909.
12. HOLMGREN, A. & F. ASLUND. 1995. Glutaredoxin. Methods Enzymol. **252:** 283–292.
13. HOLMGREN, A. 1989. Thioredoxin and glutaredoxin systems. J. Biol. Chem. **264:** 13963–13966.
14. HIROTA, K., M. MATSUI, M. MURATA, *et al.* 2000. Nucleoredoxin, glutaredoxin, and thioredoxin differentially regulate NF-κB, AP-1, and CREB activation in HEK293 cells. Biochem. Biophys. Res. Commun. **274:** 177–182.

15. MASUTANI, H., K. HIROTA, T. SASADA, *et al.* 1996. Transactivation of an inducible anti-oxidative stress protein, human thioredoxin by HTLV-I Tax. Immunol. Lett. **54:** 67–71.
16. HIROTA, K., M. MURATA, Y. SACHI, *et al.* 1999. Distinct roles of thioredoxin in the cytoplasm and in the nucleus: a two step mechanism of redox regulation of transcription factor NF-κB. J. Biol. Chem. **274:** 27891–27897.
17. UENO, M., H. MASUTANI, R. ARAI, *et al.* 1999. Thioredoxin-dependent redox regulation of p53-mediated p21 activation. J. Biol. Chem. **274:** 35809–35815.
18. KIM, Y-C., H. MASUTANI, Y. YAMAGUCHI, *et al.* 2001. Hemin-induced activation of the thioredoxin gene by Nrf2. A differential regulation of the antioxidant responsive element by a switch of its binding factors. J. Biol. Chem. **276:** 18399–18406.
19. HIROTA, K., M. MATSUI, S. IWATA, *et al.* 1997. AP-1 transcriptional activity is regulated by a direct association between thioredoxin and Ref-1. Proc. Natl. Acad. Sci. USA **94:** 3633–3638.
20. MAKINO, Y., N. YOSHIKAWA, K. OKAMOTO, *et al.* 1998. Direct association with thioredoxin allows redox regulation of glucocorticoid receptor function. J. Biol. Chem. **274:** 3182–3188.
21. EMA, M, K. HIROTA, J. MIMURA, *et al.* 1999. Molecular mechanisms of transcription activation by HLF and HIF1alpha in response to hypoxia: their stabilization and redox signal-induced interaction with CBP/p300. EMBO J. **18:** 1905–1914.
22. RUBARTELLI, A. & R. SITIA. 1991. Interleukin 1 β and thioredoxin are secreted through a novel pathway of secretion. Biochem. Soc. Trans. **19:** 255–259.
23. RUBARTELLI, A., A. BAJETTO, G. ALLAVENA, *et al.* 1992. Secretion of thioredoxin by normal and neoplastic cells through a leaderless secretory pathway. J. Biol. Chem. **267:** 24161–24164.
24. NAKAMURA, H., S. DE ROSA, M. ROEDERER, *et al.* 1996. Elevation of plasma thioredoxin levels in HIV infected individuals. Int. Immunol. **8:** 603–611.
25. NAKAMURA, H., S.C. DE ROSA, J. YODOI, *et al.* 2001. Chronic elevation of plasma thioredoxin: inhibition of chemotaxis and curtailment of life expectancy in AIDS. Proc. Natl. Acad. Sci. USA **98:** 2688–2693.
26. NAKAMURA, H., J. BAI, Y. NISHINAKA, *et al.* 2000. Expression of thioredoxin and glutaredoxin, redox-regulating proteins, in pancreatic cancer. Cancer Detect. Prev. **24:** 53–60.
27. SAHAF, B.A. SODERBERG, G. SPYROU, *et al.* 1997. Thioredoxin expression and localization in human cell lines: detection of full-length and truncated species. Exp. Cell Res. **236:** 181–192.
28. SILBERSTEIN, D.S., S. MCDONOUGH, M.S. MINKOFF & S.M. BALCEWICZ. 1993. Human eosinophil cytotoxicity-enhancing factor. Eosinophil-stimulating and dithiol reductase activities of biosynthetic (recombinant) species with COOH-terminal deletions. J. Biol. Chem. **268:** 9138–9142.
29. KAGHAD, M., F. DESSARPS, S.H. JACQUEMIN, *et al.* 1994. Genomic cloning of human thioredoxin-encoding gene: mapping of the transcription start point and analysis of the promoter. Gene **140:** 273–278.
30. TANIGUCHI, Y., Y. TANIGUCHI-UEDA, K. MORI & J. YODOI. 1996. A novel promoter sequence is involved in the oxidative stress-induced expression of the adult T-cell leukemia-derived factor (ADF)/human thioredoxin (Trx) gene. Nucleic Acids Res. **24:** 2746–2752.
31. DENKO, N., C. SCHINDLER, A. KOONG, *et al.* 2000. Epigenic regulation of gene expression in cervical cancer cells by the tumor microenvironment. Clin. Cancer Res. **6:** 480–487.
32. EJIMA, K., T. KOJI, H. NANRI, *et al.* 1999. Expression of thioredoxin and thioredoxin reductase in placentae of pregnant mice exposed to lipopolusaccharide. Placenta **20:** 561–566.
33. DAS, K.C., X.L. GUO & C.W. WHITE. 1999. Hyperoxia induces thioredoxin and thioredoxin reductase gene expression in lungs of premature baboons with respiratory distress and bronchopulmonary dysplasia. Chest **116:** 101S.

34. DAS, K.C., X.L. GUO & C.W. WHITE. 1999. Induction of thioredoxin and thioredoxin reductase gene expression in lungs of newborn primates by oxygen. Am. J. Physiol. **276:** L530–L539.
35. HAYASHI, T., Y. UENO & T. OKAMOTO. 1993. Oxidoreductive regulation of nuclear factor kappa B. Involvement of a cellular reducing catalyst thioredoxin. J. Biol. Chem. **268:** 11380–11388.
36. SACHI, Y., K. HIROTA, H. MASUTANI, *et al.* 1995. Induction of ADF/TRX by oxidative stress in keratinocytes and lymphoid cells. Immunol. Lett. **44:** 189–193
37. FUJII, S., Y. NANBU, H. NONOGAKI, *et al.* 1991. Coexpression of adult T-cell leukemia-derived factor, a human thioredoxin homologue, and human papillomavirus DNA in neoplastic cervical squamous epithelium. Cancer **68:** 1583–1591.
38. FUJII, S., Y. NANBU, I. KONISHI, *et al.* 1991. Immunohistochemical localization of adult T-cell leukaemia-derived factor, a human thioredoxin homologue, in human fetal tissues. Virchows Arch. A Pathol. Anat. Histopathol. **419:** 317–326.
39. FURUKE, K., H. NAKAMURA, T. HORI, *et al.* 1995. Suppression of adult T cell leukemia-derived factor/human thioredoxin induction by FK506 and cyclosporin A: a new mechanism of immune modulation via redox control. Int. Immunol. **7:** 985–993.
40. LEPP, S., L. PIRKKALA, S.C. CHOW, *et al.* 1997. Thioredoxin is transcriptionally induced upon activation of heat shock factor 2. J. Biol. Chem. **272:** 30400–30404.
41. AN, G. & R. WU. 1992. Thioredoxin gene expression is transcriptionally up-regulated by retinol in monkey conducting airway epithelial cells. Biochem. Biophys. Res. Commun. **183:** 170–175.
42. MARUYAMA, T., Y. KITAOKA, Y. SACHI, *et al.* 1997. Thioredoxin expression in the human endometrium during the menstrual cycle. Mol. Hum. Reprod. **3:** 989–993.
43. MARUYAMA, T., Y. SACHI, K. FURUKE, *et al.* 1999. Induction of thioredoxin, a redox-active protein, by ovarian steroid hormones during growth and differentiation of endometrial stromal cells in vitro. Endocrinology **140:** 365–372.
44. EJIMA, K., H. NANRI, M. ARAKI, *et al.* 1999. 17beta-estradiol induces protein thiol/disulfide oxidoreductases and protects cultured bovine aortic endothelial cells from oxidative stress. Eur. J. Endocrinol. **140:** 608–613.
45. YAMAMOTO, M., N. SATO, H. TAJIMA, *et al.* 1997. Induction of human thioredoxin in cultured human retinal pigment epithelial cells through cyclic AMP-dependent pathway; involvement in the cytoprotective activity of prostaglandin E1. Exp. Eye Res. **65:** 645–652.
46. DEKIGAI, H., H. NAKAMURA, J. BAI, *et al.* 2000. Geranylgeranylacetone promotes induction and secretion of thioredoxin in gastric mucosal cells and periferal blood lymphocytes. Free Radical Res. **35:** 1–8.
47. HIROTA, K., H. NAKAMURA, T. ARAI, *et al.* 2000. Geranylgeranylacetone enhances expression of thioredoxin and suppresses ethanol-induced cytotoxicity in cultured hepatocytes. Biochem. Biophys. Res. Commun. **275:** 825–830.
48. AKAMATSU, Y., T. OHNO, K. HIROTA, *et al.* 1997. Redox regulation of the DNA binding activity in transcription factor PEBP2. The roles of two conserved cysteine residues. J. Biol. Chem. **272:** 14497–14500.
49. HAYASHI, S., K. HAJIRO-NAKANISHI, Y. MAKINO, *et al.* 1997. Functional modulation of estrogen receptor by redox state with reference to thioredoxin as a mediator. Nucleic Acids Res. **25:** 4035–4040.
50. MAKINO, Y., K. OKAMOTO, N. YOSHIKAWA, *et al.* 1996. Thioredoxin: a redox-regulating cellular cofactor for glucocorticoid hormone action. Cross talk between endocrine control of stress response and cellular antioxidant defense system. J. Clin. Invest. **98:** 2469–2477.
51. XANTHOUDAKIS, S. & T. CURRAN. 1992. Identification and characterization of Ref-1, a nuclear protein that facilitates AP-1 DNA-binding activity. EMBO J. **11:** 653–665; ISSN: 0261-4189.
52. QIN, J., G.M. CLORE, W.P. KENNEDY, *et al.* 1996. The solution structure of human thiredoxin complexed with its target from Ref-1 reveals peptide chain reversal. Structure **4:** 613–620.
53. SAITO, M., H. NISHITOH, M. FUJII, *et al.* 1998. Mammalian thioredoxin is a direct inhibitor of apoptosis signal-regulating kinase (Ask) 1. EMBO J. **17:** 2596–2606.

54. NISHIYAMA, A., M. MATSUI, S. IWATA, *et al.* 1999. Identification of thioredoxin-binding protein-2/vitamin D(3) up-regulated protein 1 as a negative regulator of thioredoxin function and expression. J. Biol. Chem. **274:** 21645–21650.
55. JUNN, E., S.H. HAN, J.Y. IM, *et al.* 2000. Vitamin D3 up-regulated protein 1 mediates oxidative stress via suppressing the thioredoxin function. J. Immunol. **165:** 6287–6295.
56. YAMANAKA, H., F. MAEHIRA, M. OSHIRO, *et al.* 2000. A possible interaction of thioredoxin with VDUP1 in HeLa cells detected in a yeast two-hybrid system. Biochem. Biophys. Res. Commun. **271:** 796–800.
57. NISHIYAMA, A., T. OHNO, S. IWATA, *et al.* 1999. Demonstration of the interaction of thioredoxin with p40phox, a phagocyte oxidase component, using a yeast two-hybrid system. Immunol. Lett. **68:** 155–159.
58. WATSON, J.A., M.G. RUMSBY & R.G. WOLOWACZ. 1999. Phage display indentifies thioredoxin and superoxide dismutase as novel protein kinase C-interacting proteins: thioredoxin inhibits protein kinase C-mediated phosphorylation of histone. Biochem. J. **343:** 301–305.
59. REDL, B., P. MERSCHAK, B. ABT & P. WOJNAR. 1999. Phage display reveals a novel interaction of human tear lipocain and thioredoxin which is relevant for ligand biding. FEBS Lett. **460:** 182–186.
60. ICHIJO, H., E. NISHIDA, K. IRIE, *et al.* 1997. Induction of apoptosis by ASK1, a mammalian MAPKKK that activates SAPK/JNK and p38 signaling pathways. Science **275:** 90–94.
61. ICHIJO, H. 1999. From receptors to stress-activated MAP kinase. Oncogene **18:** 6087–6093.
62. BERTINI, R., O.M. HOWARD, H.F. DONG, *et al.* 1999. Thioredoxin, a redox enzyme released in infection and inflammation, is a unique chemoattractant for neutrophils, monocytes, and T cells. J. Exp. Med. **189:** 1783–1789.
63. NAKAMURA, H., J. VAAGE, G. VALEN, *et al.* 1998. Measurements of plasma glutaredoxin and thioredoxin in healthy volunteers and during open-heart surgery. Free Radic. Biol. Med. **24:** 1176–1786.
64. SUMIDA, Y., T. NAKASHIMA, T. YOH, *et al.* 2000. Serum thioredoxin levels as an indicator of oxidative stress in patients with hepatitis C virus infection. J. Hepatol. **33:** 616–622.
65. MAURICE, M.M., H. NAKAMURA, E.A. VAN DER VOORT, *et al.* 1997. Evidence for the role of an altered redox state in hyporesponsiveness of synovial T cells in rheumatoid arthritis. J. Immunol. **158:** 1458–1465.
66. ISOWA, N., T. YOSHIMURA, S. KOSAKA, *et al.* 2000. Human thioredoxin attenuates hypoxia-reoxygenation injury of murine endothelial cells in a thiol-free condition. J. Cell Physiol. **182:** 33–40.
67. NAKAMURA, H., L.A. HERZENBERG, J. BAI, *et al.* 2001. Circulating thioredoxin suppresses lipopolysaccharide-induced neutrophil chemotaxis. Proc. Natl. Acad. Sci. USA **98:** 15143–15148.
68. UEDA, S., H. NAKAMURA, H. MASUTANI, *et al.* 2002. Baicalin induces apoptosis via mitochondrial pathway as prooxidant. Mol. Immunol. **38:** 781–791.

Wine Polyphenols and Optimal Nutrition

FULVIO URSINI[a] AND ALEX SEVANIAN[b]

[a]*Department of Biological Chemistry, University of Padova, Padova, Italy*

[b]*University of Southern California, School of Pharmacy, Los Angeles, California, USA*

ABSTRACT: One of the key elements of Mediterranean diet is the use of wine, usually taken with foods. Besides the evidence from human experience and ancient medicine, modern experimental data support the notion that the most striking effect of wine in protecting against cardiovascular disease involves the reduction of oxidative damage to plasma lipoproteins. This oxidative damage is thought to be mediated by eating foods containing oxidized lipids. In fact, eating a meal containing oxidized lipids increases the plasma level of lipid hydroperoxides and increases the susceptibility to oxidation of LDL. The postprandial increase of LDL–, an oxidatively modified form of LDL, where apoB is unfolded and sinking in the core of the particle, is a valuable biomarker for this food-derived oxidative stress in plasma. Wine, taken with foods minimizes the postprandial rise of lipid hydroperoxides and LDL– and abolishes the increase of LDL oxidability. Among wine antioxidants, the best candidates for providing an antioxidant effect are procyanidins. These compounds are considered better antioxidants than the corresponding monomers containing catechol groups. This is due to the hydrogen transfer mechanism for the radical-scavenging reaction, which renders the reaction more specific for peroxyl radicals and pH independent. Moreover, the fast intramolecular disproportion among aroxyl radicals pulls the antioxidant reaction by both decreasing the oxidation potential and increasing the rate of the reaction. Apparently, wine procyanidins are active in preventing lipid oxidation of foods while in the digestive tract, thus preventing the postprandial plasma rise in oxidants. The likely limited bioavailability of these compounds, therefore, does not affect their relevance as key elements for optimizing nutrition and reducing risk of atherogenesis. Accordingly, studies with rabbits fed a high cholesterol diet show that grapeseed procyanidins are strongly protective not only in terms of reducing plasma lipid peroxides, but they also markedly inhibit lipid-laden foam-cell deposition. Drinking wine at meals provides this kind of protection, and the final benefits are realized by the prevention of the development of atheromatous lesions even under conditions of hypercholesterolemia.

KEYWORDS: atherosclerosis; LDL (modified); lipid hydroperoxides; cholesterol (efflux); procyanidins

INTRODUCTION

At the turn of the thirteenth century, Arnoldo da Villanova (1253–1315) collected and edited a series of rules and suggestions for maintaining human health derived from ancient tradition and wisdom. This Compendium, named *Regimen Sanitatis*

Address for correspondence: Fulvio Ursini, M.D., Department of Biochemistry, University of Padova, Viale G. Colombo 3, Padova 35121, Italy. Voice: 39-049-8276104; fax: 39-049-8073310.
ursini@mail.bio.unipd.it

Salernitani, remains a milestone in the history of medicine as the first logical and "almost" scientifically sound collection of statements about how health can be maintained by optimizing behavior and nutrition. When describing the effects of wine, the Regimen says "...During meals drink wine happily, little but often..." and also "...To avoid harming the body never drink between meals...."

The notion of the wine as a kind of "food additive" good for health only when taken with foods is intrinsic to Mediterranean dietary habits. Wine consumption apart from meals, instead, is a social, cultural and hedonistic element of cultures that use wine, where the perception of disease prevention is lost. Taking wine with foods is unquestionably a typical feature of the Mediterranean diet and this fact, although usually neglected, has to be taken in account when interpreting the epidemiological evidence for relationships between diet and risk of cardiovascular disease.

The low incidence of cardiovascular disease in relation to saturated fat intake observed in France, and referred to as "French Paradox",[1] is not so paradoxical when the integration of different foods in a diet is considered. The "French Paradox," indeed, provides epidemiological evidence for the limited value of risk assessment based on saturated fat intake and plasma cholesterol levels, when meals include significant amounts of protective factors, such as wine.

The possible mechanisms will be reviewed here supporting the hypothesis that diet plays a major role in initiation and evolution of atherosclerosis and that an optimized nutrition can actually lower the risk of disease. Wine will be discussed as a relevant nutritional factor for the optimization of nutrition.

OXIDATIVELY MODIFIED LDL AND PLAQUE FORMATION

In the modern view, atherosclerosis is a chronic inflammatory condition of arterial wall sustained by genetic and environmental factors.[2] At the molecular level, it is widely accepted that oxidatively modified LDL activates the series of biochemical events leading to the plaque formation and promoting the thrombotic complication.[3] The first event in atherosclerosis is the recruitment of monocytes and lymphocytes in the subendothelial space. This event is sparked by the interaction of minimally oxidized LDL with endothelial cells where a "response to injury" reaction is activated. The production of pro-inflammatory cytokines, adhesion molecules and colony stimulating factors accounts for the initial inflammatory reaction.[2] Under these conditions, LDL in the subendothelial space undergoes a massive oxidative degradation by both reactive oxygen species, produced by endothelial cells and macrophages, and extracellular myeloperoxidase.[4] Only this extensively oxidized LDL appears to be taken up by SR-A and CD 36 receptor in macrophages to form foam cells.[2] These receptors, and the integration of signaling events associated with cholesterol homeostasis, can be up- or down-regulated by the nature and extent of oxidative stress and modulated by specific antioxidants.[5]

In this scenario, the formation of the minimally oxidatively modified LDL emerges as the first event driving the whole process. The measurement of this event would be the ideal biomarker for a predictive risk assessment and its inhibition the goal for an efficient preventive treatment. Unfortunately, the measurement of oxidatively modified lipoproteins in plasma is not an easy task, and some authors argue against the relevance of oxidatively modified LDL in plasma on the basis of the assumption that the oxidation has to take place in the subendothelial space.[4] A 12/15 lipoxygenase activity is suggested as the most likely mechanisms for this initial oxidation.[6] However, lipoxygenase requires the presence of traces of lipid hydroperoxides in LDL to initiate peroxidation.[7] Furthermore, a common observation reveals that lipid peroxides are required for LDL oxidative modification to proceed. It is, therefore, essential that LDL be "seeded" with lipid hydroperoxides before undergoing massive oxidation in the subendothelial space.

OXIDANTS AND VASCULAR CHOLESTROL HOMEOSTASIS

Although hypercholesterolemia may cause arterial injury and oxidative stress, numerous findings argue against a direct role for cholesterol. However, there is increasing evidence that oxidants can impair cholesterol homeostasis and a more favorable oxidant-antioxidant balance brought about either by increasing the antioxidant defense or by decreasing oxidants may prevent the oxidant alteration of cholesterol trafficking. Regulatory effects on expression of HMG-CoA reductase and LDL receptors, via interaction with sterol-regulatory elements, may account for the effect of oxidative stress. Recently, sterol-regulatory elements mediating the expression of caveolin were identified that bind cholesterol and activate transcription.[8] This response is analogous to the cholesterol-mediated regulation of HMG-CoA reductase and LDL receptor protein following binding to sterol-regulatory–element binding protein. We postulate that preventing formation of oxidants and oxidative modifications during hypercholesterolemia is the means by which wine provides a protective effect at the vascular wall. By preventing the disruptive effects of oxidants such as peroxides, the normal processing of cholesterol can proceed even in the presence of high plasma cholesterol.

Emphasis has traditionally been given to the mechanisms of LDL receptor–mediated cholesterol uptake to explain cell cholesterol accumulation; however, recent studies show that free cholesterol efflux is dynamically regulated as an important component of cellular cholesterol homeostasis. Several groups have reported that oxidized LDL impairs cholesterol efflux, after challenging vascular cells with lipoprotein-cholesterol, leading to cholesterol accumulation without affecting influx rates. This effect has been attributed, at least in part, to the cholesterol oxidation products in oxidized LDL.[9] The diminished cholesterol efflux is accompanied by a reduced translocation of caveolin from intracellular membranes to the plasma membrane enriched fraction. Normally, treatment of endothelial cells with high serum-containing medium results in increased efflux of free cholestrol and translocation of caveolin to the plasma membrane, indicating a role for caveolin in mediating cholestrol efflux and maintaining a balance in the influx and efflux of cholesterol.

Caveolæ represent the terminus for cholestrol transferred from the trans-Golgi network to the plasma membrane and their formation requires the presence of caveolin, a cholestrol binding protein.[10] Caveolæ are thought to participate in the efflux of cellular cholesterol to HDL,[11] completing the cycle of cholesterol influx and efflux. Caveolin is a characteristic marker for caveolæ and required for their assembly.[12] The trans-Golgi network complex is proposed as the site where cholestrol is sorted to lysosomes or the plasma membrane.[13] Recent studies show that perturbation of this sorting process by oxidants is linked to intracellular cholesteryl esters accumulation and decreased caveolin-mediated cholestrol efflux to HDL. The effects of oxidants, including hydroperoxides associated with oxidized LDL, may be based on differential modulation of protein kinases and phosphatases that mediate the phosphorylation status and consequent translocation of caveolin.[14]

At the plasma membrane, cholesterol efflux is completed by transfer to HDL with slow bidirectional fluxes between membrane and lipid-rich HDL subclasses. If cholestrol is not removed by an acceptor, it becomes re-esterified by ACAT, thus establishing a cholestrol-ester cycle.[15] Accumulation of cholestrol esters, as a hallmark of lesion formation in humans and in cholesterol-fed rabbits, may be due to decreased ester degradation relative to its delivery by lipoproteins or synthesis in vascular cells. Similar effects on cholesterol ester synthesis and accumulation were found after treatments with oxidized lipoproteins,[16] which provoked accumulation of cholesterol, apparently by entrapment of hydrolyzed ester in lysosomes (the primary organelles to which oxidatively modified LDL is directed). Disruption of lysosomes by oxidants in LDL may not only account for the inability to complete cholesterol/cholestrol ester processing, but also may trigger apoptosis leading to the formation of the necrotic core typically seen in atheromatous lesions.[17]

LIPID HYDROPEROXIDES AND ARTERIAL MATRIX DEGRADATION

Besides the focal formation of plaque with the related thrombotic outcome, a diffuse loss of elastic properties takes place,[18] due to thickening of the wall and to degradation of matrix fibers. Degradation of matrix proteins, particularly elastin, is also involved in plaque rupture and aneurysm. Lipid hydroperoxides in LDL appear to be involved in this relevant aspect of atherosclerosis inducing the so-called vasculopathy of aging.

In a recent study, we directly visualized the effect of minimally oxidized LDL on the degradation of extracellular proteins in fresh rat aorta by means of two-photon near-infrared excitation microscopy.[19] By its deeper penetration in highly scattering tissues versus light of shorter wavelengths, the near-infrared excitation allows imaging of interior regions of intact tissues, with a penetration of about 200μm. Typical features of two-photon excitation microscopy include the simultaneous absorption of two near-infrared photons with a total energy equivalent to one photon, at half of the wavelength. Therefore, absorption in the ultraviolet region can be achieved by near-infrared excitation with the advantage of strongly reducing photodamage and photobleaching in the specimen, associated with UV excitation. We obtained for the first time high-resolution images of the intense autofluorescence arising from extracellular matrix proteins of a fresh rat aorta and we presented clear evidence that lipid hydroperoxides in LDL activate a redox-sensitive proteolytic system which degrades vascular matrix proteins. The vascular damage produced by oxidatively modified LDL is quite similar to the damage visualized in the aorta of apo-E knockout mice, a well-known model of vascular damage and atherosclerosis. The event was prevented by ascorbate and Trolox,[19] as well as by a mixture of wine polyphenols (T. Parasassi and F. Ursini, unpublished).

POST-PRANDIAL LIPOPROTEINS AND ATHROGENESIS

In 1991, Chung and Segrest,[20] proposed that lipolytic remnants of triglyceride-rich lipoproteins, whose levels are elevated during the postprandial period, present all the features needed to account for the current theories of atherogenesis. Of particular significance was the dissociation of diet risk from the risk due to elevations in plasma cholesterol, correlating postprandial lipemia with risk of cardiovascular disease. Until recently, the properties of these lipoproteins were largely regarded from the standpoint of the lipid load confronting vascular cells and the cytotoxicity of high fat-containing remnant lipoproteins. However, by considering that these lipoproteins also are enriched in peroxides and can mediate oxidative reactions in other targets, a new and critical component is added to the features ascribed to remnant particles. A recent report by Anderson, *et al.*,[21] showed that impaired endothelial function—assessed by brachial artery flow-mediated vasodilation—was directly related to triglyceride enrichment in the VLDL and LDL fractions and correlated with free radical activity and peroxide levels associated with triglycerides. The effects on vasoactivity during the postprandial phase are thought to occur by the action of triglyceride-rich remnant particles containing oxidants that impair endothelial NO production.[22]

LDL–

A subclass of oxidatively modified LDL is present in human plasma that accounts for up to 10% of LDL.[23] This LDL is more negative than native LDL and is usually referred to as LDL–. LDL– is found in the small, dense LDL subfraction that is regarded as an indicator of atherosclerosis risk.[24] In comparison to native LDL, LDL– is enriched in lipid hydroperoxides and cholesterol oxides.[23] Although the mechanism of LDL– formation has not been fully elucidated, the concentration almost doubles in the post-prandial phase (F. Ursini, unpublished) suggesting a nutritional origin. LDL– binds with diminished affinity to normal LDL receptors on human skin fibroblasts and endothelial cells, is much more cytotoxic than normal LDL,[25] is a poor ligand for scavenger receptors,[26] and induces Il-8 and monocyte chemotactic protein secretion by endothelial cells.[27] Owing to the presence of lipid hydroperoxides, LDL– is prone to copper-induced oxidation that proceeds without a lag phase.[23] The percentage of LDL– is, therefore, a major determinant of oxidability of the bulk LDL.

Using a spectroscopic approach, we found that apoB-100 secondary structure and conformation are almost completely lost in LDL–.[28] Linked to the presence of oxidized lipids, and possibly also to the presence of lysophospholipids and free fatty acids, a decreased packing of the surface lipid monolayer is evident in LDL– that is thought to favor apoB-100 unfolding and sinking into a more hydrophobic environment.

Higher percentages of LDL– have been observed in hyperlipidemia,[29] uremia[30] and diabetes.[31] The elucidation of the structure and the effects of LDL– are likely to shed further light on the well-known atherogenic effect of post-prandial lipoproteins, independently from their high TG content, and should expand the list of mechanisms by which oxidized fat is toxic. In fact, the formation of LDL– is a possible common denominator for a series of observations regarding oxidation and oxidability of plasma lipids in the post-prandial phase. In the 2–3 hours after the intake of a meal containing oxidized lipids, the plasma level of lipid hydroperoxides detectable by a chemiluminescent procedure increases,[32] the oxidation of isolated LDL proceeds faster following a shorter lag phase, and the percentage of LDL– increases. These events are prevented by either 300 mL of wine[33] or 300 mg of grapeseed extract (C. Scaccini and F. Ursini, unpublished).

WINE ANTIOXIDANTS

Wine contains a large amount of polyphenolic antioxidants.[34] Eating the usual amounts of fruit vegetables, olive oil or beer can by no means achieve the intake of antioxidants found in two to three glasses of red wine. Several procedures have been adopted for measuring the antioxidant effect. These procedures are based on the delay of the oxidation of lipids or a target molecule. In one test, a competition kinetics analysis was devised where the relative rate constant was measured for the reduction of the peroxidation-driving peroxy radicals.[34] The results obtained by different procedures are rather similar, and simple phenol analysis by the Folin-Ciocalteau reagent, although much less precise, produces comparable results. In our studies, the antioxidant capacity of white wines ranges between 0.08 and 1.22 mM equivalent of Trolox, while results from 6.45 to 41.93 were obtained for red wines. The anthocyanine monomers account for the largest part of antioxidant capacity of red wines (60%) while tannins bring the major contribution to the antioxidant capacity (51.6%) of the white wine aged in oak barrels, which showed the highest antioxidant capacity. We also observed that a decrease of antioxidant capacity does not necessarily takes place with aging of wine, apparently due to the progressive transformation of anthocyanine monomers into soluble tannins that are more efficient antioxidants.

The antioxidant capacity of procyanidines (soluble tannins) was characterized using an eletroanalytical approach.[35] These polymeric compounds are better antioxidants owing to the rapid dismutation of aroxyl radical generated in proximal chatecol moieties, which apparently drives the first oxidation step. Moreover, the antioxidant reaction of procyanidines takes place through a hydrogen transfer mechanism while an electron transfer mechanism occurs in the case of monomeric catechol-bearing compounds. This indicates a more specific reaction with peroxy radicals for procyanidines and a less favorable non-specific autoxidation.

Several studies reported an increase of plasma antioxidant capacity following the intake of wine.[36,37] We confirmed these data by our competition kinetic procedure.[38] However, there are some doubts that these data actually indicate an increased plasma concentration of polyphenols. A major discrepancy, indeed, exists between the increase of antioxidant capacity in plasma (in the range of 0.1–0.3 mM Trolox equivalents) and the actual amount of some polyphenols detectable in plasma (always in a concentration lower than 1 μM). A possible explanation is related the fact that the steady-state concentration of major plasma antioxidants (urate, ascorbate and protein thiols) may change during absorption of foods and wine and this could affect the results of the measurements of total plasma antioxidant capacity.

From the available experimental evidence, preventing the increased oxidizability of post-prandial LDL when a meal is taken with wine cannot be due simply to an increased antioxidant capacity of plasma, but, most likely, to a reduced oxidative load such as damage to lipoproteins by lipid hydroperoxides derived from foods. The decreased formation or assimilation of peroxides into plasma lipoproteins is equivalent to a decrease in oxidative load and hence increased antioxidant capacity. During digestion, oxidized lipids could promote oxidation of other lipids either in the extracellular emulsion or in the membranes of epithelial cells. Antioxidants mixed with the food during digestion could directly prevent this peroxidation, thus limiting the amount of lipid hydroperoxides reaching the blood stream.[32] The lipid hydroperoxides in the chylomicrons and their remnants are a likely candidate for the oxidative challenge that damages circulating LDL, giving rise to increases in LDL–levels and the oxidative susceptibility of the whole LDL fraction.

ANTIATHEROGENIC EFFECTS OF WINE ANTIOXIDANTS

In order to determine if wine antioxidants display antiatherogenic potencies comparable to their antioxidant activity, a study was performed using New Zealand white rabbits fed a Western diet consisting of chow supplemented with cholesterol.[39] Rabbits were fed a 0.25% cholesterol standard chow diet for eight weeks that was formulated for three study groups:

1. rabbits fed 0.25% cholesterol-chow,
2. rabbits fed 0.25% cholesterol-chow + vehicle (a proprietary matrix termed Lipoid S30 used to deliver the polyphenols), and
3. rabbits fed 0.25% cholesterol-chow + vehicle + 2% grapeseed polyphenolic extract containing a high titer of procyanidins (termed Leucoselect-Phytosome).

A fourth group of eight rabbits were maintained on cholesterol-free standard chow and served as a baseline control.

At the end of eight weeks the three cholesterol-fed groups had similar total cholesterol levels (402–455 mg/dL) whereas rabbits fed standard chow had plasma total cholesterol levels of 45 ± 7 mg/dL. This indicated that grapeseed extract has no effect on the processing of cholesterol by the liver or peripheral tissues in terms of altering plasma cholesterol levels. At the end of the study the aortas were dissected from all the animals and processed for the quantitative analysis of Sudanophilic lesions using standard procedures. The entire aorta was subdivided into three regions: (1) the aortic arch, (2) the thoracic aorta and, (3) the abdominal aorta. The extent of Sudanophilic lesions representing formation of raised fatty streaks was quantified by means of computerized imaging and planimetry and expressed on the basis of the percent of the total vessel wall area covered by lesions.

In the case of rabbits fed standard chow (baseline control), no discernible lesions were evident on any part of the aorta (i.e., less than 0.1% coverage), corresponding to no significant differences in plasma cholesterol levels over the eight-week study period. By contrast, the aortic arch was the most affected region in the cholesterol-fed animals. Aortic arches from rabbits fed 0.25% cholesterol-chow had 18.2 ± 7.6% of the vessel covered by lesions, as compared to 11.5 ± 5.1% coverage for rabbits fed 0.25% cholesterol + vehicle, and 3.0 ± 1.9 for the group receiving 0.25% cholesterol + procyanidins. The latter represented a significant ($p < 0.005$) reduction in lesions as compared to the groups not consuming procyanidins. For the 0.25% cholesterol group, 3.1 ± 2.6% of the thoracic aortae were covered by lesions, representing a significant increase in lesion coverage area as compared to the 0.25% cholesterol + vehicle-fed group (2.1 ± 1.7%), and the group receiving 0.25% cholesterol + procyanidins (0.75 ± 0.5%). Lesion severity in the abdominal aorta was intermediate to that found for the aortic arch and thoracic aortas with a significant suppression ($p < 0.01$) in lesion formation in the group receiving procyanidins. The analysis of plasma antioxidant capacity and lipid hydroperoxide content indicated that the intake of grapeseed extract abolished the oxidative stress in plasma brought about by the high cholesterol diet.

WINE AND OPTIMAL NUTRITION

The concept of essentiality in nutrition is supported by observation and experimental findings concerning the biological function of a nutrient and the occurrence of a deficiency syndrome. Dietary recommendations are, therefore, based on nutrient intakes and are consolidated in terms of recommended dietary allowances (RDA). This concept needs to be somehow revisited in light of the major insights provided by epidemiology and increased knowledge in bionutrition.[40] Some nutrients might be required at much higher dosages to decrease the risk of a given disease than for preventing the typical deficiency syndrome. Thus, this pharmacological dosage could provide a health benefit by optimizing a non-traditional function. In a broad sense, a nutrient has to be considered as a dietary factor affecting the biological functions in a way that results in health benefit. Essentially, correct nutrition is defined by nutrient intake, not food intake, and several interactions may take place among different nutrients and among nutrients and lifestyle. To add to the complexity, interactions among foods have to be considered in optimizing a dietary regimen.

The use of wine is a typical example of these concepts. Wine does not contain essential elements, thus a recommendation based on RDA is not possible; however, epidemiological data support a protective effect with respect to the incidence of cardiovascular disease. The experimental data presented here suggest that this effect could be due to lowering the harmful effect of foods containing oxidized lipids. The conclusion would be that wine can lower the risk of disease in a population exposed to a nutrition-related risk, whereas minimal if any effect can be expected for populations facing a minimal set of nutritive risk factors. This is possibly why the "French Paradox" was observed only in France. In other Mediterranean regions, specific and different dietary practices, based at least in part on the intake of fruit, vegetables, fish and lower intake of dairy products, fats and beef, the risk is already low and the protective effect of wine does not emerge from the epidemiological analysis.

REFERENCES

1. RENAUD, S. & M. DE LORGERIL. 1992. Wine, alcohol, platelets and the French paradox for coronary heart disease. Lancet **339:** 1523–1526.
2. LUSIS, A.J. 2000. Atherosclerosis. Nature **407:** 233–241.
3. STEINBERG, D., S. PARTHASARATHY, T.E. CAREW, *et al.* 1989. Beyond cholestrol: modifications of low density lipoproteins and increase of its atherogenicity. N. Engl. J. Med. **320:** 915–924.
4. CHISOLM, G. & D. STEINBERG. 2000. The oxidative modification hypothesis of atherosclerosis: an overview. Free Radical Biol. Med. **28:** 1815–1826.
5. RICCIARELLI, R., J.M. ZINGG & A. AZZI. 2000. Vitamin E reduces the uptake of oxidized LDL by inhibiting CD36 scavenger receptor expression in cultured aortic smooth muscle cells. Circulation **102:** 82–87.
6. CYRUS, T., J.L. WITZTUM, D.J. RADER, *et al.* 1999. Disruption of 12/15 lipoxygenase diminishes atherosclerosis in apoE deficient mice. J. Clin. Invest. **103:** 1597–1604.
7. SCHNURR, K., J. BELKNER, F. URSINI, *et al.* 1996. The selenoenzyme phospholipid hydroperoxide glutathione peroxidase controls the activity of the 15-lipoxygenase with complex substrates and mantains the specificity of the oxygenation products. J. Biol. Chem. **271:** 4653–4658.
8. BIST, A., P.E. FIELDING & C.J. FIELDING. 1997. Two sterol regulatory element-like sequences mediate up-regulation of caveolin gene transcription in response to low density lipoprotein free cholesterol. Proc. Natl. Acad. Sci. USA **94:** 10693–10698.
9. FIELDING, C.J., A. BIST & P.E. FIELDING. 1997. Caveolin mRNA levels are up-regulated by free cholesterol and down-regulated by oxysterols in fibroblast monolayers. Proc. Natl. Acad. Sci. USA **94:** 3753–3758.
10. MURATA, M., J. PERANEN, R. SCHREINER, *et al.* 1995. VIP21/caveolin is a cholesterol-binding protein. Proc. Natl. Acad. Sci. USA **92:** 10339–10343.
11. SMART, E.J., Y. YUN-SHU, W.C. DONZELL & R.G. ANDERSON. 1996. A role for caveolin in transport of cholesterol from endoplasmic reticulum to plasma membrane. J. Biol. Chem. **271:** 29427–29435.
12. PARTON, R.G. Caveolae and caveolins. 1996. Curr. Opin. Cell Biol. 8:542-548.
13. FIELDING, C.J. & P.E. FIELDING. 1997. Intracellular cholesterol transport. Lipid Res. **38:** 1503–1521.
14. VEPA, S., W.M. SCRIBNER & V. NATARAJAN. 1997. Activation of protein phosphorylation by oxidants in vascular endothelial cells: identification of tyrosine phosphorylation of caveolin. Free Radical Biol. Med. **22:** 25–35.
15. BROWN, M.S. & J.L. GOLDSTEIN. 1983. Lipoprotein metabolism in the macrophage: implications for cholesterol deposition in atherosclerosis. Ann. Rev. Biochem. **52:** 223–261.
16. MAOR, I. & M. AVIRAM. 1994. Oxidized low density lipoprotein leads to macrophage accumulation of unesterified cholesterol as a result of lysosomal trapping of the lipoprotein hydrolyzed cholesteryl ester. J. Lipid Res. **35:** 803–819.
17. LI, W., X.M. YUAN, A.G. OLSSON & U.T. BRUNK. 1998. Uptake of oxidized LDL by macrophages results in partial lysosomal enzyme relocalization and inactivation. Arterioscler. Thromb. Vasc. Biol. **18:** 177–184.
18. ROBERT, L., A.M. ROBERT & B. JACOTOT. 1998. Elastin-elastase-atherosclerosis revisited. Atherosclerosis **140:** 281–288.
19. PARASASSI, T., W. YU, D. DIRBIN, *et al.* 2000. Two-photon fluorescence imaging of aorta fibres reveals a new mechanism for vascular damage induced by LDL hydroperoxides. Free Radical Biol. Med. **28:** 1589–1597.
20. CHUNG, B.H. & J.P. SEGREST. 2001. Cytotoxicity of remnants of triglyceride-rich lipoproteins: an atherogenic insult? Adv. Exp. Med. Biol. **285:** 341–351.
21. ANDERSON, R.A., M.L. EVANS, G.R. ELLIS, *et al.* 2001.The relationship between postprandial lipaemia, endothelial function and oxidative stress in healthy individuals and patients with type 2 diabetes. Atherosclerosis **154:** 475–483.
22. MARCHESI, S., G. LUPATTELLI, G. SCHILLACI, *et al.* 2000. Impaired flow-mediated vasoactivity during postprandial phase in found healthy men. Atherosclerosis **153:** 397–402.

23. SEVANIAN, A., H.N. HODIS; J. HWANG, *et al.* 1997. LDL^- is a lipid hydroperoxide-enriched circulating lipoprotein. J. Lipid Res. **38:** 419–428.

24. SEVANIAN, A., J. HWANG, H.N. HODIS, *et al.* 1996. Contribution of an in vivo oxidized LDL to LDL oxidation and its association with dense LDL subpopulations. Arterioscler. Thromb. **16:** 784–793.

25. HODIS, H.N., D. KRAMSCH, P. AVOGARO, *et al.* 1994. Biochemical and cytotoxic characteristics of in vivo circulating oxidized low density lipoprotein (LDL^-). J. Lipid Res. **35:** 669–677.

26. DEMUTH, K., I. MYARA, B. CHAPPEY, *et al.* 1996. A cytotoxic electronegative subfraction is present in human plasma. Arterioscler. Thromb. **16:** 773–783.

27. DE CASTELLARNAU, C., J.L. SÁNCHEZ-QUESADA, S. BENITEZ, *et al.* 2000. Electronegative LDL from normolipemic subjects induces IL-8 and monocyte chemotactic protein secretion by human endothelial cells. Arter. Thromb. Vasc. Biol. **20:** 2281–2287.

28. PARASASSI, T., G. BITTOLO-BON, R. BRUNELLI, *et al.* 2001. Loss of apoB100 secondary structure and conformation in hydroperoxide rich, electronegative LDL–. Free Radical Biol. Med. In press.

29. SANCHEZ-QUESADA, J.M., C. OTAL-ENTRAIGAS, M. FRANCO, *et al.* 1999. Effect of simvastatin treatment on the electronegative low-density lipoprotein present in patients with heterozygous familial hypercholesterolemia. Am. J. Cardiol. **84:** 655–659.

30. ZIOUZENKOVA, O., L. ASATRYAN, M. AKMAL, *et al.* 1999. Crosslinking between low density lipoprotein ApoB and hemoglobin mediates conversion to electronegative LDL^- during the ex-vivo blood circulation. J. Biol. Chem. **274:** 18916–18924.

31. SANCHEZ-QUESADA, J.L., A. PEREZ, A. CAIXAS, *et al.* 1996. Electronegative low density lipoprotein subform is increased in patients with short-duration IDDM and is closely related to glycaemic control. Diabetologia **39:** 1469–1476.

32. URSINI, F., A. ZAMBURLINI, G. CAZZOLATO, *et al.* 1998. Postprandial plasma lipid hydroperoxides: a possible link between diet and atherosclerosis. Free Radical Biol. Med. **25:** 250–252.

33. NATELLA, F., A. GHISELLI, A. GIUDI, *et al.* 2001. Red wine with a single proxide meal reduced the increased susceptibility to peroxidation of postprandial low density lipoproteins. Free Radical Biol. Med. In press.

34. TUBARO, F., P. RAPUZZI & F. URSINI. 1999. Kinetic analysis of antioxidant capacity of wine. Biofactors **9:** 37–47.

35. URSINI, F., I. RAPUZZI, R. TONIOLO, *et al.* 2001. Characterization of the antioxidant effect of procyanidins. Methods Enzymol. **335:** 338–350.

36. MAXWELL, M., A. CRUICKSHANK & G. THORPE. 1994. Red wine and the antioxidant activity of serum. Lancet **334:** 193–194.

37. WHITEHEAD, T.P., D. ROBINSON, S. ALLAWAY, *et al.* 1995. Effect of red wine ingestion on the antioxidant capacity of serum. Clin. Chem. **41:** 32–35.

38. TUBARO, F., A. GHISELLI, P. RAPUZZI, *et al.* 1998. Analysis of plasma antioxidant capacity by competition kinetics. Free Radical Biol. Med. **24:** 1228–1234.

39. URSINI, F., F. TUBARO, J. RONG & A. SEVANIAN. 1999. Optimization of nutrition: polyphenols and vascular protection. Nutrition Rev. **57:** 241–249.

40. STRAIN, J.J. 1999. Optimal nutrition: an overview. Proc. Nutr. Soc. **58:** 395–396.

Cancer Chemopreventive Activity of Resveratrol

KRISHNA P.L. BHAT AND JOHN M. PEZZUTO

Program for Collaborative Research in the Pharmaceutical Sciences, Department of Medicinal Chemistry and Pharmacognosy, College of Pharmacy, and University of Illinois Cancer Center, University of Illinois at Chicago, Chicago, Illinois 60612, USA

ABSTRACT: Cancer chemopreventive agents are designed to reduce the incidence of tumorigenesis by intervening at one or more stages of carcinogenesis. Recently, resveratrol, a natural product found in the diet of humans, has been shown to function as a cancer chemopreventive agent. Resveratrol was first shown to act as an antioxidant and antimutagenic agent, thus acting as an anti-initiation agent. Further evidence indicated that resveratrol selectively suppresses the transcriptional activation of cytochrome P-450 1A1 and inhibits the formation of carcinogen-induced preneoplastic lesions in a mouse mammary organ culture model. Resveratrol also inhibits the formation of 12-*O*-tetradecanoylphorbol-13-acetate (TPA)–promoted mouse skin tumors in the two-stage model. The enzymatic activities of COX-1 and -2 are inhibited by resveratrol in cell-free models, and *COX-2* mRNA and TPA-induced activation of protein kinase C and AP-1–mediated gene expression are suppressed by resveratrol in mammary epithelial cells. In addition, resveratrol strongly inhibits nitric oxide generation and inducible nitric oxide synthase protein expression. NFκB is strongly linked to inflammatory and immune responses and is associated with oncogenesis in certain models of cancer, and resveratrol suppresses the induction of this transcription factor by a number of agents. The mechanism may involve decreasing the phosphorylation and degradation of IκBα. At the cellular level, resveratrol also induces apoptosis, cell cycle delay or a block in the $G_1 \rightarrow S$ transition phase in a number of cell lines. Thus, resveratrol holds great promise for future development as a chemopreventive agent that may be useful for several disorders. Preclinical toxicity studies are underway that should be followed by human clinical trials.

KEYWORDS: resveratrol; cancer chemoprevention; anti-initiation agents; antioxidants; antimutagens

INTRODUCTION

Carcinogenesis may arise as a result of chemical or biological insults to normal cells in a multistep process that involves changes at the genetic level (initiation) followed by promotion and progression that ultimately lead to malignancy.[1] Administration of agents to prevent, inhibit, or delay progression of carcinogenesis has

Address for correspondence: Dr. John M. Pezzuto, Program for Collaborative Research in the Pharmaceutical Sciences, Department of Medicinal Chemistry and Pharmacognosy, College of Pharmacy, University of Illinois at Chicago, 833 S. Wood St., Chicago, IL 60612, USA. Voice: 312-996-5967; fax: 312-996-2815.
jpezzuto@uic.edu

been termed chemoprevention.[1] Potential chemopreventive agents that have been or will be evaluated in clinical trials include micronutrients, minerals, or synthetic compounds.[2] Natural products, however, are of particular interest as chemopreventive agents.[3] One important factor is that there may be experience with human consumption. On the basis of bioassay-guided fractionation of plant extracts collected worldwide, recent discoveries of chemopreventive agents in our laboratory include brassinin,[4] deguelin,[5] sulforamate,[6] and resveratrol.[7]

Resveratrol (*trans*-3,4′,5-trihydroxystilbene), originally identified as a phytoalexin by Langcake and Pryce,[8] has attracted considerable attention due to its abundance in grapes and grape products such as wine, a long-standing component of the diet.[9] Epidemiological studies, such as with the French population, have shown an inverse correlation between intake of wine and death resulting from coronary heart disease.[10] Polyphenolics in red wine are suspected to afford these cardioprotective effects due to their spectrum of biological activities, especially as antioxidants. The discovery of resveratrol, a polyphenolic, as a cancer chemopreventive agent[7] has offered renewed interest in grapes and grape products, and dietary supplements based on resveratrol are available. We currently provide an overview of our chemoprevention studies performed with resveratrol, including inhibition of reactive oxygen species (ROS) and cyclooxygenase (COX), and efficacy in skin and mammary animal models of tumorigenesis. Relevant work by others who have investigated the chemopreventive effects of resveratrol is also discussed, as well as estrogen modulatory activities.

ANTIOXIDANT EFFECTS

Electron acceptors such as molecular oxygen react easily with free radicals, to become ROS such as $O_2^{-\cdot}$, H_2O_2, and $\cdot OH$.[11] These ROS are being continuously generated in cells exposed to an aerobic environment, and have been associated with the genesis of tumors.[12,13] The damage incurred by proteins and DNA on contact with ROS can modulate carcinogenesis at all three stages,[14,15] and the chemoprotective role of antioxidants abundant in fruits, vegetables and beverages has received considerable attention. ROS also oxidize low density lipoproteins (LDL) that interact with scavenger receptors for macrophages, inducing the formation of lipid-laden foam cells that contribute to the development of atherosclerotic lesions.[16] Thus, consumption of antioxidants such as vitamin E have been suggested to offer protection against such cardiovascular complications.[17] In a similar manner, as described in a recent review,[18] resveratrol facilitates antioxidant mechanisms. However, bearing in mind the scope of the current article, we will restrict this discussion to the cancer chemopreventive antioxidant mechanisms of resveratrol.

We have shown that resveratrol can inhibit 12-*O*-tetradecanoylphorbol-13-acetate (TPA)–induced free radical formation with cultured HL-60 cells (IC_{50}, 6.2 μg/ml).[19] In a DU145 prostate cancer cell line, resveratrol effectively inhibited growth; this was accompanied by a decrease in nitric oxide (NO) production and an inhibition of inducible nitric oxide synthase (iNOS).[20] In a related study, resveratrol was shown to suppress the formation of superoxide radical (O_2^-) and H_2O_2 produced by macrophages stimulated by lipopolysaccharide (LPS) or phorbol esters (TPA).[21]

Resveratrol also suppressed COX-2 induction, [^{3}H]arachidonic acid ([^{3}H]AA) release and prostaglandin (PG) synthesis, all stimulated by LPA or TPA.[21] Manna *et al.*[22] have shown that resveratrol is capable of inhibiting reactive oxygen intermediates (ROI) generation and lipid peroxidation induced by tumor necrosis factor (TNF) in wide variety of cells. In addition, it was reported that resveratrol inhibits unopsonized zymosan-induced oxygen radical production in murine macrophages and human monocytes and neutrophils.[23]

In vivo evidence of the antioxidant capacity of resveratrol was also illustrated by protection against renal oxidative DNA damage induced by the kidney carcinogen $KBrO_3$.[24] We have observed that pre-treatment of mouse skin with resveratrol negated several TPA-induced oxidative events in a dose-dependent manner. H_2O_2 and glutathione levels were restored to control levels, as were myeloperoxidase, oxidized glutathione reductase and superoxide dismutase activities. TPA-induced increases in the expression of *c-fos* and *TGF-β1* mRNA were also selectively inhibited.[25]

EFFECTS ON CYTOCHROME P450, ARACHIDONIC ACID, AND PROTEIN KINASE PATHWAYS

Cytochrome P450 (CYP_{450}) isozymes are a large family of constitutive and inducible heme-containing enzymes which play an important role in the metabolism of xenobiotics.[26,27] The P450s are capable of metabolizing a wide variety of carcinogens such as polycyclic aromatic hydrocarbons (PAH) and heterocyclic amines.[27,28] Greatest attention, however, has focused on CYP1A1, CYP2A6, and CYP3A4, which are selectively involved in the metabolism of these carcinogens.[29] These metabolites are generally activated forms of the pro-carcinogens that subsequently interact with the DNA of target cells. P450s are overexpressed in a variety of human tumors including breast, colon, and lung.[30–32] Patterns of tumor-specific P450 expression (such as CYP1B1) have been found in rodent liver tumors.[29] The presence of tumor-specific P450 has therapeutic implications that can offer protection against cancer.

Resveratrol has recently been shown to inhibit some CYP_{450} isozymes. Several aromatic hydrocarbons (AH) are known to induce CYP1A1 gene transcription by binding to the Ah receptor, causing translocation to the nucleus, interaction with the promoter of CYP1A1 gene, and up-regulation of CYP1A1 mRNA and protein levels.[29] It has been reported that resveratrol inhibits this Ah-induced CYP1A1 expression and activity which is mediated by several AHs such as benzo[a]pyrene (B[*a*]P), 2,3,7,8-tetrachlorodibenzo-*p*-dioxin (TCDD) and dimethylbenz[*a*]anthracene (DMBA).[33,34] With B[*a*]P and DMBA, resveratrol appears to inhibit the binding of B[*a*]P-activated nuclear Ah receptor to the xenobiotic-response element of the CYP1A1 promoter without directly binding to the receptor.[34] In contrast, Casper *et al.*[35] have shown that resveratrol serves as an antagonist for the Ah receptor. The compound promotes Ah receptor translocation to the nucleus, but inhibits transactivation of dioxin-responsive genes such as *CYP1A1* and *ILβ*. Resveratrol also inhibits other isoforms of CYP_{450} associated with the dealkylation of benzyloresorufin, ethoxyresorufin and methoxyresorufin.[36] There seems to exist some amount of selectivity with which resveratrol distinguishes and inhibits the activity of CYP1A1 over CYP1A2. Of these two isoforms, CYP1A1 is an extrahepatic enzyme that is considered relatively important for chemoprevention.[36] Nonetheless, other isoforms of CYP_{450}, such as CYP1B1, described previously to be crucial for tumor progression, are also inhibited by resveratrol. Chang *et al.*[37] have shown that resveratrol suppresses CYP1B1-catalyzed 7-ehoxyresorufin *O*-dealkylation activity and mRNA expression without any observed toxicity in MCF-7 cells. Another subtype of CYP_{450}, CYP3A4, predominantly overexpressed in colon and liver cancers, was also shown to be inactivated by resveratrol.[38]

AA is metabolized by the COX pathway to PGs, which mediate several physiological responses.[39] COX exists in two isoforms: constitutive COX-1 is important in maintaining mucosal integrity, gastric microcirculation, and motor functions, and inducible COX-2, which is triggered by cytokines and endotoxins, has been implicated in inflammatory reactions.[39] COX-2 also plays an important role in tumorigenesis as suggested by up-regulation in transformed cells and various forms of cancer.[39] Another pathway by which AA is metabolized is via lipooxygenase (LOX) to produce hydroxyeicosatetraenoic acids (HETEs) or leukotrienes.[39] The exact role of LOX is not known; however, an increased level of this enzyme has been identified in bronchitis, hepatitis, and arthritis.[40] In addition, LOX-derived metabolites have an indirect influence on the development and progression of human cancers.[40]

Early studies have shown that resveratrol, isolated from the roots of *Polygonum* species, inhibited the activity of rat peritoneal polymorphonuclear 5-LOX and COX products.[41] Subsequent work in our laboratory showed that resveratrol had cancer chemopreventive activity in assays modeling three major stages of carcinogenesis.[7] We first identified resveratrol as chemopreventive agent on the basis of its ability to inhibit the COX and hydroperoxidase activity of COX-1. On the basis of these results, we investigated the anti-inflammatory activity of resveratrol in a rat paw-edema model. Resveratrol significantly suppressed both the acute and chronic phases of edema. In addition, resveratrol was shown to suppress the development of preneoplastic lesions in DMBA-treated mouse mammary glands. No signs of toxicity were observed, as judged by morphological examination of the glands. Finally, we studied tumorigenesis in the two-stage mouse skin cancer model in which DMBA was used as initiator and TPA as promoter. During an 18-week study, mice treated with DMBA-plus TPA developed an average of two tumors per mouse with 40% tumor incidence. Application of 1, 5, 10, or 25 μmol of resveratrol together with TPA twice a week for 18 weeks reduced the number of skin tumors per mouse by 68, 81, 76, or 98%, respectively, and the percentage of mice with tumors was reduced by 50, 63, 63, or 88%, respectively.

In another study, resveratrol has been shown to inhibit the activity of the COX-1 enzyme derived from sheep seminal vesicles.[42] It was also shown in rats that resveratrol reversed mild water immersion and restraint stress (WRS)-induced gastric protection and blood flow, and attenuated the increase in PGE_2 caused by WRS, demonstrating it was a specific COX-1 inhibitor.[43] However, our recent collaborative effort has shown that resveratrol suppresses activation of *COX-2* gene expression and activity by interfering with the protein kinase C (PKC) signal transduction pathway in mammary epithelial cells.[44] Resveratrol also inhibited PKC, ERK1, and c-Jun induced *COX-2* promoter activity.[44] Further, resveratrol (in addition to other resorcin-type molecules) suppressed basal levels and TGFα-induced COX-2 promoter-dependent transcriptional activity in colon cancer cells.[45] Moreno[46] has shown that resveratrol inhibits ROS production, phosholipase A_2 (PLA_2) activity, AA release, and PGE_2 synthesis simulated by fetal calf serum (FCS) or platelet-derived growth factor (PDGF) in 3T6 fibroblasts. The COX-2 protein induced by these agents was down-regulated leading to decreased growth and DNA synthesis.[46] The same observations were extended in murine resident peritoneal macrophages, when it was shown that resveratrol inhibited LPS and PMA-induced formation of ROS, inhibited COX-2 induction, and caused marked reduction of PG synthesis and AA release.[47] Resveratrol inhibits the peroxidase but not the cyclooxygenase activity of prostaglandin H synthase-2 (PGHS-2), leading to the accumulation of PGG_2, which is a toxic endoperoxide.[48]

Promotion of initiated cells to form a population of transformed, pre-malignant cells is manifested by changes in oncogenes and tumor suppressor genes.[49] The phorbol ester tumor promoter receptor PKC belongs to an isozyme family of 11 members.[50] PKC is a well-established regulatory element in the modulation of a variety of cellular processes such as cell signaling and tumor promotion.[51] Stewart *et al.*[52] have reported that resveratrol inhibits the PKC-catalyzed phosphorylation of arginine-rich protein substrate in a non-competitive manner. Resveratrol exhibits a broad spectrum of inhibition against a variety of PKC isozymes, such as cPKC, nPKC, and αPKC.[52] More significantly, this study has attempted to explain differences in the PKC inhibition potency of resveratrol in mammalian cells versus isolated PKC, since the potency of resveratrol depends on the nature of the substrate and cofactors.[52] In an independent study, it was shown that resveratrol inhibits the activity of PKC when activated by phophatidylcholine/phosphatidylserine vesicles greater than activation by Triton X-100.[53] The authors conclude that inhibition of PKC by resveratrol is dependent on membrane effects exerted near the lipid-water interface. Stewart *et al.*[54] also found that resveratrol exhibits a more distinguished inhibitory effect on the autophosphorylation reactions of protein kinase D (PKD). Gap junctional intracellular communication (GCIC) is important for normal cell growth and suppression can lead to transformation. Many tumor promoters are known to inhibit GCIC, and Nielsen *et al.*[55] have shown that resveratrol antagonizes TPA-mediated inhibition of GCIC.

EFFECT ON CELL CYCLE AND APOPTOSIS

Apoptosis is a normal physiological process wherein cells undergo programmed cell death with considerable morphological and biochemical changes in cellular structures.[56] Apoptosis is required to maintain a balance between cell proliferation and cell loss. Since misregulation in this balance can lead to malignant transformation, induction of apoptosis in a transformed cell population suppresses the development of cancer.[57] Various phytochemicals have been shown to induce apoptosis in malignant cells and this pathway provides a promising strategy to protect against cancer.[58,59] Resveratrol induces apoptosis in HL-60 cells as demonstrated by DNA fragmentation, an increased proportion of subdiploid cell population, and a time-dependent decrease in Bcl-2 expression.[60] In the same cell line, Clement *et al.*[61] reported that resveratrol caused a dose-dependent increase in cleavage of caspase substrate poly(ADP-ribose) polymerase (PARP), and caspase inhibitors could block this effect. Recent evidence emphasizes the importance of up-regulating the CD95-CD95L system for the control of apoptosis and a number of cytotoxic drugs upregulate their expression leading to CD95-mediated signal transduction, activation of caspases, and ultimately, cell death.[62] Up-regulation of the CD95-CD95L system was also shown to be one of the mechanisms of resveratrol-induced cell death in HL-60 cells, as well as T47D breast carcinoma cells.[61]

CD95-independent mechanisms of cell death caused by cytotoxic agents have also been proposed,[63,64] and doxorubicin-induced apoptosis operates by a CD95-independent pathway.[64] Similarly, resveratrol has been shown to exhibit CD95-independent apoptosis in another monocytic leukemic cell line, THP-1.[65] It was shown that resveratrol did not cause significant changes in the expression of CD95/CD95L or induce clustering of CD95 receptors in THP-1 cells and that neutralization with anti-CD95 or anti-CD95L did not protect from resveratrol-induced apoptosis.[65] Further, it has been observed that resveratrol induced cell death in CEM-C7H2 leukemia cells in a CD95-independent manner, as judged by lack of change in apoptosis in the presence of antibodies to CD95 or CD95L.[66] Moreover, resveratrol effectively induced apoptosis in a CD95-resistant Jurkat cell line.[66]

From a different perspective, apoptosis can be induced by UV-mediated DNA damage.[67] The most important mediator of this effect is the tumor suppressor gene p53, a gene which is mutated in about 50% of tumors, and the lack of expression or function is associated with an increased risk of cancer.[68] It has been shown with JB6 C1 41 cells that resveratrol suppressed cell transformation and induced apoptosis in a p53-dependent manner.[69] Significantly, apoptosis was induced at the same concentration that was required to inhibit cell transformation.[69] Moreover, resveratrol induced apoptosis in cells expressing wild-type p53, but not in p53-deficient cells.[69] Further mechanistic work in this cell line revealed that resveratrol-mediated apoptosis and activation of p53 is mediated via a complex formation between extracellular-signal-regulated protein kinases (ERKs) and p38 kinase.[70] It was also shown that stable expression of negative mutants of ERK2 or p38 kinase or their inhibitors impaired resveratrol-mediated apoptosis in this cell line.[70] To the contrary, in erythroleukemic cells, apoptosis is a result of oxidative stress and is 5-LOX dependent.[71] Programmed cell death is induced in these cells by activation of 5-LOX, and resveratrol was shown to inhibit this effect in dose-dependent manner.[71] In addition, resveratrol inhibited the activity of 15-LOX, COX and peroxidase activity in these cells with IC_{50} values ranging from 4.5–40μM.[72]

Bax, together with the anti-apoptotic gene *bcl-2*, is a transcriptional target for p53.[73] Bax-bax homodimers act as apoptosis inducers while bcl2-bax heterodimers act as a survival signal for cells.[73] It was shown that in a rat colon carcinogenesis model resveratrol induced pro-apoptotic *bax* expression in colon aberrant cryptic foci (ACF) but not in the surrounding mucosa.[74] In addition, resveratrol treatment suppressed expression of *p21* in normal mucosa but not in ACF.[74]

An increasing body of evidence suggests that formaldehyde (HCHO) generators or capturers can play a role in cell proliferation, differentiation, and apoptosis.[75] Szende *et al.*[76] have shown that several endogenous and exogenous methylated compounds (including resveratrol in its methylated form) are potential formaldehyde generators that can induce apoptosis. Moreover, this group has reported the simultaneous occurrence of resveratrol and HCHO in white and blue grape berries, and the interaction of these substances may have a role in apoptosis.[77] Evidence for *in vivo* induction of apoptosis was obtained when it was shown that i.p. administration of resveratrol to rats inoculated with a fast growing hepatoma caused a significant decrease in tumor cell content, an increase in G_2/M accumulation, and an aneuploid peak.[78]

Resveratrol has also been shown to affect the growth and tumorigenic potential of several cancer cell lines as evidenced by inhibition of the expression and function of androgen receptor (AR) in LNCaP (prostate cancer) cells.[79] Resveratrol down-regulated the expression of androgen-induced genes such as *p21*, in addition to mediating several other effects.[79] In the same cell line, however, it was found that resveratrol neither altered the expression of nor bound to the AR, but mediated anti-androgenic effects, such as decreased intracellular and secreted PSA levels.[80] In a related study, it was found that resveratrol mediated growth inhibition and apoptosis in LNCaP cells.[81] The authors extended these observations to some androgen non-responsive cell lines, whereby resveratrol caused growth inhibition and disrupted the G_1/S phase transition of the cell cycle without causing apoptosis.[81]

There have been comparisons of the effect of resveratrol on various breast carcinoma cell lines with different metastatic potentials.[82] Resveratrol caused an accumulation of cells in S-phase with concomitant reduced expression of Rb and increased expression of p53 and bcl-2 proteins.[82] The compound was most effective against MDA-MB-435 cells, which are highly invasive.[82] Resveratrol has also been shown to suppress smooth muscle cell proliferation as seen by a $G_1 \rightarrow S$ block without induction of apoptosis.[83] In oral squamous cell carcinoma cells, resveratrol caused growth inhibition, both alone and in combination with quercetin and other polyphenolics, as shown by a decrease in DNA synthesis.[84,85] Resveratrol was found to be more potent against human gingivial epithelial cells than other cells of the oral cavity.[86] In addition, resveratrol caused a decrease in DNA synthesis and irreversible damage to cell proliferation. It is noteworthy that resveratrol did not mediate any antioxidant effects in these cells compared to quercetin and *N*-acetyl-L-cysteine.[86] Also, resveratrol significantly inhibited the growth, but not the invasion, of highly metastatic B16-BL6 melanoma cells.[87]

A considerable amount of work has been ongoing with respect to resveratrol and its cell cycle effects. It appears that resveratrol has the greatest effect on the S-phase with consequent effects on S/G_2 transition. In HL-60 cells resveratrol caused an accumulation of cells in the G_1/S phase as seen by the absence of G_2/M peaks.[88] After a 24-h treatment, resveratrol caused a significant increase in the levels of cyclins A and E along with accumulation of cdc2 in the inactive phosphorylated form.[88] Similarly, Hsieh *et al.*[89] noted that resveratrol induced NO synthase in cultured pulmonary epithelial cells with suppression of cell cycle progression through the S and G_2 phases. This was accompanied by a concomitant increase in the expression of p53 and p21 and apoptosis.[89]

Ulsperger *et al.*[90] reported that resveratrol desensitized AHTO-7 human osteoblasts to growth stimulation in response to pretreatment with carcinoma cell supernatants. Greatest inhibition was observed with pancreas, breast and renal carcinoma-derived supernatants, whereas colon and prostate had minimal effects.[90] In another study with MC3T3-E1 osteoblast cells, resveratrol stimulated proliferation and differentiation as indicated by increased DNA synthesis, alkaline phosphatase and prolyl hydroxylase activity.[91] At lower concentrations, the production of PGE_2 was reduced in these cells.

In addition to the above effects on cell proliferation, utilizing a whole cell bioassay system, resveratrol has been shown to be a potent inhibitor of DNA polymerase activity, an important enzyme required for DNA replication.[92] Ribonucleotide reductases are complex enzymes that catalyze the reduction of ribonucleotides into corresponding deoxynucleotides and are important for S-phase DNA synthesis.[93] Resveratrol was recently shown to suppress the activity of this enzyme and DNA synthesis in mammalian tumor cells thus exhibiting an antiproliferative effect.[94]

In a B[*a*]P plus 4-(methylnitrosamino)-1-(3-pyridyl)-1-butanone (NNK)–induced lung tumorigenesis A/J mice model, dietary resveratrol (500ppm) had no effect.[95] All chemopreventive agents in this study were administered during the post-initiation period. The authors conclude that since antioxidants (which act as anti-initiation agents) like resveratrol and curcumin were ineffective, the role of oxidative damage in carcinogenesis by BaP plus NNK in the mouse lung tumor model was questionable.

STUDIES ON NFκB AND IκB

NFκB is an inducible transcription factor originally identified as a heterodimeric complex consisting of a 50kDa subunit (p50) and a 65kDa subunit (p65).[96] NFκB is strongly linked to inflammatory and immune responses and is associated with oncogenesis in certain models of cancer.[96] A common feature of the regulation of transcription factors belonging to the Rel family is their sequestration in the cytoplasm as inactive complexes with a class of inhibitory molecules known as IκB.[97] Phosphorylation of IκB leads to degradation of the proteosome, allowing NFκB to translocate to the nucleus where it regulates the expression of genes involved in inflammation, cell proliferation, and apoptosis.[97,98] The effects of novel therapeutic agents that specifically target NFκB proteins are currently being assessed in experimental models of cancer.

The first evidence of resveratrol affecting the transcription factor was derived from the work of Draczynska-Lusiak *et al.*[99] In their study, it was demonstrated that oxidized low density lipoproteins (LDL) and very LDL treatment activated NFκB binding activity, and resveratrol attenuated the activation of NFκB in PC-12 cells.[99] Resveratrol mediated suppression of NFκB in mouse macrophage RAW 264.7 cells has been controversial. Tsai *et al.*[100] demonstrated that resveratrol suppressed the activity of LPS-induced inducible form of nitric oxide synthase (iNOS) as seen by a decrease in NO generation in the culture medium. The effect was mediated through down-regulation of iNOS expression at the mRNA and protein levels. Resveratrol, at a concentration of 30μM, also suppressed activation of NFκB by LPS, and inhibited phosphorylation and degradation of IκBα. However, another group report that in RAW 264.7 cells resveratrol indeed decreased LPS-mediated NO release without affecting NFκB activation.[101] It has been shown that resveratrol suppressed TNF-induced NFκB activation (phosphorylation and nuclear translocation) in a variety of cell lines, such as U-937, Jurkat, and HeLa, induced by several agents including TPA, LPS, H_2O_2, okadaic acid, and ceramide.[22] The suppression of NFκB by resveratrol coincided with the inhibition of AP-1, another transcription factor that is involved in invasiveness and tumorigenesis. Furthermore, resveratrol inhibited TNF-induced activation of AP-1, MAPK kinase, c-JNK, ROS generation, lipid peroxidation and caspase activation.[22] Recently, it has been reported that resveratrol is a potent inhibitor of NFκB nuclear translocation and IκB degradation.[102] In addition, resveratrol effects are mediated through inhibition of IKK, a regulatory key complex that phosphorylates IκB on serines 32 and 36, and blocked the expression of mRNA-encoding monocyte chemoattractant protein-1, a NFκB-regulated gene.[102]

ESTROGEN MODULATORY EFFECTS

Flavonoids and isoflavonoids, like quercetin and genistein, classified as phytoestrogens, have been reported to display both estrogenic and anti-estrogenic effects.[103] Although genistein has a 1,000-fold lower potency than estradiol (E_2), its circulating concentrations in individuals consuming a moderate amount of soyfoods is nearly 1,000-fold higher than peak levels of endogenous E_2.[104] Population-based studies have also suggested that consumption of a phytoestrogen-rich diet is protective against prostate and bowel cancer, and cardiovascular disease.[105] Hence, these phytoestrogens may function as cancer chemopreventive agents in human beings.

Resveratrol has been considered as a phytoestrogen based on its structural similarity to diethylstilbesterol. Resveratrol was shown to bind to the estrogen receptor (ER),[106] and to activate the transcription of estrogen-responsive reporter genes transfected into MCF-7 cells in a dose-dependent manner.[105] Moreover, resveratrol was reported to function as a superagonist when combined with E_2, and to increase the expression of estrogen-regulated genes in MCF-7 cells.[106] However, subsequent studies could not confirm this superagonist activity. For example, with the same cell line, Lu and Serrero[107] reported that resveratrol showed antiestrogenic activity, as demonstrated by a suppression of progesterone receptor (PR) expression induced by E_2. In addition, resveratrol down-regulated the basal levels of *TGFα1* and *IGF 1* and up-regulated *TGFβ2* mRNA. In further studies, it was reported that both isomers of resveratrol exhibited superestrogenic activity at moderate concentrations (10 and 25 μM), whereas at low concentrations (0.1 and 1 μM), antiestrogenic effects were mediated in transfected estrogen response element (ERE)-luciferase reporter experiments.[108]

In ER-positive PR1 pituitary cells, resveratrol enhanced prolactin secretion without causing growth stimulation.[109] Strikingly, the induction of prolactin secretion was blocked by a pure antiestrogen in this study. Further work with MC3T3-E1 osteoblastic cells has shown that resveratrol increased alkaline phosphatase and prolyl hydroxylase activity in these cells, indicating an estrogenic and bone loss preventive effect.[110] Both activities could be antagonized by tamoxifen, signifying an estrogenic pathway. In U2 osteogenic cancer cells transfected with ER-AF1-luciferase plasmid, resveratrol caused an estrogenic response, whereas in HepG2 liver cells, resveratrol antagonized the action of E_2 in a dose-dependent manner.[111] Resveratrol has been shown to exhibit a direct antiproliferative effect on human breast epithelial cells that was independent of estrogen receptor status.[112] Recent studies have demonstrated that resveratrol together with other polyphenolics (when administered as a red wine concentrate) could suppress the proliferation of

ER-positive and -negative mammary cancer cells at picomolar and nanomolar concentrations.[113] In ER-positive cells, such as MCF-7 and T47D, the effect was attributed to both interaction with ER as well as antioxidant effects as seen by decreased formation of ROS.[113] Previous studies have also shown that resveratrol has a direct antiproliferative effect on human breast epithelial cells that is independent of the estrogen receptor status of the cells.[114]

However, *in vivo* studies with rat models to establish the estrogenic potential of resveratrol have not confirmed suggestions provided by *in vitro* tests. Various modes of resveratrol administration have been tested. Turner *et al.*[115] administered resveratrol orally to weanling rats at concentrations ranging from 1–1,000μg/day. With lower doses, resveratrol did not affect uterine weight, uterine epithelial cell height, cortical bone histomorphometry, or serum cholesterol.[115] However, at the highest dose, resveratrol could antagonize the serum cholesterol lowering activity of E_2.[115] In another study with immature rats, resveratrol was tested by two different routes of administration (oral and s.c.) at concentrations ranging from 0.03–120mg/kg/day, and no effect on uterus weight was found.[116] In a recent study with immature Wistar rats, resveratrol was injected s.c. at three different concentrations (18, 58, and 575 mg/kg).[117] E_2 increased uterine weight, enlarged uterine lumen, and induced hypertrophy of epithelial, stromal, and myometiral cells. In contrast, resveratrol mildly decreased uterine weight, and suppressed the expression of ER-α mRNA and protein, and PR mRNA, similar to antiestrogens.[117] The authors concluded that the anti-inflammatory properties of resveratrol suppress the activation of estrogen signalling similar to other anti-inflammatory agents such as indomethacin.[118]

We performed related studies with ovariectomized female Sprague-Dawley rats that were randomized into various groups after one week of quarantine. Resveratrol (3,000mg/kg diet) was administered in the diet and estradiol (50μg/kg body weight) was dissolved in sesame oil and injected subcutaneously for 30 days. The control group received s.c injections of sesame oil only. Vaginal smears were taken every day from all rats to monitor cell morphology. At the end of the study (day 31), the rats were sacrificed, uteri were removed and cleared of intrauterine fluid, and the weights were recorded. As expected, estradiol caused approximately a three-fold increase in uterus weight compared to the vehicle control group. In the absence of estrogen, resveratrol exhibited no estrogenic activity, and in the presence of estrogen, resveratrol demonstrated no antiestrogenic activity. In support of these conclusions, there were no significant differences in uterine weights compared to respective controls (see FIGURE 1 A). Additionally, daily vaginal smears were taken to monitor estrous cytology. Cells were identified as either leukocytes or nucleated (round), or cornified (irregularly shaped, non-nucleated) epithelial cells. A raw score on a scale of 1 through 5 was assigned for cell populations ranging from entirely leukocytes (indicative of a pro-estrous stage) to entirely cornified (indicative of a diestrous stage). The control group exhibited leukocyte abundance whereas estradiol treatment resulted in predominantly cornified epithelial cells within four days. Resveratrol did not cause any significant differences between the scores compared to the respective controls (FIG. 1 B).

The only reported study where resveratrol was shown to have estrogenic properties was in stroke-prone spontaneously hypertensive rats.[119] Resveratrol administered in the diet at a concentration of 5mg/kg/day to ovariectomized rats attenuated an increase in systolic blood pressure. It also enhanced endothelin-dependent vascular

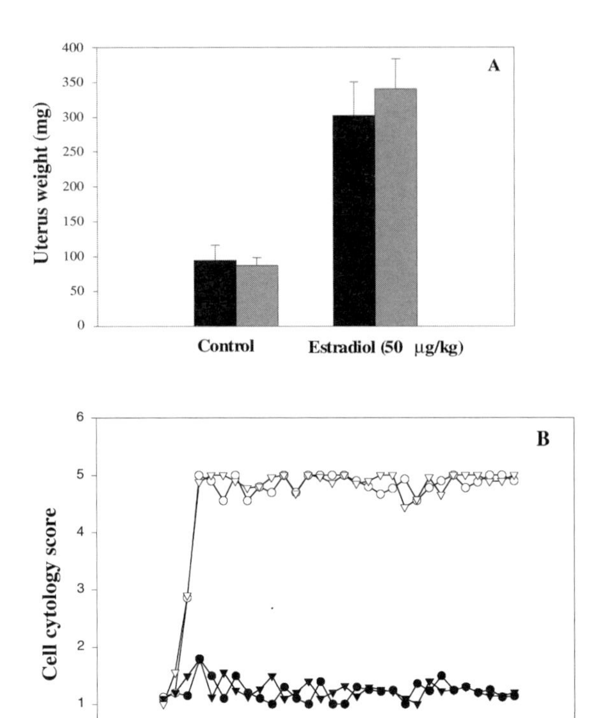

FIGURE 1. A. Effect of resveratrol on uterus weight in ovariectomized rats. Seven week-old ovariectomized female Sprague-Dawley rats were obtained from Harlan Sprague Dawley (Indianapolis, IN). All animals were placed on Teklad 4% rat/mouse chow (Harlan Teklad, Madison, WI), and maintained in accord with institutional guidelines. After one week of quarantine, animals were randomized into groups of seven. Animal cages were placed on the rack randomly to avoid variations due to environmental factors such as light and temperature that may result in a pseudo-estrous state. The groups consisted of (1) vehicle (0.1 ml sesame oil, s.c. injection) control; (2) estradiol-17β (50 μg/kg body weight in sesame oil, s.c. injection); (3) resveratrol (3,000 mg/kg diet); (4) resveratrol (3,000 mg/kg diet) plus estradiol-17β (50 μg/kg body weight in sesame oil, s.c. injection). Animals were observed twice daily and weighed twice weekly for the duration of the study (30 days). At the end of the study, the rats were sacrificed by CO_2 asphyxiation. The uteri were removed, an incision was made in each uterus to drain intrauterine fluid, they were dried between filter papers, and the dry weights recorded. Estradiol caused approximately a three-fold increase in uterus weight compared to the vehicle control group ($p < 0.01$). In the absence of estradiol, resveratrol (*gray bars*) exhibited no significant ($p = 0.32$) estrogenic activity, and in the presence of estradiol, resveratrol demonstrated no significant ($p = 0.20$) antiestrogenic activity. **B.** Effect of resveratrol on vaginal cell morphology. Resveratrol and/or estradiol were administered to ovariectomized female Sprague-Dawley rats as described above. Vaginal smears were taken daily using an eye-dropper containing 0.85% saline, placed on ringed slides on a slide tray, and observed under a light microscope using a 10× eyepiece and 10× objective. (Figure legend continued on opposite age.)

relaxation in response to acetylcholine and prevented ovariectomy-induced decreases in femoral bone strength in a manner similar to estradiol.[119] Significantly, these effects could be partially antagonized with the pure antiestrogen, ICI 182780.[119]

Recently, it was demonstrated that resveratrol could bind to both ER-α and -β with comparable affinity, but with 7,000-fold lower affinity than E_2.[120] It was also shown that the type of estrogen modulatory activity of resveratrol depends on the ERE sequence and ER subtype. Resveratrol was shown to have higher transcriptional activity when bound to ER-β than -α. Moreover, resveratrol showed antagonist activity with ER-α, but not with ER-β.[120] Consistent with this report, we have recently observed that resveratrol mediates antiestrogenic effects in endometrial cancer (Ishikawa) cells by a novel mechanism that involves selective down-regulation of ER-α, but not ER-β, manifested as suppression of estrogen-dependent alkaline phosphatase and ERE-luciferase activities, as well as PR and α1-integrin expression.[121]

As more data accumulate on the estrogen modulatory effects of resveratrol, controversy still persists with regard to its ability to serve as a chemopreventive agent in breast cancer. However, certain conclusions can be drawn based on the reports published thus far. Resveratrol indeed exhibits mixed estrogen agonist/antagonist activity with *in vitro* systems (such as reporter gene assays). However, these data could not be directly extrapolated to an *in vivo* situation using the classical uterotrophic assay. Rather, resveratrol appears to have a pure antiestrogenic effect at high doses (575 mg/kg body weight). These doses may not be relevant in a chemopreventive setting, but nevertheless provide evidence that endometrial carcinogenicity (as a result of ER-agonsim in the uterus) is not likely to be facilitated by resveratrol as with other estrogen receptor modulators such as tamoxifen. The estrogenic effect seen with hypertensive models may have alternate mechanistic pathways compared to the uterotrophic assays, and the pharmacological differences could be explained by several factors, such as selectivity, relative levels of ER-α and -β, and bioavailability in these tissues. Moreover, we have observed that resveratrol exhibits mixed estrogenic/antiestrogenic properties in some ER^+ mammary cancer cell lines (as seen by differential effects on reporter gene assays as well as natural estrogen-responsive genes such as PR), acts as a pure antiestrogen in other mammary cell lines and, in rodent models, inhibits the formation of carcinogen-induced preneoplastic mammary lesions and tumors.[122] These studies have led us to speculate that resveratrol could function as a novel selective estrogen receptor modulator (SERM).

FIGURE 1/continued. Smears were read immediately and cells were identified as either leukocytes or nucleated (round), or cornified (irregularly shaped, non-nucleated) epithelial cells. A raw score on a scale of 1 through 5 was assigned for cell populations ranging from entirely leukocytes (indicative of a pro-estrous stage) to entirely cornified (indicative of a diestrous stage), and plotted against time. Data points represent the mean of raw scores of each group. Control group (●) receiving only sesame oil had mean scores ranging from 1–2. Estradiol treatment (○) caused a significant ($p < 0.01$) increase in the score (4.5–5). Resveratrol with (▽) or without (▼) estradiol did not cause any significant ($p = 0.7$) differences in the scores, compared to the respective controls.

CONCLUSIONS AND OUTLOOK

In this article, we have briefly reviewed early and recent investigations with resveratrol in basic cancer prevention. Resveratrol represents a relatively new class of chemopreventive agent in comparison with retinoids and other diet-derived compounds. In various *in vitro* and *in vivo* models, resveratrol has proved to be capable of retarding or preventing steps of carcinogenesis, several of which are summarized in FIGURE 2. It has become apparent that resveratrol can mediate differential responses with various tissues, organs, and assay models. Thus, some activities have been or remain controversial. Nevertheless, one factor encouraging further work with resveratrol is that it is uniquely found in a soluble form in red wine, and it is virtually absent in most fruits and vegetables that form a major portion of human diet.[123] However, although wine may be considered a predominant bioavailable dietary source, ingestion of grapes or peanuts may be relevant, and dietary supplements are available. The compound also bears a simple chemical structure that is capable of interacting with a variety of receptors and enzymes, and serving as an activator or inhibitor in a number of pathways. Nonetheless, it is noteworthy that no toxicity reports have been published with respect to resveratrol in animals. In our experience, resveratrol has proven to be non-toxic, even at high doses (3,000 mg/kg diet for 120 days in rats). In addition, since resveratrol is an active ingredient of several traditional medicines used for centuries in India, China, and Japan, the general medicinal

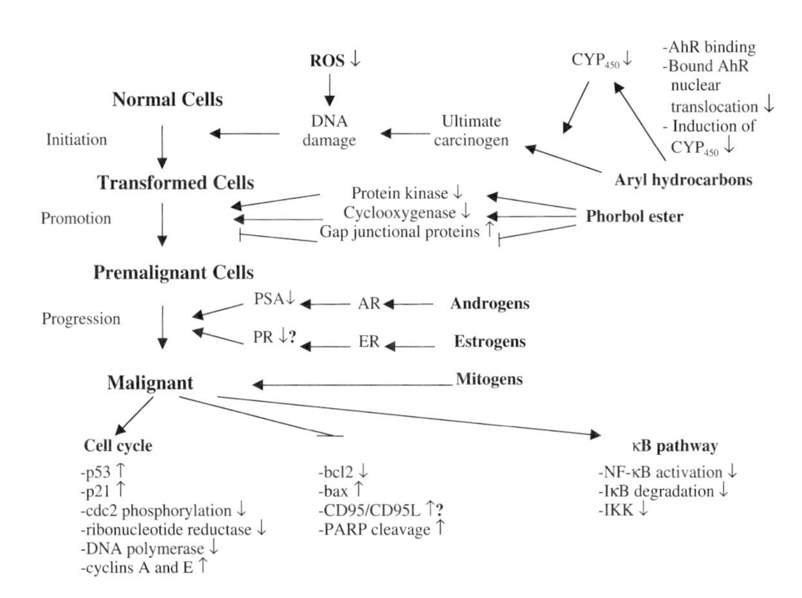

FIGURE 2. Schematic representation of the effect of resveratrol on various pathways of carcinogenesis, and cell proliferation. The upward arrow symbol (↑) indicates targets of resveratrol that are either enhanced or up-regulated, the downward arrow symbol (↓) indicates targets that either suppressed or down-regulated. A *question mark* indicates controversial data.

value and safety of this compound may be suggested.[124,125] A challenge for the future will be proper extrapolation of data from *in vitro* experiments or animal studies to the human situation. As exemplified by this article, a considerable amount of time, research, and financial resources have already been invested in the development and characterization of resveratrol. Synthesis and large-scale production have been accomplished, and preclinical toxicity studies are underway. The overall situation is unique since the compound is already consumed by human beings, so certain benefits may currently be realized, irrespective of our knowledge of mechanism. Nonetheless, full potential can only be realized on the basis of clinical trials, and it appears such trials will be performed in the future.

ACKNOWLEDGMENTS

This work was supported by P01 CA48112 awarded by the National Cancer Institute.

The authors wish to thank Pharmascience, Canada for the supply of resveratrol, Dr. Donald P. Waller, Department of Pharmaceutics and Pharmacodynamics, College of Pharmacy, University of Illinois at Chicago, for advice concerning uterotrophic studies, and Mr. Daniel Lantvit for assistance with the animal work.

REFERENCES

1. Hong, W.K. & M.B. Sporn. 1997. Recent advances in chemoprevention of cancer. Science **278:** 1073–1077.
2. Kelloff, G.J., C.C. Sigman & P. Greenwald. 1999. Cancer chemoprevention: progress and promise. Eur. J. Cancer **35:** 2031–2038.
3. Pezzuto, J.M. 1997. Plant-derived anticancer agents. Biochem. Pharmacol. **53:** 121–133.
4. Mehta, R.G., J. Liu, A. Constantinou, *et al.* 1995. Cancer chemopreventive activity of brassinin, a phytoalexin from cabbage. Carcinogenesis **16:** 399–404.
5. Udeani, G.O., C. Gerhäuser, C.F. Thomas, *et al.* 1997. Cancer chemopreventive activity mediated by deguelin, a naturally occurring rotenoid. Cancer Res. **57:** 3424–3428.
6. Gerhäuser, C., M. You, J. Liu, *et al.* 1997. Cancer chemopreventive potential of sulforamate, a novel analogue of sulforaphane that induces phase 2 drug-metabolizing enzymes. Cancer Res. **57:** 272–278.
7. Jang, M., L. Cai, G.O. Udeani, *et al.* 1997. Cancer chemopreventive activity of resveratrol, a natural product derived from grapes. Science **275:** 218–220.
8. Langcake, P. & R.J. Pryce. 1976. The production of resveratrol by *Vitis vinifera* and other members of the Vitaceae as a reponse to infection or injury. Physiol. Plant Pathol. **9:** 77–86.
9. Soleas, G.J., E.P. Diamandis & D.M. Goldberg. 1997. Wine as a biological fluid: history, production, and role in disease prevention. J. Clin. Lab Anal. **11:** 287–313.
10. Kopp, P. 1998. Resveratrol, a phytoestrogen found in red wine. A possible explanation for the conundrum of the 'French paradox'? Eur. J. Endocrinol. **138:** 619–620.
11. Halliwell, B. & J.M.C. Gutteridge. 1984. Oxygen toxicity, oxygen radicals, transition metals and disease. Biochem. J. **219:** 1–14.
12. Troll, W. & R. Wiesner. 1985. The role of oxygen radicals as a possible mechanism of tumor promotion. Annu. Rev. Pharmacol. Toxicol. **25:** 509–528.
13. Trush, M.A. & T.W. Kensler. 1991. An overview of the relationship between oxidative stress and chemical carcinogenesis. Free Radical Biol. Med. **10:** 201–209.

14. GROMADZINSKA, J. & W. WASOWICZ. 2000. The role of reactive oxygen species in the development of malignancies. Int. J. Occup. Med. Environ. Health **13:** 233–245.
15. KENSLER, T., K. GUYTON, P. EGNER, *et al.* 1995. Role of reactive intermediates in tumor promotion and progression. Prog. Clin. Biol. Res. **391:** 103–116.
16. PARTHASARATHY, S. & S.M. RANKIN. 1992. Role of oxidized low density lipoprotein in atherogenesis. Prog. Lipid Res. **31:** 127–143.
17. EICHHOLZER, M., J. LUTHY, F. GUTZWILLER & H.B. STAHELIN. 2001. The role of folate, antioxidant vitamins and other constituents in fruit and vegetables in the prevention of cardiovascular disease: the epidemiological evidence. Int. J. Vitam. Nutr. Res. **71:** 5–17.
18. BHAT, K.P.L., J.W. KOSMEDER, II & J.M. PEZZUTO. 2001. Forum review article: biological effects of resveratrol. Antioxid. Redox Signal **3:** 1047–1064.
19. LEE, S.K., Z.H. MBWAMBO, H. CHUNG, *et al.* 1998. Evaluation of the antioxidant potential of natural products. Comb. Chem. High Throughput Screen. **1:** 35–46.
20. KAMPA, M., A. HATZOGLOU, G. NOTAS, *et al.* 2000. Wine antioxidant polyphenols inhibit the proliferation of human prostate cancer cell lines. Nutr. Cancer **37:** 223–233.
21. MARTINEZ, J. & J.J. MORENO. 2000. Effect of resveratrol, a natural polyphenolic compound, on reactive oxygen species and prostaglandin production. Biochem. Pharmacol. **59:** 865–870.
22. MANNA, S.K., A. MUKHOPADHYAY & B.B. AGGARWAL. 2000. Resveratrol suppresses TNF-induced activation of nuclear transcription factors NF-κB, activator protein-1, and apoptosis: potential role of reactive oxygen intermediates and lipid peroxidation. J. Immunol. **164:** 6509–6519.
23. JANG, D.S., B.S. KANG, S.Y. RYU, *et al.* 1999. Inhibitory effects of resveratrol analogs on unopsonized zymosan-induced oxygen radical production. Biochem. Pharmacol. **57:** 705–712.
24. CADENAS, S. & G. BARJA. 2000. Resveratrol, melatonin, vitamin E, and PBN protect against renal oxidative DNA damage induced by the kidney carcinogen $KBrO_3$. Free Radical Biol. Med. **26:** 1531–1537.
25. JANG, M. & J.M. PEZZUTO. 1998. Effects of resveratrol on 12-*O*-tetradecanoylphorbol-13-acetate-induced oxidative events and gene expression in mouse skin. Cancer Lett. **134:** 81–89.
26. NELSON, D.R., L. KOYMANS, T. KAMATAKI, *et al.* 1996. P450 superfamily: update on new sequences, gene mapping, accession numbers and nomenclature. Pharmacogenetics **6:** 1–42.
27. GONZALEZ F.J. & H.V. GELBOIN. 1994. Role of human cytochromes P450 in the metabolic activation of chemical carcinogens and toxins. Drug Metab. Rev. **26:** 165–183.
28. WINDMILL, K.F., R.A. MCKINNON, X. ZHU, *et al.* 1997. The role of xenobiotic metabolizing enzymes in arylamine toxicity and carcinogenesis: functional and localization studies. Mutat. Res. **376:** 153–160.
29. MURRAY, G.I. 2000. The role of cytochrome P450 in tumor development and progression and its potential in therapy. J. Pathol. **192:** 419–426.
30. MURRAY G.I., C.O. FOSTER, T.S. BARNES, *et al.* 1991. Expression of cytochrome P450IA in breast cancer. Br. J. Cancer **63:** 1021–1023.
31. MEKHAIL-ISHAK K., N. HUDSON, M.S. TSAO & G. BATIST. 1989. Implications for therapy of drug-metabolizing enzymes in human colon cancer. Cancer Res. **49:** 4866–4869.
32. MCLEMORE T.L., S. ADELBERG, M. CZERWINSKI, *et al.* 1989. Altered regulation of the cytochrome P4501A1 gene: novel inducer-independent gene expression in pulmonary carcinoma cell lines. J. Natl. Cancer Inst. **81:** 1787–1794.
33. CIOLINO H.P., P.J. DASCHNER & G.C. YEH. 1998. Resveratrol inhibits transcription of CYP1A1 in vitro by preventing activation of the aryl hydrocarbon receptor. Cancer Res. **58:** 5707–5712.
34. CIOLINO H.P. & G.C. YEH. 1999. Inhibition of aryl hydrocarbon-induced cytochrome P-450 1A1 enzyme activity and CYP1A1 expression by resveratrol. Mol. Pharmacol. **56:** 760–767.

35. CASPER, R.F., M. QUESNE, I.M. ROGERS, *et al.* 1999. Resveratrol has antagonist activity on the aryl hydrocarbon receptor: implications for prevention of dioxin toxicity. Mol. Pharmacol. **56:** 784–790.
36. TEEL, R.W. & H. HUYNH. 1998. Modulation by phytochemicals of cytochrome P450-linked enzyme activity. Cancer Lett. **133:** 135–141.
37. CHANG, T.K., W.B. LEE & H.H. KO. 2000. Trans-resveratrol modulates the catalytic activity and mRNA expression of the procarcinogen-activating human cytochrome P450 1B1. Can. J. Physiol. Pharmacol. **78:** 874–881.
38. CHAN, W.K. & A.B. DELUCCHI. 2000. Resveratrol, a red wine constituent, is a mechanism-based inactivator of cytochrome P450 3A4. Life Sci. **67:** 3103–3112.
39. CUENDET, M. & J.M. PEZZUTO. 2000. The role of cyclooxygenase and lipoxygenase in cancer chemoprevention. Drug Metabol. Drug Interact. **17:** 109–157.
40. STEELE, V.E., C.A. HOLMES, *et al.* 2000. Potential use of lipoxygenase inhibitors for cancer chemoprevention. Expert Opin. Investig. Drugs **9:** 2121–2138.
41. KIMURA, Y., H. OKUDA & S. ARICHI. 1985. Effects of stilbenes on arachidonate metabolism in leukocytes. Biochim. Biophys. Acta **834:** 275–278.
42. SHIN, N.H., S.Y. RYU, H. LEE, *et al.* 1997. Inhibitory effects of hydroxystilbenes on cyclooxygenase from sheep seminal vesicles. Planta. Med. **64:** 283–284.
43. BRZOZOWSKI, T., P.C. KONTUREK, S.J. KONTUREK, *et al.* 2000. Expression of cyclooxygenase (COX)-1 and COX-2 in adaptive cytoprotection induced by mild stress. J. Physiol. Paris **94:** 83–91.
44. SUBBARAMAIAH, K., W.J. CHUNG, P. MICHALUART, *et al.* 1998. Resveratrol inhibits cyclooxygenase-2 transcription and activity in phorbol ester-treated human mammary epithelial cells. J. Biol. Chem. **273:** 21875–21882.
45. MUTOH, M., M. TAKAHASHI, K. FUKUDA, *et al.* 2000. Suppression of cyclooxygenase-2 promoter-dependent transcriptional activity in colon cancer cells by chemopreventive agants with a resorcin-type structure. Carcinogenesis **21:** 959–963.
46. MORENO, J.J. 2000. Resveratrol modulates arachidonic acid release, prostaglandin synthesis, and 3T6 fibroblast growth. J. Pharmacol. Exp. Ther. **294:** 333–338.
47. MARTINEZ, J. & J.J. MORENO. 2000. Effect of resveratrol, a natural polyphenolic compound, on reactive oxygen species and prostaglandin production. Biochem. Pharmacol. **59:** 865–870.
48. JOHNSON, J.L. & K.R. MADDIPATI. 1998. Paradoxical effects of resveratrol on the two prostaglandin H synthases. Prostaglandins Other Lipid Mediat. **56:** 131–143.
49. SURH, Y.H. 1999. Molecular mechanisms of chemopreventive effects of selected dietary and medicinal phenolic substances. Mutat. Res. **428:** 305–327.
50. BLOBE, G.C., S. STRIBLING, L.M. OBEID & Y.A. HANNUN. 1996. Protein kinase C isoenzymes: regulation and function. Cancer Surv. **27:** 213–248.
51. NEWTON, A.C. 1995. Protein kinase C: structure, function, and regulation. J. Biol. Chem. **270:** 28495–28498.
52. STEWART, J.R., N.E. WARD, C.G. IOANNIDES & C.A. O'BRIAN. 1999. Resveratrol preferentially inhibits protein kinase C-catalyzed phosphorylation of a cofactor-independent, arginine-rich protein substrate by a novel mechanism. Biochemistry **38:** 13244–13251.
53. GARCIA-GARCIA, J., V. MICOL, A. DE GODOS & J.C. GOMEZ-FERNANDEZ. 1999. The cancer chemopreventive agent resveratrol is incorporated into model membranes and inhibits protein kinase Cα activity. Arch. Biochem. Biophys. **372:** 382–388.
54. STEWART, J.R., K.L. CHRISTMAN & C.A. O'BRIAN. 2000. Effects of resveratrol on the autophosphorylation of phorbol ester-responsive protein kinases. Inhibition of protein kinase D but not protein kinase C isozyme autophosphorylation. Biochem. Pharmacol. **60:** 1355–1359.
55. NIELSEN, M., R.J. RUCH & O. VANG. 2000. Resveratrol reverses tumor-promoter-induced inhibition of gap-junctional intercellular communication. Biochem. Biophys. Res. Commun. **275:** 804–809.
56. TANG, D.G. & A.T. PORTER. 1996. Apoptosis: A current molecular analysis. Pathol. Oncol. Res. **2:** 117–131.
57. THOMPSON, C.B. 1995. Apoptosis in the pathogenesis and treatment of disease. Science **267:** 1456–1462.

58. KATDARE, M., H. JINNO, M.P. OSBORNE & N.T. TELANG. 1999. Negative growth regulation of oncogene-transformed human breast epithelial cells by phytochemicals. Role of apoptosis. Ann. N.Y. Acad. Sci. **889:** 247–252.
59. KATDARE, M., M.P. OSBORNE & N.T. TELANG. 1998. Inhibition of aberrant proliferation and induction of apoptosis in pre-neoplastic human mammary epithelial cells by natural phytochemicals. Oncol. Rep. **5:** 311–315.
60. SURH, Y.J., Y.J. HURH, J.Y. KANG, *et al.* 1999. Resveratrol, an antioxidant present in red wine, induces apoptosis in human promyelocytic leukemia (HL-60) cells. Cancer Lett. **140:** 1–10.
61. CLEMENT, M.V., J.L. HIRPARA, S.H. CHAWDHURY & S. PERVAIZ. 1998. Chemopreventive agent resveratrol, a natural product derived from grapes, triggers CD95 signaling-dependent apoptosis in human tumor cells. Blood **92:** 996–1002.
62. MICHEAU, O., E. SOLARY, A. HAMMANN, *et al.* 1997. Sensitization of cancer cells treated with cytotoxic drugs to fas-mediated cytotoxicity. J. Natl. Cancer Inst. **89:** 783–789.
63. MICHEAU, O., E. SOLARY, A. HAMMANN & M.T. DIMANCHE-BOITREL. 1999. Fas ligand-independent, FADD-mediated activation of the Fas death pathway by anticancer drugs. J. Biol. Chem. **274:** 7987–7992.
64. PETAK, I., D.M. TILLMAN, F.G. HARWOOD, *et al.* 2000. Fas-dependent and -independent mechanisms of cell death following DNA damage in human colon carcinoma cells. Cancer Res. **60:** 2643–2650.
65. TSAN, M.F., J.E. WHITE, J.G. MAHESHWARI, *et al.* 2000. Resveratrol induces Fas signaling-independent apoptosis in THP-1 human monocytic leukemia cells. Br. J. Hæmatol. **109:** 405–412.
66. BERNHARD, D., I. TINHOFER, M. TONKO, *et al.* 2000. Resveratrol causes arrest in the S-phase prior to Fas-independent apoptosis in CEM-C7H2 leukemia cells. Cell Death Differ. **7:** 834–842.
67. KULMS, D. & T. SCHWARZ. 2000. Molecular mechanisms of UV-induced apoptosis. Photodermatol. Photoimmunol. Photomed. **16:** 195–201.
68. HUSSAIN, S.P., M.H. HOLLSTEIN & C.C. HARRIS. 2000. p53 tumor suppressor gene: at the crossroads of molecular carcinogenesis, molecular epidemiology, and human risk assessment. Ann. N.Y. Acad. Sci. **919:** 79–85.
69. HUANG, C., W.Y. MA, A. GORANSON & Z. DONG. 1999. Resveratrol suppresses cell transformation and induces apoptosis through a p53-dependent pathway. Carcinogenesis **20:** 237–242.
70. SHE, Q.B., A.M. BODE, W.Y. MA, *et al.* 2001. Resveratrol-induced activation of p53 and apoptosis is mediated by extracellular-signal-regulated protein kinases and p38 kinase. Cancer Res. **61:** 1604–1610.
71. MACCARRONE, M., M.V. CATANI, A. FINAZZI AGRO & G. MELINO. 1997. Involvement of 5-lipoxygenase in programmed cell death of cancer cells. Cell Death Differ. **4:** 396–402.
72. MACCARRONE, M., T. LORENZON, P. GUERRIERI & A.F. AGRO. 1999. Resveratrol prevents apoptosis in K562 cells by inhibiting lipoxygenase and cyclooxygenase activity. Eur. J. Biochem. **265:** 27–34.
73. BASU, A. & S. HALDAR. 1998. The relationship between Bcl2, Bax and p53: consequences for cell cycle progression and cell death. Mol. Hum. Reprod. **12:** 1099–1109.
74. TESSITORE, L., A. DAVIT, I. SAROTTO & G. CADERNI. 2000. Resveratrol depresses the growth of colorectal aberrant crypt foci by affecting bax and $p21^{CIP}$ expression. Carcinogenesis **21:** 1619–1622.
75. TYIHAK, E., L. ALBERT, Z.I. NEMETH, *et al.* 1998. Formaldehyde cycle and the natural formaldehyde generators and capturers. Acta Biol. Hung. **49:** 225–238.
76. SZENDE, B., E. TYIHAK, L. TREZL, *et al.* 1998. Formaldehyde generators and capturers as influencing factors of mitotic and apoptotic processes. Acta Biol. Hung. **49:** 323–329.
77. KIRALY-VEGHELY, Z., E. TYIHAK, L. ALBERT, *et al.* 1998. Identification and measurement of resveratrol and formaldehyde in parts of white and blue grape berries. Acta Biol. Hung. **49:** 281–289.

78. Carbo, N., P. Costelli, M.F. Baccino, *et al.* 1999. Resveratrol, a natural product present in wine, decreases tumor growth in a rat tumor model. Biochem. Biophys. Res. Commun. **54:** 739–743.
79. Mitchell, S.H., W. Zhu & C.Y.F. Young. 1999. Resveratrol inhibits the expression and function of the androgen receptor in LNCaP prostate cancer cells. Cancer Res. **59:** 5892–5895.
80. Hsieh, T.C. & J.M. Wu. 2000. Grape-derived chemopreventive agent resveratrol decreases prostate-specific antigen (PSA) expression in LNCaP cells by androgen receptor (AR)-independent mechanism. Anticancer Res. **20:** 225–228.
81. Hsieh, T.C. & J.M. Wu. 1999. Differential effects on growth, cell cycle arrest, and induction of apoptosis by resveratrol in human prostate cancer cell lines. Exp. Cell Res. **249:** 109–115.
82. Hsieh, T.C., P. Burfeind, K. Laud, *et al.* 1999. Cell cycle effects and control of gene expression by resveratrol in human breast carcinoma cell lines with different metastatic potentials. Int. J. Oncol. 15: 245–252.
83. Zou, J., Y. Huang, Q. Chen, *et al.* 1999. Suppression of mitogenesis and regulation of cell cycle traverse by resveratrol in cultured smooth muscle cells. Int. J. Oncol. **15:** 647–651.
84. ElAttar, T.M.A. & A.S. Virji. 1999. Modulating effect of resveratrol and quercetin on oral cancer cell growth and proliferation. Anticancer Drugs **10:** 187–193.
85. ElAttar, T.M. & A.S. Virji. 1999. The effect of red wine and its components on growth and proliferation of human oral squamous carcinoma cells. Anticancer Res. **19:** 5407–5414.
86. Babich, H., A.G. Reisbaum & H.L. Zuckerbraun. 2000. In vitro response of human gingival epithelial S-G cells to resveratrol. Toxicol. Lett. **114:** 143–153.
87. Caltagirone, S., C. Rossi, A. Poggi, *et al.* 2000. Flavonoids apigenin and quercetin inhibit melanoma growth and metastatic potential. Int. J. Cancer **87:** 595–600.
88. Ragione, F.D., V. Cucciolla, A. Borriello, *et al.* 1998. Resveratrol arrests the cell division cycle at S/G_2 phase transition. Biochem. Biophys. Res. Commun. **250:** 53–58.
89. Hsieh, T.C., G. Juan, Z. Darzynkiewicz & J.M. Wu. 1999. Resveratrol increases nitric oxide synthase, induces accumulation of p53 and $p21^{WAF1/CIP1}$, and suppresses cultured bovine pulmonary artery endothelial cell proliferation by perturbing progression through S and G_2. Cancer Res. **59:** 2596–2601.
90. Ulsperger, E., G. Hamilton, M. Raderer, *et al.* 1999. Resveratrol pretreatment desensitizes AHTO-7 human osteoblasts to growth stimulation in response to carcinoma cell supernatants. Int. J. Oncol. **15:** 955–959.
91. Mizutani, K., K. Ikeda, Y. Kawai & Y. Yamori. 1998. Resveratrol stimulates the proliferation and differentiation of osteoblastic MC3T3-E1 cells. Biochem. Biophys. Res. Commun. **253:** 859–863.
92. Sun, N.J., S.H. Woo, J.M. Cassady & R.M. Snapka. 1998. DNA polymerase and topoisomerase II inhibitors from *Psoralea corylifolia.* J. Nat. Prod. **61:** 362–366.
93. Reichard, P. 1987. Regulation of deoxyribotide synthesis. Biochemistry **26:** 3245–3248.
94. Fontecave, M., M. Lepoivre, E. Elleingand, *et al.* 1998. Resveratrol, a remarkable inhibitor of ribonucleotide reductase. FEBS Lett. **421:** 277–279.
95. Hecht, S.S., P.M.J. Kenney, M. Wang, *et al.* 1999. Evaluation of butylated hydroxyanisole, myo-inositol, curcumin, esculetin, resveratrol and lycopene as inhibitors as benzo[a]pyrene plus 4-(methylnitrosamino)-1-(3-pyridyl)-1-butanone-induced lung tumorigenesis in A/J mice. Cancer Lett. **137:** 123–130.
96. Schwartz, S.A., A. Hernandez & B. Mark Evers. 1999. The role of NF-kappaB/IkappaB proteins in cancer: implications for novel treatment strategies. Surg. Oncol. **8:** 143–153.
97. Baldwin, A.S. 1999. The NF-kappa B and I kappa B proteins: new discoveries and insights. Annu. Rev. Immunol. **14:** 649–683.
98. Verma, I.M. & J. Stevenson. 1997. IkappaB kinase: beginning, not the end. Proc. Natl. Acad. Sci. USA **94:** 11758–11760.
99. Draczynska-Lusiak, B., Y.M. Chen & A.Y. Sun. 1998. Oxidized lipoproteins activate NF-kappaB binding activity and apoptosis in PC12 cells. Neuroreport **9:** 527–532.

100. TSAI, S.H., S.Y. LIN-SHIAU & J.K. LIN. 1999. Suppression of nitric oxide synthase and down-regulation of the activation of NFκB in macrophages by resveratrol. Br. J. Pharmacol. **126:** 673–680.
101. WADSWORTH, T.L. & D.R. KOOP. 1999. Effects of wine polyphenolics quercetine and resveratrol on pro-inflammatory cytokine expression in RAW 264.7 macrophages. Biochem. Pharmacol. **57:** 941–949.
102. HOLMES-MCNARY, M. & A.S. BALDWIN, JR. 2000. Chemopreventive properties of trans-resveratrol are associated with inhibition of activation of the IkappaB kinase. Cancer Res. **60:** 3477–3483.
103. DAVIS, S.R., F.S. DALAIS, E.R. SIMPSON & A.L. MURKIES. 1999. Phytoestrogens in health and disease. Recent Prog. Horm. Res. **54:** 185–210.
104. ADLERCREUTZ, C.H., B.R. GOLDIN, S.L. GORBACH, *et al.* 1995. Soybean phytoestrogen intake and cancer risk. J. Nutr. **125:** 757S–770S.
105. WISEMAN, H. 2000. The therapeutic potential of phytoestrogens. Expert Opin. Investig. Drugs **9:** 1829–1840.
106. GEHM, B.D., J.M. MCANDREWS, P.Y. CHIEN & J.L. JAMESON. 1997. Resveratrol, a polyphenolic compound found in grapes and wine, is an agonist for the estrogen receptor. Proc. Natl. Acad. Sci. U.S.A. **94:** 14138–14143.
107. LU, R. & G. SERRERO. 1999. Resveratrol, a natural product derived from grape, exhibits antiestrogenic activity and inhibits the growth of human breast cancer cells. J. Cell. Physiol. **179:** 297–304.
108. BASLY, J.P., F. MARRE-FOURNIER, J.C. LE BAIL, *et al.* 2000. Estrogenic/antiestrogenic and scavenging properties of (E)- and (Z)-resveratrol. Life Sci. **66:** 769–777.
109. STAHL, S., T-Y. CHUN & W.G. GRAY. 1998. Phytoestrogens act as estogen agonists in an estrogen-responsive pituitary cell line. Toxicol. Appl. Pharmacol. **152:** 41–48.
110. MIZUTANI, K., K. IKEDA, Y. KAWAI & Y. YAMORI. 1998. Resveratrol stimulates the proliferation and differentiation of osteoblastic MC3T3-E1 cells. Biochem. Biophys. Res. Commun. **253:** 859–863.
111. YOON, K., L. PELLARONI, K. RAMAMOORTHY, *et al.* 2000. Ligand structure-dependent differences in activation of estrogen receptor alpha in human HepG2 liver and U2 osteogenic cancer cell lines. Mol. Cell. Endocrinol. **162:** 211–220.
112. MGBONYEBI, O.P., J. RUSSO & I.H. RUSSO. 1998. Antiproliferative effect of synthetic resveratrol on human breast epithelial cells. Int. J. Oncol. **12:** 865–869.
113. DAMIANAKI, A., E. BAKOGEORGOU, M. KAMPA, *et al.* 2000. Potent inhibitory action of red wine polyphenols on human breast cancer cells. J. Cell. Biochem. **78:** 429–441.
114. MGBONYEBI, O.P., J. RUSSO & I.H. RUSSO. 1998. Antiproliferative effect of synthetic resveratrol on human breast epithelial cells. Int. J. Oncol. **12:** 865–869.
115. TURNER, R.T., G.L. EVANS, M. ZHANG, *et al.* 1999. Is resveratrol an estrogen agonist in growing rats? Endocrinology **140:** 50–54.
116. ASHBY, J., H. TINWELL, W. PENNIE, *et al.* 1999. Partial and weak oestrogenecity of the red wine constituent resveratrol: consideration of its superagonist activity in MCF-7 cells and its suggested cardiovascular protective effects. J. Appl. Toxicol. **19:** 39–45.
117. FREYBERGER, A., E. HARTMANN, H. HILDEBRAND & F. KROTLINGER. 2001. Differential response of immature rat uterine tissue to ethinylestradiol and the red wine constituent resveratrol. Arch. Toxicol. **11:** 709–715.
118. ROSENTAL, D.G., G.A. MACHIAVELLI, A.C. CHERNAVSKY, *et al.* 1989. Indomethacin inhibits the effects of oestrogen in the anterior pituitary gland of the rat. J. Endocrinol. **121:** 513–519.
119. MIZUTANI, K., K. IKEDA, Y. KAWAI & Y. YAMORI. 2000. Resveratrol attenuates ovariectomy-induced hypertension and bone loss in stroke-prone spontaneously hypertensive rats. J. Nutr. Sci. Vitaminol. (Tokyo) **46:** 78–83.
120. BOWERS, J.L., V.V. TYULMENKOV, S.C. JERNIGAN & C.M. KLINGE. 2000. Resveratrol acts as a mixed agonist/antagonist for estrogen receptors α and β. Endocrinology **141:** 3657–3667.
121. BHAT, K.P.L. & J.M. PEZZUTO. 2001. Resveratrol exhibits cytostatic and antiestrogenic properties with human endometrial adenocarcinoma (Ishikawa) cells. Cancer Res. **61:** 6137–6144.

122. BHAT, K.P.L., D. LANTVIT, K. CHRISTOV, *et al.* 2001. Estrogenic and antiestrogenic properties of resveratrol in mammary tumor models. Cancer Res. **61:** 7456–7463.
123. SOLEAS, G.J., E.P. DIAMANDIS & D.M. GOLDBERG. 1997. Resveratrol: a molecule whose time has come? and gone? Clin. Biochem. **30:** 91–113.
124. ARICHI, H., Y. KIMURA, H. OKUDA, *et al.* 1982. Effects of stilbene components of the roots of *Polygonum cuspidatum* Sieb. et Zucc. on lipid metabolism. Chem. Pharm. Bull. (Tokyo) **30:** 1766–1770.
125. PAUL, B., I. MASIH, J. DEOPUJARI & C. CHARPENTIER. 1999. Occurrence of resveratrol and pterostilbene in age-old darakchasava, an ayurvedic medicine from India. J. Ethnopharmacol. **68:** 71–76.

Resveratrol, a Component of Wine and Grapes, in the Prevention of Kidney Disease

ALBERTO A.E. BERTELLI,[a] MASSIMILIANO MIGLIORI,[b]
VINCENZO PANICHI,[c] NICOLA ORIGLIA,[b] CRISTINA FILIPPI,[b]
DIPAK K. DAS,[d] AND LUCA GIOVANNINI[a]

[a]*Department of Human Anatomy, University of Milan, 20133 Milan Italy*

Departments of [b]*Neuroscience and* [c]*Internal Medicine, University of Pisa, Pisa, Italy*

[d]*Cardiovascular Research Center, University of Connecticut School of Medicine, Farmington, Connecticut 06030-1110, USA*

ABSTRACT: Ischemia is an inciting factor in 50% of incidences of acute renal failure, and it increases the risk of organ rejection after renal transplantation. We have previously demonstrated that resveratrol (RSV) reduces ischemia-reperfusion (I/R) injury of rat kidney both by antioxidant and anti-inflammatory mechanisms. However, a clear morphological demonstration of this activity has not been made. To answer this question we have performed a new set of experiments following the experimental protocol reported below to investigate the effects of I/R injury and RSV pretreatment on kidney morphology by computerized morphometric analysis. Both renal arteries were clamped for 40 minutes in 40 male Wistar rats (b.w. 220 ± 20 g); 20 rats were pretreated with RSV 1 μM e.v. 40 minutes before clamping. All animals were reperfused for 24 hours and then sacrificed. Histological examination showed tissue conservation in treated rats. I/R-induced glomerular collapse (as revealed by mean glomerular volume and glomerular shape factor) was significantly reduced by RSV pretreatment. Capillary tuft/Bowman's capsule area ratio was enhanced in the I/R group suggesting tubular hypertension. RSV pretreatments significantly reduced this parameter to the control value. The number of platelet clots in the capillary tuft and tubular necrosis were also reduced by RSV versus I/R group. L-NAME administration worsened both functional and structural damage. Finally, cGMP urinary levels were markedly reduced from 12.1 ± 8.4 nmol/day to 0.10 ± 0.10 nmol/day in the I/R group. RSV provided cGMP (5.01 ± 1.5 nmol/day, $P < 0.05$). As expected, L-NAME administration significantly reduced cGMP in urine (0.71 ± 0.6 nmol/day). The present study confirms the protective effect of RSV pretreatment in I/R injury of rat kidney and suggests multiple mechanisms of action.

KEYWORDS: resveratrol; ischemia-reperfusion injury; renal failure

INTRODUCTION

Resveratrol (RSV), a stilbene polyphenol found in wine and grapes, has been reported to protect against coronary heart disease and to have cancer chemopreventive activity.[1] RSV is also able to inhibit platelet aggregation and to reduce oxidative stress

Address for correspondence: A.A.E. Bertelli, M.D., Ph.D., Department of Human Anatomy, University of Milan, Via Mangiagalli, 31, 20133 Milan, Italy. Fax: +39-02-58-31-61-66.

in PC 12 cells.[2–4] RSV protects LDL against peroxidative degradation by both chelating and free radical–scavenging mechanisms.[5] Recently Ray *et al.* have demonstrated that RSV reduces ischemia/reperfusion injury of heart.[6] Furthermore the same authors demonstrated that wine containing RSV and RSV alone reduces I/R injury of isolated rat hearts by antioxidant and free radical–scavenging mechanisms.[7]

To assess the ability of RSV to function as an antioxidant in I/R injury of kidney we have measured lipid peroxidation in kidney homogenates obtained from cortex and medulla of I/R rats with or without RSV treatment.[8] Furthermore, ischemic injury and the effect of RSV administration were evaluated by histological and ultrastructural examinations of the kidneys. Renal function was evaluated by monitoring creatinine clearance.

Recently, Shoskes demonstrated that the bioflavonoids quercetin and curcumin reduce I/R injury of rat kidney and the inflammatory chemokine response to that injury.[9] The author suggested that the bioflavonoids might reduce immune and non-immune renal injury.

On the basis of this observation we have also evaluated the effect of RSV on inflammatory response to I/R injury of rat kidney.[10]

Surgery and Experimental Protocol

Rats were anesthetized with an intraperitoneal injection of 3.5 mL/Kg b.w. of chloral hydrate. The abdomen was shaved and the animals were placed on a heating table kept at 39°C to maintain constant body temperature. A midline incision was made and both renal pedicles were cross-clamped. To maintain thermoregulation during surgery, the abdomen contents were replaced and the abdomen was temporarily closed with several sutures. Forty minutes later, the abdomen was reopened and the clamps were removed.

The kidneys were inspected for restoration of blood flow, and 1 mL of warm (37°C) normal saline was instilled into the abdominal cavity. The abdomen was closed in two layers and the animals were placed in a room at the constant temperature of 30°C. The animals were killed after 24 h of reperfusion. Both kidneys were harvested for histological analysis.

One milliliter of the experimental solutions containing RSV (final concentration 1 μM), vehicle (0.04% ethanol), L-NAME 15 mg/kg b.w. (as previously reported by Heineman *et al.*, 1996) or their combination was infused in femoral vein 40 minutes before ischemia.[11]

Effects of RSV Pretreatment on Mortality Rate, Renal Function and Lipid Peroxidation in I/R Rat Kidney

At the end of the experiments mortality rate induced by ischemia-reperfusion was 50% ($n = 15$); RSV administration decreased the mortality rate to 10% ($n = 15$) and serum creatinine level from 3.05 ± 0.8 mg/dL (control group) to 2.2 ± 0.5 mg/dL (resveratrol treated group). L-NAME administration worsened I/R injury. RSV administration significantly reduced TBARS level both in cortex and in medulla, to 0.226 ± 0.04 nmol/mg.prot. and to 15 ± 3.2 nmol/mg.prot., respectively, (I/R = 0.513 ± 0.12 and 36 ± 4.2 nmol/mg.prot., respectively).

Effects of RSV Pretreatment on Tissue Inflammation I/R Rat Kidney

We have found a significant organ PMNs infiltration in I/R rats both in glomeruli and in peritubular interstitium. Organ PMNs infiltration was associated with enhanced IL-6 urinary excretion, an index of kidney inflammation. RSV pretreatment have significantly reduced PMNs infiltration in I/R rats both in glomeruli and in peritubular interstitium and IL-6 urinary excretion, suggesting an anti-inflammatory activity.

Effect of I/R Injury and RSV Pretreatment on Endogenous Nitric Oxide System

All these data demonstrated a protective effect of RSV in I/R injury of rat kidney.

Aim of the Study

As demonstrated by results reported above, RSV pretreatment reduced I/R-induced kidney injury in rats. However, a clear morphological demonstration of this activity was not performed. To answer this question, we have performed a new set of experiments, following the experimental protocol reported above to investigate the effects of I/R injury and RSV pretreatment on kidney morphology by computerized morphometric analysis.

MATERIALS AND METHODS

Animals

Male Wistar rats (Charles River Italia, Calco, Italy) weighing 200 ± 40 g were housed in metabolic cages three days before surgery to sacrifice. All animals had free access to water. Animals fed with a low nitrite/nitrate diet for at least four days (Morini, S.Polo d'Enza, Italy) before the surgery. Animal care followed the criteria of the Italian Ministry of University and Scientific Research (MURST) for the care and the use of laboratory animals in research.

Chemicals

Chemicals were purchased from Sigma Chemical Co., St. Louis, MO, USA. RSV solution (1 μM) was prepared freshly and under light protection before use.

Morphometric Analysis

At the end of each experiment performed as reported above, kidneys were removed and fixed in 10% neutral buffered formalin solution and then included in paraffin. Serial sections were cut using the same microtome at 3 μm of thickness and stained with hematoxylin-eosin and periodic acid shiff (PAS).

The histological sections were examined by a Leitz microscope (Leitz, Hannover, Germany), and photographed and the images stored in a personal computer memory using a scanner (Epson Film Scan). The thickness of each section was checked as described by Weibel.[12] In these serial sections, we measured the individual volumes of 30 glomeruli using the Cavalieri principle. Histological morphometry was performed using an image analysis system consisting of a personal computer equipped with graphic tablet (Acecat II, Acecad) and scanner (Epson FilmScan 200). This system was programmed (Mocha Image Analysis Software, Jandel Scientific) to calculate mean glomerular volume and tuft area shape factor. The shape factor is a measure of how nearly circular a given object is; a perfect circle has a shape factor of 1.000 and a line has a shape factor approaching 0.000. These parameters were used to quantify glomerular collapse. Moreover, in glomeruli with area not significantly lower than the controls, we also evaluated the capillary tuft/Bowman's capsule area ratio as index of tubular hypertension.

In the same glomeruli we have also counted the number of endocapillary clots. Tubular necrosis was measured by counting the number of necrotic and non-necrotic areas in the specimen. Finally, we evaluated the percentage of tubuli with PAS+ clots.

cGMP Assay

To confirm the involvement of NO in glomerular hemodynamic as directly demonstrated in previous experiments, cyclic GMP levels in urine samples were assayed in duplicate, using an EIA kit (DRG International, Inc., USA).[8]

Statistics

Results are expressed as mean ± standard deviation; the differences between groups were studied by ANOVA and the Student-Neumann-Keuls test was used. The null-hypothesis was rejected when $p < 0.05$.

RESULTS

Histological examination showed tissue conservation in treated rats. I/R-induced glomerular collapse (as revealed by mean glomerular volume and glomerular shape factor) was significantly reduced by RSV pretreatment (see FIGURES 1 and 2). Capillary tuft/Bowman's capsule area ratio was enhanced in I/R group suggesting tubular hypertension. RSV pre-treatments significantly reduced this parameter to the control value. The number of platelet clots in the capillary tuft and tubular necrosis were also reduced in the RSV versus I/R group. L-NAME administration worsened both functional and structural damage.

Finally, cGMP urinary levels were markedly reduced from 12.1 ± 8.4 nmol/day to 0.10 ± 0.10 nmol/day in the I/R group. RSV provided cGMP (5.01 ± 1.5 nmol/day, $p < 0.05$). As expected, L-NAME administration significantly reduced cGMP in urine (0.71 ± 0.6 nmol/day)

All these results are summarized in TABLE 1.

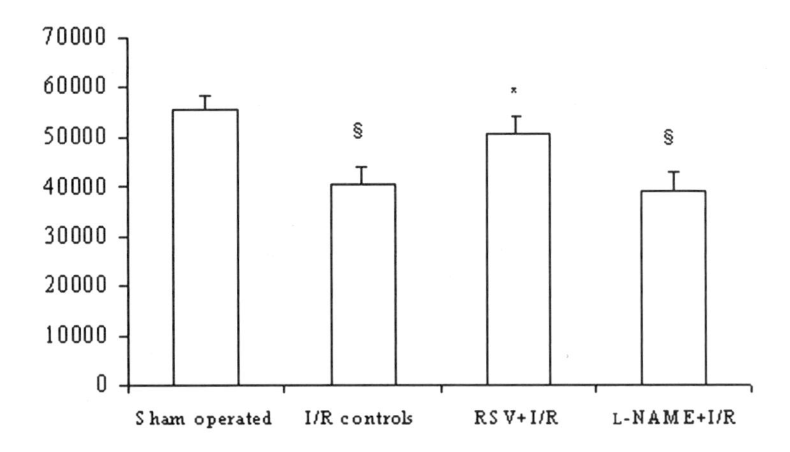

FIGURE 1. I/R induced reduction of mean glomerular volume (μm^3) was prevented by RSV pretreatment. $^*p < 0.05$ versus I/R controls, L-NAME; $^§p < 0.05$ sham operated.

DISCUSSION

As evidenced in previous reports, pretreatment with RSV reduced the mortality of I/R rats and significantly provides renal function.[8,10] Morphometric analysis supported those findings: I/R rats treated with RSV showed a significant reduction of necrosis and sloughing of the proximal tubules, and of necrotic areas in the medulla.

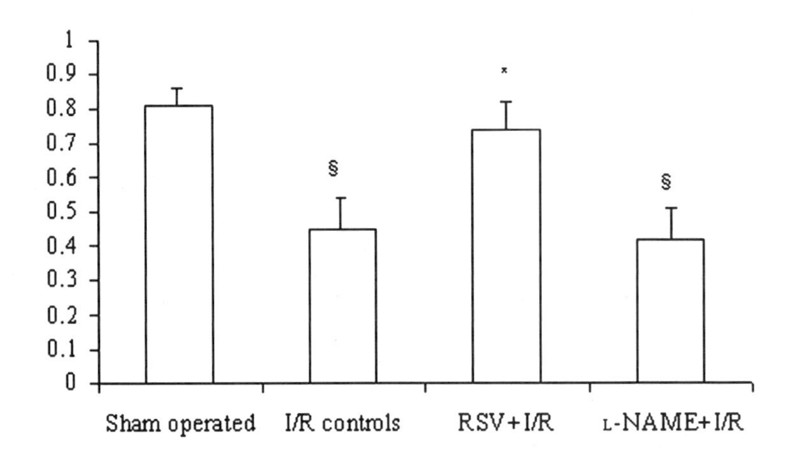

FIGURE 2. The shape factor is a measure of how nearly circular a given object is; a perfect circle has a shape factor of 1.000 and a line has a shape factor approaching 0.000. This parameter is used to quantify glomerular collapse. I/R-induced glomerular collapse was prevented by RSV. $^*p < 0.05$ versus I/R controls, L-NAME; $^§p < 0.05$ sham operated.

TABLE 1.

	Sham operated	I/R controls	RSV+I/R	L-NAME+I/R
Tubular cell necrosis (%)	n.d.	50 ± 7.5**	15 ± 9.5*	62.3 ± 11**
Tubular casts PAS+ (%)	1.2 ± 1.0	55 ± 12**	10 ± 8.4*	59 ± 10.5**
Endocapillary clots/ glomerulus	0.5 ± 0.7	12 ± 3.1**	3 ± 1.2*	9 ± 2.5**
Mean glomerular volume	55,754 ± 2,631	40,387 ± 3,684**	50,479 ± 3,847*	38,894 ± 4,265**
Glomerular shape factor	0.81 ± 0.05	0.45 ± 0.09**	0.74 ± 0.08*	0.42 ± 0.09**
Capillary tuft/Bowman's capsule area ratio	0.90 ± 0.07	0.61 ± 0.08**	0.84 ± 0.08*	0.59 ± 0.09**

*$p < 0.05$ versus I/R and I/R+L-NAME.
**$p < 0.05$ versus "sham operated".

A reduced obstruction of tubular lumen was evidenced by a lower number of PAS+ casts in the tubular lumen and by a higher capillary tuft/Bowman's capsule area ratio.

Glomerular hemodynamic was better in I/R kidney pre-treated with RSV: preserved mean glomerular volume suggested a reduction of vasoconstriction. Some Authors suggested that in I/R the combination of high NO levels with ROS and reactive nitrate intermediaries can produce lipid peroxidation and peroxinitrites, which are highly cytotoxic.[13,14] However in our study pre-treatment with L-NAME worsened I/R injury. An impairment of NO release has been proposed by other authors as one of the most important factors in the pathogenesis of I/R injury of rat kidney.[15,16] It is well known that NO_3 reflects NO production. However, the measurement of NO_2+NO_3 is not a good index of NO bioactivity in the presence of ROS; in the aqueous phase of the plasma NO is inactivated by the reaction with molecular oxygen and other ROS to form NO_2^- or with superoxide to form peroxynitrite.[17] The subsequent decomposition of peroxynitrite and thus the ratio of formed NO_2^- and NO_3^- depends on the surrounding conditions. These metabolic pathway are extensively reviewed by Kelm.[18] The results from our previous experiment using electron paramagnetic spin resonance clearly demonstrated that NO was reduced by I/R; instead RSV administration enhanced NO release.[8] Cyclic GMP is the second messenger of NO and is responsible of arterial vasodilatation. It is well established that NO released from the endothelium activates soluble guanylate cyclase causing an increase in cGMP turnover leading to the cellular effect.[19] For these reasons cGMP measurements may better reflect the vasodilatatory activity of NO. In I/R rats we found a marked reduction of cGMP. RSV administration was able to increase cGMP urinary excretion, suggesting a preserved NO bioactivity. The reduction in cGMP is an index of the well known deficiency in NO in I/R injury of the kidney that can contribute to vasoconstriction. The production of NO plays an important role in regulating vascular tone by maintaining renal (and systemic) vasodilatation.[20,21] In our experiments vasoconstriction was confirmed by glomerular collapse present in I/R kidneys, but reduced in the RSV-treated group.

In previous experiments we found that RSV reduced I/R induced polymorphonuclear cell (PMN) infiltration both in the glomerula and in the peritubular interstitium.[10] This effect may be related to modulation of TNFα-dependent ICAM-1 expression on endothelial cell membranes as demonstrated by Ferrero and coworkers.[22] However, preserved NO release may also contribute: Linas and coworkers demonstrated that NO prevents the PMN component of ischemic renal injury by blocking PMN retention and the deleterious effects of activated PMN on glomerular and tubular function and Lopez-Neblina *et al.* provided evidence that sodium nitroprusside (a NO donor) administered as late as 15 min before reperfusion has a protective effect in ischemically damaged kidneys of Sprague-Dawley rats.[23,24] They concluded that the mechanism of protection is due to the blocking of one of the steps of the interaction between leukocytes and endothelium.

The contribution of RSV to the activity of the nitric oxide pathway is still unclear. Fitzpatrick *et al.*[25] demonstrated that resveratrol didn't relax precontracted smooth muscle of intact rat aortic rings. On the contrary Chen and Pace-Asciak reported that resveratrol at a higher concentration than 3×10^{-5} M, caused relaxation of the phenylephrine-precontracted endothelium-intact aorta.[26] L-NNA reversed the relaxation, indicating an endothelium-dependent mechanism. Contrasting data was also reported about the effect of RSV on NO synthases expression.[27–29] Intriguingly, Bastianetto *et al.*[30] suggested that the protective ability of RSV against nitric oxide-related toxicity in cultured hippocampal neurons resulted from its antioxidant property rather than its inhibitory effect on intracellular enzymes such as NO synthases.

In our experiments RSV administration was able to reduce lipid peroxidation. This effect may suggest a less oxidation of NO to nitrite with a consequent preserved bioactivity as evidenced for other antioxidants.[31]

Finally, we have found an increased number of glomerular intracapillary platelets clots in I/R kidney, significantly reduced by RSV pretreatment. This effect, which may be explained by the well-known antiplatelet activity of RSV, may play a role in maintaining glomerular filtration rate.[2,3]

In conclusion, the present study confirms the protective effect of RSV pretreatment in I/R injury of rat kidney and suggests multiple mechanisms of action.

ACKNOWLEDGMENTS

The Authors wish to thank Mr. Pierluigi Frangioni (light microscopy), Mr. Leonardo Casarosa (morphometry), and Mr. Antonio Troilo (urine analysis and renal function evaluation) for their expert technical assistance.

REFERENCES

1. Jang, M., L. Cai, G.O. Udeani, *et al.* 1997. Cancer chemopreventive activity of resveratrol, a natural product derived from grapes. Science **275**(5297): 218–220.
2. Bertelli, A.A.E., L. Giovannini, D. Giannessi, *et al.* 1995. Antiplatelet activity of synthetic and natural resveratrol in red wine. Int. J. Tiss. React. **XVII**(1): 1–4.
3. Bertelli, A.A.E., L. Giovannini, R. De Caterina, *et al.* 1996. Antiplatelet activity of cis-resveratrol. Drugs Exp. Clin. Res. **XXII**(2) 61–63.
4. Chanvitayapongs, S., B. Draczynska-Lusiak, A.Y. Sun. 1997. Amelioration of oxidative stress by antioxidants and resveratrol in PC12 cells. Neuroreport **6:** 1499–1502.

5. Belguendouz, L., L. Fremont & A. Linard. 1997. Resveratrol inhibits metal iron-dependent and independent peroxidation of porcine low density lipoproteins. Biochem. Pharmacol. **53**(9): 1347–1355.
6. Ray, P.S., G. Maulik, G.A. Cordis, *et al.* 1999. The red wine antioxidant resveratrol protects isolated rat hearts from ischemia reperfusion injury. Free Radical Biol. Med. **27**(1–2): 160–169.
7. Sato, M., P.S. Ray, G. Maulik, *et al.* 2000 Myocardial protection with red wine extract. J. Cardiovasc. Pharmacol. **35:** 263–268.
8. Giovannini, L., M. Migliori, B.M. Longoni, *et al.* 2001. Resveratrol, a polyphenol found in wine, reduces ischemia reperfusion injury in rat kidney. J. Cardiovasc. Pharmacol. **37**(3): 262–270.
9. Shoskes, D.A.. 1998. Effect of bioflavonoids quercetin and curcumin on ischemic renal injury: a new class of renoprotective agents. Transplantation **66**(2): 147–152.
10. Giovannini, L., M. Migliori, B.M. Longoni, *et al.* 2000. Resveratrol pre-treatment reduces ischemia/reperfusion induced polimorphonuclear cells infiltration in rat kidneys. Int. J. Immunother. **37**(3): 262–270.
11. Heineman, A., M. Jocie, B.M. Peskar & P. Holzer. 1996. Cck-evoked hyperemia in rat gastric mucosa involves neural mechanism and nitric oxide. Am. J. Physiol. **270:** G253–G258.
12. Weibel, E.R. 1979. Point counting methods. *In* Sterological Methods, Vol. I: 101–159. E.R. Weibel, Ed. Academic Press. London.
13. Noiri, E., T. Peresleni, F. Miller, *et al.* 1996. In vivo targeting of inducible NO synthase with oligodeoxynucleotides protects rat kidney against ischemia. J. Clin. Invest. **97**(10): 2377–2383.
14. Peresleni, T., E. Noiri, W.F. Bahou, M.S. Goligorsky. 1996. Antisense oligodeoxynucleotides to inducible NO synthase rescue epithelial cells from oxidative stress injury. Am. J. Physiol. **270**(6 Pt. 2): F971–F977.
15. Lieberthal, W., E.F. Wolf & H.G. Rennke. 1989. Renal ischemia and reperfusion impair endothelium-dependent vascular relaxation. Am. J. Physiol. **256**(5 Pt. 2): F894–F900.
16. Conger, J.D., J.B. Robinette & R.W. Schrier. 1988. Smooth muscle calcium and endothelium-derived relaxing factor in the abnormal vascular responses of acute renal failure. J. Clin. Invest. **82**(2): 532–537.
17. Rubanyi, G.M. & P.M. Vanhoutte. 1986. Superoxide anions and hyperoxia inactivate endothelium-derived relaxing factor. Am. J. Physiol. **250**(5 Pt. 2): H822–H827.
18. Kelm, M.1999. Nitric oxide metabolism and breakdown. Biochim. Biophys. Acta **1411**(2-3): 273–289.
19. Murad, F., U. Forstermann, M. Nakane, *et al.* 1992. The nitric oxide-cyclic GMP signal transduction pathway in vascular smooth muscle preparations and other tissues. Jpn. J. Pharmacol. **58**(Suppl. 2): 150P–157P.
20. Deng, A. & C. Baylis. 1993. Locally produced EDRF controls preglomerular resistance and ultrafiltration coefficient. Am. J. Physiol. **264**(2 Pt. 2): F212–F215.
21. Baylis, C., K. Engels, L. Samsell & P. Harton. 1993. Renal effects of acute endothelial-derived relaxing factor blockade are not mediated by angiotensin II. Am. J. Physiol. **264**(1 Pt. 2): F74–F78.
22. Ferrero, M.E., A.E. Bertelli, A. Fulgenzi, *et al.* 1998. Activity in vit ro of resveratrol on granulocyte and monocyte adhesion to endothelium. Am. J. Clin. Nutr. **68**(6): 1208–1214.
23. Linas, S., D. Whittenburg & J.E. Repine. 1997. Nitric oxide prevents neutrophil-mediated acute renal failure. Am. J. Physiol. **272**(1 Pt. 2): F48–F54.
24. Lopez-Neblina, F., L.H. Toledo-Pereyra, R. Mirmiran, *et al.* 1996. Time dependence of Na-nitroprusside administration in the prevention of neutrophil infiltration in the rat ischemic kidney. Transplantation **61**(2):179–183.
25. Fitzpatrick, D.F., S.L. Hirschfield & R.G. Coffey. 1993. Endothelium-dependent vasorelaxing activity of wine and other grape products. Am. J. Physiol. **265**(2 Pt. 2): H774–H778.
26. Chen, C.K. & C.R. Pace-Asciak. 1996. Vasorelaxing activity of resveratrol and quercetin in isolated rat aorta. Gen. Pharmacol. **27**(2): 363–366.

27. TSAI, S.H., S.Y. LIN-SHIAU & J.K. LIN. 1999. Suppression of nitric oxide synthase and the down-regulation of the activation of NFkappaB in macrophages by resveratrol. Br. J. Pharmacol. **126**(3): 673–680.
28. CHAN, M.M., J.A. MATTIACCI, H.S. HWANG, *et al.* 2000. Synergy between ethanol and grape polyphenols, quercetin, and resveratrol, in the inhibition of the inducible nitric oxide synthase pathway. Biochem. Pharmacol. **60**(10): 1539–1548.
29. HSIEH, T.C., G. JUAN, Z. DARZYNKIEWICZ, *et al.* 1999. Resveratrol increases nitric oxide synthase, induces accumulation of p53 and p21(WAF1/CIP1), and suppresses cultured bovine pulmonary artery endothelial cell proliferation by perturbing progression through S and G2. Cancer Res. **59**(11): 2596–2601.
30. BASTIANETTO, S., W.H. ZHENG & R. QUIRION. 2000. Neuroprotective abilities of resveratrol and other red wine constituents against nitric oxide-related toxicity in cultured hippocampal neurons. Br. J. Pharmacol. **131**(4): 711–720.
31. TADDEI, S., A. VIRDIS, L. GHIADONI, *et al.* 1998. Vitamin C improves endothelium-dependent vasodilation by restoring nitric oxide activity in essential hypertension. Circulation **97**(22): 2222–2229.

Oxygen, Oxidants, and Antioxidants in Wound Healing

An Emerging Paradigm

CHANDAN K. SEN, SAVITA KHANNA, GAYLE GORDILLO, DEBASIS BAGCHI, MANASHI BAGCHI, AND SASHWATI ROY

Laboratory of Molecular Medicine, Dorothy M. Davis Heart and Lung Research Institute, Department of Surgery (CMIS), The Ohio State University Medical Center, Columbus, Ohio 43210, USA

ABSTRACT: Disrupted vasculature and high energy-demand by regenerating tissue results in wound hypoxia. Wound repair may be facilitated by oxygen therapy. Evidence supporting the mode of action of hyperbaric oxygen in promoting wound healing is sketchy, however. Topical oxygen therapy involves local administration of pure oxygen. The advantages of topical oxygen therapy include low cost, the lack of systemic oxygen toxicity, and possibility of home treatment. While this modality of wound care is of outstanding interest, it clearly lacks the support of mechanism-oriented studies. The search for mechanisms by which oxygen supports wound healing has now taken another step. Respiratory burst–derived oxidants support healing. Oxidants serve as cellular messengers to promote healing. Although this information is of outstanding significance to the practice of oxygen therapy, it remains largely unexplored. The search for "natural remedies" has drawn attention to herbals. Proanthocyanidins or condensed tannins are a group of biologically active polyphenolic bioflavonoids that are synthesized by many plants. Proanthocyanidins and other tannins facilitate wound healing. A combination of grape seed proanthocyanidin extract and resveratrol facilitates inducible VEGF expression, a key element supporting wound angiogenesis. Strategies to manipulate the redox environment in the wound are likely to be of outstanding significance in wound healing.

KEYWORDS: wound healing; antioxidants; polyphenolic bioflavonoids; oxygen therapy; oxidants

INTRODUCTION

Wound-healing abnormalities cause great physical and psychological stress to affected patients and are extremely expensive. Disrupted vasculature and high demand for energy to support processing and regeneration of wounded tissue are typical characteristics of a wound site. Low oxygen supply and high demand results in hypoxia. Oxygen delivery is a critical element for the healing of wounds.[1–3] In the

Address for correspondence: Dr. Chandan K. Sen, Laboratory of Molecular Medicine, 512 Heart & Lung Research Institute, The Ohio State University Medical Center, 473 W. 12th Avenue, Columbus, OH 43210, USA. Voice: 614-247-7786; fax: 614-247-7818.
sen-1@medctr.osu.edu

presence of poor blood flow, the availability of oxygen to the wound site is thought to be a rate-limiting step in early wound repair. Indeed, transcutaneous oxygen ($TcPO_2$) alone is able to reliably estimate probability of healing in an ischemic extremity.[4] The time line of wound healing is altered by various local conditions, such as inflammation and neuropathy; however, the most important factor regulating the regional time line of healing is blood flow. Factors that can increase oxygen delivery to the regional tissue, such as supplemental oxygen, warmth, and sympathetic blockade, can speed healing.[5,6] Intermittent oxygen therapy has been shown to promote collagen synthesis and is beneficial for producing the extracellular matrices that support wound healing.[7]

HYPERBARIC OXYGEN THERAPY

Wound repair can often be facilitated by increasing the partial pressure at which oxygen is supplied to wounds.[2] Clinical experience with adjunctive hyperbaric oxygen therapy in the treatment of chronic wounds[8] has shown that wound hyperoxia increases wound granulation tissue formation and accelerates wound contraction and secondary closure.[9,10] Nevertheless, the physiological basis for this modality remains largely unknown. Such ignorance adversely affects our ability to establish definitive criteria for the selection of patients and also to predict success in treatment. To date, there are few clinical studies that attempt to define the fundamentals underlying hyperbaric oxygen therapy. Hyperbaric studies have been criticized for the lack of well-defined wound care protocols, the absence of precise wound-healing measures, and poorly defined wound healing endpoints.[11] Evidence supporting the mode of action of hyperbaric oxygen in promoting wound healing is sketchy at best. For example, hyperbaric oxygenation above 2 atmospheres inhibits proliferation of fibroblasts and keratinocytes in cell monolayer cultures (e.g., a 10-day treatment at 3 atmospheres appeared cytostatic to keratinocytes). In contrast, hyperbaric treatment up to 3 atmospheres dramatically enhances keratinocyte differentiation, and epidermopoiesis in complete human skin equivalents.[12]

Hyperbaric oxygen therapy includes two key components: high (2–3 atm) pressure and close to 100% oxygen. What is the relative contribution of the pressure and oxygen factors? Do we need a combination of both for successful wound therapy or is normobaric oxygen treatment good enough? In the case of an exposed dermal wound, is it important to administer oxygen systematically or is topical oxygen applied locally to the wound site effective? While there are many opinions about these important questions, at present we do not have any firm evidence-based scientific conclusions. Systemic oxygen therapy is contraindicated in numerous situations and poses significant risk to organs such as the eye, brain, and lung. Under certain circumstances, negative pressure oxygen therapy has been claimed to be more effective than hyperbaric oxygen therapy.[13] A reasonable evaluation of the risk:benefit ratio of systemic oxygen therapy in the treatment of wound healing would require mechanism-oriented translational research.

TOPICAL OXYGEN THERAPY

Topical oxygen therapy represents a less explored modality in wound care.[14] Pure oxygen is locally administered to an affected region of the body at 1.03 atmospheres of pressure and can be done in the patient's own home (see FIGURE 1). It is indicated for the treatment of open wounds. The advantages of topical oxygen therapy include low cost, the lack of systemic oxygen toxicity, possibility of home treatment, and effectiveness, allowing this treatment to be prescribed for many patients early in the course of their disease rather than as a last resort.[15] Systemic hyperbaric therapy requires that patients be placed in special chambers in the presence of trained physician specialists with the delivery of oxygen in the chamber at 2–3 atmospheres of pressure. Whether topical oxygen therapy has similar efficacy as systemic hyperbaric oxygen therapy remains to be established. A few brief studies have reported the effects of topical oxygen therapy on wound healing. These studies are mostly observational and do not address underlying mechanisms.[16–18] It is claimed that topical oxygen alone or in combination with a low power laser may be useful to treat diabetic foot ulcers.[19] On the basis of prospective randomized clinical studies it has been inferred that topical oxygen therapy represents a cost-effective approach[20] to promote wound angiogenesis.[21] If indeed topical oxygen therapy emerges as a successful therapeutic modality in the treatment of wounds, it could significantly decrease the cost of caring for chronic wounds and substantially broaden the scope of patients eligible for treatment.

A RADICAL HYPOTHESIS IN SUPPORT OF OXYGEN THERAPY

The search for the mechanisms by which oxygen exerts its vital functions in wound healing has evolved another step. Reactive oxygen species (ROS, includes oxygen-derived radical as well as non-radical oxidants), often loosely termed "oxidants," are a vital part of healing.[22,23] Oxygen is the rate-limiting factor for activation of NADPH oxidase that triggers respiratory burst. Respiratory burst is a mechanism by which phagocytic cells generate oxidants from oxygen. Hyperbaric oxygen has been shown to stimulate respiratory burst activity.[24,25] Micromolar concentrations of hydrogen peroxide promote vascular endothelial growth factor (VEGF) expression in keratinocytes.[23] VEGF is an endothelial-cell–specific mitogen. The finding that VEGF was potent and specific for vascular endothelial cells and, unlike basic fetal growth factor freely diffusible, led to the hypothesis that this molecule plays a unique role in the regulation of physiological angiogenesis.[26]

Wound healing occurs in "phases." The main phases of wound healing include coagulation, which begins immediately after injury; inflammation, which initiates shortly thereafter; a migratory and proliferative process, which begins within days and includes the major processes of healing; and a remodeling process, which may last for up to a year and is responsible for scar tissue formation and development of new skin.[27] In the inflammation phase, one of the first lines of defense are migrating polymorphonuclear cells (PMNs) which locate, identify, phagocytize, kill, and digest microorganisms and eliminate wound debris. These cells, through their characteristic "respiratory burst" activity, produce $O_2^{-\bullet}$ (superoxide anion radical), which

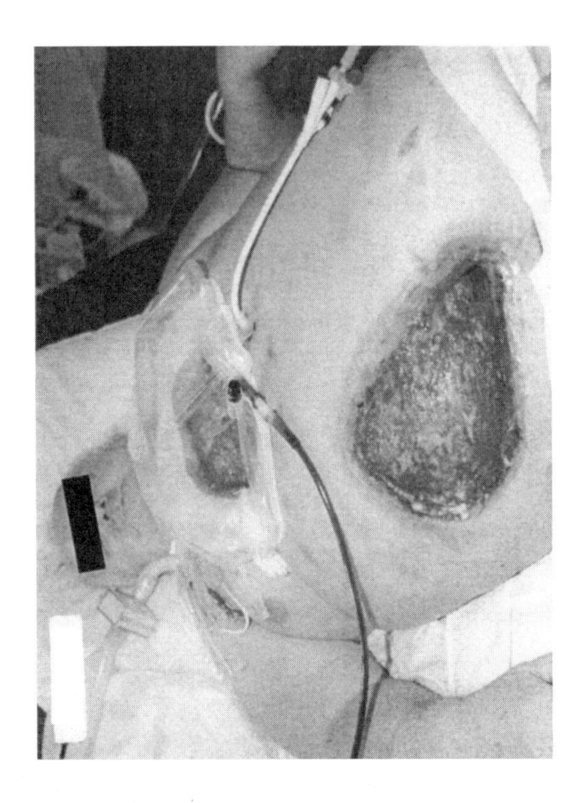

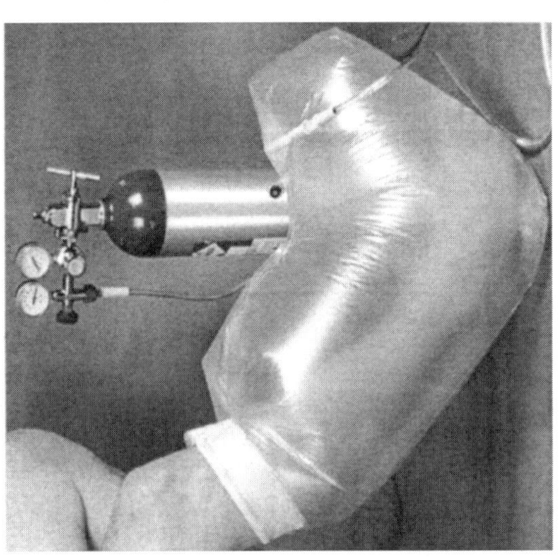

FIGURE 1. Devices for topical oxygen therapy.

is well known to be critical for defense against bacteria and other pathogens.[28] Superoxide is rapidly converted to membrane permeable form, H_2O_2, by superoxide dismutase activity or even spontaneously. Release of H_2O_2 may promote formation of other oxidants that are more stable (longer half-life) including, hypochlorous acid, chloramines, and aldehydes. The production of oxidants at the wound site is not restricted to neutrophils alone but may be also produced by macrophages, which appear and orchestrate a "long term" response to injured cells subsequent to the acute response. Taken together, this suggests that the wound site is rich in oxidants along with their derivatives such as chloramine, mostly contributed by neutrophils and macrophages. A clinically relevant model documented treatment of ischemia-induced ulcers with hydrogen peroxide cream and reported enhanced cutaneous blood recruitment not only to ulcers and adjacent sites, but also to distant sites.[29] Oxidants serve as cellular messengers that drive numerous aspects of molecular and cell biology.[30,31] While it is plausible that this information is of outstanding significance to the practice of oxygen therapy, at present it remains largely unexplored.

Consistent with the hypothesis that wound-related oxidants support the healing process, clearing oxidants from the wound environment of old rats during the early inflammation phase of the healing process decreased blood flow.[32] Exposure to mild concentrations of oxidants triggers expression of antioxidant defense proteins such as heme oxygenase 1[33] and keratinocyte growth factor[34] that are likely to protect the regenerating tissue against oxidant damage. Some would argue against the role of oxidants in wound healing. For example, Senel *et al.* have claimed that oxygen free radicals may be detrimental to ischemic skin wound healing.[35] Interpretation of results reported in this study requires careful consideration. It has been shown that treatment of the wound by allopurinol or superoxide dismutase increased tensile strength of the healing tissue. Allopurinol inhibits xanthine oxidase, a source of superoxide in endothelial cells, but does not have any effect on phagocytic or non-phagocytic oxidases that are known to be responsible for the respiratory burst phenomenon. Furthermore, superoxide dismutase accelerates the formation of hydrogen peroxide from superoxide. Hydrogen peroxide is a potent oxidant. Therefore, it is indeed plausible that the reported effects of superoxide dismutase were mediated by hydrogen peroxide. The concentration of oxidants in question is critically important. Although at micromolar concentrations oxidants such as hydrogen peroxide may favorably influence signal transduction processes that support healing, at millimolar concentrations hydrogen peroxide is likely to overwhelm the antioxidant defense system of the healing tissue[36] and trigger indiscriminate tissue damage thereby delaying healing.[37]

The effects of many growth factors and cytokines, recognized as key elements of the wound healing process, are mediated by oxidants. TGF-β1 is a pleiotropic cytokine that plays a key role in wound healing. Some fibrogenic actions of TGF-β1, necessary for extracellular matrix production, are mediated via formation of hydrogen peroxide.[38] Oxidants also promote fibroblast migration and proliferation.[39,40] Hydrogen peroxide generated by phagocytic cells in the wound site has also been shown to up-regulate endothelial-cell heparin-binding EGF mRNA, another key player in promoting wound healing.[41] Oxidants generated in response to Rac1 activation have been shown to be essential for nuclear factor κB–dependent transcriptional regulation of interleukin-1α, which, in an autocrine manner, induced collagenase-1 gene expression. Remodeling of the extracellular matrix and consequent alterations of integrin-mediated adhesion and cytoarchitecture are central to wound healing. It has been proposed that activation of Rac1 may lead to altered gene regulation and alterations in cellular morphogenesis, migration, and invasion.[42] Recent studies in our laboratory provide the first evidence that Rac1 gene transfer accelerates contraction and healing of murine excisional dermal wounds (not shown).

Platelet-derived growth factor (PDGF), commonly used in clinical wound therapy, is found as PDGF-A, AB, and BB. It exerts its effects on cells by binding to one of two membrane-bound receptors, the α-receptor or the β-receptor. Both PDGF-BB and TGF-β1 alone are more effective than hyperbaric oxygen treatment by itself in accelerating the impaired wound healing produced by ischemia. In a recent study, acutely ischemic wounds in rabbit ears were treated with saline or PDGF-BB and then animals were treated with hyperbaric air or oxygen at 2 atm abs (202.6 kPa). Hyperbaric air was without significant effect compared with control rabbits breathing air at ambient pressure. Combined treatment with hyperbaric oxygen plus PDGF-BB was synergistic in up-regulating mRNA for PDGF-β receptor. Exposure to 85% oxygen has been shown to potently increase the expression of both the PDGF-B gene and the PDGF B-type receptor.[43] These findings lay a firm rationale to test the therapeutic significance of PDGF-BB and oxygen in synergism. The results of a preliminary clinical study support the use of combined therapy using topical becaplermin (trade name for PDGF) and hyperbaric oxygen therapy as a means of successfully treating the chronic diabetic ulcer patient with deficient nitric oxide production and local wound hypoxia.[9]

The hypothesis that cytokines such as PDGF and oxygen may function synergistically to promote wound healing is in line with predictions that could be made from cell biology studies. Cytokines such as PDGF, epidermal growth factor (EGF), tumor necrosis factor (TNF)-α or interleukin (IL)-1β generate oxidants upon binding to their receptors.[44] It has been specifically demonstrated that such oxidants play a key role in driving cellular signal transduction pathways of PDGF-treated cells. Inhibitors of oxidant production inhibit PDGF-induced activation of cell signaling.[45] Consistently, in a separate study over-expression of the antioxidant-enzyme superoxide dismutase blocked the PDGF-induced expression of genes and gene products. It was shown that nitric oxide synthase induced by PDGF is mediated in part by production of superoxide.[46] Pretreatment with catalase (decomposes hydrogen peroxide) completely abrogated hydrogen peroxide-induced PDGF receptor and c-Src tyrosine phosphorylation, suggesting that PDGF receptors send mitogenic signals utilizing oxidants as messengers.[47] Endothelial cells are not only capable of sensing oxygen tension, but are also able to discriminate and respond to even small differences in oxygen tension resulting in dramatic up-regulation of the PDGF-B chain gene.[48]

Nitric Oxide

A supporting role for reactive species in wound healing has been evident from numerous studies focusing on nitric oxide. While some questions have been raised,[49] it would be fair to summarize that nitric oxide produced during the healing process clearly promotes wound repair.[50] The earliest evidence demonstrating that nitric oxide may promote wound healing was presented only five years ago when it was demonstrated that nitric oxide synthesis is critical to wound collagen accumulation and acquisition of mechanical strength.[51] Nitric oxide is expected to promote wound angiogenesis by inducing the expression of vascular endothelial growth factor.[52] Using knock-out mice and gene transfer approaches it has been established that both endothelial nitric oxide synthase[53] as well as inducible nitric oxide synthase play a key role in wound repair.[54]

HERBAL ANTIOXIDANTS IN WOUND HEALING

The search for "natural remedies" for a commonly occurring disorder such as wounds has drawn attention to herbals. From ancient times, herbals have been routinely used to treat wounds, and in many cultures their use in traditional medicine has persisted to the present. While it is possible that some time-tested herbal remedies are indeed effective, it seems to be often the case that the patient knows more about this form of medicine than the physician! In other words, lack of detailed mechanism-oriented and hypothesis-driven research poses a major drawback to the use of herbal medicine to treat wounds. For example, *Aloe vera* is commonly used for a wide range of dermatological applications including wound healing. The efficacy of *Aloe vera* in treating wound healing remains to be categorically established.[55] With the renewed interest in herbal cures, it is time to revisit the field.

There are numerous herbal derivatives that have been tried for their ability to promote wound healing. A complete discussion of these derivatives is beyond the scope of this work. While most studies are purely observational in nature, a few others have attempted to address the underlying mechanisms. For example, the polysaccharide-rich *Angelica sinensis* has a direct mucosal healing effect on gastric epithelial cells by increasing ornithine decarboxylase and c-Myc expression.[56] A *Eucommia ulmoides* Oliver leaf extract has been shown to favorably influence collagen metabolism and support wound healing. Oral administration of this herbal derivative accelerated granuloma maturation and the energy was supplied from fatty acid metabolism.[57] Eupolin extract increases fibroblast and endothelial cell growth.[58] The extract increases expression of several components of the adhesion complex and fibronectin by human keratinocytes. Eupolin reportedly stimulates the expression of many proteins of the adhesion complex and fibronectin by human keratinocytes. The adhesion complex proteins are thought to be essential to stabilize epithelium and this effect could contribute to the clinical efficacy of Eupolin in healing.[59]

Proanthocyanidins or condensed tannins are a group of biologically active polyphenolic bioflavonoids that are synthesized by many plants. Proanthocyanidins and other tannins are known to facilitate wound healing.[60,61] The mode of action, however, remains unclear. Grape seed proanthocyanidin extract, has been reported to have various clinically relevant redox-active properties.[62–66] It was recently observed that natural extracts derived from grape seeds facilitate oxidant-induced VEGF expression in keratinocytes. These results suggested that grape-seed–derived natural extracts may have beneficial effects in promoting dermal wound healing and other related skin pathologies.[23] Using a ribonuclease protection assay (RPA), the ability of GSPE to regulate oxidant-induced changes in several angiogenesis-related genes has been studied. While mRNA responses were studied using RPA, VEGF protein release from cells to the culture medium was studied using ELISA. Pretreatment of HaCaT keratinocytes with GSPE up-regulated both hydrogen peroxide as well as TNFα-induced VEGF expression and release.[23] Studies with VEGF promoter linked to a luciferase reporter showed that the herbal extract influenced the transcriptional control of inducible VEGF expression. In a murine model of dermal excisional wound, a combination of grape seed extract and 5,000ppm resveratrol markedly accelerated wound contraction and healing (not shown). In a previous section of this article, we have discussed how oxidants could support the wound healing process. Herbal extracts such as the grape seed extract are highly rich in antioxidants. This leads to an apparent paradox. How can both oxidants as well as antioxidants promote healing? While a definitive answer requires further experimentation, it should be noted that antioxidants do tend to possess signal transduction regulatory properties that may or may not be linked to their ability to detoxify oxidants.[30,31,67–69] In addition, under certain conditions such as a strong oxidizing environment lacking the support to regenerate (reduce) oxidized antioxidants, some antioxidants may assume the characteristics of a pro-oxidant.[70–73]

CONCLUSION

Recent advances in the molecular and cellular aspects of redox biology positions us well to revisit the apparently outstanding benefit of oxygen therapy in wound healing. It is likely that reactive derivatives of molecular oxygen, oxidants, for example, serve as cellular messengers to support the healing process. Strategies to manipulate the oxygen/oxidant environment in the wound are likely to be of outstanding significance.

ACKNOWLEDGMENTS

Research reported here was supported by NIH GM 27345, the Surgery Wound Healing Research Program, and US Surgical, Tyco Healthcare Group. The Laboratory of Molecular Medicine is the research wing of the Center for Minimally Invasive Surgery.

REFERENCES

1. Jonsson, K., J.A. Jensen, W.H.D. Goodson, *et al.* 1991. Tissue oxygenation, anemia, and perfusion in relation to wound healing in surgical patients. Ann. Surg. **214**: 605–613.
2. Lavan, F.B. & T.K. Hunt. 1990. Oxygen and wound healing. Clin. Plast. Surg. **17**: 463–472.
3. Wu, L., Y.P. Xia, S.I. Roth, *et al.* 1999. Transforming growth factor-beta1 fails to stimulate wound healing and impairs its signal transduction in an aged ischemic ulcer model: importance of oxygen and age. Am. J. Pathol. **154**: 301–309.
4. Padberg, F.T., T.L. Back, P.N. Thompson & R.W. Hobson, 2nd. 1996. Transcutaneous oxygen (TcPO2) estimates probability of healing in the ischemic extremity. J. Surg. Res. **60**: 365–369.
5. Hunt, T.K. & H.W. Hopf. 1997. Wound healing and wound infection. What surgeons and anesthesiologists can do. Surgical Clinics of North America **77**: 587–606.

6. SUH, D.Y. & T.K. HUNT. 1998. Time line of wound healing. Clinics Podiatr. Med. Surg. **15**: 1–9.
7. ISHII, Y., Y. MIYANAGA, H. SHIMOJO, *et al.* 1999. Effects of hyperbaric oxygen on procollagen messenger RNA levels and collagen synthesis in the healing of rat tendon laceration. Tissue Eng. **5**: 279–286.
8. BONOMO, S.R., J.D. DAVIDSON, J.W. TYRONE, *et al.* 2000. Enhancement of wound healing by hyperbaric oxygen and transforming growth factor beta3 in a new chronic wound model in aged rabbits. Arch. Surg. **135**: 1148–1153.
9. BOYKIN, J.V., Jr. 2000. The nitric oxide connection: hyperbaric oxygen therapy, becaplermin, and diabetic ulcer management. Adv. Skin Wound Care **13**: 169–174.
10. WILLIAMS, R.L. 1997. Hyperbaric oxygen therapy and the diabetic foot. J. Am. Podiatr. Med. Assoc. **87**: 279–292.
11. WILLIAMS, R.L. & D.G. ARMSTRONG. 1998. Wound healing. New modalities for a new millennium. Clin. Podiatr. Med. Surg. **15**: 117–128.
12. DIMITRIJEVICH, S.D., S. PARANJAPE, J.R. WILSON, *et al.* 1999. Effect of hyperbaric oxygen on human skin cells in culture and in human dermal and skin equivalents. Wound Repair Regen. **7**: 53–64.
13. FABIAN, T.S., H.J. KAUFMAN, E.D. LETT, *et al.* 2000. The evaluation of subatmospheric pressure and hyperbaric oxygen in ischemic full-thickness wound healing. Am. Surgeon **66**: 1136–1143.
14. FISCHER, B.H. 1969. Topical hyperbaric oxygen treatment of pressure sores and skin ulcers. Lancet **2**: 405–409.
15. HENG, M.C. 1993. Topical hyperbaric therapy for problem skin wounds. J. Dermatol. Surg. Oncol. **19**: 784–793.
16. IGNACIO, D.R., A.P. PAVOT, R.N. AZER & L. WISOTSKY. 1985. Topical oxygen therapy treatment of extensive leg and foot ulcers. J. Am. Podiatr. Med. Assoc. **75**: 196–199.
17. KAUFMAN, T., J.W. ALEXANDER & B.G. MACMILLAN. 1983. Topical oxygen and burn wound healing: a review. Burns, Including Thermal Injury **9**: 169–173.
18. UPSON, A.V. 1986. Topical hyperbaric oxygenation in the treatment of recalcitrant open wounds. A clinical report. Physical Ther. **66**: 1408–1412.
19. LANDAU, Z. 1998. Topical hyperbaric oxygen and low energy laser for the treatment of diabetic foot ulcers. Arch. Orthopaed. Trauma Surg. **117**: 156–158.
20. HENG, M.C., J. HARKER, V.B. BARDAKJIAN & H. AYVAZIAN. 2000. Enhanced healing and cost-effectiveness of low-pressure oxygen therapy in healing necrotic wounds: a feasibility study of technology transfer. Ostomy Wound Manage. **46**: 52–60.
21. HENG, M.C., J. HARKER, G. CSATHY, *et al.* 2000. Angiogenesis in necrotic ulcers treated with hyperbaric oxygen. Ostomy Wound Manage. **46**: 18–28.
22. HUNT, T.K., Z. HUSSAIN & C.K. SEN. 2001. Give me ROS or give me death. Pressure **30**: 10–11.
23. KHANNA, S., S. ROY, D. BAGCHI, *et al.* 2001. Upregulation of oxidant-induced VEGF expression in cultured keratinocytes by a grape seed proanthocyanidin extract. Free Radical Biol. Med. **31:** 38–42.
24. CLARK, L.A. & R.E. MOON. 1999. Hyperbaric oxygen in the treatment of life-threatening soft-tissue infections. Respir. Care Clin. N. Am. **5**: 203–219.
25. MADER, J.T., G.L. BROWN, J.C. GUCKIAN, *et al.* 1980. A mechanism for the amelioration by hyperbaric oxygen of experimental staphylococcal osteomyelitis in rabbits. J. Infect. Dis. **142**: 915–922.
26. FERRARA, N. & W.J. HENZEL. 1989. Pituitary follicular cells secrete a novel heparin-binding growth factor specific for vascular endothelial cells. Biochem. Biophys. Res. Commun. **161**: 851–858.
27. HUNT, T.K., H. HOPF & Z. HUSSAIN. 2000. Physiology of wound healing. Adv. Skin Wound Care **13**: 6–11.
28. BABIOR, B.M. 2000. Phagocytes and oxidative stress. Am. J. Med. **109**: 33–44.
29. TUR, E., L. BOLTON & B.E. CONSTANTINE. 1995. Topical hydrogen peroxide treatment of ischemic ulcers in the guinea pig: blood recruitment in multiple skin sites. [See comments]. J. Am. Acad. Dermatol. **33**: 217–221.
30. SEN, C.K. 2000. Cellular thiols and redox-regulated signal transduction. Curr. Topics Cell. Reg. **36**: 1–30.

31. SEN, C.K. & L. PACKER. 1996. Antioxidant and redox regulation of gene transcription [see comments]. FASEB J. **10**: 709–720.
32. KHODR, B. & Z. KHALIL. 2001. Modulation of inflammation by reactive oxygen species: implications for aging and tissue repair. Free Radical Biol. Med. **30**: 1–8.
33. HANSELMANN, C., C. MAUCH & S. WERNER. 2001. Haem oxygenase-1: a novel player in cutaneous wound repair and psoriasis? Biochem. J. **353**: 459–466.
34. BEER, H.D., M.G. GASSMANN, B. MUNZ, *et al.* 2000. Expression and function of keratinocyte growth factor and activin in skin morphogenesis and cutaneous wound repair. J. Invest. Dermatol. Symp. Proc. **5**: 34–39.
35. SENEL, O., O. CETINKALE, G. OZBAY, *et al.* 1997. Oxygen free radicals impair wound healing in ischemic rat skin. Ann. Plastic Surg. **39**: 516–523.
36. STEILING, H., B. MUNZ, S. WERNER & M. BRAUCHLE. 1999. Different types of ROS-scavenging enzymes are expressed during cutaneous wound repair. Exp. Cell Res. **247**: 484–494
37. WATANABE, S., X.E. WANG, M. HIROSE, *et al.* 1998. Rebamipide prevented delay of wound repair induced by hydrogen peroxide and suppressed apoptosis of gastric epithelial cells in vitro. Digest. Dis. Sci. **43**: 107S–112S.
38. DOMINGUEZ-ROSALES, J.A., G. MAVI, S.M. LEVENSON & M. ROJKIND. 2000. H(2)O(2) is an important mediator of physiological and pathological healing responses. Arch. Med. Res. **31**: 15–20.
39. PAPA, F., S. SCACCO, R. VERGARI, *et al.* 1998. Respiratory activity and growth of human skin derma fibroblasts. Ital. J. Biochem. **47**: 171–178.
40. YAHAGI, N., M. KONO, M. KITAHARA, *et al.* 2000. Effect of electrolyzed water on wound healing. Artificial Organs **24**: 984–987.
41. KAYANOKI, Y., S. HIGASHIYAMA, K. SUZUKI, *et al.* 1999. The requirement of both intracellular reactive oxygen species and intracellular calcium elevation for the induction of heparin-binding EGF-like growth factor in vascular endothelial cells and smooth muscle cells. Biochem. Biophys. Res. Commun. **259**: 50–55.
42. KHERADMAND, F., E. WERNER, P. TREMBLE, *et al.* 1998. Role of Rac1 and oxygen radicals in collagenase-1 expression induced by cell shape change. Science **280**: 898–902.
43. HAN, R.N., S. BUCH, B.A. FREEMAN, *et al.* 1992. Platelet-derived growth factor and growth-related genes in rat lung. II. Effect of exposure to 85% O_2. Am. J. Physiol. **262**: L140–L146.
44. SUNDARESAN, M., Z.X. YU, V.J. FERRANS, *et al.* 1996. Regulation of reactive-oxygen-species generation in fibroblasts by Rac1. Biochem. J. **318**: 379–382.
45. SIMON, A.R., U. RAI, B.L. FANBURG & B.H. COCHRAN. 1998. Activation of the JAK-STAT pathway by reactive oxygen species. Am. J. Physiol. **275**: C1640–C1652.
46. KELNER, M.J. & S.F. UGLIK. 1995. Superoxide dismutase abolishes the platelet-derived growth factor-induced release of prostaglandin E2 by blocking induction of nitric oxide synthase: role of superoxide. Arch. Biochem. Biophys. **322**: 31–38.
47. GONZALEZ-RUBIO, M., S. VOIT, D. RODRIGUEZ-PUYOL, *et al.* 1996. Oxidative stress induces tyrosine phosphorylation of PDGF alpha-and beta-receptors and pp60c-src in mesangial cells. Kidney Int. **50**: 164–173.
48. KOUREMBANAS, S., R.L. HANNAN & D.V. FALLER. 1990. Oxygen tension regulates the expression of the platelet-derived growth factor-B chain gene in human endothelial cells. J. Clin. Invest. **86**: 670–674.
49. SHUKLA, A., A.M. RASIK & R. SHANKAR. 1999. Nitric oxide inhibits wounds collagen synthesis. Mol. Cell. Biochem. **200**: 27–33.
50. EFRON, D.T., D. MOST & A. BARBUL. 2000. Role of nitric oxide in wound healing. Curr. Opin. Clin. Nutr. Metab. Care **3**: 197–204.
51. SCHAFFER, M.R., U. TANTRY, S.S. GROSS, *et al.* 1996. Nitric oxide regulates wound healing. J. Surg. Res. **63**: 237–240.
52. FRANK, S., B. STALLMEYER, H. KAMPFER, *et al.* 1999. Nitric oxide triggers enhanced induction of vascular endothelial growth factor expression in cultured keratinocytes (HaCaT) and during cutaneous wound repair. FASEB J. **13**: 2002–2014.
53. LEE, P.C., A.N. SALYAPONGSE, G.A. BRAGDON, *et al.* 1999. Impaired wound healing and angiogenesis in eNOS-deficient mice. Am. J. Physiol. **277**: H1600–H1608.

54. YAMASAKI, K., H.D. EDINGTON, C. MCCLOSKY, *et al.* 1998. Reversal of impaired wound repair in iNOS-deficient mice by topical adenoviral-mediated iNOS gene transfer. J. Clin. Invest. **101**: 967–971.
55. VOGLER, B.K. & E. ERNST. 1999. Aloe vera: a systematic review of its clinical effectiveness. Br. J. Gen. Practice **49**: 823–828.
56. YE, Y.N., E.S. LIU, V.Y. SHIN, *et al.* 2001. A mechanistic study of proliferation induced by *Angelica sinensis* in a normal gastric epithelial cell line. Biochem. Pharmacol. **61**: 1439–1448.
57. LI, Y., K. METORI, K. KOIKE, *et al.* 2000. Granuloma maturation in the rat is advanced by the oral administration of *Eucommia ulmoides* Oliver leaf. Biol. Pharmaceut. Bull. **23**: 60–65.
58. PHAN, T.T., M.A. HUGHES & G.W. CHERRY. 1998. Enhanced proliferation of fibroblasts and endothelial cells treated with an extract of the leaves of *Chromolaena odorata* (Eupolin), an herbal remedy for treating wounds. Plastic & Reconstructive Surgery **101**: 756–765.
59. PHAN, T.T., J. ALLEN, M.A. HUGHES, *et al.* 2000. Upregulation of adhesion complex proteins and fibronectin by human keratinocytes treated with an aqueous extract from the leaves of *Chromolaena odorata* (Eupolin). Eur. J. Dermatol. **10**: 522.
60. HUPKENS, P., H. BOXMA & J. DOKTER. 1995. Tannic acid as a topical agent in burns: historical considerations and implications for new developments. Burns **21**: 57-61.
61. ROOT-BERNSTEIN, R.S. 1982. Tannic acid, semipermeable membranes, and burn treatment [letter]. Lancet **2**: 1168.
62. BAGCHI, D., A. GARG, R.L. KROHN, *et al.* 1998. Protective effects of grape seed proanthocyanidins and selected antioxidants against TPA-induced hepatic and brain lipid peroxidation and DNA fragmentation, and peritoneal macrophage activation in mice. Gen. Pharmacol. **30**: 771–776.
63. BAGCHI, D., A. GARG, R.L. KROHN, *et al.* 1997. Oxygen free radical scavenging abilities of vitamins C and E, and a grape seed proanthocyanidin extract in vitro. Res. Commun. Mol. Pathol. Pharmacol. **95**: 179–189.
64. BAGCHI, M., J. BALMOORI, D. BAGCHI, *et al.* 1999. Smokeless tobacco, oxidative stress, apoptosis, and antioxidants in human oral keratinocytes [published erratum appears in Free Radical Biol. Med. 1999 Jun. **26**(11–12):1599]. Free Radical Biol. Med. **26:** 992–1000.
65. RAY, S.D., M.A. KUMAR & D. BAGCHI. 1999. A novel proanthocyanidin IH636 grape seed extract increases in vivo Bcl- XL expression and prevents acetaminophen-induced programmed and unprogrammed cell death in mouse liver. Arch. Biochem. Biophys. **369**: 42–58.
66. YE, X., R.L. KROHN, W. LIU, *et al.* 1999. The cytotoxic effects of a novel IH636 grape seed proanthocyanidin extract on cultured human cancer cells. Mol. Cell. Biochem. **196**: 99–108.
67. SEN, C.K. 1998. Redox signaling and the emerging therapeutic potential of thiol antioxidants. Biochem. Pharmacol. **55**: 1747–1758.
68. SEN, C.K., S. KHANNA, S. ROY & L. PACKER. 2000. Molecular basis of vitamin E action. Tocotrienol potently inhibits glutamate-induced pp60(c-Src) kinase activation and death of HT4 neuronal cells. J. Biol. Chem. **275**: 13049–13055.
69. SEN, C.K., L. PACKER & O. HANNINEN, Eds. 2001. Handbook of oxidants and antioxidants in exercise. Elsevier. Amsterdam, 1207 pp.
70. BAGNATI, M., C. PERUGINI, C. CAU, *et al.* 1999. When and why a water-soluble antioxidant becomes pro-oxidant during copper-induced low-density lipoprotein oxidation: a study using uric acid. Biochem. J. **340**: 143–152.
71. BLAND, J.S. 1998. The pro-oxidant and antioxidant effects of vitamin C. Alternative Med. Rev. **3**: 170.
72. LIEHR, J.G. & D. ROY. 1998. Pro-oxidant and antioxidant effects of estrogens. Methods Mol. Biol. **108**: 425–435.
73. OHSHIMA, H., Y. YOSHIE, S. AURIOL & I. GILIBERT. 1998. Antioxidant and pro-oxidant actions of flavonoids: effects on DNA damage induced by nitric oxide, peroxynitrite and nitroxyl anion. Free Radical Biol. Med. **25**: 1057–1065.

Protective Effects of a Novel Niacin-Bound Chromium Complex and a Grape Seed Proanthocyanidin Extract on Advancing Age and Various Aspects of Syndrome X

HARRY G. PREUSS,[a] DEBASIS BAGCHI,[b] AND MANASHI BAGCHI[b]

[a]*Department of Physiology Georgetown University Medical Center, Washington, D.C. 20007, USA*

[b]*School of Pharmacy and Allied Health Professions, Creighton University, Omaha, Nebraska 68178, USA*

ABSTRACT: Aging is the progressive accumulation of changes with time that are responsible for the ever-increasing likelihood of disease and death. The precise cascade of pathological events mainly responsible for aging are still not clearly understood, but enhanced production of free radicals and its deleterious effects on proteins, nucleic acids, and fats, as well as enhanced glycosylation of proteins and DNA are prevalent during aging. Insulin resistance may be a common etiology, at least in part, behind the pathobiological alterations of advancing age. Prevalent age-related disorders such as cardiovascular diseases, obesity, and cancer have been associated with impaired glucose/insulin metabolism and its consequences. This leads to future strategies to combat the aging process and chronic disorders such as the components of syndrome X associated with aging. Increasing the intake of antioxidants and/or substances recognized to enhance insulin sensitivity is a natural means of combatting the glucose/insulin perturbations and free radical damage. Accordingly, ingestion of niacin-bound chromium and natural antioxidants such as grape seed proanthocyanidin extract has been demonstrated to improve insulin sensitivity and/or ameliorate free radical formation and reduce the signs/symptoms of chronic age-related disorders including syndrome X. These natural strategies possess a highly favorable risk/benefit ratio.

KEYWORDS: aging; insulin resistance; free radicals; grape seed extract

INTRODUCTION

Insulin resistance often develops with advancing age, and may play a prominent role in the aging process through its association with enhanced glycosylation of proteins and DNA, and increased free radical formation.[1] Insulin is responsible for transporting glucose (blood sugar) into the body's cells, where it is used for energy and stored for future use. When a person is resistant to insulin, glucose transport is

Address for correspondence: Harry G. Preuss M.D., M.A.C.N., C.N.S., Department of Physiology, Georgetown University Medical Center, Med-Dent Building, Room 103 SE, 3900 Reservoir Road, NW, Washington, DC 20007, USA. Voice: 202-687-1441; fax: 202-687-8788.
preusshg@georgetown.edu

impaired. In turn, the pancreas, where insulin is manufactured, compensates by releasing more and more insulin, which in turn leads to elevated insulin levels, or hyperinsulinemia. Common age-related disorders such as type 2 diabetes mellitus, cardiovascular diseases, hypertension, dyslipidemia, obesity, polycystic ovary disease, sleep apnea, and certain hormone-sensitive cancers have also been associated with impaired glucose/insulin metabolism, strengthening the association of insulin resistance with aging.[2] Pathological mechanisms behind the perturbations of glucose/insulin metabolism and glycosylation theory of aging have been extensively reviewed. Further strategies for countering the aging process and chronic disorders associated with it through natural means may be directed toward enhancing sensitivity of the glucose-insulin system and/or ameliorating formation of free radicals[3–5] (see FIGURE 1).

Whether the pathological events associated with insulin resistance are a significant contributing factor to illness found in aging populations is not a trivial question. In modern acculturated countries, non-communicable, age-related diseases such as cardiovascular disorders, tumors, and diabetes mellitus account for over two-thirds of deaths.[6–8] This occurs despite medical knowledge sufficient to control the extent of these disorders. The appearance of the common chronic disorders seems to be related, at least in part, to lifestyle choices[7–9]—choices which often affect the status of the glucose/insulin system.[10] Obesity is a highly prevalent and unabating public health problem in the United States and has been demonstrated as an epidemic. Approximately 61% of U.S. adults are obese primarily because they eat imprudently and exercise negligibly. Involuntary weight gain adversely affects all elements of the atherogenic cardiovascular risk profile, including blood lipids, blood pressure, and glucose tolerance. The metabolically linked cluster of risk factors associated with obesity have been characterized as an insulin resistance syndrome (syndrome X).[11]

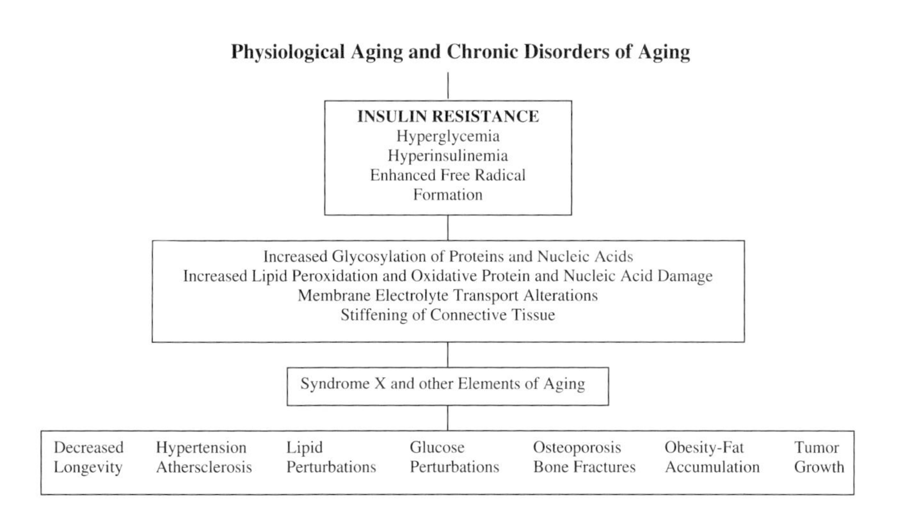

FIGURE 1. Hypothesis associating insulin resistance and its manifestations with aging.

Syndrome X is a common metabolic disorder affecting more than 75 million Americans. One of the main symptoms of syndrome X is a condition called "insulin resistance," where the body becomes resistant to its own insulin. When insulin is not functioning properly, the body compensates by releasing more insulin, which results in hyperinsulinemia. Hyperinsulinemia is a primary indicator of syndrome X, which can lead to increased risk of diabetes, heart disease, and obesity.

Clinical observations support the concept that elevated glucose and insulin levels secondary to insulin resistance contribute to the aging process.[11,12] Diabetics and end stage renal disease patients undergoing hemodialysis characteristically show disturbances in glucose/insulin metabolism.[13,14]

PERTURBATIONS IN THE GLUCOSE/INSULIN SYSTEM

The best direct evidence of maintaining healthy levels of insulin during aging is limited largely to animal studies. Masoro *et al.*[15] used a rat model to demonstrate that even slight elevations in insulin and/or glucose levels might influence, at least to some extent, various chronic disorders associated with aging. The ability of caloric restriction to slow many manifestations of the aging process and to augment the life span in rats supports a prominent role to improve the glucose/insulin system and reduce the chronic disorders of aging. Caloric restriction improves insulin sensitivity also.[15] Over the years, rodent studies that have used caloric restriction as a vehicle to prolong life-span, have been duplicated in primates.[16] The best evidence that caloric restriction may benefit human longevity lies in the experience of individuals in the well-publicized biosphere study during which many cardiovascular risk factors improved with caloric restriction.[17] This "insulin resistance hypothesis" would be strengthened considerably if it were found that natural means other than caloric restriction can also improve insulin sensitivity. However, the definitive data necessary to make such determinations are not available. Therefore, more extensive studies, similar to caloric restriction, must be carried out using agents capable of enhancing insulin sensitivity and reducing non-enzymatic glycosylation and oxidative stress.[18–21]

INTERACTION BETWEEN ANTIOXIDANTS AND INSULIN SENSITIZERS

An interesting association exists between insulin sensitizers that increase insulin sensitivity and antioxidants that decrease free radical damage. Recent research by Cunningham (1998) and Chausmer (1998) indicates that products characteristically classified as antioxidants can also enhance insulin sensitivity.[22,23] Many disease states associated with free radicals are the same as those found with insulin resistance.[24] In earlier studies, it was demonstrated that supplementation with a novel niacin-bound chromium complex promoted normal insulin function and lowered blood pressure in humans.[25] It was also noted that the niacin-bound chromium complex had antioxidant properties.[26] Preliminary studies suggest that antioxidants and enhancers of insulin sensitivity increase the average life span.[27–29] Earlier research showed a profile in which high levels of other antioxidants such as vitamin A and vitamin E were important for human longevity.[30] Therefore, it is reasonable to argue that the predominant pathological mechanism behind aging, cardiovascular disorders, cancer, and some eye disorders relates to the destructive influence of insulin resistance and oxidative tissue damage.

ACUTE EFFECTS OF A NOVEL NIACIN-BOUND CHROMIUM COMPLEX (NBC) AND A GRAPE SEED PROANTHOCYANIDIN EXTRACT (GSPE)

Using both laboratory and clinical studies over the last few years, we have concentrated on the therapeutic effects of two powerful natural products, a novel niacin-bound chromium complex (NBC, commercially known as ChromeMate™), which is a well-recognized insulin sensitizer and a proven supplement for aging and age-associated disorders such as the components of syndrome X, and a grape seed proanthocyanidin extract (GSPE), which is a broad spectrum antioxidant.

Our initial examination of the acute effects of NBC and GSPE focused on rats.[32] Measuring elevation in blood pressure is a sensitive way to assess changes in insulin sensitivity in rats[26,31](see TABLE 1). The effects of NBC and GSPE, alone and in combination, were examined in two separate investigations carried out on normotensive Brown Norway Fischer 344 rats (BN/F344) and hypertensive (SHR) rats. In the first study, NBC and GSPE decreased SBP significantly at three weeks in normotensive hybrid BN/F344 rats, −3 and −6 mm Hg, respectively, as compared to control, suggesting an improvement in insulin sensitivity. Although circulating glucose levels and HbA1C were essentially similar, insulin levels were significantly lower in rats consuming either NBC or GSPE. Renal TBARS, an estimate of oxidative damage to fatty membranes, were also significantly lower in BN/F344 rats ingesting NBC or GSPE. In a second study, changes in the respective systolic blood pressure in SHR were found after four weeks with NBC, GSPE, or the combination compared to control. Results from a losartan challenge group demonstrated decreased activity of the renin-angiotensin system in each test group of SHR. The lowered levels of creatinine in the test SHR suggest decreased activity of angiotensin 2 secondary to the ingestion of the two dietary supplements. From these two acute studies, we concluded that oral NBC and GSPE alone and in combination can significantly lower systolic blood pressure in normotensive and hypertensive rats and can affect the glucose-insulin and renin-angiotensin systems without producing any obvious toxicity.

EFFECTS OF LONG-TERM SUPPLEMENTATION WITH A NOVEL NIACIN-BOUND CHROMIUM COMPLEX (NBC), A GRAPE SEED PROANTHOCYANIDIN EXTRACT (GSPE), AND A ZINC MONOMETHIONINE COMPLEX (ZM)

The hypothesized benefits of NBC and GSPE are largely based on acute studies. In a chronic study, we determined the use of a combination of NBC, GSPE, and ZM in protecting against insulin resistance during a prolonged period of time (exceeding

TABLE 1. Various biochemical and metabolic parameters of BN/F344 rats and SHR at 21–30 days

Group	Delta BW (g)	Delta SBP (mm Hg)	Glucose (mg/dL)	Insulin (μU/mL)	HbA1C (%)	Triglycerides (mg/dL)	Cholesterol (mg/dL)	Creatinine (mg)
BN/F344 (21 days)								
Control	6.3 ± 2.1	+4.0 ± 1.4	169 ± 15.2	5.94 ± 1.6	4.9 ± 0.1	228 ± 11.0	141 ± 6.8	—
NBC	5.0 ± 3.5	+1.0 ± 1.2*	157 ± 7.3	2.91 ± 0.4*	4.9 ± 0.1	219 ± 25.0	165 ± 6.0	—
GSPE	10.0 ± 2.1	−2.0 ± 1.1*	187 ± 23.0	3.13 ± 0.5*	4.9 ± 0.1	223 ± 23.5	159 ± 8.5	—
SHR (30 days)								
Control	43 ± 2.5	+26 ± 5.5	86 ± 3.9	183 ± 12.3	4.6 ± 0.2	120 ± 5.6	81.2 ± 2.0	1.0 ± 0.06
NBC	36 ± 2.3	+7.9 ± 2.1*	116 ± 20.5	190 ± 9.0	4.7 ± 0.1	131 ± 11.9	85.3 ± 1.6	0.6 ± 0.04*
GSPE	43 ± 5.1	−2.1 ± 4.9*	116 ± 25.1	201 ± 13.3	4.8 ± 0.1	116 ± 8.0	81.2 ± 1.5	0.6 ± 0.04*
NBC + GSPE	41 ± 1.7	−2.9 ± 5.9*	110 ± 5.9	187 ± 7.6	5.0 ± 0.2	108 ± 4.9	83.8 ± 3.5	0.5 ± 0.04*

Each value is the mean ± SEM from eight rats; *significantly different from control; BW, bodyweight; SBP, systolic blood pressure.

one year) in Brown Norway Fischer 344 rats. In this study, 104 BN/F344 rats were divided into two groups: a control group receiving a basic diet and a test group receiving the same diet with added NBC (5 ppm), oligomeric proanthocyanidin rich GSPE (250 ppm) and ZM (18 ppm).[21] Initial mean systolic blood pressures of both control and test groups were 122 mm Hg. During the first seven months, the systolic blood pressure of the control animals steadily increased to 140 mm Hg and remained at this level for the next 7–8 months. In contrast, the systolic blood pressure of the test animals initially decreased over the first four months to as low as 110–114 mm Hg. The systolic blood pressure then increased over the following months, essentially reaching the starting value of 120 mm Hg. This was still significantly lower than control ($p < 0.001$). In 12 control and 12 test rats, hepatic TBARS formation, an estimate of lipid peroxidation/free radical formation, was significantly lower after one year of ingesting the test diet ($p < 0.04$); and HbA1C was also statistically significantly lower in the test group (4.9% vs. 4.6%, $p < 0.003$). Circulating levels of cholesterol, HDL, and triglycerides were essentially similar between the two groups (see TABLE 2). Body, kidney, and liver weights were not different after one year of ingesting the different diets; but epididymal fat pad weight was less in the group receiving supplements. It was observed after prolonged supplementation of a combination of insulin sensitizers and antioxidants (NBC, GSPE and ZM) that they can markedly lower systolic blood pressure in normotensive rats. The combination also helped to lessen oxidative damage to fats as suggested by decreased TBARS formation, and lower HbA1C without showing signs of toxicity.[21]

Thus, both acute and chronic rat studies suggested that NBC and GSPE could lower blood pressure and sensitize the glucose/insulin system. Beneficial effects were also noted on oxidative stress and the renin/angiotensin and nitric oxide systems. Our earlier studies demonstrate that NBC supplementation caused significant loss of body fat and sparing of muscle in overweight African-American women.[33]

TABLE 2. Serum chemistry following one year of supplementation

Parameter	Control	Test	p
Insulin (ng/mL)	5.2 ± 0.5	4.1 ± 0.6	0.15
Glucose (mg/dL)	200 ± 22	250 ± 18	0.08
Insulin/Glucose	3.0 ± 0.5	1.7 ± 0.2	0.03*
HbA1C (ng/mL)	4.9 ± 0.1	4.6 ± 0.1	0.04*
Cholesterol (mg/mL)	132 ± 4.6	133 ± 5.8	NS
HDL (mg/dL)	35 ± 2.0	33 ± 1.3	NS
Triglycerides (mg/dL)	284 ± 17	310 ± 18	NS
Uric Acid (mg/dL)	4.3 ± 0.8	4.0 ± 0.6	NS
BUN (mg/dL)	12.0 ± 0.3	12.0 ± 0.3	NS

Each value is the mean ± SEM from eight rats; *significantly different from control.

HUMAN CLINICAL STUDY

Hypercholesterolemia, a significant cardiovascular risk factor, is prevalent in the American population. Although many pharmaceuticals provide adequate therapeutic benefits, many drugs also cause significant adverse reactions, such as liver problems with the "statins." Accordingly, ways of lowering cholesterol levels using safe natural products would be beneficial and welcomed. In a randomized, double-blind, placebo-controlled study on humans it was observed that the combination of NBC and GSPE could favorably influence circulating cholesterol, LDL, and autoantibodies to oxidized LDL[34]. We examined 40 hypercholesterolemic patients (220–300 mg/dL) in a randomized, double-blind, placebo-controlled study. The four groups of 10 patients each received placebo b.i.d., NBC 200 μg b.i.d., GSPE 100 mg b.i.d., or a combination of NBC plus GSPE at the same dosage bid. Over 2 months, the change in total cholesterol from baseline among groups was: placebo −3.5% ± 4, GSPE −2.5% ± 2, NBC −10% ± 5, and NBC+GSPE −16.5% ± 3. The decrease in the last group was significantly different from placebo ($p < 0.01$). The major change was in the LDL levels: placebo −3.0% ± 4, GSPE −1.0% ± 2.0, NBC −14% ± 4.0, NBC+GSPE −20% ± 6.0 (see FIGURE 2). Again, the combination of NBC plus GSPE

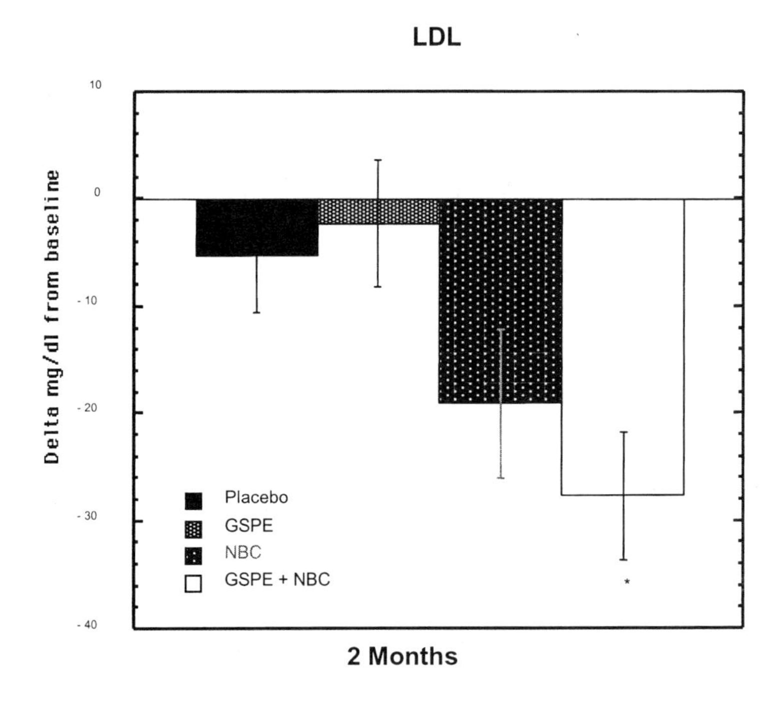

FIGURE 2. Difference between LDL at two months and baseline in the four groups: (1) the placebo group; (2) grape seed proanthocyanidin extract (GSPE)–fed group; (3) niacin-bound chromium complex (NBC)–treated group; and (4) GSPE+NBC–treated group. *Significantly different.

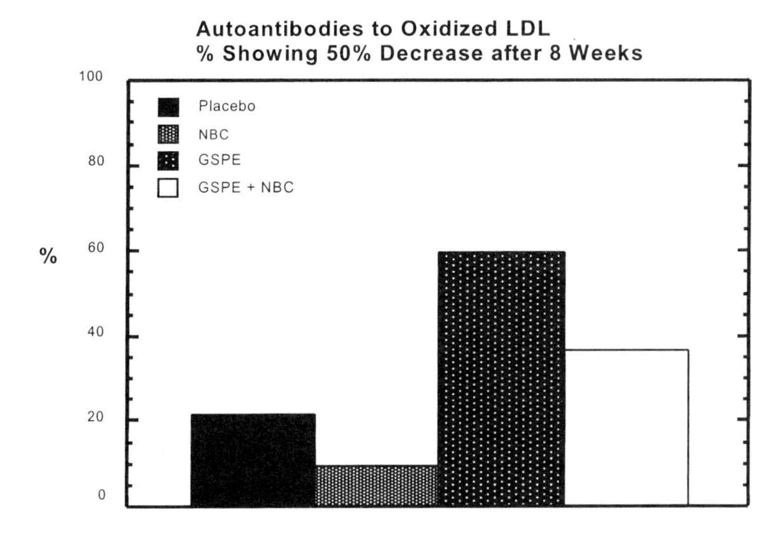

FIGURE 3. Percent of individuals in the four groups showing a greater than 50% reduction in autoantibodies to oxidized LDL.

significantly decreased LDL when compared to placebo ($p < 0.01$). HDL levels did not essentially change among the groups. Also, there was no significant difference for the triglyceride concentrations among the groups. No statistically significant differences were observed in the level of autoantibodies to oxidized LDL. Groups receiving GSPE had greater decreases in this parameter, that is, –17.3% and –10.4% in the placebo and NBC group in comparison to –30.7% and –44.0% in the GSPE and NBC+GSPE groups. There was also a trend for those groups receiving GSPE to decrease autoantibodies to oxidized LDL greater than 50% over eight weeks: placebo 2/9, NBC 1/10, GSPE 6/10, and NBC+GSPE 3/8 (see FIGURE 3). No significant changes occurred in the levels of circulating homocysteine among the four groups. However, only five subjects showed high levels outside the range of normal (5–15 μmole/L). Four of these occurred in the GSPE group. Three out of the four showed a marked decrease. One subject from the combination group showed no change. In this basically normotensive population, the blood pressures were not significantly different. Nevertheless, the one hypertensive subject on GSPE had a marked decrease in systolic (–42 mm Hg) and diastolic (–13 mm Hg) BP. From the above study, we conclude that a combination of natural products specifically NBC and GSPE can decrease LDL levels significantly.

SUMMARY AND CONCLUSIONS

Over the last century, humans have made great strides toward increased longevity. Much of this has come about through better hygiene, availability of better nutrients, and medical advancements. Our present information suggests that substances known to ameliorate insulin resistance are active in lowering glycosylation, free radical formation, and membrane transport alterations. These are all beneficial changes. Unfortunately, it is generally recognized that the number of type 2 diabetics is increasing, and this is mainly attributed to the increase in the overweight state and obesity. The latter is attributed to increased consumption of calories and less exercise in today's busy life schedule. In summary, the current paper will encourage people to seek a better life style and use of appropriate dietary supplements, which may favorably affect life-span and reduce the incidence of advancing age–induced chronic disorders and improve deleterious symptoms of syndrome X.

REFERENCES

1. REAVEN, G.M., N. CHEN, C. HOLLENBECK & Y.D.I. CHEN. 1989. Effect of age on glucose tolerance and glucose uptake in healthy individuals. J. Am. Ger. Soc. **37:** 735–740.
2. DEFRONZO, R. 1981. Glucose intolerance and aging. Diabetes Care **4:** 493–501.
3. BROUGHTON, D.L. & R.L. TAYLOR. 1991. Review: deterioration of glucose tolerance with age: the role of insulin resistance. Age Aging **20:** 221–225.
4. DEFRONZO, R.A. & E. FERINIMMI. 1991. Insulin resistance: a multifaceted syndrome responsible for NIDDM, obesity, hypertension, dyslipidemia, and atherosclerotic cardiovascular disease. Diabetes Care **14:** 173–194.
5. BAGCHI, M., D. BAGCHI, E.B. PATTERSON, *et al.* 1996. Age-related changes in lipid peroxidation and antioxidant defense in Fischer 344 rats. Ann. Acad. Sci. **793:** 449–452.
6. BRESLOW, L. 1982. Prevention and control of noncommunicable diseases. World Health Forum **3:** 429–431.
7. EDITORIAL. 1982. Mass strategies of prevention; the swings and roundabouts. Lancet **2:** 1256–1257.
8. BOOTH, F.W., S.E. GORDON, C.J. CARLSON & M.T. HAMILTON. 2000. Waging war on modern chronic diseases: primary prevention through exercise biology. J. Appl. Physiol. **88:** 774–787.
9. WILLIAMS, D.E., A.T. PREVOST, M.J. WHICHELOW, *et al.* 2000. A cross-sectional study of dietary patterns with glucose intolerance and other features of the metabolic syndrome. Br. J. Nutr. **83:** 257–266.
10. ZIMMET, P.Z. 1988. Primary prevention of diabetes mellitus. Diabetes Care **11:** 258–262.
11. REAVEN, G.M. 1988. Role of insulin resistance in human disease (Banting Lecture 1988). Diabetes **37:** 1595–1607.
12. CERAMI, A., H. VLASSARE & M. BROWNLEE. 1987. Glucose and aging. Sci. Am. **256:** 90–96.
13. MAKITA, Z., S. RADOFF, E.J. RAYFIELD, *et al.* 1991. Advanced glycosylation end products in patients with diabetic nephropathy. N. Engl. J. Med. **325:** 836–842.
14. PARFREY, P.S. & J.D. HARNETT. 1994. Long-term cardiac morbidity and mortality during dialysis therapy. Adv. Nephrol. **23:** 311–330.
15. MASORO, E.J., R.J.M. MCCARTER, M.S. KATZ & C.A. MCMAHAN. 1992. Dietary restriction alters characteristics of glucose fuel use. J. Gerontol. **47:** B202–B208.
16. HOPKIN, K. 1995. Aging in focus: caloric restriction may put brakes on aging. J. Nat. Inst. Health **7:** 47–50.
17. WALFORD, R.L., S.B. HARRIS & M.W. GUNION. 1992. The calorically restricted low-fat nutrient-dense diet in Biosphere 2 significantly lowers blood glucose, total leucocytes count, cholesterol and blood pressure in humans. Proc. Natl. Acad. Sci. USA **89:** 11533–11537.
18. GONDAL, J.A., P. MACARTHY, A.K. MYERS & H.G. PREUSS. 1996. Effects of dietary sucrose and fibers on blood pressure in spontaneously hypertensive rats. Clin. Nephrol. **45:** 163–168.
19. PREUSS, H.G., S.T. JARRELL, N. BUSHEHRI, *et al.* 1997. Nutrients and trace elements as they affect blood pressure in the elderly. Ger. Nephrol. Urol. **6:** 169–179.

20. PREUSS, H.G. 1997. Effects of glucose/insulin perturbations on aging and chronic disorders of aging: the evidence. J. Am. Coll. Nutr. **16:** 397–403.
21. PREUSS, H.G., S. MONTAMARRY, B. ECHARD, *et al.* 2001. Long-term effects of chromium, grape seed extract, and zinc on various metabolic parameters of rats. Mol. Cell. Biochem. **223:** 95–102.
22. CUNNINGHAM, J.J. 1998. The glucose/insulin sysytem and vitamin C: implications in insulin-dependent diabetes mellitus. J. Am. Coll. Nutr. **17:** 105–108.
23. CHAUSMER, A.B. 1998. Zinc, insulin, and diabetes. J. Am. Coll. Nutr. **17:** 109–115.
24. DIPLOCK, A.T. 1991. Antioxidant nutrients and disease prevention: an overview. Am. J. Clin. Nutr. **53:** 189S–193S.
25. MERTZ ,W. 1993. Chromium in human nutrition: a review. J. Nutr. **123:** 626–633.
26. PREUSS, H.G., P.L. GROJEC, S. LIEBERMAN & R.A. ANDERSON. 1997. Effects of different chromium compounds on blood pressure and lipid peroxidation in spontaneously hypertensive rats. Clin. Nephrol. **47:** 325–330.
27. MCCARTY, M.F. 1993. Homologous physiological effects of phenformin and chromium picolinate. Med. Hypotheses **41:** 316–334.
28. EVANS, G.W. & L.K. MEYER. 1994. Life span is increased in rats supplemented with a chromium-pyridine 2 carboxylate complex. Adv. Sci. Res. **1:** 19–23.
29. LEDVINA, M. & M. HODANOVA. 1980. The effect of simultaneous administration of tocopherol and sunflower oil on the life-span of female mice. Exp. Gerontol. **15:** 67–71.
30. MECOCCI, P., M.C. POLIDORI, L. TROIANO, *et al.* 2000. Plasma antioxidants and longevity: a study on healthy centenarians. Free Radical Biol. Med. **28:** 1243–1248.
31. PREUSS, H.G., J.A. GONDAL, E. BUSTOS, *et al.* 1995. Effect of chromium and guar on sugar-induced hypertension in rats. Clin. Nephrol. **44:** 170–177.
32. TYSON, D.A., N.A. TALPUR, B.W. ECHARD, *et al.* 2000. Acute effects of grape seed extract on the systolic blood pressure of normotensive and hypertensive rats. Res. Commun. Pharmacol. Toxicol. **5:** 91–106.
33. TYSON, D.A., N.A. TALPUR, B.W. ECHARD, *et al.* 2000. Acute effects of grape seed extract on the systolic blood pressure of normotensive and hypertensive rats. Res. Commun. Pharmacol. Toxicol. **5:** 91–106.
34. CRAWFORD, V., R. SCHECKENBACH, J. BELLANTI, *et al.* 1999. Effects of niacin-bound chromium supplementation on body composition in overweight African-American women. Diabetes, Obesity Metab. **1:** 331–337.
35. PREUSS, H.G., D. WALLERSTEDT, N. TALPUR, *et al.* 2000. Effects of Chromium and Grape Seed Extract on the Lipid Profile of Hypercholesterolemic Subjects: A Pilot Study. J. Med. **31:** 227–246.

Cellular Protection with Proanthocyanidins Derived from Grape Seeds

DEBASIS BAGCHI,[a,b] MANASHI BAGCHI,[a] SIDNEY J. STOHS,[a] SIDHARTHA D. RAY,[b] CHANDAN K. SEN,[c] AND HARRY G. PREUSS[d]

[a]*Creighton University School of Pharmacy & Allied Health Professions, Omaha, Nebraska 68178, USA*

[b]*Department of Pharmacology, Toxicology and Medicinal Chemistry, AMS College of Pharmacy and Health Sciences, Long Island University, Brooklyn, New York 11201, USA*

[c]*Laboratory of Molecular Medicine, Department of Surgery, The Ohio State University Medical Center, Columbus, Ohio 43210, USA*

[d]*Department of Physiology and Biophysics, Georgetown University Medical Center, Washington, DC 20007, USA*

ABSTRACT: Grape seed proanthocyanidins have been reported to possess a broad spectrum of pharmacological and medicinal properties against oxidative stress. We have demonstrated that IH636 proanthocyanidin extract (GSPE) provides excellent protection against free radicals in both *in vitro* and *in vivo* models. GSPE had significantly better free radical scavenging ability than vitamins C, E and β-carotene and demonstrated significant cytotoxicity towards human breast, lung and gastric adenocarcinoma cells, while enhancing the growth and viability of normal cells. GSPE protected against tobacco-induced apoptotic cell death in human oral keratinocytes and provided protection against cancer chemotherapeutic drug-induced cytotoxicity in human liver cells by modulating cell cycle/apoptosis regulatory genes such as bcl2, p53 and c-myc. Recently, the bioavailability and mechanistic pathways of cytoprotection by GSPE were examined on acetaminophen-induced hepatotoxicity and nephrotoxicity, amiodarone-induced pulmonary toxicity, doxorubicin-induced cardiotoxicity, DMN-induced immunotoxicity and MOCAP-induced neurotoxicity in mice. Serum chemistry changes, integrity of genomic DNA and histopathology were assessed. GSPE pre-exposure provided near complete protection in terms of serum chemistry changes and DNA damage, as well as abolished apoptotic and necrotic cell death in all tissues. Histopathological examination reconfirmed these findings. GSPE demonstrated concentration-/dose-dependent inhibitory effects on the drug metabolizing enzyme cytochrome P450 2E1, and this may be a major pathway for the anti-toxic potential exerted by GSPE. Furthermore, GSPE treatment significantly decreased TNFα-induced adherence of T-cells to HUVEC by inhibiting VCAM-1 expression. These results demonstrate that GSPE is highly bioavailable and may serve as a potential therapeutic tool in protecting multiple target organs from structurally diverse drug- and chemical-induced toxicity.

KEYWORDS: proanthocyanidins; free radicals; antioxidants

Address for correspondence: Debasis Bagchi, Ph.D., FACN, C.N.S., Department of Pharmacy Sciences, Creighton University School of Pharmacy & Allied Health Professions, 2500 California Plaza, Omaha, NE 68178, USA. Voice: 402-280-2950; fax: 402-280-1883.
debsis@creighton.edu

INTRODUCTION

Free radicals have been implicated in more than one hundred disease conditions in humans, including arteriosclerosis, AIDS, arthritis, brain and cardiovascular dysfunctions, carcinogenesis, cataracts, diabetes, ischemia and reperfusion injury of many tissues, ocular dysfunction and tumor promotion.[1–3] Antioxidants/free radical scavengers function as inhibitors at both initiation and promotion/propagation/transformation stages of tumor promotion/carcinogenesis, and protect cells against oxidative damage.[1,2] Proanthocyanidins are powerful naturally occurring polyphenolic antioxidants widely available in fruits, vegetables, seeds, nuts, flowers and bark.[4,5] Proanthocyanidins are known to possess antibacterial, antiviral, antiinflammatory, antiallergic and vasodilatory actions.[4–6] They have also been shown to inhibit lipid peroxidation, platelet aggregation, capillary permeability and fragility.[4] Proanthocyanidins have been shown to modulate the activity of regulatory enzymes including cyclooxygenase, lipooxygenase, protein kinase C, angiotensin-converting enzyme, hyaluronidase enzyme and cytochrome P450 activities.[4–6] In our laboratory, concentration- and dose-dependent free radical scavenging abilities of a novel IH636 grape seed proanthocyanidin extract (GSPE) were assessed both *in vitro* and *in vivo* models, and compared with vitamins C, E and β-carotene. GSPE exhibited significantly better protection than vitamin C, E and β-carotene.[7-9] GSPE demonstrated selective cytotoxicity towards cultured human breast, lung and gastric adenocarcinoma cells, while enhancing the growth and viability of normal cells.[10]

To understand the mechanistic pathways of cytoprotection and organ-specific bioavailability of GSPE, a series of experiments were conducted. Protective ability of GSPE was assessed against smokeless tobacco–induced oxidative stress, genomic DNA fragmentation, and apoptotic cell death in a primary culture of normal human oral keratinocytes, and compared with vitamins C and E, singly and in combination. The protective ability of GSPE was also assessed against chemotherapeutic drug(s) (4-hydroxyperoxycyclophosphamide or idarubicin)–induced cytotoxicity towards cultured normal human liver cells. Furthermore, organ-specific bioavailability and protective ability of GSPE was investigated against a broad spectrum of drug- and chemical-induced multiorgan toxicity in mice. The chemoprotective ability of GSPE was assessed against acetaminophen (APAP)-induced hepatotoxicity, amiodarone (AMI)-induced pulmonary toxicity, doxorubicin (DOX)-induced cardiotoxicity, dimethylnitrosamine (DMN)-induced splenotoxicity and O-ethyl-S,S-dipropyl phosphorodithioate (MOCAP)-induced neurotoxicity. Serum chemistry changes, integrity of genomic DNA and histophathologic assessment were conducted. Furthermore, the regulatory effect of GSPE and drug metabolizing enzyme cytochrome P450 2E1 and cell adhesion molecules were determined.

IN VITRO FREE RADICAL SCAVENGING ABILITY

The free radical scavenging abilities of GSPE, vitamin E, and vitamin C against biochemically generated superoxide anion and hydroxyl radical were assessed *in vitro* at varying concentrations via cytochrome *c* reduction and chemiluminescence response. A concentration-dependent inhibition was demonstrated by GSPE. At a

100 mg/L concentration, GSPE exhibited 78–81% inhibition of superoxide anions and hydroxyl radicals. Under identical conditions, vitamin C inhibited these two free radicals by approximately 12–19%, while vitamin E inhibited these two oxygen radicals by 36–44%.[7] GSPE also demonstrated superior RSA against peroxyl radicals as compared to Trolox.[9]

Concentration-dependent protective ability of GSPE was assessed against hydrogen peroxide–induced oxidative stress in cultured J774A.1 macrophage and neuroactive PC-12 adrenal pheochromocytoma cells using a laser scanning confocal microscopy. The overall intracellular oxidized states of these cells following incubation with hydrogen peroxide was assessed at an excitation wavelength of 513nm using 2,7-dichlorofluorescein diacetate as fluorescent probe. Approximately 5.8- and 4.5-fold increases in fluorescence intensity were observed following incubation of J774A.1 and PC-12 cells with 0.50mM hydrogen peroxide for 24 h, respectively. Pretreatment of the J774A.1 cells with 50 and 100 mg/L GSPE decreased hydrogen peroxide-induced fluorescent intensity by 36% and 70%, while under these same conditions approximately 50% and 70% in fluorescence intensities were observed in PC-12 cells, respectively. Thus, GSPE provided significant protection against oxidative stress induced by hydrogen peroxide in these cells.[11]

DIFFERENTIAL CYTOTOXICITY TOWARDS HUMAN NORMAL AND MALIGNANT CELLS

The cytotoxicity of GSPE was assessed towards selected human cancer cells, including cultured MCF-7 breast cancer, A-427 lung cancer, and CRL-1739 gastric adenocarcinoma cells at 25 and 50 mg/L concentrations for 0–72 h using cytomorphology and MTT cytotoxicity assay, and these effects were compared with those on two normal cells, including normal human gastric mucosal and J774A.1 macrophage cells. Concentration- and time-dependent cytotoxic effects were induced by GSPE in human breast, lung, and gastric adenocarcinoma cells, while under these same conditions GSPE enhanced the growth and viability of the normal cells.[10]

SMOKELESS TOBACCO–INDUCED OXIDATIVE STRESS AND APOPTOTIC CELL DEATH AND PROTECTION BY ANTIOXIDANTS

The protective ability of GSPE was assessed against smokeless tobacco–induced oxidative damage and programmed cell death (apoptosis) in a primary culture of human keratinocytes. Approximately 9, 29, and 25% apoptotic cell death were observed in these cells following treatment with 100, 200, and 300μg/mL tobacco-treated cells, respectively. Pretreatment of the 300 μg/mL tobacco-treated cells with 100 mg GSPE/L reduced tobacco-induced apoptotic cell death by 85% in oral cells, while a combination of vitamins E and C (75μM each) reduced tobacco-induced apoptotic cell death by 46%.[12]

PROTECTION AGAINST CANCER CHEMOTHERAPY DRUG-INDUCED TOXICITY IN HUMAN LIVER CELLS

Anticancer chemotherapeutic agents are effective in inhibiting growth of cancer cells *in vitro* and *in vivo*, however, toxicity to normal cells is a major problem in this therapeutic intervention. The effect of GSPE was assessed to ameliorate chemotherapy drug–induced toxic effects in cultured Chang epithelial cells, established from nonmalignant human liver tissue. These cells were treated *in vitro* with idarubicin (Ida) (30 nM) or 4-hydroxyperoxycyclophosphamide (4-HC) (1 μg/mL) with or without GSPE (25 μg/mL). The cells were grown *in vitro* and the growth rate of the cells was determined using the MTT [3-(4,5-dimethylthiazol-2-yl)-2,5-diphenyltetrazolium bromide; thiazolyl blue] assay. Results showed that GSPE decreased the growth inhibitory and cytotoxic effects of Ida as well as 4-HC on Chang epithelial cells. Because these chemotherapeutic agents are known to induce apoptosis in the target cells, the Chang epithelial cells were analyzed for apoptotic cell population by flow cytometry. There was a significant decrease in the number of cells undergoing apoptosis following treatment with GSPE. An increased expression of the anti-apoptotic protein bcl2 was observed in GSPE-treated cells. These results indicate that GSPE can be a potential candidate to ameliorate the toxic effects associated with chemotherapeutic agents and one of the mechanisms of action of GSPE includes upregulation of bcl2 expression.[13]

ACETAMINOPHEN (APAP)-INDUCED HEPATO- AND NEPHROTOXICITY, AMIODARONE (AMI)-INDUCED PULMONARY TOXICITY AND DOXORUBICIN (DOX)-INDUCED CARDIOTOXICITY *IN VIVO*, AND PROTECTION BY GSPE

The short and long term protective effects of GSPE were examined on APAP overdose-induced lethality and liver toxicity. Mice were administered nontoxic doses of GSPE (three or seven days, 100 mg/kg, p.o.) followed by hepatotoxic doses of APAP (400 or 500 mg/kg, i.p.). GSPE dramatically decreased APAP-induced mortality, serum alanine aminotransferase (ALT) activity, a biomarker of hepatotoxicity, and hepatic DNA damage. APAP caused a massive elevation in ALT activity, which exceeded control value (45 ± 2 U/L) by approximately 663-fold (29,813 ± 463 U/L). In contrast, ALT activity did not change following oral administration of GSPE alone for seven days (27 ± 2 U/L). However, this mode of GSPE preexposure followed by APAP administration showed excellent hepatoprotection. GSPE+APAP combination also showed a dramatic decrease in ALT activity (2,792 ± 78 U/L). Histopathological evaluation of liver and kidney sections showed a remarkable interference of GSPE against APAP toxicity and substantial inhibition of apoptotic and necrotic liver cell death. APAP was also shown to phosphorylate (deactivate) the bcl-X_L gene, a death inhibitor gene and a positive regulator of the bcl2 family of genes. In contrast, GSPE alone enhanced the expression of the bcl-X_L gene and significantly reduced APAP-induced phosphorylation of bcl-X_L gene. Thus, GSPE significantly attenuates APAP-induced lethality, liver toxicity, hepatic DNA damage, ALT activity, apoptotic cell death and positively influences gene expression.[14–16]

This high dose of APAP (500mg/kg, p.o.) also caused a significant increase in serum BUN level (3.2-fold increase). Vehicle alone (control) or GSPE alone (100 mg/kg/day) did not alter kidney function at all, rather BUN levels remained the same.[16,17] However, GSPE preexposure for seven days significantly reduced APAP-induced BUN elevation. A 3.2-fold BUN increase by APAP alone was reduced to 1.5-fold by GSPE+APAP. GSPE preexposure was very effective in minimizing injury inflicted by APAP in both liver and kidneys. DNA damage was also assessed in the variously treated tissues. GSPE treatment alone for seven days did not alter the integrity of genomic DNA in the kidneys. Approximately 4.8- and 2.6-fold increases in DNA fragmentation in the liver and kidney tissues were observed following APAP treatment (500 mg/kg, p.o.). However, GSPE preexposure exhibited a significant effect on APAP-induced DNA fragmentation in the liver (157% of control) and kidney (106% of control) tissues. Similar to serum chemistry parameters, GSPE preexposure did not fully reverse APAP-induced DNA fragmentation, reflecting the residual damage in the heavily injured cells associated with ongoing recovery, repair, and healing processes.[16,17]

The potential of GSPE to defend the lungs from AMI-induced changes were evaluated biochemically and histopathologically. Four consecutive doses of AMI (a total of 200mg/kg in four days) caused approximately a 9.0-fold increase in serum CK (creatine kinase) activity, and a 5-fold increase in serum ALT activity. Control and GSPE alone sera showed normal ranges of BUN and CK indicating AMI's failure to influence these two organs. However, GSPE preexposure prior to and during AMI treatment considerably reduced pulmonary injury, and brought down the serum chemistry changes close to normal. The integrity of lung DNA was determined quantitatively by sedimentation assay and qualitatively by agarose gel electrophoresis. AMI alone induced moderate DNA fragmentation (152%), and the GSPE+AMI group showed near total countering of AMI-induced pulmonary genomic DNA damage. This effect was due to bioavailability of GSPE in the lung to defend against AMI-related toxic effects. Agarose gel electrophoresis of genomic DNA isolated from lung tissue mirrored the quantitative assay. Histopathological evaluation supported biochemical data.[16,17]

The cytoprotective efficacy of GSPE was tested on DOX-induced cardiotoxicity in mice. DOX alone induced a 6.0-fold increase in serum CK activity. Coupled with CK activity, a 6.0-fold increase in serum ALT activity in the absence of any increase in BUN was observed in DOX-exposed animals. In this context, a 6.0-fold increase in CK activity is considered toxicologically significant and may threaten animal survival, whereas a 6.0-fold increase in ALT activity is totally a reversible change and does not pose any danger to animal health. Despite significant alterations induced by DOX, 7-day GSPE preexposure totally abolished DOX-effects and brought down the CK/ALT activities close to normal. These data also suggests that GSPE was bioavailable in both the heart and the liver in order to counteract DOX-effects.[16,17]

12-O-TETRADECANOYLPHORBOL-13-ACETATE (TPA), DIMETHYLNITROSAMINE (DMN) AND MOCAP-INDUCED MULTIORGAN TOXICITY, AND PROTECTION BY GSPE

The protective abilities of GSPE, vitamin E, vitamin C, β-carotene and a combination of vitamins E plus C against TPA-induced lipid peroxidation and DNA fragmentation in the brain and liver tissues of mice, as well as against production of reactive oxygen species (ROS) in the peritoneal macrophages of mice, were assessed *in vivo*.[8] TPA is a well-known inducer of ROS and tumor promotion *in vivo*. Pretreatment of mice with GSPE (100mg/kg), vitamin E (100mg/kg), vitamin C (100mg/kg), β-carotene (50mg/kg) and a combination of vitamins E plus C (100mg/kg each) decreased TPA-induced production of ROS in peritoneal macrophages by 71, 43, 16, 17, and 51%, respectively, via chemiluminescence response, and 69, 32, 15, 18, and 47%, respectively, via cytochrome *c* reduction as compared to controls. Pretreatment of mice with the same dosages of GSPE, vitamin E, vitamin C, β-carotene and a combination of vitamins E plus C decreased TPA-induced DNA fragmentation by 50, 31, 14, 11, and 40% in brain tissues, and 47, 30, 10, 11 and 38% in liver tissues, respectively, while lipid peroxidation was reduced by 61, 45, 13, 8, and 48% in brain tissues, and 46, 36, 12, 7, and 39% in liver mitochondria and 59, 47, 14, 12, and 53% in liver microsomes, respectively, as compared to controls. Pretreatment of mice with GSPE (25, 50 and 100mg/kg) resulted in a significant dose-dependent inhibition of TPA-induced production of ROS in peritoneal macrophage cells, and lipid peroxidation and DNA fragmentation in brain and liver tissues compared to controls. These results demonstrate that GSPE is bioavailable to the target organs, and provides significantly greater protection against ROS and free radical–induced lipid peroxidation and DNA damage than vitamins E, C and β-carotene, as well as a combination of vitamins E plus C.[8]

DMN is a potent immunotoxin as well as a carcinogen. The spleen is one of the primary targets of DMN besides many other organs in the body. Unfortunately, splenotoxicity lacks a serum chemistry marker. A 10mg/kg i.p. dose of DMN induced a 52.3-fold increase in ALT activity, while GSPE preexposure provided a 94% protection against DMN-induced ALT leakage. Compared with the control and GSPE-exposed animals, DMN alone caused a massive fragmentation of genomic DNA in the spleen and induced massive cell death by apoptosis in addition to necrosis. On a quantitative basis DMN caused a 2.5-fold increase in DNA fragmentation over control tissues. However, the damage induced by DMN was completely abolished by GSPE preexposure. A near total recovery from the damage was evident in GSPE+DMN exposed mice. Qualitatively, an identical pattern of changes in the integrity of DNA, was observed. GSPE provided dramatic protection against DMN-induced splenotoxicity in hispathological examination.[16,18] The dramatic protective effects of GSPE on DMN-induced splenotoxicity, raises several possibilities:

- *(i)* interference with DMN metabolism;
- *(ii)* detoxification of DMN metabolites including biological reactive intermediates;
- *(iii)* interference with endonuclease activity, and
- *(iv)* interference with DNA methylation.

Whether GSPE influenced DMN-induced changes in bcl-X_L expression in the spleen remains open to future investigation.

O-Ethyl-S,S-dipropyl phosphorodithioate (MOCAP) induced a 3.5-fold increase in serum CK levels along with marginally altering serum ALT and BUN levels. GSPE+MOCAP treated animals showed a serum chemistry profile, which closely resembled control animals or those treated with GSPE alone. Following treatment with MOCAP, most animals were lethargic and showed loss of control in their activities. Some animals died because of non-specific neurotoxicity. MOCAP did not alter the integrity of genomic DNA significantly. Similarly, GSPE had no adverse effects on the brain DNA. In histopathological evaluation, MOCAP did not induce any overt histopathological changes except mild periventricular leukomalacia, mineralization and apoptotic cell death. These effects were partially reversed in GSPE+MOCAP exposed brains. Apoptotic cell death was observed microscopically in the absence of massive genomic DNA fragmentation.[16,18]

HUMAN CLINICAL TRIALS

Hypercholesterolemia, a significant cardiovascular risk factor, is prevalent in the American population. Many drugs lower circulating cholesterol levels, but they are not infrequently associated with severe side effects. We examined 40 hypercholesterolemic subjects (total cholesterol 210–300mg/dL) in a randomized, double-blind, placebo-controlled study.[19] The four groups of 10 subjects received either placebo b.i.d., chromium polynicotinate (CM) 200μg b.i.d., GSPE 100mg b.i.d., or a combination of CM and GSPE at the same dosage b.i.d. Over two months, the average percent change ± SEM in total cholesterol from baseline among groups was: placebo −3.5% ± 4, GSPE −2.5% ± 2, CM −10% ± 5, and combination −16.5% ± 3. The decrease in the last group was significantly different from placebo ($p < 0.01$). The major decrease in cholesterol concentration was in the LDL levels: placebo −3.0% ± 4, GSPE −1.0% ± 2.0, CM −14% ± 4.0, and combination −20% ± 6.0. Again, the combination of CM and GSPE significantly decreased LDL when compared to placebo ($p < 0.01$). HDL levels did not change among the groups. Also, there was no significant difference in the triglyceride concentrations among the groups; and no statistically significant differences were seen in the levels of autoantibodies to oxidized LDL. However, the trend was for the two groups receiving GSPE to have greater decreases in the latter parameter, i.e., −30.7% and −44.0% in the GSPE and combined groups in contrast to −17.3% and −10.4% in the placebo and chromium groups. We determined the number of subjects in each group who decreased autoantibodies to oxidized LDL greater than 50% over 8 weeks and found these ratios among groups: placebo, 2/9; CM, 1/10; GSPE, 6/10; and combined, 3/8. Thus, 50% of subjects (9/18) receiving GSPE had a greater than 50% decrease in autoantibodies compared to 16% (3/19) in the two groups not receiving GSPE. No significant changes occurred in the levels of circulating homocysteine and blood pressure among the four groups. These results demonstrate that a combination of CM and GSPE can decrease total cholesterol and LDL levels significantly. Furthermore, there was a trend toward decreased circulating autoantibodies to oxidized LDL in the two groups receiving GSPE.[19]

In another human clinical study, GSPE supplementation ameliorated symptoms of chronic pancreatitis in patients after traditional therapy had failed. GSPE 100 mg t.i.d. provided effective symptom control by reducing both pain index and incidence of vomiting in these patients significantly.[20]

MECHANISMS OF CYTOPROTECTION BY GSPE

GSPE preexposure significantly demonstrated profound antiendonucleolytic, antilipoperoxidative, antiapoptogenic and antinecrogenic potential. A series of *in vivo* studies demonstrated the bioavailability of GSPE to multiple target organs including liver, kidney, heart, spleen and the brain. Orally administered GSPE for 7–10 consecutive days followed by exposure to toxic doses of diverse organotropic chemicals protected select organ toxicities. Furthermore, these studies addressed antinephrotoxic, anti-immunotoxic and antineurotoxic potential of GSPE. Data presented here show that GSPE administration for several days is clearly bioavailable in multiple target organs.[14–18]

Previous studies from our laboratories have linked the protective abilities of GSPE with inactivation of antiapoptotic gene bcl-X_L, and modification of several other critical molecular targets such as DNA-damage/DNA-repair, lipid peroxidation and intracellular Ca^{2+} homeostasis.[14] Especially, GSPE provided dramatic protection against APAP-induced hepato- and nephrotoxicity, significantly increased bcl-X_L expression in the liver tissue and antagonized both necrotic and apoptotic deaths of liver cells *in vivo*. However, it was not clear from this study whether the antiapoptotic and antinecrotic effects of GSPE were:

(*i*) due to its interference with endonuclease activity,
(*ii*) due to its antioxidant effect, or,
(*iii*) due to its ability to inhibit microsomal drug metabolizing enzyme(s), such as cytochrome P450 2E1.

Since cytochrome P450 2E1 primarily metabolizes acetaminophen and other drugs/chemicals in mice and rats, we assessed cytochrome P450 2E1's catalytic activity in both *in vitro* and *in vivo* models. Overall this investigation compared the *in vitro* aniline hydroxylation patterns of:

(*i*) *in vivo* GSPE-exposed and unexposed (control) mouse liver microsomes,
(*ii*) induced (1% acetone in drinking water for 3 days) and uninduced rat liver microsomes in the presence and absence of GSPE *in vitro*, and
(*iii*) control rat liver microsomes in the presence of an anti-APAP agent 4-AB *in vitro*.

For the *in vivo* assessment, male B6C3F1 mice were fed GSPE diet (ADI 100 mg/kg) for four weeks, and liver microsomes were isolated from both control and GSPE-fed mice for aniline hydroxylation, a specific marker of cytochrome P450 2E1 activity. Data showed that hydroxylation was 40% less in microsomes from GSPE-exposed livers compared to control microsomes. Similarly, when rat liver microsomes were incubated with various concentrations of GSPE *in vitro* (100 µg/mL), aniline hydroxylation was inhibited to various degrees (uninduced: 40 and 60% and induced: 25 and 50%, respectively, with 100 and 250μg/mL). Influence of GSPE on hydroxylation patterns were compared with another hepatoprotective agent 4-AB, a well-known modulator of nuclear enzyme poly(ADP-ribose)polymerase, and the data shows that 4-AB did not alter aniline hydroxylation at all. Collectively, these results may suggest that GSPE has the ability to inhibit cytochrome P450 2E1, and this additional cytoprotective attribute is in conjunction with its novel antioxidant and/or antiendonucleolytic potential.[21]

Altered expression of cell adhesion molecule expression has been implicated in a variety of chronic inflammatory conditions. Regulation of adhesion molecule expression by specific redox sensitive mechanisms has been reported. The effects of GSPE was evaluated on the expression of TNFα-induced ICAM-1 and VCAM-1 expression in human umbilical vein endothelial cells (HUVEC). GSPE at low concentrations (1–5μg/ml), down-regulated TNFα-induced VCAM-1 expression but not ICAM-1 expression in HUVEC. Such regulation of inducible VCAM-1 by GSPE was also observed at the mRNA expression level. A cell-cell co-culture assay was performed to verify whether the inhibitory effect of GSPE on the expression of VCAM-1 was also effective in down-regulating actual endothelial cell/leukocyte interaction. GSPE treatment significantly decreased TNFα-induced adherence of T-cells to HUVEC.[22] The potent inhibitory effect of low concentrations of GSPE on agonist-induced VCAM-1 expression suggests therapeutic potential of this extract in inflammatory conditions and other pathologies involving altered expression of VCAM-1.[22]

CONCLUSION

A number of epidemiological studies have exhibited that high antioxidant status is directly linked to low risk of degenerative disease. Increased consumption of fresh fruits and green vegetables can promote antioxidant status in the human body. On the other hand, cigarette smoking, physical stress, and over-processed foods, laden with hydrogenated fat, sugar, refined flours and chemicals, can potentiate enhanced free radical production and oxidative stress in the human body leading to a broad spectrum of degenerative diseases. Unfortunately, while the physical, emotional, environmental and chemical stressors are growing, the essential nutrition in our increasingly processed and pre-packaged food supply is decreasing. Furthermore, while coping with our hectic lifestyle we have great difficulty finding time to consume a good amount of pesticide-free fresh fruits and green vegetables. Thus, a novel, natural and bioavailable antioxidant may improve our lifestyle by enhancing our antioxidant status. Taken together, our research studies demonstrate that GSPE is highly bioavailable and may serve as a potential therapeutic tool in protecting multiple target organs from structurally diverse drug-, environmental-, and chemical-induced toxic assaults. Further, mechanistic and clinical studies are in progress to unveil the benefits of this novel, natural antioxidant.

REFERENCES

1. HERMAN, D. 1982. The free radical theory of aging. *In* Free Radicals in Biology. W.A. Pryor, Ed.: 255–275. Academic Press. New York.
2. HALLIWELL, B., J.M.C. GUTTERIDGE & C.E. CROSS. 1992. Free radicals, antioxidants and human disease: where are we now? J. Lab. Clin. Med. **199:** 598–620.
3. AMES, B.N. 1992. Pollution, pesticides and cancer. J. AOAC Int. **75:** 1–5.
4. RICE-EVANS, C.A. & L. PACKER, Eds. 1997. Flavonoids in Health and Disease. Marcel Dekker, Inc. New York.
5. HANEFELD, M. & K. HERRMANN. 1976. On the occurrence of proanthocyanidins, leucoanthocyanidins and catechins in vegetables. Z. Lebensm. Unters. Forsch. **161:** 243–248.
6. CHEN, Z.Y., P.T. CHAN, K.Y. HO, *et al.* 1996. Antioxidative activity of natural flavonoids is governed by number and location of their aromatic hydroxyl groups. Chem. Phys. Lipids **79:** 157–163.
7. BAGCHI, D., A. GARG, R.L. KROHN, *et al.* 1997. Oxygen free radical scavenging abilities of vitamins C and E, and a grape seed proanthocyanidin extract in vitro. Res. Commun. Mol. Pathol. Pharmacol. **93:** 179–189.
8. BAGCHI, D., A. GARG, R.L. KROHN, *et al.* 1998. Protective effects of grape seed proanthocyanidins and selected antioxidants against TPA-induced hepatic and brain lipid peroxidation and DNA fragmentation, and peritoneal macrophage activation in mice. Gen. Pharmacol. **30:** 771–776.
9. SATO, M., G. MAULIK, P.S. RAY, *et al.* 1999. Cardioprotective effects of grape seed proanthocyanidin against ischemic reperfusion injury. J. Mol. Cell. Cardiol. **31:** 1289–1297.
10. YE, X., R.L. KROHN, W. LIU, *et al.* 1999. The cytotoxic effects of a novel IH636 grape seed proanthocyanidin extract on cultured human cancer cells. Mol. Cell. Biochem. **196:** 99–108.
11. BAGCHI, D., C.A. KUSZYNSKI, J. BALMOORI, *et al.* 1998. Hydrogen peroxide-induced modulation of intracellular oxidized states in cultured macrophage J774A.1 and neuroactive PC-12 cells, and protection by a novel IH636 grape seed proanthocyanidin extract. Phytotherapy Res. **12:** 568–571.
12. BAGCHI, M., J. BALMOORI, D. BAGCHI, *et al.* 1999. Smokeless tobacco, oxidative stress, apoptosis, and antioxidants in human oral keratinocytes. Free Radical Biol. Med. **26:** 992–1000.
13. JOSHI, S.S., C.A. KUSZYNSKI, E.J. BENNER, *et al.* 1999. Amelioration of cytotoxic effects of idarubicin and 4-hydroxyperoxycyclophosphamide on Chang liver cells by a novel grape seed proanthocyanidin extract. Antiox. Redox Signal. **1:** 563–570.
14. RAY, S.D., M.A. KUMAR & D. BAGCHI. 1999. A novel proanthocyanidin IH636 grape seed extract increases in vivo bcl-XL expression and prevents acetaminophen-induced programmed and unprogrammed cell death in mouse liver. Arch. Biochem. Biophys. **369:** 42–58.
15. RAY, S.D., H. PARIKH, E. HICKEY, *et al.* 2001. Differential effects of IH636 grape seed proanthocyanidin extract and a DNA repair modulator 4-aminobenzamide on liver microsomal cytochrome P4502E1-dependent aniline hydroxylation. Mol. Cell. Biochem. **218:** 27–33.
16. BAGCHI, D., S.D. RAY, D. PATEL & M. BAGCHI. 2001. Protection against drug- and chemical-induced multiorgan toxicity by a novel grape seed proanthocyanidin extract. Drugs Exp. Clin. Res. **XXVII:** 3–15.
17. RAY, S.D., D. PATEL, V. WONG & D. BAGCHI. 2000. In vivo protection of DNA damage associated apoptotic andnecrotic cell deaths during acetaminophen-induced nephrotoxicity, amiodarone-induced lung toxicity and doxorubicin-induced cardiotoxicity by a novel IH636 grape seed proanthocyanidin extract. Res. Commun. Mol. Pathol. Pharmacol. **107:** 137–166.
18. RAY, S.D., V. WONG, A. RINKOVSKY, *et al.* 2000. Unique organoprotective properties of a novel IH636 grape seed proanthocyanidin extract on cadmium chloride-induced nephrotoxicity, dimethylnitrosamine (DMN)-induced splenotoxicity and MOCAP-induced neurotoxicity in mice. Res. Commun. Mol. Pathol. Pharmacol. **107:** 105–128.

19. PREUSS, H., D. WALLERSTEDT, N. TALPUR, *et al.* 2000. Effects of niacin-bound chromium and grape seed proanthocyanidin extract on the lipid profile of hypercholesterolemic subjects: a pilot study. J. Med. **31:** 227–246.
20. BANERJEE, B. & D. BAGCHI. 2001. Beneficial effects of a novel IH636 grape seed proanthocyanidin extract in the treatment of chronic pancreatitis. Digestion **63:** 203–206.
21. RAY, S.D., H. PARIKH, E. HICKEY, *et al.* 2001. Differential effects of IH636 grape seed proanthocyanidin extract and a DNA repair modulator 4-aminobenzamide on liver microsomal cytochrome P450 2E1-dependent aniline hydroxylation. Mol. Cell. Biochem. **218:** 27–33.
22. SEN, C.K. & D. BAGCHI. 2001. Regulation of inducible adhesion molecule expression in human endothelial cells by grape seed proanthocyanidin extract. Mol. Cell. Biochem. **216:** 1–7.

Polyphenols and Red Wine as Peroxynitrite Scavengers

A Chemiluminescent Assay

SILVIA ALVAREZ, TAMARA ZAOBORNYJ, LUCAS ACTIS-GORETTA, CÉSAR G. FRAGA, AND ALBERTO BOVERIS

Physical Chemistry - PRALIB, School of Pharmacy and Biochemistry, University of Buenos Aires, Buenos Aires, Argentina

ABSTRACT: A novel chemiluminescent assay for evaluating peroxynitrite ($ONOO^-$)-scavenging capacity was developed. The experimental protocol ensures sensitivity and reproducibility of measurements. The addition of 0−500 μM $ONOO^-$ to rat liver homogenate generated a luminous signal that was analyzed by chemiluminescence in a LKB Wallac liquid scintillation counter. The obtained optimal conditions were: 1–2 mg/mL of homogenate protein in 120 mM KCl, 30 mM phosphate buffer (pH 7.4), and 220 μM $ONOO^-$ at 30°C. As polyphenols we used (+)-catechin, (−)-epicatechin, and myricetin. The most efficient of the compounds tested was myricetin with an IC_{50} of 20 μM. The effectiveness of this method was verified by evaluating the antioxidant ability of three red wine samples to decrease peroxynitrite-initiated chemiluminescence. The $ONOO^-$-scavenging activity of wines measured by this assay was related to the phenolic level of the samples. The quickness and reliability of this assay makes it particularly suitable for a large-scale screening of watery food extracts.

KEYWORDS: peroxynitrites; polyphenols; red wine; chemiluminescence

INTRODUCTION

The antioxidant activity of red wine and of its major polyphenols have been demomstrated in many experimental systems spanning the range from *in vitro* studies (isolated human LDL, liposomes, macrophages, cultured cells) to investigations in human subjects.[1–3] Flavonoids are a large class of polyphenolic compounds, ubiquitous in plants, containing some hydroxyl groups in positions that constitute the chemical basis of their antioxidant activity. The relatively high reactivity of flavonoids with oxygen radicals such as superoxide anion and hydroxyl radical is well known. It has recently been reported that wine polyphenols effectively scavenge nitric oxide and peroxynitrite.[4,5] Evaluating the capacity of antioxidants to prevent peroxynitrite ($ONOO^-$) damage in biological systems is of great interest. $ONOO^-$ is an oxidizing and nitrating molecule produced *in vivo* in mammals as both intracellular and extracellular metabolite through the diffusion-controlled reaction of NO

Address for correspondence: Silvia Alvarez, Fisicoquímica, Facultad de Farmacia y Bioquímica, Universidad de Buenos Aires, Junín 956, 1113, Buenos Aires, Argentina.Voice/fax: 5411-4508-3646.
salvarez@ffyb.uba.ar

and O_2^- ($k = 1.9 \times 10^{10}\ M^{-1}s^{-1}$).[6] Peroxynitrite is able to initiate lipid peroxidation, which may involve the participation of low concentrations of $\cdot NO_2$ or NO produced in this process. Taking this into account, the $ONOO^-$-initiated chemiluminescence seems to afford a good basis to develop an assay to evaluate the capacity of compounds to prevent ONOO-damage in biological systems.

The aim of this work is to standardize the $ONOO^-$-initiated chemiluminescence as a microassay to evaluate the ability of antioxidants to prevent $ONOO^-$ damage in biological systems. Three Argentinian red wines and the polyphenols (+)-catechin, (–)-epicatechin, and myricetin were tested as antioxidant compounds using the present methodology.

MATERIALS AND METHODS

Peroxynitrite was prepared by reacting 2 M $NaNO_2$ and 2.11 M H_2O_2 in HNO_3, and stabilized with 4 M NaOH. Peroxynitrite-initiated chemiluminescence was measured in a LKB Wallac, 1209 Rackbeta liquid scintillation counter (Turku, Finland) in the out-of-coincidence mode. The results are expressed as total counts registered during 30 s (counts) or counts per mg of homogenate protein present in the assay (counts/mg prot).

For chemiluminescence measurements, polyphenols were used in a range concentration of 0–50 mM, and for wine samples a volume of 10 mL of (dilution 1:10, v:v) was used.

The optimal experimental conditions obtained were: 0.5 to 1.2 mg/mL of homogenate protein in a reaction medium containing 120 mM KCl, 30 mM phosphate buffer (pH 7.4) and 200 mM $ONOO^-$, in a final volume of 2 mL.

RESULTS AND DISCUSSION

Maximal emission was linearly related to the protein concentration in the range 0.5 to 1.2 mg/mL. Progressive decrease of photoemission was observed when protein concentration was higher than 1.2 mg/mL, probably due to turbidity quenching. Maximal emission values also depended on temperature. The optimal temperature seemed to be 30°C.

The polyphenols (–)-epicatechin ($IC_{50} = 6.5\,\mu M$) and (+)-catechin ($IC_{50} = 13\,\mu M$), were more potent than myricetin ($IC_{50} = 20\,\mu M$). The difference can be due to different chemical structure. It is worth noting that myriricetin shows higher antioxidant potency when assayed in chemical systems (ability to inhibit NADH oxidation originated by $ONOO^-$).[7]

The effectiveness of the present method was verified by evaluating the antioxidant ability of three red wine samples to decrease peroxynitrite-initiated chemiluminescence. The $ONOO^-$-scavenging capacity of wines measured by this assay was related to the phenolic content of the samples (TABLE 1). Cabernet sauvignon was the red wine with the highest antioxidant capacity (measured as percentage of chemiluminescence inhibition), and it also presented the highest phenolic content.

TABLE 1. Phenolic content and inhibition of $ONOO^-$-initiated chemiluminescence inhibition for wine samples

Wine sample	Phenolic content (GAE, mg/mL)	CL inhibition (%)
Cabernet sauvignon	2011 ± 41	13.9 ± 1.0
Malbec	1441 ± 49	4.81 ± 0.5
Blended[a]	814 ± 50	1.18 ± 0.3
White	250 ± 37	1.00 ± 0.3

NOTES: For CL measurements, a volume of 10 μL of the wine sample (dilution 1:10) was used. GAE, gallic acid equivalents; CL, chemiluminescence.

[a]Blended: Cabernet sauvignon, Malbec, and Merlot.

Accordingly, the white wine sample had both the lowest antioxidant capacity and phenolic content.

The experimental protocol of this novel assay for evaluating $ONOO^-$-scavenging capacity using a chemiluminescent method ensures sensitivity and reproducibility of measurements. The quickness and reliability of this assay makes it particularly suitable for a large-scale screening of watery food extracts.

REFERENCES

1. SERAFINI, M., G. MAIANI & A. FERRO-LUZZI. 1998. Alcohol-free red wine enhances plasma antioxidant capacity in humans. J. Nutr. **128:** 1003–1007.
2. LOTITO, S.B. & C.G. FRAGA. 1999. (+)-Catechin as antioxidant: mechanisms preventing human plasma oxidation and activity in red wines. BioFactors. **10:** 125–130.
3. BURNS, J., P.T. GARDNER, J. O'NEIL, *et al.* 2000. Relationship among antioxidant activity, vasodilation capacity, and phenolic content of red wines. J. Agric. Food Chem. **48:** 220–230.
4. HAENEN, G.R., J.B. PAQUAY, R.E. KORTHOUWER & A. BAST. 1997. Peroxynitrite scavenging by flavonoids. Biochem. Biophys. Res. Commun. **236:** 591–593.
5. PAQUAY, J.B., G.R. HAENEN, R.E. KORTHOUWER & A. BAST. 1997. Peroxynitrite scavenging by wines. J. Agric. Food Chem. **45:** 3357–3358.
6. RADI, R., A. DENICOLA, B. ALVAREZ, *et al.* 2000. The biological chemistry of peroxynitrite. *In* Nitric Oxide. L.J. Ignarro, Ed.: 57–82. Academic Press. San Diego, CA.
7. BOVERIS, A., S. ALVAREZ, S. LORES ARNAIZ & L. VALDEZ. 2001 Peroxynitrite scavenging by mitochondrial reductants and plant polyphenols. *In* Handbook of Antioxidants. E. Cadenas & L. Packer, Eds.: 351–370. Marcel Dekker, Inc. New York.

Polyphenols in Red Wines Prevent NADH Oxidation Induced by Peroxynitrite

LAURA B. VALDEZ, LUCAS ACTIS-GORETTA, AND ALBERTO BOVERIS

Laboratory of Free Radical Biology, School of Pharmacy and Biochemistry, University of Buenos Aires, Buenos Aires, Argentina

ABSTRACT: It is known that red wines have antioxidant properties owing to the presence of polyphenols and hydroxycinnamates. It has been reported that these compounds are efficient scavengers of peroxynitrite ($ONOO^-$). We assayed the $ONOO^-$-scavenging activity of plant polyphenols, hydroxycinnamates, and red wines using a fluormetric assay that involves the participation of $ONOO^-$ (200 μM) as oxidant and NADH (50–100 μM) as a target molecule. NADH oxidation was prevented by pure polyphenols and hydroxycinnamates. With the wines, the Cabernet Sauvignon showed more potent $ONOO^-$-scavenging ability than the Malbecs and the blended wines. These results correlate well with the total phenol content determined by Folin-Ciocalteau assay.

KEYWORDS: **polyphenols; red wines; peroxynitrite; NADH oxidation**

INTRODUCTION

Red wine is a rich source of phenolic compounds (flavonoids and non-flavonoids) and its antioxidant effect has been shown in different *in vitro* and *in vivo* systems. The most common flavonoids in red wine are the flavonols (myricetin, quercetin and kaempferol), flavan-3-ols ((+)-catechin and (−)-epicatechin) and the anthocyanins.[1] The major non-flavonoids found in wines are the hydroxycinnamates caffeic and *p*-coumaric acids.[1] Ferulic acid has also been reported in wines.[1] Polyphenols can act as reducing agents and antioxidants by the hydrogen-donating property of their hydroxyl groups[2] as well as by their metal-chelating abilities.[3] A relatively high reactivity of flavonoids with radicals such as superoxide anion (O_2^-),[4] hydroxyl radical (·OH),[2] and nitric oxide (NO)[5] has been reported. In addition, flavonoids and simple phenolics are efficient scavengers of peroxynitrite ($ONOO^-$).[6] Peroxynitrite is an oxidizing and nitrating molecule produced *in vivo* through the diffusion-controlled reaction between NO and O_2^-, both as an intracellular and as an extracellular metabolite. Peroxynitrite is a powerful one- and two-electron oxidant ($E^{o\prime}$ (ONOOH, $H^+/\cdot NO_2$, H_2O) = 1.6–1.7 V and $E^{o\prime}$ (ONOOH, H^+/NO_2^-, H_2O) = 1.3–1.37 V), that efficiently oxidizes the sulfhydryl group of cysteine and glutathione (GSH),[7] the sulfur atom of methionine[8] and ascorbate.[9]

Address for correspondence: Laura B. Valdez, Cátedra de Fisicoquímica, Facultad de Farmacia y Bioquímica, Universidad de Buenos Aires, Junín 956, 1113 Buenos Aires, Argentina. Voice\fax: (54-11)-4508-3646.
lbvaldez@ffyb.uba.ar

Ann. N.Y. Acad. Sci. **957: 274–278 (2002).**

In this work, the $ONOO^-$ scavenging activity of plant flavonoids (myricetin, (+)-catechin and (–)-epicatechin), hydroxycinnamates (caffeic, ferulic and chlorogenic acid), plant extracts (grape seed and *Ginkgo biloba*), and red wines was assayed.

METHODS

The $ONOO^-$ scavenging activity of flavonoids, hydroxycinnamates, plant extracts and red wines, was assayed at 37°C, using a fluorometric assay ($\lambda_{exc} - \lambda_{em} = 340–463$ nm) that involves the participation of $ONOO^-$ as oxidant and NADH as a target molecule. The reaction medium consisted of 100mM phosphate buffer (pH 7.0), 0.1mM DTPA, 100μM NADH and 200μM $ONOO^-$. Reduced glutathione (0–150μM), flavonoids (0–200μM), hydroxycinnamates (0–100 μM), plant extracts (0–100μg/mL) and red wines (0–50 μL of a 1/5 dilution) were used in competition with NADH. This indirect fluorometric technique was used to estimate the rate constants of the reactions of $ONOO^-$ with phenolic compounds, and the concentrations inhibiting 50% NAD formation (IC_{50}) of the different extracts and red wines, according to Valdez *et al.*[10]

Total phenols were analyzed according to the Folin-Ciocalteau method, as described by Fogliano *et al.*,[11] and the results were expressed as gallic acid equivalents (GAE).

RESULTS AND CONCLUSIONS

Exposure of NADH to $ONOO^-$ causes the oxidation of the reduced pyridine nucleotide. The addition of flavonoids and hydroxycinnamates to the reaction medium decreased the extent of NADH oxidation by $ONOO^-$.

The rate constants of the reactions between $ONOO^-$ and flavonoids or hydroxycinnamates were estimated from the rate constant for the reaction of NADH and $ONOO^-$ (see TABLE 1). The highest value of the rate constants corresponded to the reaction of $ONOO^-$ with myricetin; the catechins and ferulic acid had the lowest values, and the other two hydroxycinnamates had intermediate values. The IC_{50} were in agreement with the given second order reaction constants.

According to Bors *et al.*[2] the following structural group are important determinants for the antioxidant potential of flavonoids:

(*a*) the O-dihydroxy (catechol) structure in the B ring;
(*b*) the 2,3-double bond in conjugation with a 4-oxo function; and
(*c*) the additional presence of 3- and 5- hydroxyl groups.

Our results on NADH oxidation by $ONOO^-$ showed that the $ONOO^-$ scavenging activity was markedly higher for myricetin than for the others two catechins and the hydroxycinnamates. This finding is consistent with the structural requirements for the antioxidant capacity described above, since myricetin meets all three criteria, whereas the catechins meet only two. On the other hand, the number of phenolic

TABLE 1. Apparent second-order rate constants and IC_{50} of the reactions of $ONOO^-$ with polyphenols and hydroxycinnamates

	Reductants	$k \times 10^{-3}$ (M^{-1} s^{-1})	IC_{50} (units)
Flavonoids[a]	Myricetin	4.9 ± 0.3	35 ± 1 μM
	(+)-Catechin	0.65 ± 0.05	275 ± 23 μM
	(−)-Epicatechin	0.59 ± 0.05	313 ± 23 μM
Hydroxycinnamates[a]	Chlorogenic acid	1.1 ± 0.2	173 ± 36 μM
	Caffeic acid	1.3 ± 0.3	144 ± 29 μM
	Ferulic acid	0.63 ± 0.05	507 ± 15 μM
Plant extracts[b]	Grape seed	—	118 ± 16 μg/mL
	Ginkgo biloba	—	85 ± 11 μg/mL
Red wines[c]	Cabernet Sauvignon	—	22 μl/mL
	Malbec	—	28 μl/mL
	Blended wine	—	44 μl/mL

[a]The corresponding rate constants for the reactions between $ONOO^-$ and the phenolic compounds were estimated taking into account the rate constant for the reaction of GSH and $ONOO^-$ at 37°C ($k_c = 1.4 \times 10^3$ M^{-1} s^{-1})[12] and the calculated rate constant for the reaction of NADH with $ONOO^-$ ($k_b = (1.8 \pm 0.1) \times 10^3$ M^{-1} s^{-1}).

[b]The units are roughly comparable to μM, after considering a content of 31% of flavonoids in the extracts and a mean relative mass of 320 kDa.

[c]The IC_{50} were expressed as the volume of red wine (μL) per mL of medium reaction in the curvette that inhibits NADH oxidation at 50%.

hydroxyl groups seems to be important for the prevention of $ONOO^-$-mediated NADH oxidation by hydroxycinnamates: chlorogenic and caffeic acid were more effective than ferulic acid. This observation agrees with the fact that ferulic acid has only one available phenolic hydroxyl group per molecule, compared with the two potential active groups that chlorogenic and caffeic acid have. Besides, the similarity of IC_{50} for these two hydroxycinnamates supports that the phenolic hydroxyl group is the only one active.

The extracts from *Ginkgo biloba* and grape seed also inhibited NADH oxidation induced by $ONOO^-$. The estimated IC_{50} was lower for *Ginkgo biloba* than for grape seed extract. The amount of whole extract needed for half inhibition of NADH oxidation, indicates that these two extracts are rich in polyphenols able to scavenge $ONOO^-$ (TABLE 1).

All the red wines assayed were able to protect NADH oxidation induced by $ONOO^-$. Cabernet Sauvignon showed a more potent $ONOO^-$ scavenging effect (22 μL/mL) than Malbec (28 μL/mL) and a blended wine (44 μL/mL) to prevent 50% of the NADH oxidation induced by $ONOO^-$ (see FIGURE 1 A). The peroxynitrite antioxidant capacity of wines was correlated to the amount of phenolic compounds: 2011 ± 41 mg GAE/L for Cabernet Sauvignon, 1441 ± 49 mg GAE/L for Malbec, and 814 ± 50 mg GAE/L for the blended wine (FIG. 1 B).

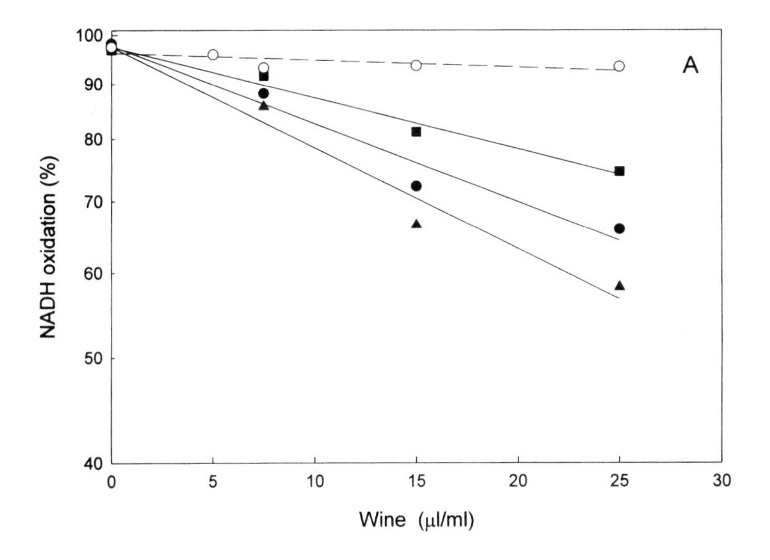

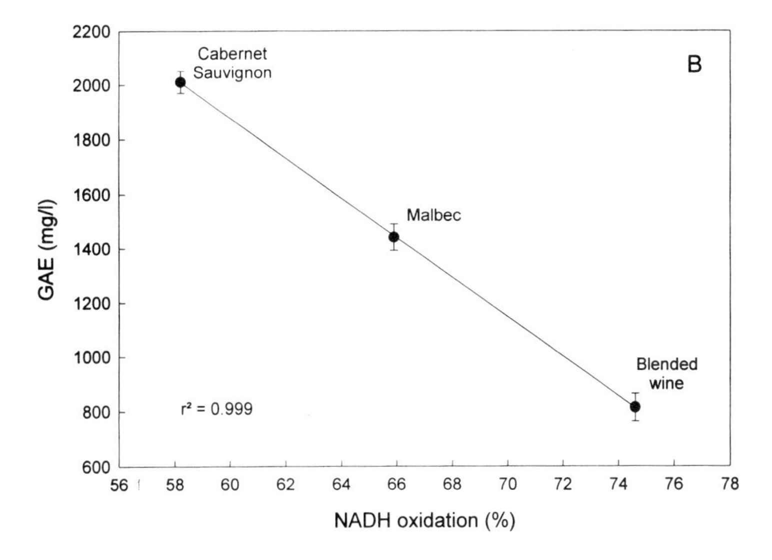

FIGURE 1. (**A**) Effect of a Cabernet Sauvignon wine (▲), a Malbec wine (●), a blended wine (■) and a white wine (○) on NADH oxidation induced by peroxynitrite. Plot of Log of NADH oxidation as a function of wine volume (μL) of a 1/5 dilution per ml of medium reaction in the curvette. (**B**) Relationships between the total phenols content determined by Folin-Ciocalteau assay, expressed as mg GAE/L wine, and the percentage of NADH oxidation induced by peroxynitrite in the presence of the red wines (25 μL of a 1/5 dilution per mL of the medium reaction).

Our results are in agreement with the view that flavonoids are potent scavengers of $ONOO^-$.[6] Moreover, red wines also exerted $ONOO^-$-scavenging capacity due to the presence of flavonoids and hydrocycinnamates. In addition, this technique is useful for evaluating the $ONOO^-$ antioxidant capacity of different types of wines and others watery extracts containing phenolic compounds.

REFERENCES

1. Crozier, A., J. Burns, A.A. Aziz, *et al.* 2000. Antioxidant flavonols from fruit, vegetables and beverages: measurements and bioavailability. Biol. Res. **33:** 79–88.
2. Bors, W., W. Heller, C. Michel & M. Saran. 1990. Flavonoids as antioxidants: determination of radical-scavenging efficiencies. Meth. Enzymol. **186:** 343–355.
3. Brown, J.E., H. Khodr, R.C. Hider & C.A. Rice-Evans. 1998. Structural dependence of flavonoid interactions with Cu^{2+} ions: implications for their antioxidant properties. Biochem. J. **330:** 1173–1178.
4. Sichel, G., C. Corsaro, M. Scallia, *et al.* 1991. In vitro scavenger activity of some flavonoids and melanins against O_2^-. Free Radical Biol. Med. **11:** 1–7.
5. Haenen, G.R.M.M. & A. Bast. 1999. Nitric oxide radical scavenging of flavonoids. Meth. Enzymol. **301:** 490–503.
6. Haenen, G.R.M.M., J.B.G. Paquay, R.E.M. Korthouwer & A. Bast. 1997. Peroxynitrite scavenging by flavonoids. Biochem. Biophys. Res. Commun. **236:** 591–593.
7. Radi, R., J.S. Beckman, K.M. Bush & B.A. Freeman. 1991. Peroxynitrite oxidation of sulfhydryls: the cytotoxic potential of superoxide and nitric oxide. J. Biol. Chem. **266:** 4244–4250.
8. Pryor, W.A., X. Jin & G.L. Squadrito. 1994. One- and two-electron oxidations of methionine by peroxynitrite. Proc. Natl. Acad. Sci. USA **91:** 11173–11177.
9. Bartlett, K., D.F. Church, P.L. Bounds & W.H. Koppenol. 1995. The kinetics of the oxidation of L-ascorbic acid by peroxynitrite. Free Radical Biol. Med. **18:** 85–92.
10. Valdez, L.B., S. Alvarez, S. Lores Arnaiz, *et al.* 2000. Reactions of peroxynitrite in the mitochondrial matrix. Free Radical Biol. Med. **29:** 349–356.
11. Fogliano, V., V. Verde, C. Randazzo & A. Ritieni. 1999. Method for measuring antioxidant activity and its application to monitoring the antioxidant capacity of wines. J. Agric. Food Chem. **47:** 1035–1040.
12. Quijano, C., B. Alvarez, R.M. Gatti, *et al.* 1997. Pathways of peroxynitrite oxidation of thiol groups. Biochem. J. **322:** 167–173.

Comparative Study on the Antioxidant Capacity of Wines and Other Plant-Derived Beverages

LUCAS ACTIS-GORETTA,[a] GERARDO G. MACKENZIE,[b]
PATRICIA I. OTEIZA,[b] AND CÉSAR G. FRAGA[a]

[a]Physical Chemistry–PRALIB, [b]Biological Chemistry–IQUIFIB (UBA-CONICET), School of Pharmacy and Biochemistry, University of Buenos Aires, Buenos Aires, Argentina

ABSTRACT: Consistent epidemiological data point to a reduced morbidity and mortality from coronary heart disease and atherosclerosis in people consuming plant-derived beverages such as tea or wine. We studied the antioxidant capacity of three red wines (W) and compared it those of tea and herbal "mate" tea infusions. The antioxidant capacity was evaluated measuring: (1) the inhibition of the luminol-induced chemiluminescence assay (TRAP); (2) the inhibition of 2.2´-thiobarbituric–reactive substances (TBARS) formation in liposomes by fluorescence; (3) the protection of Jurkat cells from AMVN-induced oxidation, measuring the oxidation of 5-(and-6)-carboxy-2´7´-dichlorodihydrofluorescein diacetate to a fluorescent derivative. The polyphenolic content was estimated spectrophotometrically and by HPLC with electrochemical detection. All three beverages provided antioxidant protection in the three assays in a dose-dependent manner. Significant and positive correlations were found between antioxidant capacity and total polyphenol content, especially in the Jurkat cell oxidation assay (r: 0.96, p < 0.01). Results suggest that these dietary components could be a source of antioxidants that protect from oxidative stress. Further studies of absorption and metabolism of the active compounds will judge the physiological relevance of these results for human health.

KEYWORDS: antioxidants; red wine; tea; "mate" tea

INTRODUCTION

There are consistent epidemiologic data pointing to reduced morbidity and mortality from coronary diseases in people consuming fruits, vegetables, and plant-derived foods. The health benefits ascribed to the consumption of red wines may be related to the high content of polyphenols in this beverage. Since coronary heart disease and cancer have been long associated with conditions of oxidative stress, the protective actions of red wine have been attributed to the presence of polyphenolic compounds, especially flavonoids.

Address for correspondence: Dr. César G. Fraga, Fisicoquímica, Facultad de Farmacia y Bioquímica, Junín 956, 1113-Buenos Aires, Argentina. Voice/fax: 54-11-4508-3646.
cfraga@ffyb.uba.ar

In this study, we characterized the antioxidant capacity exerted by three polyphenol-containing beverages: wine, black tea (tea), and herbal "mate" tea (mate). The antioxidant capacity was evaluated using three different methodologies. The results associate the concentration of total phenolics and catechins in the beverages with their antioxidant capacity.

MATERIALS AND METHODS

Beverages

Three Argentinean wines [a Cabernet Sauvignon (CS), a Malbec, and one generic], tea and mate from commercial sources were evaluated. Tea and mate were prepared using 2 g of herb in 250 mL of boiling water as infusion (tea) or decoction (mate). Tea and mate beverages were prepared in the way normally employed in households. Only one brand of tea and mate were used, since preliminary experiments showed no significant differences in the antioxidant capacity exerted by various commercial brands of these herbs.

Phenolic Content

Total phenolic content was evaluated spectrophotometrically using the Folin-Ciocalteau assay.[1] Results were expressed as gallic acid equivalents (GAE). (+)-Catechin, (−)-epicatechin, chlorogenic acid, and galleate were evaluated by HPLC with electrochemical detection.[2]

Luminol-Chemiluminescence Assay (TRAP)

The antioxidant capacity of the beverages was evaluated by the inhibition of the chemiluminescence generated by the reaction of luminol with 2,2′-azobis(amidinopropane) chlorhydrate (AAPH).[3] The antioxidant capacity values were calculated as Trolox (a water-soluble analogue of vitamin E) equivalents.

2-Thiobarbituric Acid-Reactive Substances (TBARS)

Phosphatidylcholine liposomes prepared as described,[4] were incubated with 10 mM AAPH (60 min, 37°C) in the presence or absence of the beverages. TBARS were evaluated fluorometrically and expressed as malondialdehyde equivalents.

Cell Oxidation

Jurkat cells (human leukemia T cells) were obtained from the American Cell Type Culture Collection (Rockville, MD, USA). Cell viability was assessed by exclusion of the dye Trypan Blue. For the experiments, 6×10^4 cells in 0.2 mL PBS were incubated for 5 min at 4°C in the presence or absence of the different beverages. After the addition of 1 mM 2,2′-azobis (2,4-valeronitrile) (AMVN), cells were incubated for 60 min at 37°C with continuous shaking. Following a brief centrifugation, cells were incubated in the presence of 10 μM of 5(or 6)-carboxy-2′7′-dichlorodihydrofluorescein diacetate for 30 min at 37°C.[5] Cells were centrifuged, and the pellet was resuspended in PBS containing 0.1% (v/v) Igepal and incubated for 30 min at room temperature. The fluorescence (RF) in the samples was measured at 525 nm (λ_{exc} = 475 nm).

RESULTS AND DISCUSSION

Wines, tea, and mate were studied for their antioxidant capacity using three different assays (see TABLE 1). We consider that evaluating the antioxidant capacity of a substance using several methodological approaches will permit a more complete understanding of the mechanisms by which the substance exerts its antioxidant action. In the present study we evaluated the inhibition of the oxidation using: (a) hydrophilic substance in a pure chemical system (TRAP); (b) lipids forming synthetic membranes (liposomes); and (c) living cells.

TABLE 1 shows the total phenolic content in the samples, and their antioxidant capacity measured by the three assays. The spectrophotometric evaluation of the total phenolic content showed that wines presented the highest values (2 to 5 times higher than tea and mate, $p < 0.01$). Among the three wines, differences in total phenolic content were also significant ($p < 0.01$). When the antioxidant capacity was evaluated by the TRAP assay, CS wine showed the highest antioxidant capacity followed by mate, malbec, generic wine, and tea. A significant and positive correlation ($p < 0.001$) was found between TRAP values and the total phenolic content, except for mate. An interpretation of these results is that part of the total antioxidant capacity of mate is related to non-phenolic compounds. The second methodology used for estimating the antioxidant capacity of the beverages, was the inhibition of AAPH-induced TBARS formation in liposomes. The highest inhibition was exerted by CS and Malbec wines, followed by mate and tea. This assay provides information about the ability of the beverages to inhibit the oxidation of lipids by radicals generated in the aqueous phase and supports a surface-related action of polyphenols. In the third methodology used, we determined the ability of the beverages to protect cells from

TABLE 1. Total phenolics content and antioxidant capacity in red wines, tea and mate

Samples	Phenolics (mg/L)	TRAP[a] (nL)	Liposome oxidation[b] (nL)	Cell oxidation[c] (nL)
CS wine	2,011 ± 41	130 ± 35	150 ± 65	26 ± 5
Malbec wine	1,441 ± 49	310 ± 82	170 ± 17	53 ± 7
Generic wine	814 ± 50	500 ± 31	380 ± 69	140 ± 9
Tea	352 ± 33	1,360 ± 180	250 ± 30	23 ± 3
mate	451 ± 18	248 ± 10	300 ± 39	29 ± 5

Total phenolics content is expressed as gallic acid equivalents (mg/L). Antioxidant capacity is expressed as the volume that inhibited the oxidation in the different systems assayed (see MATERIALS AND METHODS).

[a]TRAP, amount that was equivalent to the inhibition exerted by 0.5 mM Trolox.

[b]Liposome oxidation, volume of beverage that inhibited 50% of TBARS production.

[c]Cell oxidation, volume that inhibited 50% of AMVN-induced cell fluorescence.

TABLE 2. (+)-catechin, (−)-epicatechin, gallate, and chlorogenic acid in wines, tea, and mate

Samples	(+)-Catechin	(−)-Epicatechin	Gallate	Chlorogenic acid
		(μM)		
CS wine	108 ± 8	32 ± 3	98 ± 1	nd
Malbec wine	91 ± 12	46 ± 1	61 ± 7	nd
Generic wine	72 ± 1	34 ± 7	29 ± 9	nd
Tea	25 ± 13	46 ± 21	98 ± 23	22 ± 1
mate	14 ± 5	nd	nd	609 ± 50

Tea infusion and mate decoction were prepared as indicated in MATERIALS AND METHODS. nd, below detection limit.

AMVN-mediated oxidation. Tea and CS wine were the most effective beverages, followed by mate.

The content of various phenolic compounds in the beverages is shown in TABLE 2. Wines presented a higher content of free (+)-catechin (CS > Malbec > generic), compared to tea and mate. By contrast, levels of free (−)-epicatechin were similar in wines and tea. It is well known that in tea, catechins are mainly conjugated with gal-

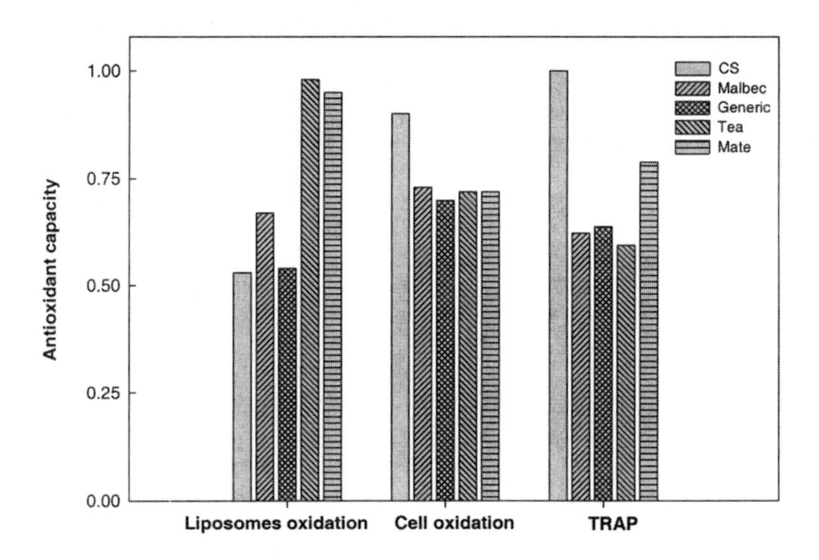

FIGURE 1. Antioxidant capacity of wines, tea, and mate evaluated by three different assays. The beverages were tested at the same concentration of total phenolics (350 μM). Antioxidant capacity is expressed relative to the value of oxidation obtained in the absence of the tested beverages. Values are means of at least three independent experiments (ESM were under 10%).

late residues. The fact that we did not hydrolize the samples could lead to an underestimation of the total amount of catechins present in the samples.

The matrix effect of the different beverages was evaluated by comparing the antioxidant capacity of the beverages in a dilution that had equivalent amounts of phenolic compounds (see FIGURE 1). In this condition, the antioxidant capacity evaluated by both the TRAP assay and the inhibition of cell oxidation, was significantly higher ($p < 0.01$) for the CS (0.7 mM Trolox) than for the others beverages (around 0.4 mM). On the contrary, when the antioxidant capacity was evaluated by the ability of the beverages to inhibit liposome oxidation, tea and mate were the most potent (100% inhibition).

The present results indicate that: (a) the three beverages assayed, wine, tea, and mate can provide antioxidant protection at tissue levels attained in normal human diets; (b) not all the methods used to evaluate the antioxidant capacity of the beverages are equivalent; (c) the comparison among different beverages underscored the capacities of certain types of wines (CS in this study) over infusions of tea and decoctions of mate; (d) in populations consuming high daily amounts of mate (Argentina, Uruguay, Brazil), the contribution of mate to the antioxidant defenses in the body is significant; e) the food matrix may be influencing the antioxidant capacity of phenolic compounds. Further studies of absorption and metabolism of the active compounds will enable judgment of the relevance of these results for human health.

ACKNOWLEDGMENTS

Supported with grants from UBA, CONICET and Ministry of Health (Ramon Carrillo-Arturo Oñativia Fellowship).

REFERENCES

1. FOGLIANO, V., V. VERDE, G. RANDAZZO & A. RITIENI. 1999. Method for measuring the antioxidant activity and its application to monitoring the antioxidant capacity of wines. J. Agric. Food. Chem. **47:** 1035–1040.
2. LOTITO, S.B. & C.G. FRAGA. 1999. (+)-catechin prevents human plasma oxidation. Free Radical Biol. Med. **24:** 435–441.
3. LISSI, E., M. SALIM-HANNA, C. PASCUAL & M.D. DEL CASTILLO. 1995. Evaluation of total antioxidant potential (TRAP) and total antioxidant reactivity from luminol-enhanced chemilumenescence measurements. Free Radical Biol. Med. **18:** 153–158.
4. VERSTRAETEN, S.V. & P.I. OTEIZA. 1995. Sc^{3+}, Ga^{3+}, In^{3+}, Y^{3+}, and Be^{2+} promote changes in membrane physical properties and facilitate Fe^{2+}-initiated lipid peroxidation. Arch. Biochem. Biophys. **322:** 284–290.
5. Oteiza P.I., M.S. Clegg, M.P. Zago & C.L. Keen. 2000. Zinc deficiency induces oxidative stress and AP-1 activation in 3T3 cells. Free Radical Biol. Med. **28:** 1091–1099.

Assessing the Antioxidant Capacity in the Hydrophilic and Lipophilic Domains

Study of a Sample of Argentine Wines

SILVINA B. LOTITO, M. LOURDES RENART, AND CÉSAR G. FRAGA

Physical Chemistry/PRALIB, School of Pharmacy and Biochemistry, University of Buenos Aires, Buenos Aires, Argentina

ABSTRACT: The antioxidant capacity of different types of red and white wines was assessed by different assays. Two assays were used to evaluate the antioxidant capacity in aqueous phase: (1) inhibition of the generation of 2,2′-azinodi(3-etilbencenotiazolin-6-sulfonate) (ABTS)-derived radical; and (2) protection of vitamin E in human plasma. The results indicated that red wines were not all equally effective, and a comparison could be made among different types. The ABTS assay showed that Cabernet Sauvignon wines had the highest antioxidant capacity followed by Malbec wines. Red wines showed a protective capacity of 70–90% in preventing SH-groups oxidation, and, in addition, they were also effective in preventing the oxidative damage in lipid domains (30–97% protection to liposomes, 20–70% to vitamin E). The antioxidant capacity of the white wines was significantly lower, as evaluated by all the assays. Significant correlations were found for phenolics and catechin content with the antioxidant capacity of the studied wines.

KEYWORDS: antioxidants; red wines; Argentine wines

INTRODUCTION

Consistent epidemiological data suggest that the regular intake of fruits, vegetables, and their derivatives may reduce the risk of several chronic diseases.[1] The content of phenolics in red wine may explain the apparent compatibility of a high fat diet and a low incidence of coronary atherosclerosis in populations that drink red wine daily (French Paradox).[2] Since those pathologies have been long associated to free radical–mediated damage, it is possible to theorize that the phenolics present in wine could have a role as antioxidants *in vivo*. A well balanced characterization of the antioxidant capacity of wines is therefore required for studies on health effects of wines.

In this study, we used several assays to evaluate the antioxidant capacity of some Argentine wines, both in aqueous phase and lipid phase. The results are associated to the phenolics, (+)-catechin, and gallic acid content measured in the wines.

Address for correspondence: Dr. César G. Fraga, Fisicoquímica, Facultad de Farmacia y Bioquímica, Junín 956, 1113-Buenos Aires, Argentina. Voice/fax: 54-11-4508-3646.
cfraga@ffyb.uba.ar

EXPERIMENTAL

The sample consisted of seven red [Cabernet Sauvignon (CS), Malbec (MB), Generic (A, B, C, D, E)] and two white wines (F, G).

The antioxidant capacity was evaluated by:

(*a*) *ABTS method.* Wines (dilution 1/25) were studied for their ability to inhibit the formation of 2,2′-azinodi(3-etilbencenotiazolin-6-sulfonate) (ABTS)–derived cation radical.[3] This radical is formed by the interaction of ABTS (305 μM) with ferrylmyoglobin radical species, which were generated by a mixture of metmyoglobin (3.05 μM) and H_2O_2 (250 μM). The reaction was followed at 600 nm, for three minutes. Results are referred to a control of maximal oxidation in the absence of wines, and expressed as percentage of inhibition of ABTS oxidation.

(*b*) *Protection of thyols (SH) groups.* Bovine albumin (45 g/L) was oxidized by incubation (60 min, 37°C) with a 50 mM water-soluble generator of peroxyl radical [2,2′-azobis(amidinopropane) hydrochloride (AAPH)]. The exposed SH groups of albumin were determined, by a spectrophotometric technique,[4] in the absence or the presence of wines (dilution 1/100). Results are referred to a control of maximal oxidation in the absence of wines, and expressed as percentage of inhibition of SH groups oxidation.

(*c*) *Protection of liposomes.* Liposomes made of L-α-phosphatydilcholine (0.5 g/L) were incubated with 10 mM AAPH (60 min, 37 ºC) in the absence or the presence of wines (dilution 1/25). Lipid oxidation was evaluated as substances reactive to 2-thiobarbituric acid (TBARS), fluorometrically.[5] Results are referred to a control of maximal oxidation in the absence of wines, and expressed as percentage of inhibition of lipid oxidation.

(*d*) *Protection of vitamin E.* Human plasma was incubated with 50 mM AAPH (4 h, 37°C) in the absence or the presence of wines (dilution 1/10). Vitamin E (as α-tocopherol) was evaluated by HPLC (reverse phase), with electrochemical detection.[6] Results are referred to a control of maximal oxidation in the absence of wines, and expressed as percentage of inhibition of α-tocopherol oxidation.

Phenolics content was evaluated using the Folin-Ciocalteau assay, as described by Fogliano *et al.*[7] Results are expressed as gallic acid equivalents (GAE). (+)-Catechin and gallic acid were evaluated by HPLC, reverse phase, with electrochemical detection.[6]

RESULTS AND DISCUSSION

We studied four different assays to evaluate the antioxidant capacity of a sample of Argentine wines. The antioxidant capacity was evaluated by the protection to targets of oxidation present in hydrophilic or lipophilic domains. Red wines were able to prevent the oxidation in both the aqueous and lipid phase (see TABLE 1). The results indicated that the assayed red wines were not all equally effective. In the ABTS method, the antioxidant activity is indicated by the ability of wines to inhibit the reaction of formation of ABTS-derived radicals, in the aqueous phase. The ABTS assay showed that CS had the highest antioxidant activity, followed by MB. Generic wines exerted an intermediate antioxidant activity as measured by this method. Results for white wines indicated a 30-times lower antioxidant capacity (TABLE 1). The second method used to evaluate the antioxidant capacity in aqueous phase was the protection to SH groups of albumin. The maximal oxidation of the albumin SH groups reached under the present conditions was about 50–60%. The highest protection to SH-groups was observed for MB. For generic red wines, it was observed that some had antioxidant capacities similar to CS or better, but others were as low as white wines (TABLE 1).

The assayed wines were also highly effective preventing oxidative damage in lipid domains, with red ones being the most protective. The protection to liposome membranes reached more than 90% for MB and CS (dilution 1/25). Considering the entire sample of red wines, they were shown to exert an average of 60% protection, compared to a lower antioxidant capacity for white wines (average 11%). The assessment of the prevention of vitamin E oxidation in plasma is of high physiological relevance. CS wine exhibited the highest ability to protect plasma vitamin E from oxidation. The best red wines exerted a 40–70% protection to vitamin E, and some generic wines showed an antioxidant capacity almost similar to white wines.

The correlations between the different methods are shown in TABLE 2. To assess the substances responsible for the antioxidant capacity, we correlated the phenolics, (+)-catechin, and gallic acid content, with the results obtained using the various assays to evaluate antioxidant capacity (TABLE 2).

Our results shows a higher antioxidant capacity of red wines compared to white wines in both the aqueous and lipid phases. It is interesting to point out that although the polyphenols present in wines are water soluble, they also account for the antioxidant capacity in lipid environments. That effect in lipid domains can result from preventing important oxidative events, such as the oxidation of membranes, or the

TABLE 1. Antioxidant capacity of wines and total phenols in wines

Wine	ABTS	SH groups	Liposomes	Vitamin E	Phenolics
	(%)				(GAE, mg/mL)
CS	38.3	58	90	71	2788
MB	27.5	77	92	43	1780
Generic A	12.5	49	62	37	1696
Generic B	11.5	32	47	38	1176
Generic C	18.5	1	58	32	1209
Generic D	14.0	1	49	21	1411
Generic E	6.0	43	31	32	890
White F	0.6	1	2.5	20	279
White G	0.4	17	19	35	238

NOTES: Wines of different types and varieties were evaluated for the antioxidant capacity. Antioxidant capacity is expressed as percentage of inhibition (%) referred to the value obtained in the absence of wine; GAE: gallic acid equivalents.

TABLE 2. *R* values for the correlations between antioxidant capacity assays and phenolic content

	ABTS	SH groups	Liposomes	Vitamin E
ABTS		0.63^{c}	0.94	0.80^{a}
SH groups			0.77^{a}	0.71^{b}
Liposomes				0.71^{b}
Phenolics	0.94	0.62^{b}	0.88	0.73^{b}
(+)-Catechin	0.89	0.69^{b}	0.69	0.89
Gallic acid	0.84	0.65^{c}	0.79	0.77

NOTES: The correlations were significant at $p < 0.001$, except for those indicated as *a*: $p < 0.03$; *b*: $p < 0.05$, and *c*: $p < 0.07$.

depletion of lipid soluble antioxidants as vitamin E inside the plasma lipoproteins. The methods applied set the significance of wines as antioxidants not only in hydrophilic phases (which is expected from the aqueous nature of the wines) but also in lipid environments.

REFERENCES

1. HERTOG, M.G.L. *et al.* 1993. Dietary antioxidant flavonoids and risk of coronary heart disease: the Zutphen Elderly Study. Lancet **342:** 1007–1011.
2. RENAUD, S. & M. DE LORGERIL. 1992. Wine, alcohol, platelets, and French paradox for coronary heart disease. Lancet **339:** 1523–1526.
3. MILLER, N.J. *et al.* 1993. A novel method for measuring antioxidant capacity and its application to monitoring the antioxidant status in premature neonates. Clin. Sci. **84:** 407–412.
4. MURPHY, M.E. & J.P. KEHRER. 1989. Oxidation state of tissue thiol groups and content of protein carbonyl groups in chickens with inherited muscular dystrophy. Biochem. J. **260:** 359–364.
5. VERSTRAETEN, S.V. & P.I. OTEIZA. 1995. Sc^{3+}, Ga^{3+}, In^{3+}, Y^{3+}, and Be^{2+} promote changes in membrane physical properties and facilitate Fe^{2+}-initiated lipid peroxidation. Arch. Biochem. Biophys. **322:** 284–290.
6. LOTITO, S.B. & C. G. FRAGA. 1998. (+)-Catechin prevents human plasma oxidation. Free Radical Biol. Med. **18:** 435–442.
7. FOGLIANO, V. *et al.* 1999. Method for measuring antioxidant activity and its application to monitoring the antioxidant capacity of wines. J. Agric. Food Chem. **47:** 1035–1040.

Physiological Model of the Stimulative Effects of Alcohol in Low-to-Moderate Doses

ANATOLY G. ANTOSHECHKIN

Nulab, Inc., Los Angeles, California 90039, USA

ABSTRACT: A physiological model of alcohol use is proposed, based on the observation that ethanol is a natural intracellular metabolite. It is suggested that ethanol synthesis is an intermediate step in the metabolic pathway of elimination of the excess of energy-releasing substrates from mitochondria and the cytosol when the cell does not require additional energy. The process is first of all important for elimination of acetyl-CoA from mitochondria. Ingested alcohol passes across plasma and mitochondrial membranes by diffusion and follows the endogenous ethanol pathway but in the reverse direction. This pathway is used when the cell needs additional energy. In that case, exogenous ethanol supplies the respiratory chain with electrons via NADH on the path of oxidation to acetaldehyde and then to acetate by alcohol and aldehyde dehydrogenases. After conversion to acetyl-CoA, it goes into the citrate cycle and is thus used for energy production entirely. If citrate cycle activity is decreased constitutionally, acetyl-CoA enters into citrate cycle too slowly and energy deficit arises, ethanol compensates for the energy deficit. Acetate then diffuses out of mitochondria. The less energy production in mitochondria, the more the cell is ready to accept alcohol as an energy source. Citrate cycle activity can be depressed as a result of single-nucleotide polymorphisms of genes controlling some enzymes involved in the cycle, which influences the alcohol requirement.

KEYWORDS: redox shuttle; intracellular ethanol formation; energy-producing pathway

The action of alcohol on central nervous system has a dose-dependent biphasic character. The stimulative effects are characteristic for the first phase, while in the second phase, toxic or pharmacological effects predominate with increased dose.[1] The toxic effects of alcohol are probably determined by such well-established pharmacological mechanisms as the blockade of neurotransmitter receptors and signaling pathways, and some others.[2] To advance our understanding of the biochemical basis of health benefits of mild-to-moderate alcohol consumption, it seems advisable to search for physiological mechanisms of alcohol action that participate in normal biochemical reactions without disturbing them.

The proposed model of physiological action of ingested alcohol is based on an observation that ethanol is not a foreign compound to human cells, but is a natural intracellular metabolite.[3–5] It is suggested, that ethanol has two functions. First, it acts as a substrate shuttle ethanol ↔ acetaldehyde for the transmembrane transfer of redox equivalents between mitochondria and cytosol.[6,7] Second, it is an intermediate

Address for correspondence: Anatoly G. Antosheckin, M.D., Ph.D., 3504 Candlewood Ct., Fairfield, CA 94533, USA. Voice/fax: 707-434-1640.
antoshechkin@earthlink.net

metabolite in the pathway of elimination of excess energy-releasing substrates from mitochondria and the cytosol. Such a substrate is first of all acetyl-CoA, a surplus of which can arise in mitochondria when the cell requirement for production of adenosine triphosphate (ATP) is lowered. The metabolic pathway consists of the following sequence of conversions: acetyl-CoA → acetate → acetaldehyde → ethanol. The reactions acetate → acetaldehyde → ethanol are catalyzed by NAD-dependent aldehyde dehydrogenase and alcohol dehydrogenase. These reduction reactions are accompanied by energy elimination from mitochondria in the form of redox equivalents.

Upon the entrance into the cytosol, ethanol can esterify excess of some carboxylic acids, including fatty acids—another energy-releasing substrate—and is excreted from the cell in the form of esters or in free form.[5] Esterification of the carboxylic acids inactivates their carboxylic groups and facilitates their transport through membranes. Ethyl esters formation is probably carried out by glutathione *S*-transferases.[8]

Because all reactions of the intracellular ethanol pathway are reversible, ingested alcohol follows this pathway, but in the reverse direction (see FIGURE 1). By simple diffusion, it passes from the blood stream across plasma and mitochondrial membranes of the cell and then follows the pathway: ethanol → acetaldehyde → acetate → acetyl-CoA → citrate cycle. This pathway is used especially when the cell needs additional energy and NAD+ exceeds NADH in mitochondria. In this case, exogenous alcohol carries out the function of an energy-releasing substrate, supplying electrons via NADH into the respiratory chain at the stages of oxidation to acetaldehyde and the latter to acetate. Under subsequent conversion of acetate into acetyl-CoA and its entering into the citrate cycle, ethanol is oxidized completely.

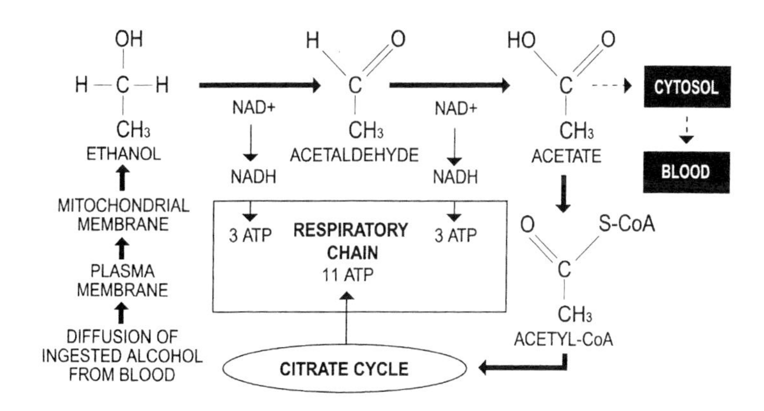

FIGURE 1. Energy-producing pathway of ingested ethanol. Oxidation of one molecule of ethanol on the stages of conversion to acetaldehyde and then to acetate supplies the respiratory chain with redox equivalents via NADH, which is enough to produce six molecules of ATP. Further conversion of acetate into acetyl-CoA and its oxidation in the citrate cycle gives additional energy for production of 11 ATP molecules. Therefore, the energetic efficiency of ethanol oxidation to acetate is only two times lower than the efficiency of the citrate cycle, the key process of energetic metabolism.

However, if activity of the citrate cycle is constitutionally depressed, acetyl-CoA goes into the citrate cycle too slowly, and an energy deficit arises; in this case alcohol conversion does not proceed to the acetyl-CoA stage, but stops at the stage of acetate, which diffuses into the cytosol. In this case, alcohol compensates the citrate cycle insufficiency by delivery of electrons via NADH into the respiratory chain at the stages of oxidation to acetaldehyde and acetate. Oxidation of ethanol during these two reactions can significantly substitute for the energetic function of citrate cycle.

Based on the proposed model, motivation for alcohol consumption by moderate drinkers is determined by the reduction of energy production in cells of the human body, first of all, in neurons. Transient physiological reduction of ATP formation is evoked by temporal deficit of nutrients in the body. More constant decline of ATP production in mitochondria of neurons may be a result of genetically determined variations of carbohydrate metabolism and citrate cycle activity. The lower the energy production in mitochondria, the more the neuron is ready to accept alcohol as an energy source.

There are many observations favoring an assumption that ingested alcohol interferes in energy metabolism. For instance, in our observations the addition of ethanol to partially depleted culture medium to a concentration of 10mg/mL promotes fibroblast growth *in vitro*. Mild doses of alcohol enhance man's physical performance, which is why alcohol is included in the list of banned substances by the International Olympic Committee. It is known that cells under stress conditions waste more energy and require its reinforcement.[9] Depending on the nature of their work and their psychoemotional status, many people feel psychological tension and fatigue at the end of the workday, which is a manifestation of light stress. Intake of mild-to-moderate doses of alcohol eliminates these stress manifestations, which provides at least partial motivation for the consumption of alcoholic beverages.[10]

REFERENCES

1. LUNDQUIST, F. 1975. Interference of ethanol in cellular metabolism. Ann. N.Y. Acad. Sci. **252:** 11–20.
2. U.S. DEPARTMENT OF HEALTH AND HUMAN SERVICES. 1997. Ninth Special Report to the U.S. Congress on Alcohol and Health. NIH Publication No. 97-4017: 65–98.
3. ANTOSHECHKIN, A.G. 2001. On intracellular formation of ethanol and its possible role in energy metabolism. Alcohol Alcohol. **36:** 608.
4. ANTOSHECHKIN, A.G., *et al.* 1988. Experimental evidence of the intracellular formation of ethanol and its role in the cell as an intermediate metabolite. Izv. Acad. Nauk SSSR, Biol. **(1):** 139–142.
5. ANTOSHECHKIN, A.G., *et al.* 1988. Determination of human fibroblast metabolism in vitro by gas chromatography-mass spectrometry of cell-excreted metabolites. Anal. Biochem. **169:** 33–44.
6. GRUNNET, N. 1973. Oxidation of acetaldehyde by rat liver mitochondria in relation ethanol oxidation and the transport of reducing equivalents across the mitochondrial membrane. Eur. J. Biochem. **35:** 236–243.
7. BAKKER, B.M., *et al.* 2000. The mitochondrial alcohol dehydrogenase Adh3p is involved in a redox shuttle in *Saccharomyces cerevisiae*. J. Bacteriol. **182:** 4730–4737.
8. BORA, P.S. & L.G. LANGE. 1993. Molecular mechanism of ethanol metabolism by human brain of fatty acids ethyl esters. Alcohol. Clin. Exp. Res. **17:** 28–30.

9. TOUSSAINT, O., A. HOUBION & J. REMACLE. 1994. Effect of modulation of the energetic metabolism on the mortality of cultured cells. Biochim. Biophys. Acta **1186:** 209–220.
10. SWENDSEN, J.D., *et al.* 2000. Mood and alcohol consumption: an experience sampling test of the self-medication hypothesis. J. Abnorm. Psychol. **109:** 198–204.

A Queue Paradigm Formulation for the Effect of Large-Volume Alcohol Intake on the Lower Urinary Tract

C.P. ARUN

Department of Urology, James Paget Hospital, Great Yarmouth, Norfolk NR31 6LA, England

ABSTRACT: We examine the urologic consequences of large-volume alcohol intake in light of recent advances in hollow viscera biomechanics and urinary tract pharmacology. Recent studies have shown that alcohol is depressive on the isolated rabbit detrusor. Patients teetering on or having manifest pelvi-ureteric junction obstruction or bladder outflow obstruction have been known to demonstrate symptoms on alcohol consumption. In men over 50, the cause is usually an enlarged prostate. Loin pain is precipitated in the case of the former and urinary retention in the latter case after consuming alcohol. Causation is difficult to prove for ethical reasons as well as the practical difficulties in running a prospective trial. It appears that a combination of rapid fill (due to diuretic effects), a weakened pump (due to the depressive effect of alcohol at least with the detrusor) and outlet obstruction lead to the loin pain and urinary retention seen in clinical practice. People liable to urinary tract obstruction would be well advised to avoid large volume alcohol intakes.

KEYWORDS: queue paradigm; lower urinary tract; large-volume alcohol intake

Recently, an intuitive model based on probability theory, the queue paradigm (the sequence: arrival-service-departure) was proposed by the author[1] for modeling physiological systems that could be conceived of as "service systems." Organ systems and subsystems that could be considered to be "physiological service systems" include the cardiovascular system,[2] the gastrointestinal system and the toxin disposal system.[1] The present work outlines a conceptual framework based on the queue paradigm for studying the effect of alcohol intake on the lower urinary tract (LUT).

In lower animals, frequent voiding of small amounts of urine is normal and even of value as for example, in marking one's territory. Among humans however, civilization demands that the act of micturition be performed at a socially convenient time and place. Under normal circumstances, the lower urinary tract unit expels its contents to the external environment on initiating the micturition reflex. Although the function of urine storage is of social value, that of urine expulsion is the more important from a physiological standpoint. A bladder full of urine can be very painful even

Address for correspondence: C.P. Arun, FRCS, Department of Urology, James Paget Hospital, Lowestoft Road, Gorleston, Great Yarmouth, Norfolk NR31 6LA, England. Voice: 44.1493.452.731; fax: 44-1493-452-666.
arunpeter@yahoo.com

for short periods of time (experienced easily enough if one should have difficulty in finding a toilet in time). In the clinical condition termed acute retention of urine (ARU), one is unable to pass urine even on initiating a micturition reflex. In the indexed literature, evidence for the possible role of alcohol in ARU is accumulating. The mechanisms involved are worth exploring for the insights they can provide into the pharmacological effects of alcohol as well as for the dynamics of the LUT.

The value of considering a physiological system as a stochastic service system is that one can then employ the rich resources of the mathematically founded theory of stochastic processes to study its dynamics (stochastic processes can be roughly described as those whose outcome cannot be predicted in advance). Using the queue paradigm, we can deal with such entities as arrivals to the system, the service level (λ) of the server and the departures from the system of interest. Although arrivals and departures are implicit to the so-called input-output principle, the concept of "service level" is new to physiological system modeling. Service level provides us with a measure of organ (or sub-organ) function. Moreover, the concept of congestion that is an important subject of study in stochastic processes can be useful in describing the dynamics of physiological systems: the term pathophysiological gridlock is self-explanatory. For example, what we term "congestive" cardiac failure, can be understood to be "congestion" of the cardiovascular queueing system.[2] FIGURE 1 is a schematic representation of the LUT as a queueing system. The concept of congestion (analogous to traffic gridlock) is also of value in describing the LUT's response to alcohol intake. Equation **(1)**, states the simple relation between arrival rate, service level (λ) and the departure rate for a physiological service system.

$$\text{Service level } (\lambda) = \text{arrivals/departures} \tag{1}$$

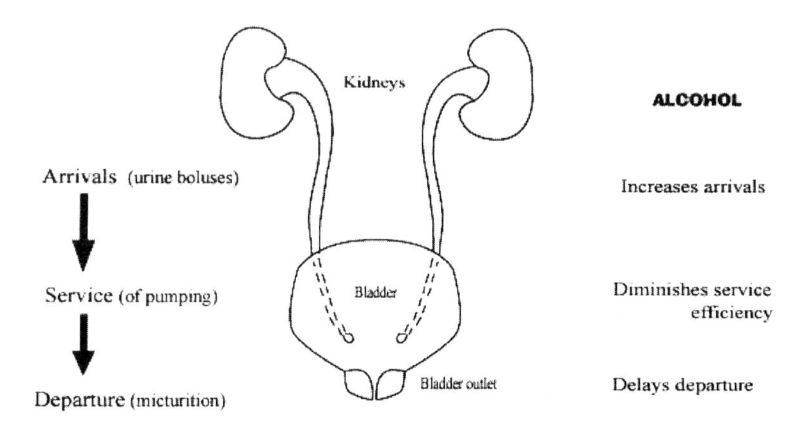

FIGURE 1. Diagram showing how the lower urinary tract may be formulated in terms of the queue paradigm to study the effects of alcohol. The self-explanatory text to the left details this. To the right, we can visualize the effect of alcohol in terms of the sequence: arrivals-service-departure. Here, we are interested in the service of pumping; if we were interested in studying the dynamics of bladder storage (for another problem), we could represent this equally well using the queue paradigm.

For our purposes, the arrival rate relates to boluses of urine that enter the bladder from the kidneys. The departure rate is the rate at which urine leaves the bladder at micturition. The service level is the efficiency with which the bladder is able to expel its contents. Delivered service—like the probability of any given event—cannot possibly exceed unity. Clearly, if the demanded service level is greater than unity (i.e., the amount of urine exceeds that which can be expelled) congestion will ensue. With the LUT, this manifests itself as retention.

THE EFFECT OF ALCOHOL, ESPECIALLY LARGE-VOLUME INTAKES AS A QUEUEING PROBLEM

It transpires that alcohol can act at all three levels, arrivals, service and departure of the queue paradigm to produce congestion of the LUT. The arrivals of urine can be increased by both the water load of certain alcoholic beverages as well as by the diuretic effect of its alcohol content. Alcohol has been demonstrated to be inhibitory to the muscle (the server) of the bladder, both *in vitro*[3] and *in vivo*[4] studies and hence the service of urine expulsion suffers. Furthermore, voiding of urine (departure) can be compromised since alcohol intake can reduce one's ability to respond to the sense of bladder distension. Unfortunately, with viscoelastic muscular structures such as the urinary bladder, the window of optimal function is a narrow one and can easily be overwhelmed. In men who are over fifty years of age, enlargement of the prostate increases outlet resistance and hence the work that the bladder has to perform in expelling urine. In addition, over-distension (inevitable if micturition is delayed), can overwhelm the bladder's ability to cope with a urine load and result in retention.

Worldwide, alcohol is probably the most commonly consumed, socially acceptable, non-therapeutic drug. The queue paradigm provides us with a useful conceptual framework for describing the effects of alcohol on the lower urinary tract and drawing some useful lessons. In the future, it may be possible to obtain quantitative relations for predicting the response of the LUT to various combinations of arrival, service level, and departure using techniques such as heavy-traffic approximation. Large-volume alcohol intakes are best avoided especially in men over fifty years of age. After consuming alcohol, it is probably unwise to postpone micturition.

REFERENCES

1. ARUN, C.P. 1999. Queueing Theory in Clinical Practice: Application to the Cardiovascular System. Proceedings of the 1999 Summer Computer Simulation Conference. M.S. Obaidat, A. Nicanci and B. Sadoun, Eds.: 481–484. SCS Press, San Diego.
2. ARUN, C.P. 2000. Queueing and inventory theory in clinical practice: application to clinical toxicology. Ann. N.Y. Acad. Sci. **919:** 284–287.
3. OHMURA, M., A. KONDO, *et al.* 1997. Effects of ethanol on responses of isolated rabbit urinary bladder and urethra. Int. J. Urol. **4**(3): 295–299.
4. YOKOI, K., M. OHMURA, *et al.* 1996. Effects of ethanol on in vivo cystometry and in vitro whole bladder. J. Urol. **156**(4): 1489–1491.

Oxidative Stress and Inflammatory Reaction Modulation by White Wine

ALBERTO A.E. BERTELLI,[a] MASSIMILIANO MIGLIORI,[b] VINCENZO PANICHI,[c] BIANCAMARIA LONGONI,[d] NICOLA ORIGLIA,[b] AGNESE FERRETTI,[d] MARIA GIUSEPPA CUTTANO,[d] AND LUCA GIOVANNINI[b]

[a]*Department of Human Anatomy, University of Milan, 20133, Milan, Italy*

[b]*Neuroscience Department, Pharmacology Section,* [c]*Internal Medicine Department, and* [d]*Department of Oncology, Transplants and Advanced Technologies in Medicine University of Pisa, Pisa, Italy*

ABSTRACT: Wine and olive oil, essential components of the Mediterranean diet, are considered important factors for a healthy life style. Tyrosol (T) and caffeic acid (CA) are found in both extra virgin olive oil and in white wine. Three white wines from the northeast Italy and four white wines from Germany were analyzed for their content of T and CA. These compounds were tested for their antioxidant activity and their capacity to modulate three different cytokines: IL-1β, IL-6, and TNF-α, which are currently considered to be the major cytokines influencing the acute phase of the inflammatory response. Furthermore, the antioxidant activity of T and CA was analyzed by monitoring the oxidation of a redox-sensitive probe by using laser scanning confocal microscopy. T and CA, applied at nanomolar range, were found to significantly reduce the generation of oxidants induced by azobis-amidinopropane-dihydrochloride. Peripheral blood mononuclear cells (PBMC) from healthy volunteers were incubated at 37°C for 12 hours with 100 ng LPS (*E. coli* and *P. maltofilia*). Increasing doses of T and CA (150 nM to 300 μM) were added and cell-associated IL-1β and TNF-α were determined by immunoreactive tests after three freeze-thaw cycles. IL-6 release was also determined in cell surnatants. LPS-stimulated PBMC showed a significant increase in cytokine release, while T and CA, used at nanomolar concentrations, were able to modulate their expression. Taken together, these results suggest a remarkable effect of white wine non-alcoholic compounds on oxidative stress and inflammatory reaction.

KEYWORDS: cytokines; white wine; inflammatory reaction; antioxidants

INTRODUCTION

The Mediterranean diet is correlated with a lower incidence of coronary heart disease[1,2] and in this diet wine and olive oil are essential components. Tyrosol (T) and caffeic acid (CA) are found both in extra virgin olive oil and in white wine. It is well known that experimental feeding of polyphenolics compounds of olive oil increase the resistance of low density lipoprotein to oxidation in *ex vivo*[3,4] underlying the capacity of this substances to act as antioxidants;[3,5] furthermore they exert

Address for correspondence: A.A.E. Bertelli, M.D., Ph.D., Department of Human Anatomy, University of Milan, Via Mangiagalli, 31, 20133 Milan, Italy. Fax: +39-02-58-31-61-66.

other biological activities including inhibition of platelet aggregation and suppression of cytokine-stimulated NO production.[6,7] Three white wines from the Collio area, Friuli, Northeast Italy (Tocai, Verduzzo, and Tocai late harvest) and four white wines from Germany (Riesling), were analyzed for their content of T (11 to 37.6 mg/L) and CA (3.3 to 23.1 mg/L). The aim of this study is to evaluate the activity of T and CA for their antioxidant activity and their capacity of modulating three different cytokines: interleukin-1β (IL-1β), interleukin-6 (IL-6) and tumor necrosis factor alpha (TNF-α), which are currently considered to be the major cytokines influencing the acute phase of the inflammatory response.

Furthermore the antioxidant activity of T and CA was also analyzed by monitoring the oxidation of a redox-sensitive probe by using laser scanning confocal microscopy.

MATERIAL AND METHODS

Monocytes production of IL-1β, TNFα, and IL6 was evaluated in 18 healthy volunteers as follows.

Cell Isolation

Monocyte-enriched PBMC were isolated by Ficoll (Histopaque-1077, Sigma Aldrich Co.) gradient centrifugation from heparinzed blood. Cells were washed in HBSS (Sigma Aldrich Co, UK) counted and cultured in RPMI 1640 medium supplemented with fetal calf serum, glutamine and antibiotics. Purity of monocyte fraction was confirmed by FACS analysis using an anti-CD 14 monoclonal antibody. Cells were incubated with lipopolysaccaride (LPS) from *E. coli* and *P. maltofilia* 100 ng/mL and simultaneously with increasing doses of T and CA (150 nM to 300 μM). Monocytes were incubated with: LPS alone; LPS and T; LPS and CA; LPS, CA, and T.

Cytokine Assay

After the incubation, cells were lysed by three freeze (−40°C) and thawing (37°C) cycles. Intracellular cytokine concentrations were determined by ELISA for IL-1β and TNF-α (Medgenix Diagnostics, Brussels, Belgium). The percentages of intra- and inter-assay variability were 11.9% and 12.5% for TNF-α and 10.8% and 10.4% for IL-1β. Samples were assayed in duplicate. Results were normalized to the monocytes concentration of 2.5×10 monocytes and expressed in pg/2.5×10^6 cells. IL-6 release was also determined in cells surnatant by ELISA and expressed in pmol/mL.

Human Umbilical Vein Endothelial Cell (HUVEC) Cultures

HUVEC were isolated from human umbilical vein as described.[8] HUVEC were used for experiments in the third subpassage. HUVEC were identified for their typical cobblestone appearance and positive indirect immunofluorescence for factor VIII. All agents used for cells dissociation and cultures were obtained from Sigma (Milan, Italy).

Monitoring of Reactive Oxygen Species Production

For microscopic observation, HUVEC were plated on gelatin-coated glass-plastic dishes until they reached subconfluence. HUVEC were loaded with 10μM dihydrofluorescin (DHF) (Molecular Probes, Eugene, Oregon, USA), an oxidation sensitive probe, for 30 min at 37°C in culture medium. The dish was positioned for imaging on the stage of an inverted microscope (Nikon Eclipse TE300) equipped with a laser confocal scanning system (Radiance Plus, Biorad). Fluorescence images were collected every three minutes using the 488nm excitation wavelength from an argon laser and the 515nm emission filter. In order to minimize photo-oxidation of the probe, the laser beam was attenuated to 50% of maximal illumination, and exposure of cells to light was limited to the image acquisition intervals (about 2 sec every 3 min) via the acquisition software.

After a stable baseline was attained, the free radical–generating system consisting of either 0.5M of the initiator 2.2′-azobis(2-amidinopropane) dihydrochloride (ABAP) (Wako Chemicals USA, Inc., Dallas, Texas) was added to control cells and to cells treated with CA and T. The decrease in the level of oxidative stress was monitored and recorded in the hard disk of the computer for further analysis.

RESULTS

LPS-stimulated PBMC showed a significant increase in release of all cytokines examined, whereas T and CA were able to modulate their expression at different concentrations.

Significant ($p < 0.05$) inhibition of IL-1β cell–associated LPS-induced production was determined by T at 200μM and CA at 150μM; TNF-α intracellular level were significantly decreased by CA at 150 nM.

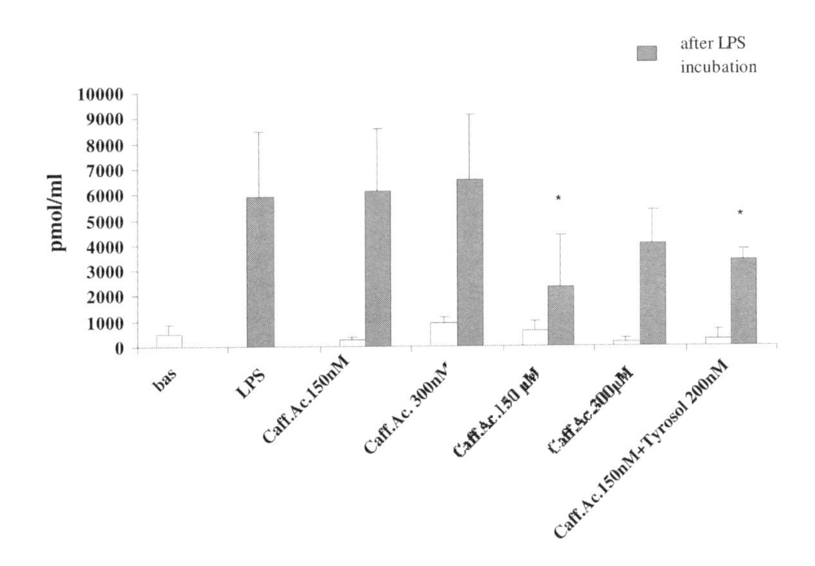

FIGURE 1. Effect of caffeic acid on PBMC IL-6 release. $*p < 0.05$ versus LPS.

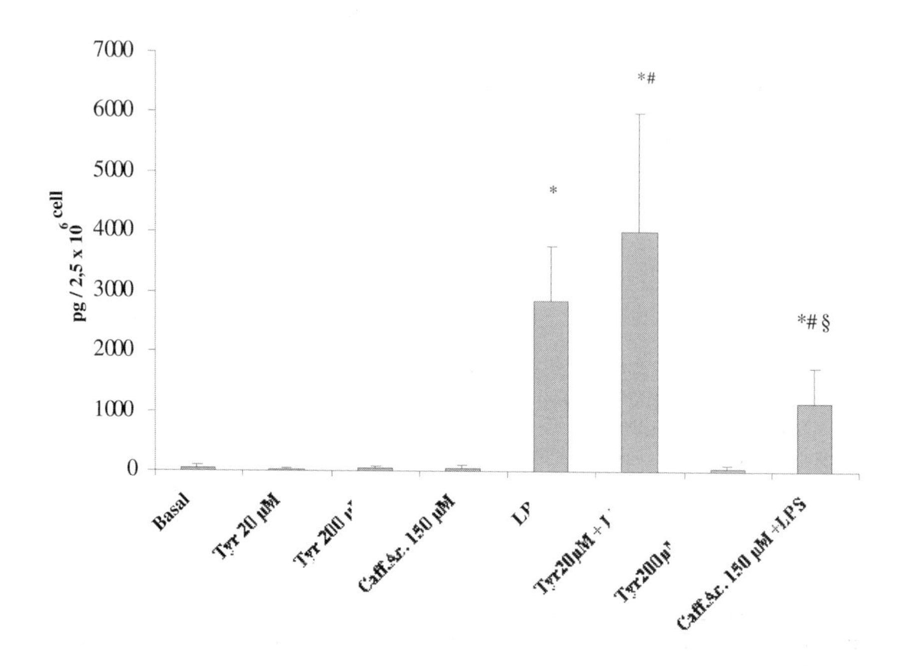

FIGURE 2. Effect of tyrosol and caffeic acid on PBMC-associated production of IL-1β. $^*p < 0.001$ versus control; $^{\#}p < 0.01$ versus LPS; $^{\S}p < 0.01$ versus Tyr 20 μM + LPS.

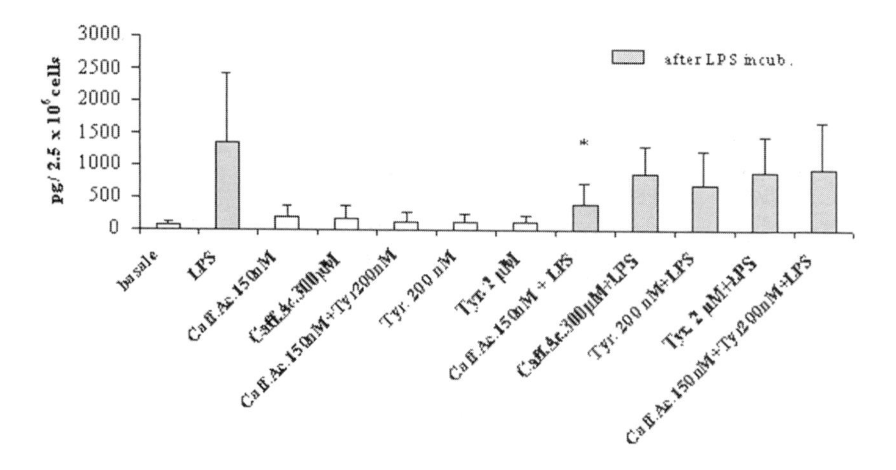

FIGURE 3. Effect of tyrosol and caffeic acid on PBMC-cell associated production of TNF-α. $^*p < 0.05$ versus LPS.

The IL-6 LPS induced release was significantly ($p < 0.05$) reduced by T at 20μM and CA at 150μM; furthermore a complementary effect of CA and T was noted at the concentrations of 150nM and 200nM respectively ($p < 0.05$).

The analytical results are shown in FIGURES 1 to 3.

As shown in FIGURE 4 CA and T clearly reduces the fluorescent signal due to oxidative stress of endothelial cells. The oxidant generator ABAP was able to significantly increase the florescent signal, indicating an alteration in the redox state of the cell.

DISCUSSION

Data obtained in our study showed a significant antinflammatory and antioxidant activity of some polyphenolic compounds found in white wine, namely tyrosol and caffeic acid.

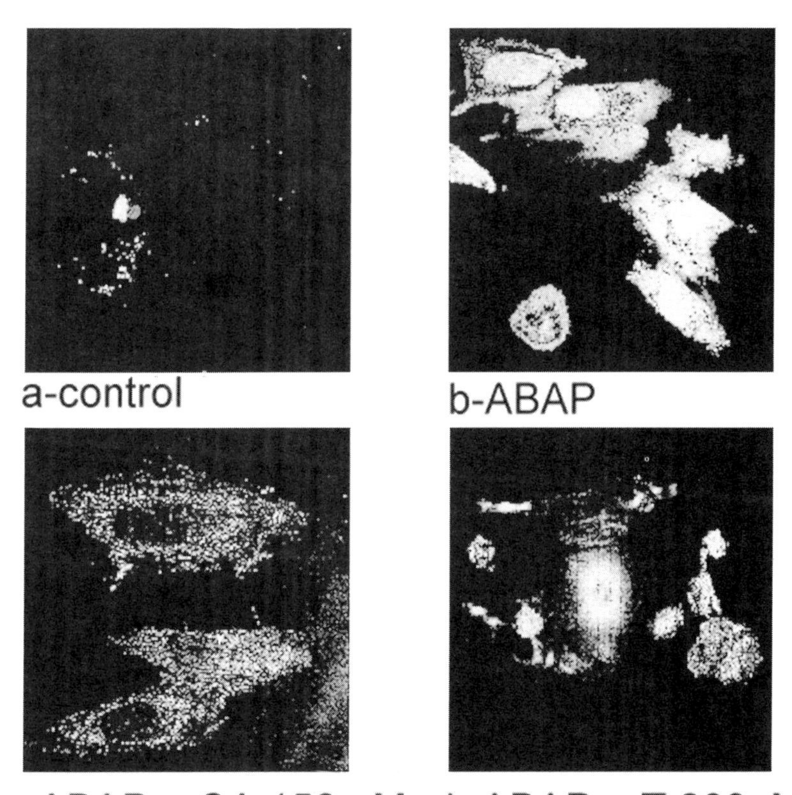

FIGURE 4. Effect of CA and T on fluorescent signal induced by oxidative stress (ABAP) in HUVEC.

During the inflammatory response, bacterial-derived fragments stimulate the monocyte/macrophage system to produce cytokines. Among these, IL1 and TNF-α are released initially in the site of inflammation and have a potent local and systemic effects. These cytokines, followed later by IL-6, are able to initiate and maintain the inflammatory response regulating the transcription of acute phase response proteins. Freshly isolated human peripheral blood mononuclear cells, cultured in absence of stimuli, did not produce IL-6 in most cases; however after incubation with different stimuli, such as LPS, an important intracellular synthesis of IL-1 and TNF-α was detected while IL-6 was almost entirely secreted into the extracellular fluid.[9]

In this paper we demonstrated an inhibitory activity of T and CA on the LPS-induced TNF-α, IL-1β and IL-6 production from PBMC of healthy volunteers; moreover a synergistic effect was noted when these substances were tested together even at lower concentrations.

The antioxidant activity of T and CA has been described by de la Puerta *et al.*[10] on intact rat peritoneal leukocytes stimulated with calcium ionophore. These compounds were also able to inhibit leukotriene B4 generation at the lipoxygenase level. In our experiments pre-treatment of endothelial cells with CA and T, led to a remarkable decrease in the cell fluorescence confirming that these compounds found in white wine are potential candidate as antioxidants *in vivo*.

These data are in agreement with other authors that showed how some antioxidants, including butylated hydroxyanisole and tetrahydropapaveroline, inhibit the production of TNF-α, IL-1β and IL-6 in PBMC stimulated by LPS, and this antioxidant-mediated inhibition of cytokine production was correlated with low levels of the corresponding messenger RNAs.[11]

On this evidence we suppose that the inhibition of cytokine production observed for T and CA may be related to a modulatory effect on the transcription factors NF-κ B and AP-1.

We conclude that the phenolics found in white wine (tyrosol and caffeic acid) can produce different antinflammatory and antioxidant activities, which opens up the possibility that some white wine may provide enhanced health benefits.

REFERENCES

1. de Lorgeril, M., P. Salen & F. Paillard. 2000. Diet and medication for heart protection in secondary prevention of coronary heart disease. New concepts. Nutr. Metab. Cardiovasc. Dis. **10**(4): 216–222.
2. Willett, W.C., F. Sacks, A. Trichopoulou, *et al.* 1995. Mediterranean diet pyramid: a cultural model for healthy eating. Am. J Clin. Nutr. **61**(6 Suppl.):1402S-1406S.
3. Giovannini, C., E. Straface, D. Modesti, *et al.* 1999. Tyrosol, the major olive biophenol, protects againist oxidized-LDL-induced injury in Caco-2 cells. J Nutr. **129**(7): 1269–1277.
4. Abu-Amsha, R., K.D. Croft, I.B. Puddey, *et al.* 1996. Phenolic content of various beverages determines the extent of inhibition of human serum and low density lipoprotein oxidation in vitro: identification and mechanism of action of some cinnamic acid derivates from red wine. Clin. Sci. **91**(4): 449–458.
5. Nardini, M., F. Patella, V. Gentili, *et al.* 1997. Effect of caffeic acid dietary supplementation on the antioxidant defense system in rat: an in vivo study. Arch. Biochem. Biophys. **342**(1): 157–160.
6. Koshihara, Y., T. Neichi, S. Murota, *et al.* 1984. Caffeic acid is a selective inhibitor for leukotriene biosynthesis. Biochim. Biophys. Acta **792**(1): 92–97.

7. SOLIMAN, K.F. & E.A. MAZZIO. 1998. In vitro attenuation of nitric oxide production in C6 astrocyte cell culture by various dietary compounds. Proc. Soc. Exp. Biol. Med. **218**(4): 390–397.
8. JAFFE, E.A., R.L. NACHMANN, C.G. BECKER, *et al.* 1973. Culture of human endothelial cells derived from umbilical veins. Identification by morphologic and immunologic criteria. J Clin. Invest. **52:** 2745–2756.
9. SCHINDLER, R., J. MANCILLA & C.A. DINARELLO. 1990. Correlations and interactions in the production of interleukin-6 (IL-6), IL-1, and tumor necrosis factor (TNF) in human blood mononuclear cells: IL-6 suppresses IL-1 and TNF. Blood **75**(1): 40–47
10. DE LA PUERTA, R., V. RUIZ GUTIERREZ & J.R. HOULT. 1999. Inhibition of leukocyte 5-lipoxygenase by phenolics from virgin olive oil. Biochem. Pharmacol. **57**(4): 445–449.
11. EUGUI, E.M., B. DELUSTRO, S. ROUHAFZA, *et al.* 1994. Some antioxidants inhibit, in a co-ordinate fashion, the production of tumor necrosis factor-alpha, IL-beta, and IL-6 by human peripheral blood mononuclear cells. Int. Immunol. **6** (3):409–422.

Reduction of Myocardial Ischemia Reperfusion Injury with Regular Consumption of Grapes

JIANHUA CUI, GERALD A. CORDIS, ARPAD TOSAKI,
NILANJANA MAULIK, AND DIPAK K. DAS

Cardiovascular Research Center, University of Connecticut School of Medicine, Farmington, Connecticut 06030-1110, USA

ABSTRACT: Recently several polyphenolic antioxidants derived from grape seeds and skins have been implicated in cardioprotection. This study was undertaken to determine if the grapes were equally cardioprotective. Sprague Dawley male rats were given (orally) standardized grape extract (SGE) for a period of three weeks. Time-matched control experiments were performed by feeding the animals 45 µg/100 of glucose plus 45 µg/100 g fructose per day for three weeks. After 30 days, rats were sacrificed, hearts excised and perfused via working-mode. Hearts were made ischemic for 30 min followed by two hours of reperfusion. At 100 mg/kg and at 200 mg/kg, SGE provided significant cardioprotection as evidenced by improved post-ischemic ventricular recovery and reduced amount of myocardial infarction. No cardioprotection was apparent when rats were given grape samples at a dose of 50 mg/100 g/day. *In vitro* studies demonstrated that the SGE could directly scavenge superoxide and hydroxyl radicals which are formed in the ischemic reperfused myocardium. The results demonstrate that the heats of the rats fed SGE reduced myocardial ischemia reperfusion injury by functioning as *in vivo* antioxidant.

KEYWORDS: cardioprotection; proanthocyanidin; resveratrol; antioxidants

INTRODUCTION

Recent studies demonstrated the cardioprotective abilities of various grape components. For example, grape seed proanthocyanidin were found to provide significant cardioprotection against ischemia/reperfusion injury.[1,2] Resveratrol, which is present in grape skin, was found to ameliorate myocardial ischemic reperfusion injury by reducing cell death due to necrosis and apoptosis.[3] This study was, therefore, designed to show whether grapes could also provide cardioprotection as the grapes contain the same polyphenolic components as red wine.

The results of our study demonstrated that the hearts of the rats fed Standardized Grape Extract (SGE) for three weeks were resistant to ischemic reperfusion injury. Three different doses were used; the results showed that SGE at 100 mg/kg/day and 200 mg/kg/day were cardioprotective while at 50 mg/kg/day SGE was not effective.

Address for correspondence: Dipak K. Das, Ph.D., Cardiovascular Research Center, University of Connecticut, School of Medicine, Farmington, CT 06030-1110, USA. Voice: 860-679-3687; fax: 860-679-4606.
ddas@neuron.uchc.edu

MATERIALS AND METHODS

Sprague Dawley male rats of about 300g body weight were used for our study. Rats were fed *ad libitum* regular rat chow with free access to water. A group of rats were given (orally) for three weeks three different doses of Standardized Table Grape Preparation (50, 100, or 200mg/kg body weight/day) (California Table Grape Commission) consisting of extracts of red, green, and blue-black California grapes, seeded and seedless varieties, in a freeze-dried powder form. The preparation was processed by the California Table Grape Commission, and stored to preserve the integrity of the biologically active compounds found in fresh grapes. Time matched control experiments were performed by feeding the animals 45µg/100 of glucose plus 45µg/100 g fructose per day for 3 weeks. After 21 days, the rats were properly anesthetized with pentobarbital (65mg/kg). After intravenous administration of heparin (500IU/kg), the chests were opened, the hearts were rapidly excised and mounted on a non-recirculating Langendorff perfusion apparatus.[9] Retrograde perfusion is established at a pressure of 100 cm H_2O with an oxygenated normothermic Krebs-Henseleit bicarbonate (KHB) buffer with the following ion concentrations (in mM): 118.0 NaCl, 24.0 $NaHCO_3$, 4.7 KCL, 1.2 KH_2PO_4, 1.2 $MgSO_4$, 1.7 $CaCl_2$, and 10.0 glucose. The KHB buffer has been previously equilibrated with 95% O_2/5% CO_2, pH 7.4 at 37°C. After perfusing the heart via the Langendorff mode for 10 min, the pulmonary vein is cannulated and the Langendorff perfusion discontinued for subsequent working heart perfusion as described previously.[9] It is essentially a left-heart preparation in which oxygenated KHB at 37°C enters the cannulated pulmonary vein and left atrium at a filling pressure of 17cm H_2O. The perfusion fluid then passes to the left ventricle from which it is spontaneously ejected through the aortic cannula against a pressure of 100 cm H_2O. Aortic flow can be measured by a calibrated rotameter while coronary flow is measured by the timed collection of the coronary perfusate dripping from the heart. The aortic flow is recirculated while coronary effluent can be collected or recirculated. Heart rate, aortic developed pressure, and its first derivative (dp/dt_{max}) are acquired and recorded. Coronary flow is terminated for 15 min to induce global ischemia which is followed by two hours of reperfusion. Aortic pressure was measured using a Gould P23XL pressure transducer connected to a sidearm of the aortic cannula, the signal was amplified using a Gould 6600 series signal conditioner and monitored on a real-time data acquisition and analysis system (CORDAT II.). At the end of 10 minutes, after the attainment of steady state cardiac function, baseline functional parameters was recorded and coronary effluent collected for biochemical assays. The circuit was then switched back to the retrograde mode and heart perfused for 15 min with KHB buffer. Hearts was then subjected to global ischemia for 30 min followed by two hours reperfusion.

At the end of reperfusion, a 10% (w/v) solution of triphenyl tetrazolium in phosphate buffer was infused into the aortic cannula.[10] The hearts were excised and stored at −70°C. Sections (0.8mm) of frozen heart were fixed in 2% paraformaldehyde, placed between two cover slips and digitally imaged using a Microtek ScanMaker 600z. To quantitate the areas of interest in pixels, a NIH Image 5.1 (a public-domain software package) was used. The infarct size (transmural) was quantified in pixels.

Evaluation of SGE for Free Radical Scavenging Activities

Since grapes contain several polyphenolic antioxidants, we examined whether SGE could directly scavenge reactive oxygen species. Free radical scavenging activities were determined by chemiluminescence technique using an EG & G Berthold Lumat LB 9507 luminometer.[10] Superoxide anions were generated by the action of xanthine oxidase (7 mU) on xanthine (100 μM) in a 500 μL reaction mixture containing 10 mM sodium carbonate, pH 9.0 and 28 μM luminol. The reaction was started by luminometer injection of xanthine. The hydroxyl radical assay mixture containined xanthine oxidase (0.6 mU), 10 μM sodium carbonate, pH 9.0, 28 μM luminol, 100 μM xanthine, 100 μM EDTA, and 100 μM ferric chloride. The scavenging activities were compared with 10 mU/mL SOD for superoxide anion and dimethyl thiourea for hydroxyl radical.

STATISTICAL ANALYSIS

For statistical analysis, a two-way analysis of variance (ANOVA) followed by Scheffe's test was first carried out to test for any differences between groups. If differences were established, the values were compared using Student's *t*-test for paired data. The values were expressed as mean ± SEM. The results were considered significant for $p < 0.05$.

RESULTS

Reduction of Myocardial Ischemic/Reperfusion Injury with SGE

The hearts of the rats given SGE orally significantly improved post-ischemic contractile function as compared to those given control diets. Three different doses were used. SGE had no effect on heart rate and coronary flow as shown in TABLE 1. Aortic flow was reduced in all groups during the postischemic reperfusion. However, SGE at 100 mg/kg and 200 mg/kg doses significantly improved the aortic flow during the reperfusion as compared to control and 50 mg/kg dose groups. Postischemic LVDP and LV_{max}dp/dt also showed significant improvement for the hearts of the rats given 100 mg/kg and 200 mg/kg SGE.

Myocardial infarct size expressed as the per cent infarct of the entire risk area was 33.8% for the control hearts subjected to 30 min ischemia followed by two hours of reperfusion. There was no change in the infarct size for the hearts of the rats fed 50 mg/kg SGE. However, there was a significant reduction in the infarct size for the heart of the animals given either 100 mg/kg or 200 mg/kg SGE.

In Vitro *Oxygen Free Radicals Scavenging by SGE*

To confirm the antioxidant effects of grapes, the scavenging activities of SGE were measured *in vitro*. The results for the chemiluminescence response of the chemically generated superoxide and hydroxyl radicals in the presence of SGE and luminol are shown in FIGURE 1. Free radical scavenging efficiency was calculated as compared to that of SOD for superoxide anion and of DMTU for hydroxyl radicals. The results demonstrate that SGE was able to scavenge both superoxide and hydroxyl radicals efficiently.

TABLE 1. **Effects of SGE on postischemic ventricular function and myocardial infarction**

	Group	Baseline	Reperfusion		
			30 min	60 min	120 min
Heart Rate (beats/min)	control	313 ± 7	287 ± 6	279 ± 6	277 ± 7
	SGE50	305 ± 8	281 ± 6	284 ± 6	276 ± 5
	SGE100	307 ± 8	283 ± 7	274 ± 6	276 ± 5
	SGE200	316 ± 5	306 ± 8	300 ± 8	280 ± 11
Aortic Flow (ml/min)	control	52.3 ± 2.5	11.7 ± 1.1	9.5 ± 1.6	7.5 ± 1.5
	SGE50	52.6 ± 3.6	12.8 ± 1.1	10.9 ± 1.6	9.7 ± 1.2
	SGE100	50.7 ± 3.5	23.4 ± 1.7*	24.8 ± 1.7*	17.9 ± 1.6*
	SGE200	50 ± 2	24.1 ± 4	26.8 ± 2.6*	19.9 ± 2.3*
Coronary Flow (ml/min)	control	26.5 ± 1.1	17.6 ± 0.5	17.1 ± 1.5	16.3 ± 1.3
	SGE50	27.1 ± 1.2	184 ± 1.9	18.6 ± 1.2	16.9 ± 1.6
	SGE100	27.3 ± 2.0	18.3 ± 2.2	18.2 ± 1.7	17.2 ± 2.0
	SGE200	27 ± 1	20 ± 2.6	20 ± 1.8	18.0 ± 1.9
LVDP (left ventricle developed pressure (mmHg)	control	107 ± 5	34 ± 5	36 ± 3	23 ± 5
	SGE50	105 ± 6	45 ± 5	44 ± 6	38 ± 6
	SGE100	103 ± 7	65 ± 4*	69 ± 7	51 ± 5
	SGE200	111 ± 5	87 ± 4*	79 ± 5*	64 ± 6*
DP/DT (mmHg/sec)	control	3,469 ± 148	1,650 ± 116	1,759 ± 104	1,568 ± 65
	SGE50	3,436 ± 201	1,815 ± 137	1,876 ± 129	1,675 ± 141
	SGE100	3,513 ± 124	1,988 ± 88*	2,049 ± 92*	1,839 ± 63*
	SGE200	3,600 ± 260	2,471 ± 183*	2,264 ± 131*	1,899 ± 74*
Infarct Size (size %)	control				34.5 ± 1.5
	SGE50				32.5 ± 0.8
	SGE100				25.8 ± 0.5*
	SGE200				25.2 ± 0.7*

NOTES: Results are expressed as mean ± SEM from six rats per group; SGE 50, 100, 200; standardized grape extract 50, 100, 200 mg/kg body weight; $^*p < 0.05$ compared with control.

DISCUSSION

In this study, we demonstrate that the hearts of the rats fed SGE (100mg/kg and 200mg/kg) were resistant to myocardial ischemic reperfusion injury. The SGE-fed rat hearts displayed significant improvement in post-ischemic left ventricular function and reduced myocardial infarct size. *In vitro* results suggested that at least a part of the cardioprotective effects of SGE was due to its ability to function as an antioxidant.

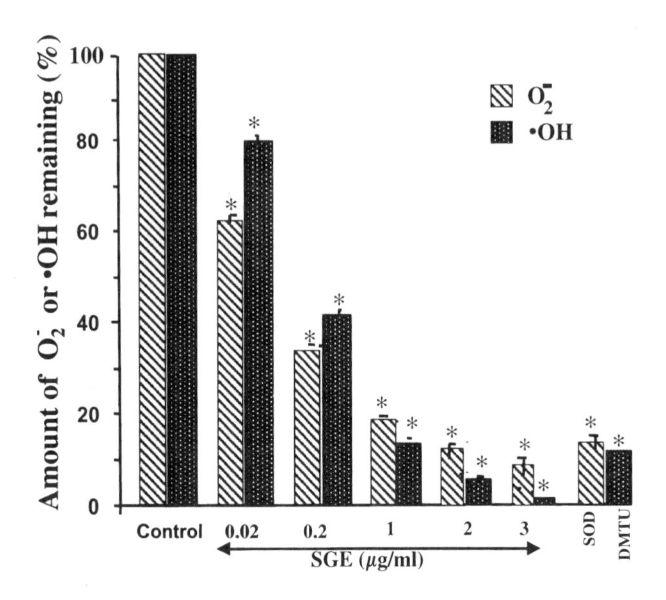

FIGURE 1. Superoxide anion and hydroxyl radical scavenging activities of SGE. Superoxide anions and hydroxyl radicals were generated chemically, and their scavenging was determined with SGE. Scavenging activities were compared against SOD for superoxide anion and DMTU for hydroxyl radicals.

Substantial evidence exists to support the notion that ischemia and reperfusion generate superoxide and hydroxyl radicals among other cytotoxic free radicals.[11] The presence of reactive oxygen species were confirmed directly by estimating free radical formation and indirectly by assessing lipid peroxidation and DNA breakdown products.[11,12] Among the oxygen free radicals, superoxide anion (O_2^-) is the most innocent free radical while the hydroxyl radical (·OH) possesses the most deleterious effects to the cells. Virtually all the biomolecules including lipids, proteins, and DNAs are potential targets for ·OH radical attack. The results of our study demonstrated that SGE used in our study can directly scavenge both superoxide and hydroxyl radicals. These results were further supported by a significant reduction of MDA content in the SGE-treated hearts.

Antioxidants have long been known to protect against the damaging effects of free radical–mediated tissue injury especially ischemia reperfusion injury of the heart and other organs. Amidst the intense interest generated in light of the various findings which support red wine as being a plausible preventive intervention against coronary heart disease, we were interested to see whether the grapes from which wines are manufactured also possess the same cardioprotective properties. Similar to wine, grapes contain polyphenolic antioxidants including resveratrol, anthocyanins, catechin, quercetin, flavans, and several phenolics. All of these polyphenolic compounds have been shown to possess potent antioxidant properties.

Grapes as opposed to other fruits and vegetables as sources of polyphenols and antioxidants are unique in a number of ways. First of all, resveratrol and a few other polyphenols are only present in grapes and are virtually absent from commonly consumed fruits and vegetables, and as such the consumption of grapes would constitute their only source in the diet. Resveratrol, a stilbene polyphenol, has recently been found to protect the hearts from ischemic reperfusion injury by its ability to reduce both necrosis and apoptosis.[3] The grape seed-derived proanthocyanidins have been found to be cardioprotective by their ability to reduce ventricular arrhythmias and cardiomyocyte apoptosis.[1,2] Grapes appear to possess the same degree of cardioprotective properties. The cardioprotective properties of grapes may be attributed at least in part to the presence of several unique polyphenolic compounds like resveratrol and proanthocyanidins.

ACKNOWLEDGMENTS

This study was supported in part by NIH HL 22559, HL 33889, HL 34360, and HL 56803 and a grant from the California Table Grape Commission.

REFERENCES

1. Sato, M., G. Maulik, P.S. Ray, *et al.* 1999. Cardioprotective effects of grape seed proanthocyanidin against ischemic reperfusion injury. J. Mol. Cell. Cardiol. **31:** 1289–1297.
2. Sato, M., D. Bagchi, A. Tosaki & D.K. Das. 2001. Grape seed proanthocyanidin reduces cardiomyocyte apoptosis by inhibiting ischemia/reperfusion-induced activation of JNK-1 and c-Jun. Free Radical Biol. Med. **16:** 729–739.
3. Ray, P.S., G. Maulik, G.A. Cordis, *et al.* 1999. The red wine antioxidant resveratrol protects isolated rat hearts from ischemia reperfusion injury. Free Radical Biol. Med. **27:** 160–169.
4. Engelman, D., M. Watanabe, R.M. Engelman, *et al.* 1995. Hypoxic preconditioning preserves antioxidant reserve in the working rat heart. Cardiovasc. Res. **29:** 133–140.
5. Yoshida, T., N. Maulik, R.M. Engelman, *et al.* 1997. Glutathione peroxidase knockout mice are susceptible to myocardial ischemia reperfusion injury. Circulation **96**(II): 216–220.
6. Maulik, G., N. Maulik, V. Bhandari, *et al.* 1997. Evaluation of antioxidant effectiveness of a few herbal plants. Free Radical Res. **27:** 221–228.
7. Das, D.K. & N. Maulik. 1995. Protection against free radical injury in the heart and cardiac performance. *In* Exercise and Oxygen Toxicity. C.K. Sen, L. Packer, O. Hanninen, Eds.: 359–388. Elsevier Science. Amsterdam.
8. Das, D.K. & N. Maulik. 1994. Evaluation of antioxidant effectiveness in ischemia reperfusion tissue injury. Methods Enzymol. **233:** 601–610.
9. Tosaki, A., D. Bagchi, D. Hellegouarch, *et al.* 1993. Comparisons of ESR and HPLC methods for the detection of hydroxyl radicals in ischemic/reperfused hearts. A relationship between the genesis of oxygen-free radicals and reperfusion-induced arrhythmias. Biochem. Pharmacol. **45:** 961–969.
10. Saija, A., M. Scalese, M. Laiza, *et al.* 1995. Flavonoids as antioxidant agents: importance of their interaction with biomembranes. Free Radical Biol. Med. **19:** 481–486.
11. Das, D.K. & N. Maulik. 1994. Evaluation of antioxidant effectiveness in ischemia reperfusion tissue injury. Methods Enzymol. **233:** 601–610.
12. Das, D.K. & N. Maulik. 1995. Protection against free radical injury in the heart and cardiac performance. *In* Exercise and Oxygen Toxicity. C.K. Sen, L. Packer, O. Hanninen, Eds.: 359–388. Elsevier Science. Amsterdam.

Cardioprotective Abilities of White Wine

JIANHUA CUI,[a] ARPAD TOSAKI,[a] GERALD A. CORDIS,[a]
ALBERTO A.E. BERTELLI,[b] ALDO BERTELLI,[c]
NILANJANA MAULIK,[a] AND DIPAK K. DAS[a]

[a]*Cardiovascular Research Center, University of Connecticut School of Medicine, Farmington, Connecticut 06030-1110, USA*

[b]*Institute of Anatomy and* [c]*Department of Pharmacology, University of Milan, Milan, Italy*

ABSTRACT: To study if white wines, like red wine, can also protect the heart from ischemia reperfusion injury, ethanol-free extracts of three different white wines (WW1, WW2 and WW3) (100 mg/100 g body weight) were given orally to Sprague Dawley rats (200 g body weight) for three weeks. Control rats were given water only for the same period of time. After three weeks, rats were anesthetized and sacrificed, and the hearts excised for the preparation of isolated working rat heart. All hearts were subjected to 30 min global ischemia followed by two hours of reperfusion. The results demonstrated that among the three different white wines, only WW2 showed cardioprotection as evidenced by improved post-ischemic ventricular recovery compared to control. The amount of malonaldehyde production in white wine–fed rat hearts were lower compared to that found in control hearts indicating reduced formation of the reactive oxygen species. *In vitro* studies using chemiluminescence technique revealed that these white wines scavenged both superoxide anions and hydroxyl radicals. The results of our study demonstrated that only WW2 white wine provided cardioprotection as evidenced by the improved the post-ischemic contractile recovery and reduced myocardial infarct size. The cardioprotective effect of this white wine may be attributed, at least in part, from its ability to function as an *in vivo* antioxidant.

KEYWORDS: cardioprotection; white wine; antioxidants

INTRODUCTION

A large number of epidemiological studies suggest that mild-to-moderate consumption of red wine is associated with a reduced incidence of mortality and morbidity from coronary heart disease.[1] This gave rise to what is now popularly termed the French Paradox.[2] Amidst the intense interest generated by the various findings that support red wine's being a plausible preventive intervention against coronary heart disease, we were interested in what role, if any, it plays in the face of an acute ischemic insult and whether it can exert any beneficial effects in the context of ischemia reperfusion injury. The results of our study demonstrated potent cardioprotective abilities of red wine and the cardioprotection was attributed at least in part to

Address for correspondence: Dipak K. Das, Ph.D., Cardiovascular Research Center, University of Connecticut School of Medicine, Farmington, CT 06030-1110, USA. Voice: 860-679-3687; fax: (860) 679-4606.
ddas@neuron.uchc.edu

the polyphenolic antioxidants, especially resveratrol and proanthocyanidins present in the red wine.[3]

Red wine is rich in several typical polyphenols including resveratrol (3,5,4′-trihydroxystilbene), a naturally occurring antioxidant phytoalexin and proanthocyanidins. These polyphenols which are mostly present in the seeds and skins of the grapes possess many biological functions including protection against atherosclerosis, lipid peroxidation and cell death.[4] Both proanthocyanidins and resveratrol possess potent cardioprotective properties.[5,6] Grape seed proanthocyanidins have recently found to potentiae anti-death signaling events.[11,12]

White wine, on the other hand, contains many of the polyphenolic antioxidants which are also present in the red wine. However, compared to red wine, white wine contains minimal amount of proanthocyanidins, and resveratrol is present in only insignificant amount in white wine. On the other hand, white wines are rich in certain polyphenols. Whether these polyphenols present in white wine contribute to cardioprotection is not known. The present study was undertaken to determine if white wine can also provide cardioprotection similar to red wine.

MATERIALS AND METHODS

White Wine

Three different white wines (WW1 [Verduzzo Friulano Isonzo, 1999], WW2 [Tocai Friulano Collio 1999], and WW3 [Passito di Tocai Friulano Collio 1999]) from Cantina Produttori Cormons, Italy, were chosen for this study. Ethanol-free extracts of the wines were prepared by vacuum evaporation. Wine was frozen at −40°C for 12 hours, and then lyophilized bringing the product from −35°C to −5°C under vacuum (−180mm Hg) for eight hours. After 12 hours, the temperature was raised from 5°C to 35°C at 180mm Hg for 24 hours. The jelly-like resulting product was harvested and transferred into dark vials. The filling operation was performed under UV light to avoid contamination.

Animals

Male Sprague-Dawley rats (275–300g, Harlan Sprague-Dawley) were used in this study. They were provided with food and water *ad libitum* until the start of the experimental procedure. Rats were randomly assigned to one of four groups, A (Control), B (white wine #1, WW1), C (white wine #2, WW2) and D (white wine #3, WW3). The animals were fed *ad libitum* with free access to water. They were given orally (once a day) either water (control) or the white wine extracts for three weeks. All animals received care in compliance with the principles of laboratory animal care formulated by the National Society for Medical Research and *Guide for the Care and Use of Laboratory Animals* prepared by the National Academy of Sciences and published by the National Institute of Health (publication no. NIH 80-23, revised 1978).

The perfusion buffer used in this study consisted of a modified Krebs-Henseleit bicarbonate buffer (KHB) (in mM: 118 NaCl, 4.7 KCl, 1.2 $MgSO_4$, 1.2 KH_2PO_4, 25 $NaHCO_3$, 10 Glucose and 1.7 $CaCl_2$, gassed with 95% O_2-5% CO_2, pH 7.4) which was maintained at a constant temperature of 37°C and was gassed continuously for the entire duration of the experiment. Rats were anesthetized with sodium pentobarbital (80 mg/kg b.w., i.p. injection, Abbott Laboratories), anticoagulated with heparin sodium (500 IU/kg b.w., i.v. injection, Elkins-Sinn Inc.). After ensuring sufficient depth of anesthesia, thoracotomy was performed, hearts were excised and immersed in ice-cold perfusion buffer. Both aorta and pulmonary veins were cannulated as quickly as possible and hearts were perfused in the retrograde Langendorff mode at a constant perfusion pressure of 100cm H_2O for a stabilization period of 10 minutes. Then the circuit was switched to the antegrade working mode as described previously.[9] The hearts pumped against an afterload of 100cm H_2O perfused with a constant preload of 17cm H_2O. Aortic pressure was measured using a Gould P23XL pressure transducer connected to a sidearm of the aortic cannula, the signal was amplified using a Gould 6600 series signal conditioner and monitored on a real-time data acquisition and analysis system (CORDAT II.). Heart rate, developed pressure (defined as the difference of the maximum systolic and diastolic aortic pressures) and the first derivative of developed pressure were all derived or calculated from the continuously obtained pressure signal. Aortic flow was measured using a calibrated flowmeter (Gilmont Instruments, Inc.) and coronary flow was measured by timed collection of the coronary effluent dripping from the heart. At the end of 10 min, after the attainment of steady state cardiac function, baseline functional parameters were recorded and coronary effluent collected for biochemical assays. The circuit was then switched back to the retrograde mode and hearts were perfused with either KHB (Group A), Red wine extract (Group B) or *trans*-resveratrol (Group C), the latter at a concentration of 10μM, for a duration of 15 minutes. At the end of this period, hearts were subjected to global ischemia for 30 min followed by a two-hour reperfusion. The first 10 minutes of reperfusion were in the retrograde mode to allow for post-ischemic stabilization and recordings were made in the antegrade working mode at 15 minutes, 30 minutes, 60 minutes, and 120 minutes into reperfusion to allow for assessment of functional recovery. Coronary effluent samples for biochemical assays were also collected at the above time points and stored at −20°C.

Hearts to be used for infarct size calculations ($n = 6$ per group) were taken upon termination of the experiment and immersed in 1% Triphenyl tetrazolium solution in Phosphate buffer (Na_2HPO_4 88mM, NaH_2PO_4 1.8mM) preheated to 37°C for 20 minutes and then stored at −70°C for later processing. Frozen hearts (including only ventricular tissue) were sliced in a plane perpendicular to the apico-basal axis into approximately mm thick sections, blotted dry, placed in between microscope slides and scanned on a single pass flat bed scanner (Hewlett-Packard Scanjet 5p).[10] Using the NIH Image 1.6.1 image processing software, each image was subjected to background subtraction, contrast enhancement and grayscale conversion for improved clarity and distinctness. Total as well as infarct zones of each slice were traced and the respective areas were calculated in terms of pixels and cm^2, the respective volumes being obtained in cm^3. The weight of each slice was then recorded to facilitate the expression of total and infarct masses of each slice in grams. The total and infarct masses of each heart were then calculated by taking a weighted average. Infarct size was taken to be the percent infarct mass of total mass for any one heart.

White Wine Composition with HPLC

The white wine extracts were dissolved in a mixture of water/ethanol (95:5, v/v). The clear solutions were separated with High Performance Liquid Chromatography (HPLC).[11] Separation of the phenolic compounds was achieved on a Hyersil-2-column with 3μ material at room temperature. The analytes were monitored with a Spectroflow 757 variable wavelength detector at 280nm. The compounds were identified and quantitated against authentic standards.

In Vitro *Superoxide and Hydroxyl Radical Assay*

Since white wines contain several polyphenolic antioxidants, we examined if these white wine extracts could directly scavenge reactive oxygen species. Free radical scavenging activities were determined by chemiluminescence technique using an EG & G Berthold Lumat LB 9507 luminometer.[12] Superoxide anions were generated by the action of xanthine oxidase (7mU) on xanthine (100μM) in a 500μL reaction mixture containing 10mM sodium carbonate, pH 9.0 and 28μM luminol. The reaction was started by luminometer injection of xanthine. The hydroxyl radical assay mixture contained xanthine oxidase (0.6 mU), 10μM sodium carbonate, pH 9.0, 28μM luminol, 100μM xanthine, 100μM EDTA, and 100μM ferric chloride. The scavenging activities were compared with 10 mU/ml SOD for superoxide anion and dimethyl thiorea for hydroxyl radical.

STATISTICAL ANALYSIS

For statistical analysis, a two-way analysis of variance (ANOVA) followed by Scheffe's test was first carried out to test for any differences between groups. If differences were established, the values were compared using Student's *t*-test for paired data. The values were expressed as mean ± SEM. The results were considered significant for $p < 0.05$.

RESULTS

Effects of White Wine on Cardioprotection

The effects of the white wine extract on post-ischemic left ventricular function are summarized in TABLE 1. All data are presented as the mean ± the standard error of mean. As expected, except for heart rate and coronary flow, postischemic ventricular functions were depressed in all four groups. For example, after 60 min of reperfusion AF, LVDP, and LVdp/dt were 15.2 ± 2.1 mL/min, 37.3 ± 2.7 mm Hg, and 1,155 ± 137 mm Hg/min, respectively, in the control group compared to those at the baseline levels (52.8 ± 2.4 mL/min, 100.6 ± 1.8 mm Hg, and 3687 ± 201 mm Hg/sec, respectively). Although similar trends were noticed for other groups, these values were significantly higher for the WW2 group (26.2 ± 3.0 mL/min, 71.8 ± 1.2 mm Hg, and 1796 ± 211 mm Hg/min, respectively) compared to control group and WW1 and WW3 groups. Thus, WW2 displayed significant recovery of post-ischemic myocardial function as compared to control group. This was also evidenced by significantly higher values in the aortic flow readings throughout the reperfusion period, which were maintained closer to baseline values. Developed pressure and LVdp/dt were also markedly higher in these groups from R-30 onwards.

In this particular study, hearts were arrested by global ischemia for 30 min, therefore, whole ventricle was considered as the area of risk. Control hearts subjected to ischemia and reperfusion had an infarct size of 32.7 ± 3.1%. Percent infarct size was

TABLE 1. Effects of white wines on postischemic ventricular function and myocardial infarction[a]

	Group	Baseline	Reperfusion at 30 min	Reperfusion at 60 min	Reperfusion at 120 min
Heart rate (beats/min)	Control	313 ± 7	287 ± 6	279 ± 6	277 ± 7
	WW1	315 ± 8	285 ± 7	278 ± 7	272 ± 8
	WW2	312 ± 7	290 ± 9	280 ± 8	285 ± 8
	WW3	317 ± 6	295 ± 7	284 ± 6	280 ± 7
Aortic flow (mL/min)	Control	52.3 ± 2.6	11.7 ± 1.1	9.5 ± 1.6	7.6 ± 1.5
	WW1	49.1 ± 2.8	11.0 ± 0.9	9.6 ± 1.1	7.9 ± 1.3
	WW2	52.9 ± 3.1	12.0 ± 1.8	10.2 ± 1.5	8.8 ± 2.0
	WW3	51.7 ± 2.5	24.8 ± 2*	28.3 ± 1.8*	22.9 ± 1.4*
Coronary flow (mL/min)	Control	26.5 ± 1.1	17.6 ± 0.5	17.1 ± 1.5	16.3 ± 1.3
	WW1	26.0 ± 1.6	18.9 ± 1.8	17.9 ± 0.8	16.8 ± 1.1
	WW2	27.0 ± 1.3	17.9 ± 0.8	15.9 ± 1.2	16.0 ± 0.8
	WW3	26.3 ± 1.1	16.9 ± 1.2	19.2 ± 1.6	18.2 ± 1.1
LVDP (left ventricle developed pressure) (mm Hg)	Control	107 ± 5	34 ± 5	36 ± 3	23 ± 5
	WW1	111 ± 8	39 ± 6	43 ± 7	29 ± 4
	WW2	101 ± 8	38 ± 3	42 ± 5	31 ± 2
	WW3	102 ± 8	60 ± 7*	68 ± 8*	49 ± 6*
dp/dt (mm Hg/sec)	Control	3,469 ± 148	1,650 ± 116	1,759 ± 104	1,568 ± 65
	WW1	3,601 ± 109	1,800 ± 174	1,834 ± 132	1,589 ± 93
	WW2	3,437 ± 97	1,738 ± 185	1,679 ± 105	1,492 ± 93
	WW3	3,581 ± 98	2,012 ± 88*	2,194 ± 113*	2,000 ± 123*
Infarct size (%)	Control				34.5 ± 1.5
	WW1				33.6 ± 1.1
	WW2				31.4 ± 1.0
	WW3				23.5 ± 0.8*

[a]Results are expressed as mean ± S.E.M. of six rats per group; ww, white wine; $*p < 0.05$ compared with control.

significantly reduced for WW2 only (21.5 ± 1.5). There were no differences in infarct size between the control group and WW1 and WW3 groups.

In Vitro *Radical Scavenging Assay*

To further confirm that lowering of oxidative stress in the white wine–fed rat hearts was indeed due to oxygen free radical scavenging activities of white wines, the scavenging activities were measured *in vitro*. The results for the chemiluminescence response of the chemically generated superoxide and hydroxyl radicals in the presence of white wines and luminol are shown in FIGURE 1. Inhibition efficiency was calculated as compared to that of SOD for superoxide anion and of DMTU for hydroxyl radicals. The results demonstrate that all white wines were able to scavenge both superoxide and hydroxyl radicals efficiently. Again, WW2 was more efficient in scavenging both superoxide and hydroxyl radicals compared to other white wines.

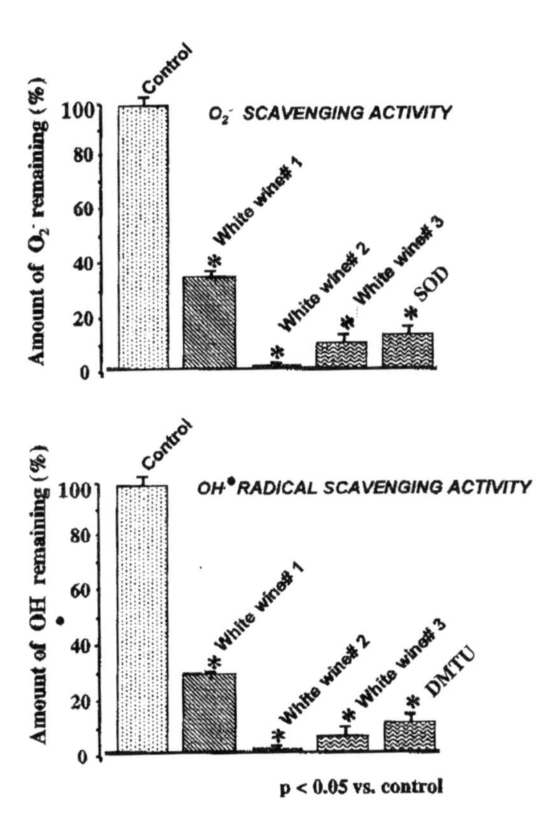

FIGURE 1. Superoxide anion and hydroxyl radical scavenging activities of white wines. Superoxide anions and hydroxyl radicals were generated chemically, and their scavenging was determined with white wines. Scavenging activities were compared against SOD for superoxide anion and DMTU for hydroxyl radicals.

Polyphenolic Composition of White Wines

The phenolic composition of the white wines was determined by HPLC. As shown in TABLE 2, the tyrosol content of white wines 1 and 2 were identical. WW3 had about one-third of the tyrosol present in WW1 and WW2. White wine 2 was unique in the sense that it was the only white wine that contained significant amount of caffeic acid, catechin, and resveratrol.

DISCUSSION

The results of this study demonstrated that white wine can also protect the hearts from detrimental effects of ischemia reperfusion injury as evidenced by improved post-ischemic ventricular function and reduced myocardial infarction. Among the three different white wines used in the study, only one group, white wine 2, was found to be cardioprotective. In contrast, all the white wines reduced the amount of oxidative stress in the heart as indicated by decreased MDA formation. These wines were also found to be highly effective in directly scavenging superoxide and hydroxyl radicals. The results indicate that reduction of the oxidative stress may not be the only criteria for the wines to be cardioprotective. The phenolic composition of WW2 was completely different from those found in the other two groups of wines. Although it was not shown, the presence of significant amount of caffeic acid, catechin and resveratrol in the WW2 group could play a role in its ability to protect the myocardium from ischemic reperfusion injury.

The polyphenol fraction of wine includes phenolic acids (p-coumaric, cinnamic, caffeic, gentisic, ferulic, and vanillic acids), trihydroxy stilbenes (resveratrol and polydatin), and the flavonoids (catechin, epicatechin and quercetin).[13] The free radical–scavenging property of various flavonoids has been observed in relation to hydroxyl and superoxide as well as peroxyl and alkoxyl radicals.[14] Flavonoids also possess the ability to chelate iron ions known to catalyze many free radical–generating processes. This property probably also contributes to their antioxidant effectiveness. Antioxidants have long been known to protect against the damaging effects of free radical–mediated tissue injury especially ischemia reperfusion injury of the heart and other organs.[15] Substantial evidence exists to support the notion that

TABLE 2. Composition of white wines

	WW1	WW2	WW3
Tyrosol	28.1 ± 5.4	30.4 ± 1.9	10.6 ± 1.2
Caftaric acid (as caffeic acid)	<0.1	13.6 ± 0.9	<0.1
Catechin	17.4 ± 2.8	55.4 ± 2.4	<0.5
t-Resveratrol	<1.0	1.04	<1.0
Quercitin	<0.5	<0.5	<0.5

NOTES: The white wine extracts were dissolved in a mixture of water/ethanol (95:5, v/v). The clear solutions were separated with HPLC and detected at 280nm. A Hyersil-2-column with 3 μ material was used. The compounds were identified and quantitated against authentic standards.

ischemia and reperfusion generate oxygen free radicals which contribute to the pathogenesis of ischemic reperfusion injury.[16] The presence of reactive oxygen species were confirmed directly by estimating free radical formation and indirectly by assessing lipid peroxidation and DNA breakdown products.[15,16] Virtually, all the biomolecules including lipids, proteins and DNA molecules are the potential targets for attack by the ·OH radical. Thus, free radical–scavenging ability of white wine 2 may at least in part be responsible for its cardioprotective properties. However, it should be noted that all the white wines effectively scavenged hydroxyl radicals, although the antioxidant action was found to be the highest for the WW2. This would suggest that there is some uniqueness in the WW2. Indeed, WW2 contained significant amount of resveratrol, caffeic acid and catechin compared to those present in other white wines. Resveratrol, a stilbene polyphenol, has recently been found to protect the hearts from ischemic reperfusion injury by its ability to reduce both necrosis and apoptosis.[5] Proanthocyanidins were shown to be cardioprotective by its ability to reduce ventricular arrhythmias and cardiomyocyte apoptosis.[6–8] Catechin is also cardioprotective and mostly found in green tea.[17] The role of caffeic acid in cardioprotection is not known. It is also possible there is as yet unknown compound present in white wine 2 which contributes to cardioprotective properties of this wine.

To the best of our knowledge, this is the first report to indicate the cardioprotective effects of white wine. The cardioprotective effects appear to be mediated by some unique antioxidant present in this wine which may include resveratrol, catechin, and caffeic acid. Further research is needed to determine the factors responsible for cardioprotection in the white wines.

ACKNOWLEDGMENTS

This study was supported by NIH HL 34360, HL 22559, HL 33889, HL 56803, and a Grant from the California Table Grape Association.

REFERENCES

1. Fuhrman, B., A. Lavy & M. Aviram. 1995. Consumption of red wine with meals reduces the susceptibility of human plasma and low-density lipoprotein to lipid peroxidation. Am. J. Clin. Nutr. **61:** 549–554.
2. Renaud, S. & M. De Lorgeril. 1992. Wine, alcohol, platelets and the French Paradox for coronary heart disease. Lancet **339:** 1523–1526.
3. Sato M., P. Ray, G. Maulik, *et al.* 2000. Myocardial protection with red wine extract. J. Cardiovasc. Pharmacol. **35:** 263–268.
4. Bertelli, A.A.E., L. Giovannini, D. Giannessi, *et al.* 1995. Antiplatelet activity of synthetic and natural resveratrol in red wine. Int. J. Tissue Reac. **17:** 1–3.
5. Ray, P.S., G. Maulik, G.A. Cordis, *et al.* 1999. The red wine antioxidant resveratrol protects isolated rat hearts from ischemia reperfusion injury. Free Radical Biol. Med. **27:** 160–169.
6. Sato, M., G. Maulik, P.S. Ray, *et al.* 1999. Cardioprotective effects of grape seed proanthocyanidin against ischemic reperfusion injury. J. Mol. Cell. Cardiol. **31:** 1289–1297.
7. Pataki, T., I. Bak, P. Kovacs, *et al.* 2002. Grape seed proanthocyanidins reduced ischemia/reperfusion-induced injury in isolated rat hearts. Am. J. Clin. Nutrition. **75:**.

8. SATO, M., D. BAGCHI, A. TOSAKI & D.K. DAS. 2001. Grape seed proanthocyanidin reduces cardiomyocyte apoptosis by inhibiting ischemic/reperfusion-induced activation of Jnk-1 and c-Jun. Free Radical Biol. Med. **16:** 729–739.
9. ENGELMAN, D., M. WATANABE, R. ENGELMAN, *et al.* 1995. Hypoxic preconditioning preserves antioxidant reserve in the working rat heart. Cardiovasc. Res. **29:** 133–140.
10. YOSHIDA, T., N. MAULIK, R.M. ENGELMAN, *et al.* 1997. Glutathione peroxidase knockout mice are susceptible to myocardial ischemia reperfusion injury. Circulation **96**(II): 216–220.
11. ANDLAUER, W., C. STUMPF & P. FURST. 2000. Influence of the acetification process on phenolic compounds. J. Agric. Food Chem. **48:** 3533–3536.
12. MAULIK, G., N. MAULIK, V. BHANDARI, *et al.* 1997. Evaluation of antioxidant effectiveness of a few herbal plants. Free Radical Res. **27:** 221–228.
13. SOLEAS, G.J., E.P. DIAMANDIS & D.M. GOLDBERG. 1997. Wine as a biological fluid: history, production and role in disease prevention. J. Clin. Lab. Anal. **11:** 287–313.
14. RICE-EVANS, C.A., N. MILLER, P.G. BOLWELL, *et al.* 1995. The relative antioxidant activities of plant-derived polyphenolic flavonoids. Free Radical Res. **22:** 375–383.
15. DAS, D.K. & N. MAULIK. 1994. Evaluation of antioxidant effectiveness in ischemia reperfusion tissue injury. Methods Enzymol. **233:** 601–610.
16. TOSAKI, A., D. BAGCHI, D. HELLEGOUARCH, *et al.* 1993. Comparisons of ESR and HPLC methods for the detection of hydroxyl radicals in ischemic/reperfused hearts. A relationship between the genesis of oxygen-free radicals and reperfusion-induced arrhythmias. Biochem. Pharmacol. **45:** 961–969.
17. HERTOG, M.G., E.J. FESKENS, P.C. HOLLMAN & M.B. KATAN. 1993. Dietary antioxidant flavonoids and risk of coronary heart disease: the Zutphen Elderly study. Lancet **342:** 1007–1011.

Doctor, Should I Have a Drink?

An Algorithm for Health Professionals

ROGER R. ECKER[a] AND ARTHUR L. KLATSKY[b]

[a]*Summit Medical Center, Oakland, California 94609, USA*

[b]*Kaiser Permanente Medical Center, Oakland, California, USA*

KEYWORDS: alcohol; coronary heart disease

Patients are deluged with media advice, sometimes accurate but often commercial, self-serving and confusing. They may also seek or receive advice from the internet, nurses, pharmacists, dieticians, and practitioners of non-traditional medicine. The physician is often the person best qualified to synthesize the relevant information and give sound advice to his or her patient. Prevention as well as treatment of disease has always been a prime objective. Enhancement of the quality of life is also important, and often crucial to compliance.

Physicians find themselves between Scylla and Charybdis regarding alcohol consumption, conflicted by information about benefits of moderate drinking and the manifest misery which alcoholism causes. A judgment about who might benefit and who might be harmed requires a careful history and a considered explanation to the patient. Many decide simply to ignore the subject. Others choose a "one size fits all" course, advising reduced drinking or abstinence because of alcohol's potential for harm. Such approaches might potentially be harmful to the health of some individuals. They are inadequate in light of current epidemiological data about health effects of alcohol. We have devised an algorithm (see FIGURE 1) to assist health care professionals in advising patients about drinking.

Many studies have shown reduction of risk of fatal and nonfatal cardiovascular disease, mostly coronary heart disease (CHD) and ischemic stroke, in light/moderate drinkers.[1–3] This reduced risk has been observed in a wide variety of patient populations, including those with diabetes, hypertension and prior myocardial infarction.[1,2] Plausible mechanisms for alcohol's benefit include increased high-density lipoprotein cholesterol (HDL-C), several antithrombotic actions, and increased insulin sensitivity. Non-alcohol ingredients in some alcoholic beverages, especially red wine, offer hypothetical additional benefit, but observational data are conflicting about the role of beverage choice (wine, liquor or beer). It appears likely that all alcoholic beverage types decrease CHD risk.[1–3] The optimal amount of alcohol for lowest risk of CHD or death is not entirely clear, but net harm is seen in some studies when in excess of two drinks/day in men and above 1 drink/ day in women. A special consideration in women is evidence that moderate drinking increases her risk of breast cancer.[4]

Address for correspondence: Roger R. Ecker, M.D., Summit Medical Center, 301 Harbor Light Rd, Alameda, CA 94501-5965, USA. Voice: 510-522-6622; fax: 510-522-6616.
r.ecker@att.net

Ann. N.Y. Acad. Sci. **957: 317–320 (2002).**

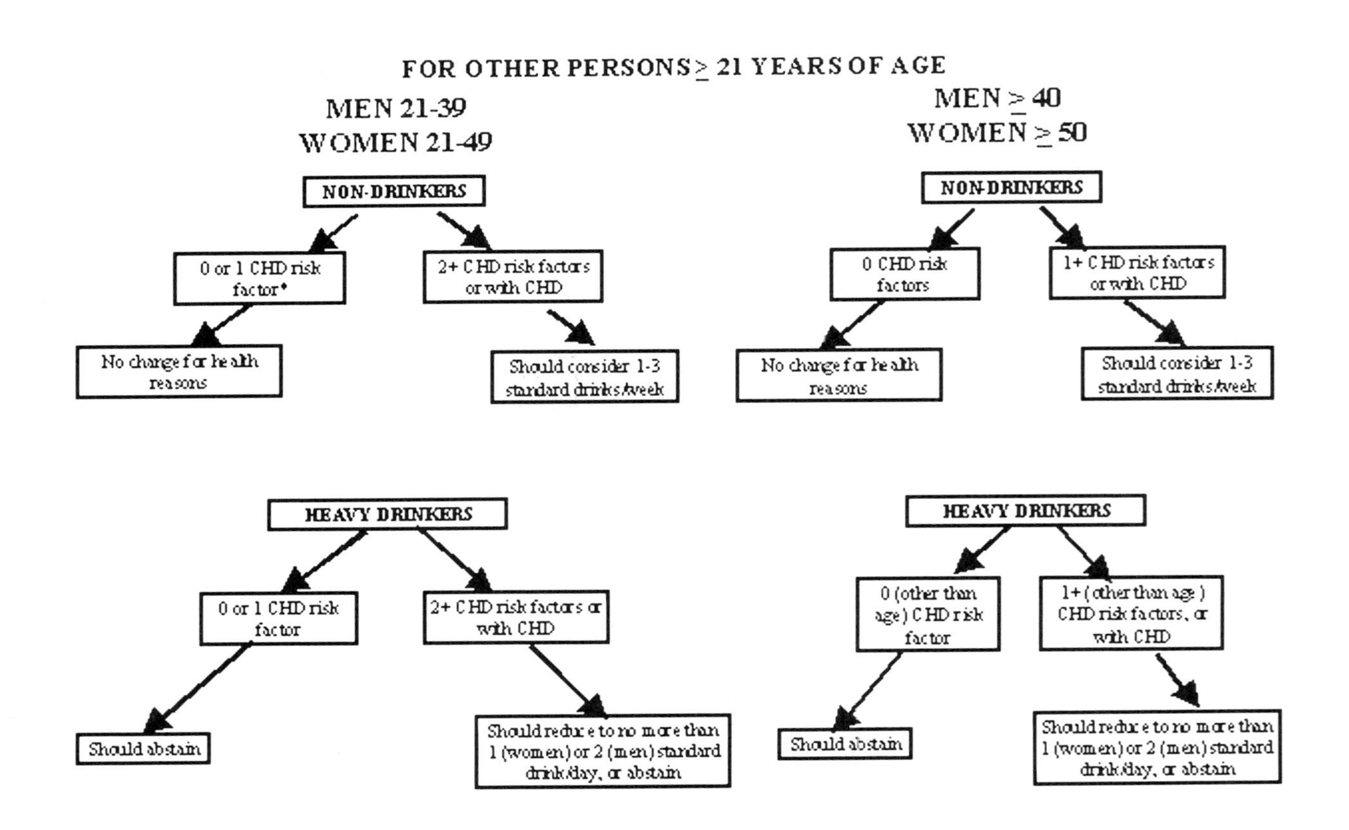
FOR OTHER PERSONS ≥ 21 YEARS OF AGE
MEN 21-39
WOMEN 21-49
NON-DRINKERS
0 or 1 CHD risk factor*
2+ CHD risk factors or with CHD
No change for health reasons
Should consider 1-3 standard drinks/week
HEAVY DRINKERS
0 or 1 CHD risk factor
2+ CHD risk factors or with CHD
Should abstain
Should reduce to no more than 1 (women) or 2 (men) standard drink/day, or abstain
MEN ≥ 40
WOMEN ≥ 50
NON-DRINKERS
0 CHD risk factors
1+ CHD risk factors or with CHD
No change for health reasons
Should consider 1-3 standard drinks/week
HEAVY DRINKERS
0 (other than age) CHD risk factor
1+ (other than age) CHD risk factors, or with CHD
Should abstain
Should reduce to no more than 1 (women) or 2 (men) standard drink/day, or abstain

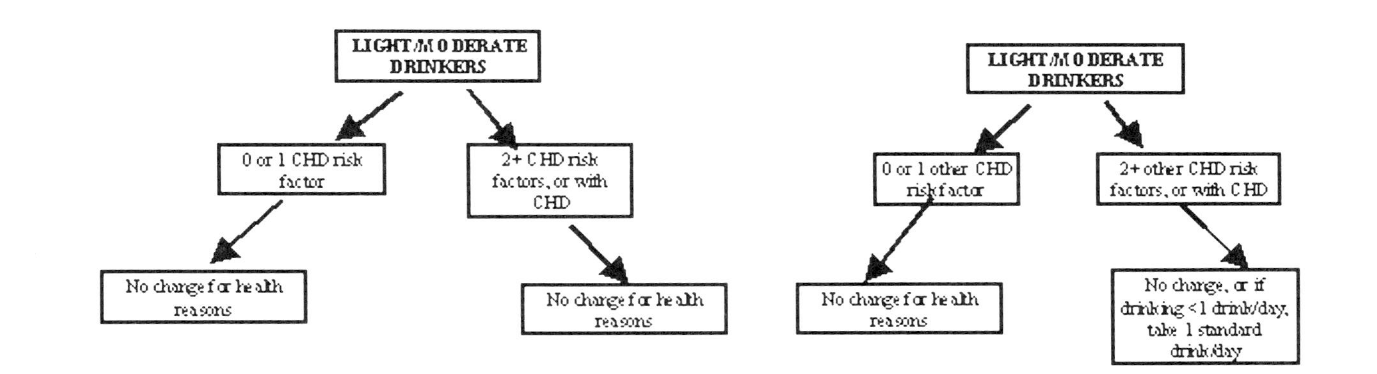

FIGURE 1. An algorithm for health professionals for advice on moderate alcohol drinking for health. **Exclusions:** (1) persons less than 21 years of age; (2) pregnant women; (3) nondrinkers avoiding alcohol owing to family history (FH) of alcoholism or religious/moral reasons; (4) nondrinkers with personal history of alcohol problems; (5) nondrinkers with known organ damage from alcohol, any chronic liver disease, or evident genetic risk of breast/ovarian cancer. "Moderate" is defined as not more that one standard drink/day for women and not more than two standard drinks/day for men. "Heavy" is in excess of three drinks/day for men and in excess of two drinks/day for women. A standard drink is a 12 oz. bottle/can of beer, a 4 oz glass of table wine, or a 1 1/2 oz of distilled spirits. *Coronary Heart Disease (CHD) Risk Factors (RF)—National Cholesterol Education Program (NCEP) Guidelines: (1) FH of CHD, parent or sibling ($M < 55$ or $F < 65$); (2) smoker; (3) hypertension; (4) diabetes; (5) total cholesterol at least 200; HDL cholesterol less than 35 (if greater than 60, subtract one RF); (7) age (already included).

What advice should be given the patient? Cessation of smoking, control of weight, hypertension and diabetes, lipid management by diet, exercise and drugs remain the cornerstones of CHD prevention. After these basics, alcohol should be considered. For persons at above average CHD risk, alcohol abstinence, except for special reasons, is not best. The special reasons include high risk of alcoholism (always in need of individual assessment), pregnancy, liver disease, increased genetic risk of breast cancer,[4] certain medications, and religious/moral reasons for abstinence. Heavy drinkers judged to be addicted should be counseled to abstain. Some heavy drinkers who can control their drinking might be advised to decrease to one or two drinks per day. Non-drinkers who were former light drinkers with no addiction problems should be advised to resume drinking for health reasons if they have at least two risk factors for CHD and no reason for exclusion, and those with 0–1 CHD risk factors should be told the options and to drink at their discretion. Non-drinkers older than 40 years (men) or 50 (women) who have 1+ CHD risk factors should be advised to consider taking 1–3 standard drinks per week, as should younger men and women who have at most 2+ CHD risk factors. Some will feel that these recommendations are controversial, but we believe them justified by the epidemiological data.

We recommend taking the patient through the algorithm (FIG. 1) either in person or utilizing the patient's history to reach a decision on advice. Note exclusions, definitions of levels of drinking, and the definition of a standard drink. How and where one drinks are also important. Drinking before driving or operating machinery is clearly unwise. Drinking alcohol with food may enhance its healthful effects. Reduction of CHD is optimal when one or two standard drinks are consumed on five or six days of the week.[5]

We believe that we have reached a time when physicians and other health professionals can offer their patients objective, sound, evidence-based information on alcohol and health.

REFERENCES

1. PAOLETTI, R., A.L. KLATSKY, A. POLI & S. ZAKHARI, Eds. 2000. Moderate Alcohol Consumption and Cardiovascular Disease. Kluwer Academic Publishers. Dordrecht, the Netherlands.
2. KLATSKY, A.L. 2001. Editorial: Should patients with heart disease drink alcohol? JAMA **285:** 2004–2006.
3. CORRAO, G., L. RUBBIATI, V. BAGNARDI, *et al.* 2000. Alcohol and coronary heart disease: a meta-analysis. Addiction **95:** 1505–1523.
4. LONGNECKER, M.P. 1994. Alcoholic beverage consumption in relation to risk of breast cancer: meta-analysis and review. Cancer Causes Control **5:** 73–82.
5. MCELDUFF, P., & A.J. DOBSON. 1997. How much alcohol and how often? Population based case-control study of alcohol consumption and risk of a major coronary event. Br. Med. J. **314:** 1159–1164.

Preservation of Paraoxonase Activity by Wine Flavonoids

Possible Role in Protection of LDL from Lipid Peroxidation

BIANCA FUHRMAN AND MICHAEL AVIRAM

Lipid Research Laboratory, Technion Faculty of Medicine, The Rappaport Family Institute for Research in the Medical Sciences, and Rambam Medical Center, Haifa, Israel

ABSTRACT: Paraoxonase is an esterase physically associated with HDL, and its activity has been shown to be inversely related to the risk of cardiovascular diseases. We have shown that paraoxonase can hydrolyze specific lipid peroxides in oxidized lipoproteins and in atherosclerotic lesions. Paroxonase was shown to be inactivated by oxidative stress. Consumption of wine flavonoids was shown to preserve paraoxonase activity by reducing the oxidative stress in apolipoprotein E–deficient mice, thereby contributing to paraoxonase hydrolytic activity on lipid peroxides in oxidized lipoproteins and atherosclerotic lesions.

KEYWORDS: red wine; flavonoids; polyphenols; antioxidants; paraoxonase; atherosclerosis

INTRODUCTION

Human serum paraoxonase activity has been shown to be inversely related to the risk of cardiovascular disease.[1–3] HDL-associated serum paraoxonase (PON 1) has recently been shown to protect LDL, as well as the HDL itself, against oxidation induced by copper ions or by free radicals generators,[4,5] and this effect could be related to the hydrolysis of the lipoproteins' oxidized lipids.[6] Thus, an effort is being made to identify natural dietary components that are potent antioxidants with the ability to preserve or increase PON 1 activity and as a result to promote hydrolysis of LDL-associated lipid peroxides and thereby attenuate the progression of atherosclerosis.

The intake of flavonoids, which constitute one of the largest groups of antioxidant phytochemicals, was shown to be inversely related to morbidity and mortality from coronary heart disease,[7] and this phenomenon could be associated with the inhibition of LDL oxidation.[8,9] Red wine is a rich dietary source for polyphenols, such as the flavonols quercetin and myricetin, and also the 3-flavanols catechin and epi(gallo) catechin, and it was shown to inhibit LDL oxidation in humans[10] and to attenuate development of atherosclerosis in animal models.[11]

In the present study we investigated the effect of red wine consumption and its polyphenols quercetin and catechin on the serum paraoxonase activity of the atherosclerotic apolipoprotein E–deficient (E^0) mice.

Address for correspondence: Dr. Bianca Fuhrman, The Lipid Research Laboratory, Rambam Medical Center, Haifa, Israel, 31096. Voice: 972-4-8295278; fax: 972-4-8520076.
fuhrman@tx.technion.ac.il

EXPERIMENTAL DESIGN

Apolipoprotein E–deficient mice (E^0 mice) were kindly provided by Dr. Jan Breslow, the Rockefeller University, New York City. At 4 weeks of age, 40 E^0 mice were assigned randomly to four groups, 10 mice in each group. The mice were supplemented for six weeks in their drinking water with:

1. placebo (alcoholized water, 1.1% alcohol);
2. red wine (Cabernet Sauvignon) containing 1.1% alcohol, 0.5mL/day/mouse (50μg equivalents of catechin);
3. catechin, 50μg/day/mouse in 1.1% alcoholized solution;
4. quercetin 50μg/day/mouse in 1.1% alcoholized solution.

Blood was collected from the retroorbital plexus under anesthesia with ether into ependorf tubes with 1mM of Na_2 (EDTA), after six weeks of treatment.

Serum paraoxonase activity was measured as aryl esterase using phenylacetate as the substrate. Initial rates of hydrolysis were determined spectrophotometrically at 270nm. The assay mixture includes 1.0mM phenylacetate and 0.9mM $CaCl_2$ in 20mM Tris HCl, pH 8.0. Nonenzymatic hydrolysis of phenylacetate is subtracted from the total rate of hydrolysis. The E_{270} for the reaction is 1,310 M^{-1} cm^{-1}. One unit of aryl esterase activity is equal to 1 μmol of phenylacetate hydrolyzed/min/mL.

RESULTS AND DISCUSSION

Serum paraoxonase activity, measured as arylesterase activity, in E^0 mice (whose plasma is oxidized and highly susceptible to oxidation) was found to be reduced by 27%, in comparison to the activity observed in control mice (see FIGURE 1A). PON measurements in serum derived from E^0 mice after 6 weeks of polyphenols consumption revealed 14%, 113%, and 75% higher paraoxonase activity after the consumption of catechin, quercetin or red wine, respectively in comparison to paraoxonase activity in serum from the placebo group (FIG. 1B).

These results are in accordance with other recent studies, showing that several antioxidants, including licorice root–derived glabridin or pomegranate juice, preserve human serum paraoxonase (PON1) activity as they decrease the content of lipid peroxides, which inactivates PON 1.[12,13]

The increased levels of serum PON in E^0 mice that consumed red wine polyphenols can contribute to the reduction in LDL oxidative state by PON action on LDL-associated lipid peroxides. High serum PON activity in these mice may have also resulted from the reduced oxidative stress in the presence of the red wine-derived polyphenolic antioxidants.

CONCLUSION

As paraoxonase protects against lipid peroxidation, the beneficial effect of red wine-polyphenols on paraoxonase activity can be considered as an additional anti-atherogenic property of red wine.

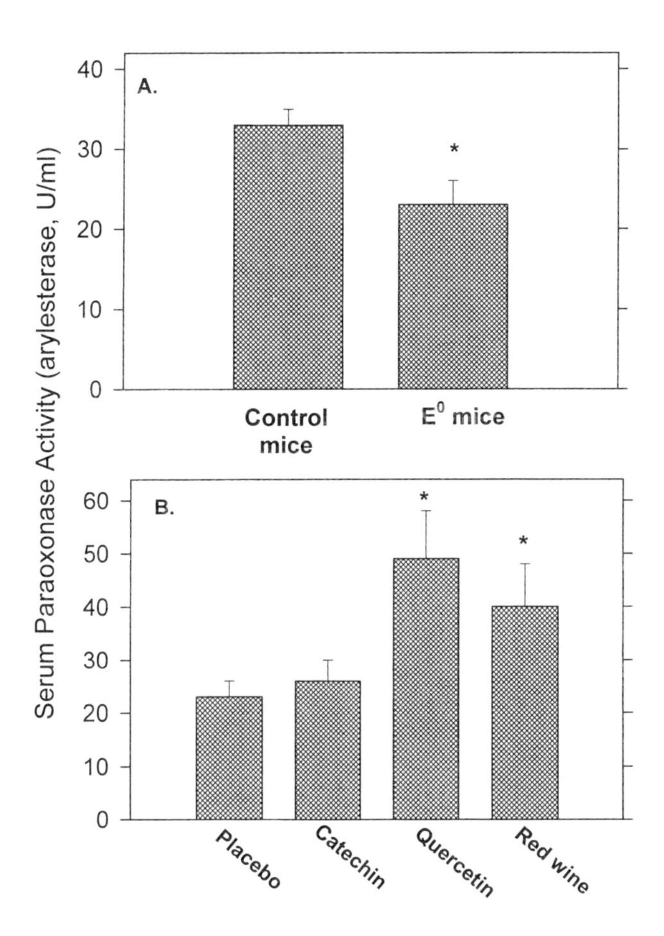

FIGURE 1. The effect of catechin, quercetin, or red wine consumption by E^0 mice on serum paraoxonase activity. Paraoxonase activity was measured as arylesterase activity in: (**A**) serum derived from E^0 mice in comparison to control mice; (**B**) serum derived from E^0 mice that consumed placebo, catechin, quercetin or red wine for 6 weeks. Results are expressed as mean ± S.D. of three separate determinations. $^*p < 0.01$ versus placebo.

REFERENCES

1. La Du, B.N., S. Adkins, C.L. Kuo & D. Lipsig. 1993. Studies on human serum paraoxonase/arylesterase. Chem. Biol. Interact. **87:** 25–34.
2. La Du, B.N., M. Aviram, S. Billecke, *et al.* 1999. On the physiological role(s) of the paraoxonases. Chem. Biol. Interact. **119/120:** 379–388.
3. Aviram, M., 1999. Does paraoxonase play a role in susceptibility to cardiovascular disease? Mol. Med. **5:** 381–386.
4. Aviram, M., M. Rosenblat, C.L. Bisgaier, *et al.* 1998. Paraoxonase inhibits high density lipoprotein (HDL) oxidation and preserves its functions: a possible peroxidative role for paraoxonase. J. Clin. Invest. **101:** 1581–1590.

5. AVIRAM, M., S. BILLECKE, R. SORENSON, *et al.* 1998. Paraoxonase active site required for protection against LDL oxidation involves its free sulfhydryl group and is different from that required for its arylesterase/paraoxonase activities: selective action of human paraoxonase allozymes Q and R. Arterioscler. Thromb. Vasc. Biol. **18:** 1617–624.
6. AVIRAM, M., E. HARDAK, J. VAYA, *et al.* 2000. Human serum paraoxonases (PON1) Q and R selectively decrease lipid peroxides in human coronary and carotid atherosclerotic lesions. PON 1 esterase has peroxidase-like activities. Circulation **101:** 2510–2517.
7. HERTOG, M.G., D. KROMHOUT, C. ARAVANIS, *et al.* 1995. Flavonoid intake and long-term risk of coronary heart disease and cancer in the seven countries study. Arch. Intern. Med. **155:** 381–386.
8. AVIRAM, M. & B. FUHRMAN. 1998. Polyphenolic flavonoids inhibit macrophage-mediated oxidation of LDL and attenuate atherogenesis. Atherosclerosis. **137**(Suppl.): S45–S50.
9. FUHRMAN, B. & M. AVIRAM. 2001. Flavonoids protect LDL from oxidation and attenuate atherosclerosis. Curr. Opin. Lipidol. **12:** 41–48
10. FUHRMAN, B., A. LAVY & M. AVIRAM. 1995. Consumption of red wine with meals reduces the susceptibility of human plasma and LDL to undergo lipid peroxidation Am. J. Clin. Nutr. **61:** 549–554.
11. HAYEK, T., B. FUHRMAN, J. VAYA, *et al.* 1997. Reduced progression of atherosclerosis in the apolipoprotein E deficient mice following consumption of red wine, or its polyphenols quercetin or catechin, is associated with reduced susceptibility of LDL to oxidation and aggregation. Arterioscler. Thromb. Vasc. Biol. **17**: 2744–2752.
12. AVIRAM, M., M. ROSENBLAT, S. BILLECKE, *et al.* 1999. Human serum paraoxonase (PON 1) is inactivated by oxidized low density lipoprotein and preserved by antioxidants. Free Radical Biol. Med. **26:** 892–904.
13. AVIRAM, M., L. DORENFELD, M. ROSENBLAT, *et al.* 2000. Pomegranate juice consumption reduces oxidative stress, low density lipoprotein modifications and platelet aggregation: studies in the atherosclerotic apolipoprotein E deficient mice and in humans. Am. J. Clin. Nutr. **71:** 1062–1076.

Absorption and Metabolism of Antioxidative Polyphenolic Compounds in Red Wine

SHINJI YAMASHITA,[a] TOSHIYASU SAKANE,[a] MASAMI HARADA,[b] NAMINO SUGIURA,[b] HIROFUMI KODA,[b] YOSHINOBU KISO,[b] AND HITOSHI SEZAKI[a]

[a]*Faculty of Pharmaceutical Sciences, Setsunan University, 45-1 Nagaotouge-cho, Hirakata City Osaka 573-0101, Japan*

[b]*Institute for Health Care Science, Suntory Ltd., Osaka 618-8503 Japan*

ABSTRACT: We have shown that drinking red wine reduces oxidation of LDL. This reduction in oxidation has been attributed to the polyphenolic compounds in red wine, but the mechanisms of absorption and metabolism of these compounds has been unclear. We therefore investigated the absorption and metabolism of polyphenols using rats to identify their active forms in biological fluids. We also investigated the effect of tartaric acid (TA), a major organic acid in wine, on the absorption of polyphenols. Our results suggested that low molecular weight polyphenols are absorbed in the intestine and metabolized to their glucuronide conjugates, which exhibit antioxidative activity in plasma, and that TA can enhance the bioavailability of wine polyphenols.

KEYWORDS: polyphenols; antioxidants; red wine; tartaric acid

INTRODUCTION

We have reported that intake of red wine enhanced the resistance of LDL to oxidation in humans. This enhanced antioxidative activity of LDL has been ascribed to the polyphenolic compounds in red wine.[1] However, the mechanism on absorption and metabolism of these polyphenols has been unclear. Therefore, we investigated the absorption and metabolism of the polyphenols in wine using rats to identify their active form in biological fluids. The effect of tartaric acid (TA),a major organic acid in wine, on the absorption of the polyphenols was also investigated.

MATERIALS AND METHODS

Male Wistar rats (280–320g) were fasted for 15 h prior to the administration of polyphenols. Under diethyl ether anesthesia, the rats had placed cannulae surgically placed in the left femoral aorta. After the surgery, the animals were placed in restraining cages. Polyphenols suspended in water were administered orally at a dose of 100mg/kg after the rats awoke.

Address for correspondence: Hirofumi Koda, Ph.D., Institute for Health Care Science, Suntory Ltd., 1-1-1 Wakayamadai, Shimamoto-cho, Mishima-gun, Osaka 618-8503 Japan. Voice: 075-962-8800; fax: 075-962-3791.
Hirofumi_Koda@suntory.co.jp

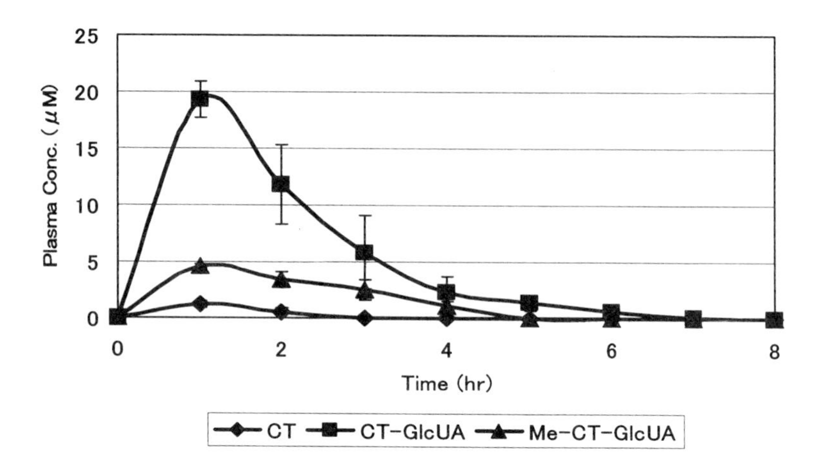

FIGURE 1. (+)-Catechin metabolites in rat after oral administration. CT: (+)-catechin; CT-GlcUA: (+)-Catechin5-β-*O*-glucuronide; Me-CT-GlcUA: Methyl (+)-catechin5-β-*O*-glucuronide.

300μl of blood was collected from each rat in a heparinized tube at two-hour intervals for eight hours. Plasma concentration of polyphenols and their metabolites were determined using HPLC. Structural characteristics of metabolites of polyphenols were analyzed using NMR spectra and mass spectra (MS).

Antioxidative activity of each sample as superoxide anion radical scavenger was determined using the ESR method.[2]

RESULTS

Absorption and Metabolism of (+)-Catechin and (−)-Epicatechin

The major metabolites of (+)-catechin (CT) and (−)-epicatechin (EC) administered orally were identified as CT 5-*O*-β-glucuronide and EC 5-*O*-β-glucuronide, respectively, in plasma by MS and NMR analysis. The plasma concentrations of

TABLE 1. Superoxide anion radical scavenging activity of (+)-catechin and (−)-epicatechin metabolites

Compounds	EC_{50} (μM)
(+)-Catechin	17
(+)-Catechin-GlcUA	15
Me-(+)-catechin-GlcUA	1640
(−)-Epicatechin	19
(−)-Epicatechin-GlcUA	16
Me-(−)-Epicatechin-GlcUA	1730

these glucuronide conjugates were higher than those of their parent compounds (see FIGURE 1). In ESR these glucuronide conjugates exhibited high antioxidative activity as superoxide anion radical scavengers. The potencies of antioxidative activities of these glucuronide conjugates were equivalent to those of their parent compounds (TABLE 1).

Absorption and Metabolism of Phenolic Acids

When phenolic acids such as gallic acid (GA) and protocatechuic acid (PA) were fed to rats, the antioxidative parent compounds appeared together with the unidentified metabolites. Plasma concentrations of GA and PA were higher than those of their metabolites.

Effect of Tartaric Acid on Absorption of (+)-Catechin

Tartaric acid (1%, 2%), a major organic acid in red wine, slowed the intestinal absorption of CT and increased total AUC of CT-GlcUA in plasma (see FIGURE 2).

CONCLUSION

Low-molecular weight antioxidative polyphenols in red wine were absorbed from the intestine and appeared in the plasma mainly as glucuronide conjugates (in the cases of CT and EC) or intact structure (in the cases of GA and PA) possessing antioxidative ability, which were exerted to inhibit LDL oxidation. In addition, the coexistence of 1% or 2% TA increased the AUC of CT metabolites in plasma. This result suggests that TA can enhance absorption and the bioavailability of polyphenolics in red wine.

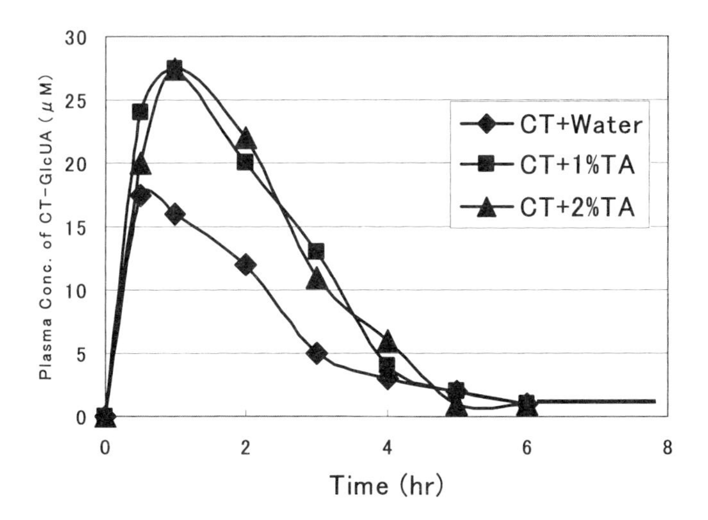

FIGURE 2. Effect of tartaric acid on intestinal absorption of (+)-catechin in rat. CT: (+)-catechin; CT-GlcUA: (+)-Catechin5-β-*O*-glucuronide; TA: tartaric acid.

REFERENCES

1. KONDO, K., A. MATSUMOTO, H. KURATA, *et al.* 1994. Inhibition of oxidation of low-density lipoprotein with red wine. Lancet. **344:** 1152.
2. HATANO, T., R. EDAMATSU, M. HIRAMATSU, *et al.* 1989. Effects of the interaction of tannins with co-existing substances. VI. Effects of tannins and related polyphenols on superoxide anion radical, and on 1,1-diphenyl-2-picrylhydrazyl radical. Chem. Pharm. Bull. **37:** 2016–2021.

The *in Vivo* Antithrombotic Effect of Wine Consumption on Human Blood Platelets and Hemostatic Factors

ERNA P.G. MANSVELT,[a] DAVID P. VAN VELDEN,[b] ELBA FOURIE,[a] MARIETJIE ROSSOUW,[a] SUSAN J. VAN RENSBURG,[c] AND C. MARIUS SMUTS[d]

[a]*Department of Haematological Pathology,* [b]*Department of Family Medicine and Primary Care, and* [c]*Department of Chemical Pathology, Faculty of Health Sciences, University of Stellenbosch, Tygerberg 7505, South Africa*

[d]*Nutrional Intervention Research Unit, MRC, Tygerberg 7505, South Africa*

ABSTRACT: We compared the *in vivo* effect of red vs. white wine consumption on platelet aggregation, responsiveness and membrane viscosity, plasma total antioxidant status, thromboxane B_2 levels, and fibrinolysis. Diet and red wine had a synergistic effect in decreasing platelet aggregation. Red wine did not have a significantly more favorable effect on the fibrinolytic factors than white wine. The reduction in platelet membrane viscosity after red wine, which could contribute to the protective antithrombotic role of red wine, needs further explanation.

KEYWORDS: red wine; white wine; aggregation; platelet membrane; microviscosity; oxidation; fibrinolysis

INTRODUCTION

The anti-aggregatory effect of high levels of alcohol on blood platelets via phospholipase A_2 inhibition has been proved.[1] Several investigators reported that red wine has a more pronounced anti-aggregatory effect than white wine.[2] The phenolic compounds in red wine have an antioxidative effect on the poly-unsaturated fatty acids in the platelet lipid membranes that makes them less reactive.[3] Platelet membrane microviscosity is related to its cholesterol/phospholipid ratio[4] or can be an indicator of lipid membrane oxidative status.[5]

OBJECTIVES

We investigated the *in vivo* effect of regular and moderate wine consumption and compared the effect of red and white wine on platelet aggregation and responsiveness, plasma thromboxane B_2 (TxB_2), platelet membrane microviscosity, platelet

Address for correspondence: Erna P.G. Mansvelt, Department of Haematological Pathology, Faculty of Health Sciences, University of Stellenbosch, PO Box 19063, Tygerberg 7505, South Africa. Tel +27 21 938 5348; Fax +27 21 938 4609.
epgm@gerga.sun.ac.za

membrane cholesterol/phospholipid ratio, plasma total antioxidant status (TAS), and fibrinolysis: plasma fibrinogen, tissue plasminogen activator (tPA), plasminogen activator inhibitor-1 (PAI-1).

The relation between platelet aggregation and platelet membrane microviscosity, which is a measurement of membrane cholesterol/phospholipid ratio or fatty acid oxidation, was examined as a possible explanation for the antiatherogenic effect of alcohol in the form of wine on blood platelets.

DESIGN AND METHODS

Six female and seven male volunteers, aged 25–56 years, on no medication, abstained from alcohol for 21 days. They then consumed red and then white wine for 28 days respectively, interspersed by 21 days of abstinence. Daily intake of alcohol was 23 grams for females, 32 grams for males. Fasting blood and urine specimens were taken before (baseline 1 and 2) and after each wine consumption period (postred, postwhite) and finally after 21 days washout. All test subjects adhered to their habitual lifestyle and diet, recording their daily red meat and fat intake.

Platelet aggregation and responsiveness were determined on a Coulter Chronolog aggregometer using standard agonists. Platelet membrane microviscosity was measured as a function of fluorescence polarization. tPA and PAI-1 were determined by ELISA (Stago, France) and TxB_2 by EIA (Biotrak, Amersham, UK). Platelet lipids were extracted and determined by standard techniques. Plasma TAS was determined by TAS Kit (Randox Laboratories, Crumlin, UK).

RESULTS

Platelet aggregation with low concentration of collagen (1 µg/mL) and plasma TxB_2 levels decreased significantly after both red and white wine consumption compared to baseline values (see FIGURE 1). Platelet aggregation decreased significantly with addition of high concentration of collagen (5 µg/mL) only after red wine consumption in five of the subjects maintaining a Mediterranean diet ($p < 0.05$). Collagen lag times, however, decreased after both wines, indicating that the platelets became more responsive after wine consumption. An exception was found after red wine in the five subjects following a Mediterranean diet.

Plasma fibrinogen levels decreased after both wines, but only significantly after white wine consumption compared to baseline ($p < 0.05$). PAI-1 levels decreased after red wine and significantly increased after abstinence from red wine ($p < 0.05$). PAI-1 levels decreased significantly after white wine consumption compared to the preceding baseline levels ($p < 0.05$). tPA levels remained unaltered after both wines. Platelet membrane microviscosity decreased significantly after red wine ($p < 0.02$) compared to baseline and significantly increased after white wine ($p < 0.05$) compared to the preceding baseline value (see FIGURE 2).

A significant decrease in membrane cholesterol/phospholipid ratio was found only after white wine consumption compared to baseline ratio ($p < 0.02$). Plasma

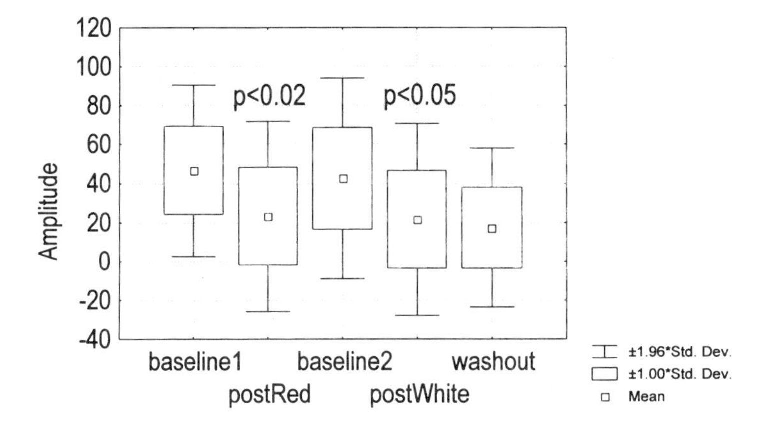

FIGURE 1. Platelet aggregation with 1 μg/mL collagen measured as percentage amplitude compared to 100% aggregation. Decrease in aggregation after red wine compared to baseline 1 ($p < 0.02$) and after white wine compared to baseline 2 ($p < 0.05$). Mean values for 13 volunteers □. (Baseline 1: after three weeks abstention, before red wine; baseline 2: after three weeks abstention, before white wine consumption).

TAS tended to increase after red wine; however, white wine consumption had a pro-oxidant effect on TAS ($p < 0.02$). No significant correlations were found between platelet aggregation and membrane microviscosity or between membrane cholesterol /phospholipid ratio and microviscosity or between TAS and microviscosity.

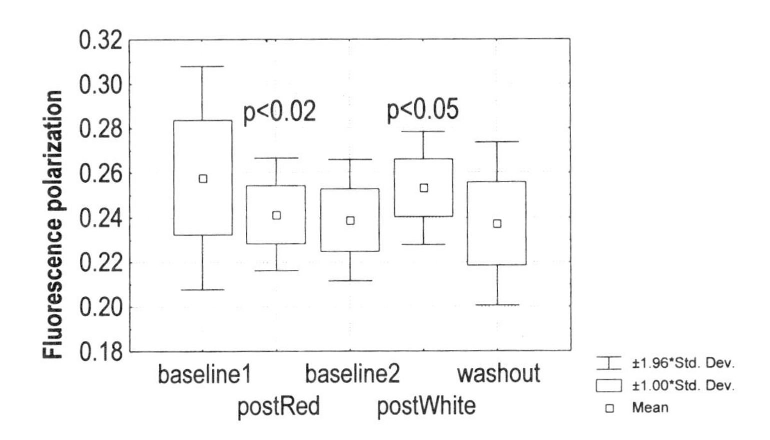

FIGURE 2. Platelet membrane microviscosity decreased after red wine compared to baseline 1 ($p < 0.02$); increased after white wine compared to baseline 2 ($p < 0.05$). Mean values for 13 volunteers □. (Baseline 1: after three weeks abstention, before red wine; baseline 2: after three weeks abstention, before white wine consumption).

CONCLUSIONS

This study indicates that diet and red wine may have a synergistic effect in decreasing platelet aggregation. Red wine did not have a significantly more favorable effect on the fibrinolytic factors than white wine. The decrease in platelet membrane microviscosity after red wine consumption could not be related to change in platelet membrane cholesterol/phospholipid ratio. A similar decrease in membrane microviscosity after red but not after white wine consumption has been found on red cells.[6] This effect needs further explanation and could contribute to the protective antithrombotic role of red wine.

REFERENCES

1. RUBIN, R. 1989. Ethanol interferes with collagen-induced platelet activation by inhibition of arachidonic acid mobilization. Arch. Biochem. Biophys. **270:** 99–113.
2. RENAUD, S. & M. DE LORGERIL. 1992. Wine, alcohol, platelets and the French paradox for coronary heart disease. Lancet **339:** 1523–1526.
3. POLETTE, A, D. LEMAITRE, M. LAGARDE, *et al.* 1996. N-3 fatty acid-induced lipid peroxidation in human platelets is prevented by catechins. Thromb. Haemost. **75:** 945–949.
4. SHATTIL S.J. & R.A. COOPER. 1976. Membrane microviscosity and human platelet function. Biochemistry **59:** 4832–4837.
5. VAN RENSBURG, S.J., W.M.U. DANIELS, J. VAN ZYL, *et al.* 1994. Lipid peroxidation and platelet membrane fluidity – implications for Alzheimer's disease? NeuroReport **5:** 2221–2224.
6. SHARPE, P.C., L.T. MCGRATH, E. MCCLEAN, *et al.* 1995. Effect of red wine consumption on lipoprotein (a) and other risk factors for atherosclerosis. Quart. J. Med. **88:** 101–107.

Corrupt Captains and Convicts

PHILIP NORRIE

Elanora Heights, N.S.W. Australia 2101

ABSTRACT: **Australia is unique in that so many of its wine companies were founded by members of the medical profession. The First Fleet, which brought the first convicts and settlers to Australia in 1787, was delayed until wine was provided as a medicine for the long voyage from England. Australia's wine doctors were advocating the use of wine as a medicine 200 years before the French Paradox.**

KEYWORDS: **wine production; Australian wine; wine doctors**

"Wine in moderation—the thinking person's health drink" Philip Norrie

Have you ever wondered why so many vineyards in Australia were established by Doctors? Australia is unique amongst wine producing countries in that 60% of the fruit from any vintage is processed by wine companies established by Australia's 180 or more wine doctors.

It all began in 1787 when the First Fleet was preparing to sail from England for New South Wales. Surgeon John White, the surgeon in charge of the health of all convicts, sailors, soldiers, and free settlers, was not satisfied with the standard rations issued to the fleet. On 7th February 1787 White wrote to Governor Phillip, the Governor of the proposed new colony in Australia, asking for what he called "necessaries."[1] This is a term he used to describe extra items of food not included in the standard ration such as sugar, currants, rice, sago, barley, soup, tea, spices, and wine.

White used wine as a medicine throughout the voyage to Australia to prevent malnutrition and diseases. For example on 20th December, 1787 White wrote the following in his diary—"On those days the scurvy began to show itself in the Charlotte (one of the convict transport ships), mostly among those who had the dysentery to a violent degree; but I was pretty well able to keep it under by a liberal use of the essence of malt and some good wine, which ought not to be classed among the most indifferent antiscorbutics. For the latter we were indebted to the humanity of Lord Sydney and Mr. Nepean, principal and under secretary of state."[2]

So good was Surgeon White's care that only 24 of the 775 convicts in the First Fleet died. It was a different story with the Second Fleet. Conditions were totally different in the Second Fleet and 274 of the 965 convicts died during the voyage out to Australia. When the Second Fleet finally arrived in Port Jackson, a similar number

Address for correspondence: Philip Norrie, M.B., B.S., M.Sc., M.Soc.Sc. (Hons.), General Practitioner, 2/50 Kalang Road, Elanora Heights, N.S.W. Australia 2101. Voice: 61-2-9913-1088/61-2-9974-5794; fax: 61-2-9970-6152.
winedoc@sneaker.net.au

of convicts as had died on the way out were so sick they had to be hospitalized in the infant colony's basic hospital, so only about one third of the original convict population were fit for work upon arrival—a sad state of affairs for the colony and the convicts.

After the Third Fleet, convict transport ships came out to Australia on their own and not in groups or fleets. They were a mixture of converted Royal Navy ships, ex-slaving vessels or ships especially built for the purpose. Whether the ship had a doctor on board or not for the long voyage to Sydney Town was largely up to the Transport Company which owned the ship.

The turning point in the medical treatment of convicts, during transportation out to Australia came in 1814 with the voyage of the Surrey. The Surrey was loaded with 200 male convicts, marine guards and her crew. She left England on 22nd February, 1814 in company with another convict transport the Broxbornebury.[3] During the voyage her master, Captain James Patterson, kept the convicts closely confined with poor ventilation because he feared the convicts would take over his ship. The bedding was not cleaned or aired and the cells were not properly cleaned or fumigated. After leaving Rio on April 21, "gaol fever" or typhus took hold, not only of the convicts but also the guards and crew. By the time the Surrey reached the east coast of Australia the death toll had reached 51 including 36 convicts. That was a pity, but what was really bad for business was that the Captain of the ship, the First Mate, the Second Mate, the boatswain, the ship's surgeon, six seamen and four soldiers had also died. So there was no one to navigate the ship. Fortunately the Broxbornebury passed by and transferred a brave man on board the fever ridden ship to navigate it into Port Jackson. Once inside the Heads of Sydney harbor the ship was quarantined on the northern shore of the harbor and the many remaining sick treated in tents erected as a temporary hospital. So began the use of North Head as a quarantine station.

Governor Macquarie, the Governor of New South Wales, ordered an investigation and appointed Dr. William Redfern to do the job. Redfern was Sydney's leading doctor as well as Australia's first wine doctor, when he established his vineyard "Campbellfields," south west of Sydney, in 1818. Redfern had been a convict transported to Australia on board the Minerva in 1801, so he knew both sides of the problem. It was this investigation, along with Redfern's findings and recommendations, that was to have a marked impact on Australia's wine industry. Redfern found out that the Captain had withheld rations from the convicts, including their wine ration, so that he could sell them at various ports on the way out, such as Rio and Capetown, not to mention the high prices he would have commanded for them in Sydney town, to make extra money for himself. So the convicts became weak and more susceptible to disease. He also found that the ship's surgeon had no authority over the ship's master, so that his recommendations for better ventilation, bedding cleansing and fumigation of cells, were overruled and ignored. Redfern also found fault with other things such as inadequate clothing and withholding of soap so it could be sold later in Sydney. Such basic things as adequate clothing, food, water, wine and cleanliness meant the difference between life and death for these people.

In a letter to Macquarie dated 30th September, 1814, Redfern made his eleven recommendations about how to prevent further tragedies such as the Surrey's 1814 voyage.[4] Redfern recommended, amongst other things, that one-quarter of a pint of wine with added lime juice be given to each convict each day—out in the open so no one could withhold their ration—to prevent malnutrition and scurvy. He also recommended that every transport ship have a properly qualified doctor on board, preferably Royal Naval surgeons used to shipboard life, and that these surgeons be given more power and authority by being commissioned so that they were not subordinate to the ship's master, but solely in charge of and responsible for the convict's welfare while the ship's master was solely in charge of the safety of the ship.

After the defeat of Napoleon in 1815, there was no need for a large standing army or navy because the threat of Bonaparte was gone and many surgeons were retired on half pay. These men found their prospects better in Australia, and because Australia was the only British colony taking convicts last century, out they came.

Redfern's recommendations were acted upon immediately, so Australia found itself host to many naval surgeons doing convict transport and later migrant transport service. These doctors knew the benefits of wine which would, as Redfern put it, "maintain the Vigor of the System"; and "Dispel Despondency" from being kept confined below decks in prison cells.[5]

The better wines from Europe would have been kept by the English wine merchants and the poorer wines shipped out to Australia. By the time the wine had spent six months in a leaking oak cask in the bilge of a ship it would have run the risk of oxidation and seawater contamination. So these doctors retired to Australia and established their vineyards to make wine to help their patients, thus avoiding the potential problems associated with transporting wine to Australia.

That is why the Australian medical profession's symbol should not be the traditional snake caduceus, but a glass of wine and a set of convict leg irons, because Australia's medical profession began with convict transport doctors maintaining their convict patient's health with wine.

Redfern's recommendations were later published as "Instructions for Surgeons,– Superintendent on board convict ships bound for New South Wales or Van Diemens Land and for the Masters of those ships." This book was standard issue to all transports, their masters and surgeons and was one of Australia's first public health documents.

Today Australia has over 180 wine doctors, but the Australian medical profession in general is up to its stethoscopes in wine not only as consumers, but also as prescribers and producers—it is an integral part of our unique medical heritage. If it wasn't for the Surrey's corrupt captain, Australia may have had a shortage of doctors.

Later vineyards were established in Australia's leading lunatic asylums to help prevent malnutrition, dysentery and to provide activity for the patients. Dr. Frederick Norton Manning established Australia's first lunatic asylum vineyard at the Tarban Creek Asylum (now the Gladesville Psychiatric Hospital in Sydney) in N.S.W. in the 1870's; Dr. Beattie-Smith established the Ararat Asylum vineyard and Dr. Watkins the Sunbury Asylum vineyard, both in Victoria, in the 1880's; while Dr. William Lennox Cleland did the same at the Adelaide lunatic asylum (now the Glenside Psychiatric Hospital) in South Australia also in the 1880's.

Besides Dr. Redfern, other Australian Wine Doctors wrote about the health benefits of wine long before today's new age of pro–wine-in-moderation research. Dr. Henry John Lindeman wrote a letter to the editors of the New South Wales Medical Gazette in 1871 "Wine as a Therapeutic Agent and Why it Should become our national beverage" extolling the virtues of pure wine over "King Rum" and other adulterated alcoholic beverages.[6] In 1880 Dr. William Lennox Cleland the medical superintendent of the Adelaide Lunatic Asylum wrote a paper "Some remarks upon wine as a food and its production"[7] while Dr. Thomas Fiaschi addressed the Australasian Trained Nurses' Association in Sydney in 1906 with a detailed and extensive lecture entitled "The Various Wines used in Sickness and Convalescence."[8]

It can been seen that without the puritanical bias featuring in some countries such as the U.S.A., Australia's medical profession were free to prescribe wine as a medicine for their patients from the time of the first European settlement. Not only did they prescribe it but they also made it! Australia's Wine Doctors were up to their stethoscopes in wine not only as consumers but also as prescribers and producers. Thus Australia can be regarded as the first extensive modern day research trial about the medicinal virtues of wine as a medicine, conducted by its wine doctors.

REFERENCES

1. WHITE, J. 1962. Journal of a Voyage to N.S.W. originally published 1790. Published in Association with R.A.H.S. Angus and Robertson, p. 18.
2. Ibid, p. 103.
3. BATESON, C. 1969. The Convict Ships 1787–1868. Brown, Son and Ferguson Ltd. Glasgow. Pages 340 and 382
4. THE LIBRARY COMMITTEE OF THE COMMONWEALTH PARLIAMENT. 1919. Commonwealth of Australia Historical Records of Australia Series 1. Governors' Despatches to and from England Volume VIII, July 1813—December 1815, p. 275.
5. Ibid, p. 287.
6. NORRIE, P.A. 1993. The History of Dr. H.J. Lindeman and Lindeman's Wines. Appollo Books. Sydney, p. 30.
7. NORRIE, P.A. 1994. Australia's Wine Doctors, Vol. 3. Past South Australian Wine Doctors Organon (Aust.) Pty. Ltd.,p. 14.
8. NORRIE, P.A. 1990. Vineyards of Sydney, Cradle of the Australian Wine Industry. Horwitz. Grahame. Sydney, p. 204.

The Cardioprotective Effect of Wine on Human Blood Chemistry

DAVID P. VAN VELDEN,[a] ERNA P.G. MANSVELT,[b] ELBA FOURIE,[c] MARIETJIE ROSSOUW,[c] AND A. DAVID MARAIS[d]

[a]*Department of Family Medicine and Primary Care,* [b]*Department of Haematological Pathology,* [c]*Department of Chemical Pathology, Faculty of Health Sciences, University of Stellenbosch, Tygerberg 7505, South Africa*

[d]*Lipid Laboratory, Cape Heart Centre and MRC Cape Heart Group, University of Cape Town Health Sciences Faculty, Anzio Road, Observatory 7925, South Africa*

ABSTRACT: We investigated the *in vivo* effects of regular consumption of red and white wine on the serum lipid profile, plasma plasminogen activator-1, homocysteine levels, and total antioxidant status. This study confirmed that moderate consumption of wine, red more than white, exerts cardioprotective effects through beneficial changes in lipid profiles and plasma total antioxidant status.

KEYWORDS: red wine; white wine; lipoprotein profile; small dense LDL particle size; total antioxidant status

INTRODUCTION

Epidemiological studies generally showed a strong inverse association between wine consumption and mortality for coronary heart disease (CHD). Antiatherogenic alteration in plasma lipoproteins, particularly increase in high-density lipoprotein (HDL) cholesterol, is considered as the most plausible mechanism of the protective effect of alcohol consumption on CHD. Current clinical guidelines[1] accept that the risk of vascular disease correlates with LDL and HDL cholesterol levels. Low HDL cholesterol concentration, mild hypertriglyceridemia, and a smaller and denser species of LDL constitute a common atherogenic phenotype[2] which is associated with other risk factors for atherosclerosis: insulin resistance, defects in coagulation, including hyper-fibrinogenemia,[3] defects in fibrinolysis, including raised plasminogen activator inhibitor-1 (PAI-1),[4] hypertension, and hyperuricemia. Wine consumption may have an influence on a number of these cited risk factors for CHD.

Address for correspondence: David P. van Velden, Department of Family Medicine and Primary Care, Faculty of Health Sciences, University of Stellenbosch, P.O. Box 19063, Tygerberg 7505, South Africa. Voice: +27-21-938-9233; fax: +27-21-938-9153.
dpvv@gerga.sun.ac.za

OBJECTIVES

Because small LDL can be modulated by lifestyle interventions associated with the Mediterranean diet,[5] we investigated the *in vivo* effects of regular and moderate red and white wine consumption on the serum lipid profile and antioxidant status as another plausible mechanism of the protective effect of wine on coronary heart disease. The procedure was repeated on the same subjects, substituting white wine for red wine.

DESIGN AND METHODS

Thirteen healthy volunteers (7 male, 6 female) aged 25–56 years, on no medication, abstained from alcohol for 21 days prior to and 21 days after the study. During the two 28-day study periods, the subjects consumed first red, then white wine (equivalent to a daily alcohol intake of 32 grams for males, and 23 grams of alcohol for females) interspersed by 21 days of abstention. No other alcoholic beverages and no medication were allowed during the study period. They continued with their normal activities and normal diet, but were requested to record their daily diet. At base line, after 28 days of wine consumption, and after 21 days of washout fasting, blood samples were taken for serum lipogram, uric acid, albumin, γ-glutamyl transferase, total bilirubin, PAI-1 and homocysteine levels, and total antioxidant status (TAS) determination. Cholesterol, HDL and TG were determined on an ADVIA 1650 (Bayer), PAI-1 was determined by ELISA (Stago, France), total plasma homocysteine by HPLC, plasma TAS by TAS kit (Randox Laboratories, Crumlin, UK), small dense LDL particles by non-denaturating polyacrylamide gradient gel electrophoresis (GGE).

RESULTS

The lipid profiles responded significantly more anti-atherogenic on red compared to white wine with a decrease in total cholesterol, increase in HDL cholesterol (see FIGURE 1), and decrease in LDL cholesterol, with minimal effect on triglyceride levels. A trend was observed for an increase in the size of small, dense LDL particles ($p = 0.082$).

Plasma TAS increased with red wine consumption, with a significant decrease during white wine consumption ($p < 0.02$) (see FIGURE 2).

A non-significant decrease in PAI-1 levels after red wine was followed by a significant increase ($p < 0.05$) after successive abstinence, and levels decreased after white wine consumption ($p < 0.05$) compared to preceding baseline.

None of the test subjects had raised fasting homocysteine levels. Homocysteine levels were unaffected by wine consumption.

No significant changes were found in liver and renal functions.

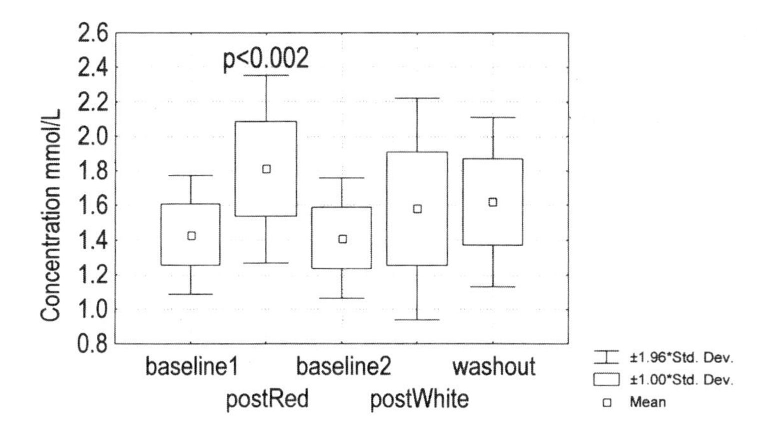

FIGURE 1. Mean values of HDL for 13 volunteers increased after red wine compared to baseline 1 ($p < 0.002$). (Baseline 1: after three weeks abstention, before red wine; baseline 2: after three weeks abstention, before white wine consumption).

CONCLUSIONS

This study confirmed that regular and moderate consumption of wine, red more than white, exert cardioprotective effects through beneficial changes in lipid profiles. The trend towards an increase in the size of small dense particle LDL during wine consumption is a positive finding which has a protective effect for cardiovascular disease.

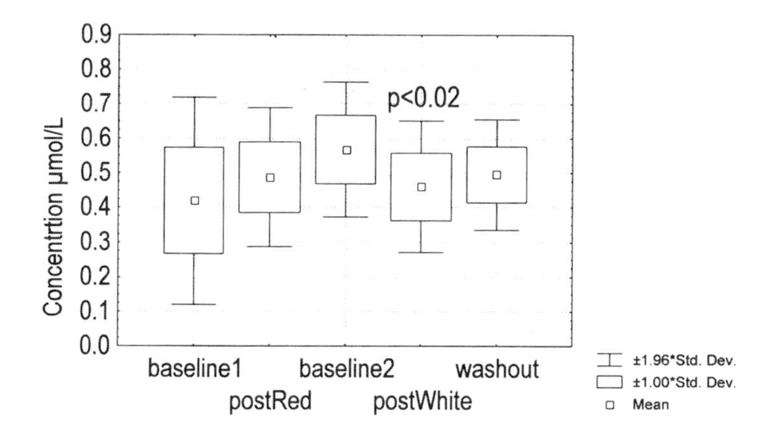

FIGURE 2. Plasma TAS increased after red wine consumption compared to baseline 1 ($p > 0.05$), and decreased after white wine ($p < 0.02$) compared to baseline 2. (Mean values for 13 volunteers). (Baseline 1: after three weeks abstention, before red wine; baseline 2: after three weeks abstention, before white wine consumption).

Plasma antioxidant status increased after red wine, but decreased significantly after white wine consumption. It is known that alcohol has a pro-oxidant effect, and this finding can be explained by the significantly lower antioxidant content in the non-alcoholic components of white wine compared to that of red wine.

It seems that homocysteine levels are unaffected by wine consumption of short duration.

REFERENCES

1. Wood, D.A. & G. De Backer, O. Faergeman, *et al.* & the Task Force, Prevention of Coronary Heart Disease in Clinical Practice. 1998. Recommendations of the second joint task force of the European society of Cardiology, European Atherosclerosis Society and European Society of Hypertension. Eur. Heart J. **19:** 1434–1503.
2. Austen, M.A. & M.C. Kin, K.M.Vranizan & R.M. Kraus. 1990. Atherogenic lipoprotein phenotype: a proposed genetic marker for coronary heart disease. Circulation **82:** 495–506.
3. Maki, A.C. & M.H. Davidson, P. Marx, *et al.* 2000. Association between elevated fibrinogen and the small, dense low-density lipoprotein phenotype among postmenopausal women. Am. J. Cardiol. **85:** 451–456.
4. Festa, A. & R. D'Agostino, Jr., L. Mykären, *et al.* 1999. Low-density lipoprotein particle size is inversly related to plasminogen activator inhibitor-1 levels. The Insulin Resistance Atherosclerosis Study. Arterioscler. Thromb. Vasc. Biol. **19:** 605– 610.
5. Marais, A.D. 2000. Therapeutic modulation of low-density lipoprotein size. Curr. Opin. Lipidol. **11:** 597–602.

Melatonin

An Antioxidant in Edible Plants

RUSSEL J. REITER AND DUN-XIAN TAN

Department of Cellular and Structural Biology,
The University of Texas Health Science Center,
San Antonio, Texas 78229-3900, USA

ABSTRACT: Melatonin, a molecule with antioxidant properties that is widely distributed in the animal kingdom, has now been shown to exist in the plant kingdom, including edible plants. Our findings show that melatonin is not only an endogenously produced antioxidant, but that it is also consumed in the diet. Since melatonin concentrations in the blood correlate with the total antioxidant status of this fluid, it is likely that dietary melatonin could be important in protecting against oxidative damage.

KEYWORDS: melatonin; antioxidant; free radicals; oxidative damage; edible plants

INTRODUCTION

Melatonin, *N*-acetyl-5-methoxytryptamine, has long been known to be a secretory product of the vertebrate pineal gland. This product, which is synthesized in and secreted from the pineal gland in a circadian manner with highest values at night, is related to the regulation of other 24 hour cycles, mediates annual reproductive changes in seasonal breeding animals, and has oncostatic actions.

Besides these functions, melatonin was recently discovered to be a free radical scavenger[1,2] and antioxidant.[2–4] As to its direct interactions, melatonin has been shown to detoxify the hydroxyl radical ($\cdot OH$), hydrogen peroxide (H_2O_2), peroxynitrite anion ($ONOO^-$), nitric oxide ($\cdot ON$) and hypochlorous acid ($HOCl$). The resulting products which are produced when melatonin reacts with these toxic agents are listed in TABLE 1.

In addition to these actions, melatonin, possibly via receptor-mediated mechanisms, has been shown to have indirect antioxidative actions as well. Thus, the indole promotes the activity of several enzymes which metabolically remove toxic reactants; these enzymes include superoxide dismutase, catalase, glutathione peroxidase, and glutathione reductase. Also, melatonin inhibits the potentially proxidative enzyme, nitric oxide synthase.[2–4] Both the direct scavenging actions of melatonin as well as its indirect effects on the enzymes that metabolize toxic reactants are induced with both physiological and pharmacological levels of melatonin.

Address for correspondence: Russel J. Reiter, Ph.D. Department of Cellular and Structural Biology, The University of Texas Health Science Center, 7703 Floyd Curl Drive, Mail Code 7762, San Antonio, TX 78229-3900, USA. Voice: 210-567-3859; fax: 210/567-6948.
Reiter@uthscsa.edu

TABLE 1. A summary of the free radicals and other reactive species reported to be directly scavenged by melatonin along with the products that are formed as a consequence of these interactions

Reactant scavenged	Product identified
$^1O_2$	N^1-Acetyl-N^2-formyl-5-methoxytryptamine
H_2O_2	N^1-Acetyl-N^2-formyl-5-methoxytryptamine
·OH	Cyclic 3-hydroxymelatonin
HOCl	2-Hydroxymelatonin
·NO	*N*-Nitrosomelatonin
$ONOO^-$/ONOOH	6-Hydroxymelatonin; cyclic 2-hydroymelatonin; cyclic 3-hydroxymelatonin; 1-nitromelatonin; 1-hydroxymelatonin

NOTES: $^1O_2$, singlet oxygen; H_2O_2, hydrogen peroxide; ·OH, hydroxyl radical; HOCl, hypochlorous acid; ·NO, nitric oxide; $ONOO^-$, peroxynitrite anion; $ONOO^-$, peroxynitrous acid.

MELATONIN IN EDIBLE PLANTS

The existence of melatonin in plants was initially theorized after the indole was found in algae, species that have characteristics of both the plant and animal kingdoms.[5] This discovery was followed by identification of melatonin in a variety of different plants, including mono- and dicotyledonous angiosperms. The highest melatonin levels reported in plants to date are those measured in so-called medicinal plants,[6] i.e., fever few and St. John's wort; in these plants melatonin concentrations are in the range of μg/g tissue. In the seeds of a number of plants and in the cherry fruit, melatonin levels are in the ng/g tissue range (TABLE 2).[7]

Several facts have come to light since melatonin was discovered in plants: (a), the indole is found in roots, stems, leaves, fruits and seeds; (b), in different plants, the concentrations of melatonin vary widely; (c), and melatonin in unequally distributed within a given plant. While melatonin has been found in a wide variety of plant tissues, few foodstuffs have been examined as to their melatonin levels.[8]

CONCLUDING REMARKS

Melatonin, consumed in the diet, is absorbed by the gut and significantly alters circulating levels of the indole. Since the concentration of melatonin in the blood correlates positively with the total antioxidant status of this fluid,[9] the implication is that consuming foodstuffs containing melatonin would increase the antioxidative capacity of the organism. Melatonin readily crosses both the placenta and blood-brain barrier and gets into every cell and, so far as is known, into all subcellular compartments including the mitochondria.[2–4] Melatonin's presence in mitochondria may be of special importance since these organelles are a major source of free radicals and oxidative damage.

TABLE 2. Melatonin concentrations in a some of the plant products that have been analyzed

Common name	Scientific name	Melatonin concentration (ng/g)
Milk thistle seed	*Silybum marianum*	2
Tart cherry fruit (Balaton)	*Prunus cerasus*	2
Poppy seed	*Popaver somniferum*	6
Anise seed	*Pimpinela anisum*	7
Coriander seed	*Coriandrum sativum*	7
Celery seed	*Apium graveolens*	7
Flax seed	*Linum usitatissimum*	12
Green cardamom seed	*Elettaria cardamomum*	15
Tart cherry fruit (Montmorency)	*Prunus cerasus*	15
Alfalfa seed	*Medicago sativum*	16
Fennel seed	*Foeniculum vulgare*	28
Sunflower seed	*Helianthus annuus*	29
Fenugreek seed	*Trigonella foenum-graecum*	43
Wolf berry seed	*Lycium barbarum*	103
Black mustard seed	*Brassuca nigra*	129
White mustard seed	*Brassica hirta*	189
St. John's wort, leaf	*Hypericum perforatum*	1,750
Fever few, gold leaf	*Tanacetum parthenium*	1,920
Fever few, green leaf	*Tanacetum parthenium*	2,450
St. John's wort, flower	*Hypericum perforatum*	4,390
Huang-qin	*Scutellaria biacalensis*	7,110

NOTES: Clearly, the amount of melatonin plants contain varies widely, and melatonin within a given plant is not uniformly distributed.

ACKNOWLEDGMENTS

D-X.T. is supported by NIH training Grant T32AG00165-13.

REFERENCES

1. TAN, D.X., L.D. CHEN, B. POEGGELER, *et al.* 1993. Melatonin: a potent, endogenous hydroxyl radical scavenger. Endocrine J. **1:** 57–60.
2. TAN, D.X., L.C. MANCHESTER, R.J. REITER, *et al.* 2000. Significance of melatonin in antioxidative defense system. Biol. Signals Recept. **9:** 137–159.
3. REITER, R.J., D.X. TAN, C. OSUNA & E. GITTO. 2000. Actions of melatonin in the reduction of oxidative stress. J. Biomed. Sci. **7:** 444–458.

4. REITER, R.J., D.X. TAN, D. ACUÑA-CASTROVIEJO, *et al.* 2000. Melatonin: mechanisms and actions as an antioxidant. Curr. Top. Biophys **24:** 171–183.
5. POEGGELER, B. & R. HARDELAND. 1994. Detection and quantification of melatonin in a dinoflagellate, Gonyaulax polyedra: solutions to the problem of methoxyindole destruction in non-vertebrate material. J. Pineal Res. **17:** 1–10.
6. MURCH, S.J., C.B. SIMMONS & P.K. SAXENA. 1997. Melatonin in fever few and other medicinal plants. Lancet **350:** 1598–1599.
7. MANCHESTER, L.C., D.X. TAN, R.J. REITER, *et al.* 2000. High levels of melatonin in the seeds of edible plants: possible function in germ tissue protection. Life Sci. **67:** 3023–3029.
8. REITER, R.J. 1999. Phytochemicals: melatonin. *In* Encyclopedia of Food Science and Technology. F.J. Frances, Ed.: 1918–1922. John Wiley. New York.
9. BENOT, S., R. GOBERNA, R.J. REITER, *et al.* 1999. Physiological levels of melatonin contribute to the antioxidant capacity of human serum. J. Pineal Res. **27:** 59–64.

EPR, Free Radicals, Wine, and the Industry

Some Achievements

GORDON J. TROUP[a] AND CHARLES R. HUNTER[b]

[a]*School of Physics and Materials Engineering, Monash University, Clayton, Victoria 3168, Australia*

[b]*Anatomy Deapartment, Monash University, Clayton, Victoria 3168, Australia*

ABSTRACT: The achievements and uses of electron paramagnetic resonance in the wine industry are pointed out. For example, it first detected stable free radicals in red and white wines and showed that the radical concentration, and therefore the antioxidant action, of white wines was increased by skin and oak exposure. It is expected that EPR will be used more in the future.

KEYWORDS: EPR; free radicals; wine; wine industry

EPR (electron paramagnetic resonance), a well-known detector of free radicals and paramagnetic ions, seems to be the forgotten elder sister of NMR in the wine industry, so we would like to present some of its research achievements. It first detected stable free radicals in red and white wines, and showed that the radical concentration, and therefore the antioxidant action, of white wines was increased by skin and oak exposure.[1] FIGURE 1 shows the first observations of free radicals in red wine, and in red grape juice extract. The wine free radical sits in the middle of the Mn^{2+} spectrum found in all wines. In order to detect these signals readily by EPR, it was necessary to cold evaporate the wines to (1/10)–(1/15) of the original sample volume. A Varian E-12 ~9.1 Ghz spectrometer was used, with standard quartz sample tubes, and specimens frozen using liquid N_2 to avoid microwave polar losses. Free radicals were observed in all red wines, but only in those whites with significant skin/oak exposure. Cu^{2+} is not observed. The radicals are mainly on the phenolics.[1]

Wine fractionation into anthocyanines, flavonoids, and non-flavonoids showed the associated free radicals to be in the ratio 5 : 1 : undetectable, thus showing the antioxidant action of the anthocyanines. The terms are somewhat vague, and the exclusion of one category from another may not have been perfect. The flavonoid component also had a Cu^{2+} signal: this is due to Cu^{+} in the whole wine being changed to Cu^{2+} by pH change, and the effect of the column, during the separation.

Bottle aging potential may be shown by the growth of the (whole) wine free radical signal in the laboratory with time. Bailey's "classic" cabernet 1988 and 1991 wines were made by the same winemaker to the same "recipe." Coincidentally, both wines (cold evaporated) had the same initial free radical signal level. The behavior of the signal was observed over some days, as the specimens were left in the

Address for correspondence: Gordon J. Troup, School of Physics and Materials Engineering, Monash University, Clayton, Victoria 3168, Australia. Voice: +61-39-9053639; fax: +61-39-9053637.

gordon.troup@spme.monash.edu.au

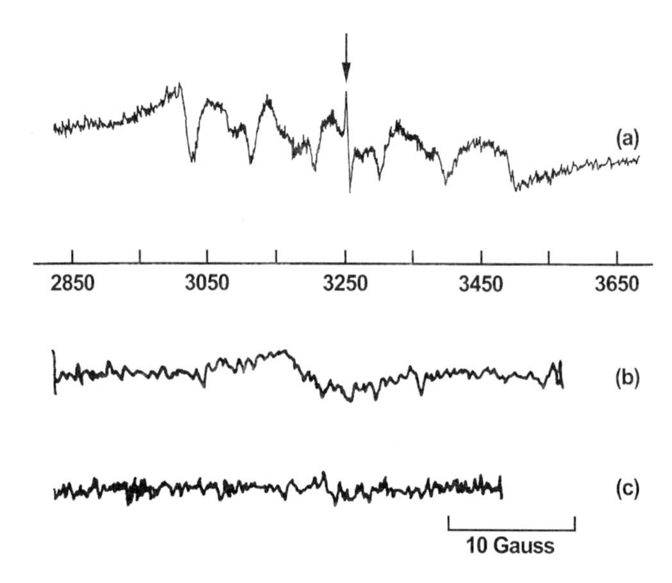

FIGURE 1. Curve (**a**): free radical signal from a red wine in the center of the Mn^{2+} signal. Curves (**b**) and (**c**): EPR signal from red and from white grapejuice extracts, respectively.

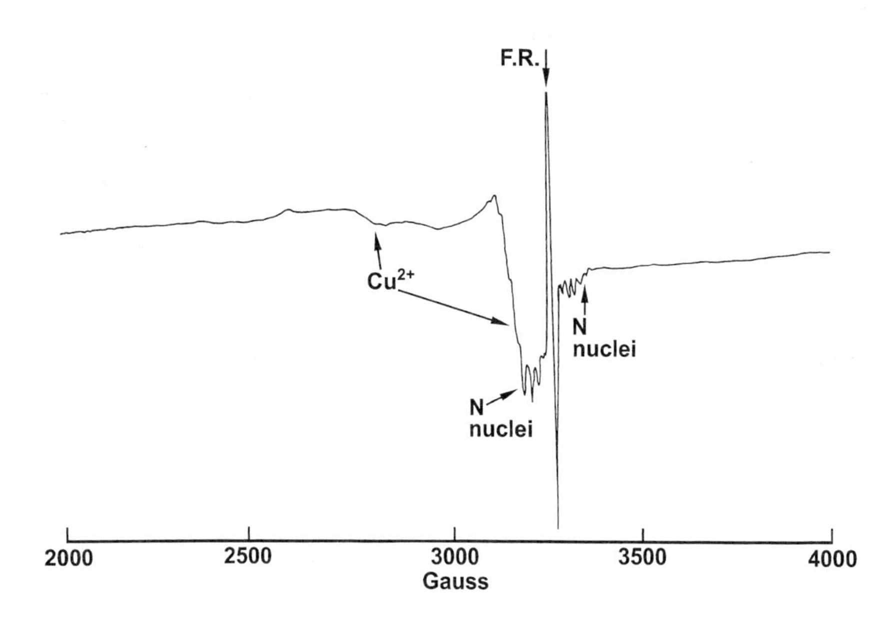

FIGURE 2. EPR signal from the waxy bottle deposit thrown by some South Australian red wines.

air-conditioned laboratory (at about 20°C). The 1991 wine signal started to rise immediately, while the 1988 one "lagged" and never reached the maximum reached by the 1991 wine. A band of experienced tasters agreed that the older wine would take little, if any, further bottle aging, while the younger would continue to improve with age.

Some South Australian red wines throw a waxy red bottle deposit. This was known to be an anthocyanin-protein compound prior to EPR investigation. The EPR spectrum is shown in FIGURE 2, and displays a large free radical signal (expected), and a Cu^{2+} signal with N superhyperfine structure. Therefore it is possible that Cu is involved in the cross-linking of the compound, since Cu–amine bonds are well known. Artificially aged (40°C for about three months) white wines without antioxidant addition, turn brown and throw a fluffy brown precipitate. The EPR spectrum of the precipitate also shows a Cu^{2+} signal without hyperfine structure and a free radical signal. The composition of the precipitate has yet to be determined.

Grapeseed polyphenols decrease in concentration prior to ripening according to a recent theory.[2]

This has been verified by EPR as well as by HPLC,[3] and we suspect that this is the first time that EPR has been used as a "routine" tool in the wine industry. More such studies are planned.

"Powdery mildew" attacks grape vine leaves. It is treated by adding materials, presumed to produce free radicals under sunlight, to the leaves in powder form, or in "solution," with a soap. The following materials, in the solid state, currently being tested, already show a free radical signal: yeast extract, milk powder, whey powder, lactose, and a methionine-vitamin B mixture.This work is still in progress.

The amount of information gathered by applying EPR to these problems in the wine industry clearly shows that EPR should be used much more than it has been previously, which will certainly be of benefit to the industry.

ACKNOWLEDGMENTS

Thanks to the following sponsors of travel to the NYAS Wine and Alcohol conference: the School of Physics and Materials Engineering; the Science Faculty; and the School of Economics and Business Studies (MBA: Prof. R. Willis), Monash University; Mrs. Kath Byer, Notting Hill Hotel, Clayton, Victoria; Nutri-Green, Swan Hill, Victoria; and Evelyn Faye Health and Beauty, Melbourne, Victoria.

REFERENCES

1. TROUP, G.J., *et al.* 1994. Free radicals in red wine but not in white? Free Radicals Res. **20:** 63–69.
2. KENNEDY, J.A., *et al.* 2000. Changes in grapeseed polyphenols during fruit ripening. Phytochemistry **55:** 77–85.
3. KENNEDY, J.A., G.J. TROUP, *et al.* 2000. Development of seed polyphenols in berries from *Vitis vinifera* L. cv. Shiraz. Aust. J. Grape and Wine Res. **6:** 244–254.

Index of Contributors

